Computation, Logic, Philosophy

A collection of Essays

Mathematics and Its Application (China Series)

Computation, Logic, Philosophy

A Collection of Essays

Professor Wang Hao

Famous Mathematical Logician, Mathematician and Philosopher
Rockefeller University, New York, USA

Science Press
Beijing, China

Kluwer Academic Publishers
Dordrecht/Boston/Lancaster/Tokyo

Responsible Editors Yang Xianying Mei Lin

ISBN-13: 978-94-010-7561-9 e-ISBN-13: 978-94-009-2356-0
DOI: 10.1007/978-94-009-2356-0

Distribution rights throughout the world,
excluding The People's Republic of China,
granted to Kluwer Academic Publishers,
P.O.Box 17/3300 AA Dordrecht, Holland

Sold and distributed in the U.S.A. and Canada
by Kluwer Academic Publishers,
101 Philip Drive, Assinippi Park, Norwell, MA 02061, U.S.A.

Sold and distributed in the People's Republic of China
by Science Press, Beijing

In all other countries, sold and distributed
by Kluwer Academic Publishers Group,
P.O. Box 322, 3300 AH Dordrecht, Holland

Softcover reprint of the hardcover 1st edition 1990

SERIES EDITOR'S PREFACE

'Et moi, ..., si j'avait su comment en revenir, je n'y serais point allé.'
Jules Verne

The series is divergent; therefore we may be able to do something with it.
O. Heaviside

One service mathematics has rendered the human race. It has put common sense back where it belongs, on the topmost shelf next to the dusty canister labelled 'discarded non-sense'.
Eric T. Bell

Mathematics is a tool for thought. A highly necessary tool in a world where both feedback and non-linearities abound. Similarly, all kinds of parts of mathematics serve as tools for other parts and for other sciences.

Applying a simple rewriting rule to the quote on the right above one finds such statements as: 'One service topology has rendered mathematical physics ...'; 'One service logic has rendered computer science ...'; 'One service category theory has rendered mathematics ...'. All arguably true. And all statements obtainable this way form part of the raison d'être of this series.

This series, *Mathematics and Its Applications*, started in 1977. Now that over one hundred volumes have appeared it seems opportune to reexamine its scope. At the time I wrote

> "Growing specialization and diversification have brought a host of monographs and textbooks on increasingly specialized topics. However, the 'tree' of knowledge of mathematics and related fields does not grow only by putting forth new branches. It also happens, quite often in fact, that branches which were thought to be completely disparate are suddenly seen to be related. Further, the kind and level of sophistication of mathematics applied in various sciences has changed drastically in recent years: measure theory is used (non-trivially) in regional and theoretical economics; algebraic geometry interacts with physics; the Minkowsky lemma, coding theory and the structure of water meet one another in packing and covering theory; quantum fields, crystal defects and mathematical programming profit from homotopy theory; Lie algebras are relevant to filtering; and prediction and electrical engineering can use Stein spaces. And in addition to this there are such new emerging subdisciplines as 'experimental mathematics', 'CFD', 'completely integrable systems', 'chaos, synergetics and large-scale order', which are almost impossible to fit into the existing classification schemes. They draw upon widely different sections of mathematics."

By and large, all this still applies today. It is still true that at first sight mathematics seems rather fragmented and that to find, see, and exploit the deeper underlying interrelations more effort is needed and so are books that can help mathematicians and scientists do so. Accordingly MIA will continue to try to make such books available.

If anything, the description I gave in 1977 is now an understatement. To the examples of interaction areas one should add string theory where Riemann surfaces, algebraic geometry, modular functions, knots, quantum field theory, Kac-Moody algebras, monstrous moonshine (and more) all come together. And to the examples of things which can be usefully applied let me add the topic 'finite geometry'; a combination of words which sounds like it might not even exist, let alone be applicable. And yet it is being applied: to statistics via designs, to radar/sonar detection arrays (via finite projective planes), and to bus connections of VLSI chips (via difference sets). There seems to be no part of (so-called pure) mathematics that is not in immediate danger of being applied. And, accordingly, the applied mathematician needs to be aware of much more. Besides analysis and numerics, the traditional workhorses, he may need all kinds of combinatorics, algebra, probability, and so on.

In addition, the applied scientist needs to cope increasingly with the nonlinear world and the extra mathematical sophistication that this requires. For that is where the rewards are. Linear models are honest and a bit sad and depressing: proportional efforts and results. It is in the non-linear world that infinitesimal inputs may result in macroscopic outputs (or vice versa). To appreciate what I am hinting at: if electronics were linear we would have no fun with transistors and computers; we would have no TV; in fact you would not be reading these lines.

There is also no safety in ignoring such outlandish things as nonstandard analysis, superspace and anticommuting integration, p-adic and ultrametric space. All three have applications in both electrical engineering and physics. Once, complex numbers were equally outlandish, but they frequently proved the shortest path between 'real' results. Similarly, the first two topics named have already provided a number of 'wormhole' paths. There is no telling where all this is leading - fortunately.

Thus the original scope of the series, which for various (sound) reasons now comprises five subseries: white (Japan), yellow (China), red (USSR), blue (Eastern Europe), and green (everything else), still applies. It has been enlarged a bit to include books treating of the tools from one subdiscipline which are used in others. Thus the series still aims at books dealing with:

- a central concept which plays an important role in several different mathematical and/or scientific specialization areas;
- new applications of the results and ideas from one area of scientific endeavour into another;
- influences which the results, problems and concepts of one field of enquiry have, and have had, on the development of another.

The present volume, one of the first in the 'Chinese subseries' of MIA, also, appropriately enough, one dealing with fundamental issues: interrelations between logic and computer science. The advent of computers has sparked off revived interest in a host of fundamental issues in science and mathematics such as computability, recursiveness, computational complexity and automated theorem proving to which latter topic the author has made seminal contributions for which he was awarded the ATP prize in 1982.

It is a pleasure to welcome this volume in this series.

The shortest path between two truths in the real domain passes through the complex domain.

J. Hadamard

La physique ne nous donne pas seulement l'occasion de résoudre des problèmes ... elle nous fait pressentir la solution.

H. Poincaré

Never lend books, for no one ever returns them; the only books I have in my library are books that other folk have lent me.

Anatole France

The function of an expert is not to be more right than other people, but to be wrong for more sophisticated reasons.

David Butler

Bussum, August 1989

Michiel Hazewinkel

CONTENTS

INTRODUCTION xi

PART ONE. BROAD ISSUES 1

1. ON FORMALIZATION 3
   1.1 Systematization [1955(53)] 3
   1.2 Communication 4
   1.3 Clarity and consolidation 5
   1.4 Rigour 6
   1.5 Approximation to intuition 7
   1.6 Application to philosophy 8
   1.7 Too many digits 9
   1.8 Ideal language 10
   1.9 How artificial language? 11
   1.10 The paradoxes 12

2. THE CONCEPT OF COMPUTABILITY [(1953)] 13
   2.1 Formalizing intuitive concepts 13
   2.2 The intuitive concept of computability 13
   2.3 Computation by theoretical machines 15
   2.4 General recursive functions 15
   2.5 Constructive proofs 17
   2.6 Effective methods 19
   2.7 Speed functions 20
   2.8 Transfinite recursions 22
   2.9 The indeterminate domain of computable functions 25

3. PROCESS AND EXISTENCE IN MATHEMATICS [1961(60)] 30

4. LOGIC, COMPUTATION AND PHILOSOPHY [1971(66)] 47
   4.1 Logic and logical positivism 47
   4.2 What is mathematics? 48
   4.3 Logic and computation 50
   4.4 Relatively undecidable propositions and absolutely unsolvable problems 51
   4.5 Foundations of set theory 53
   4.6 What is mathematics? (continued) 56

PART TWO. AUTOMATED THEOREM PROVING(ATP) .61

5. COMPUTER THEOREM PROVING AND ARTIFICIAL INTELLIGENCE [1984(82)] .63

APPENDIX: CITATION FOR HAOWANG AS WINNER OF "MILESTONE" AWARD IN AUTOMATIC THEOREM-PROVING .75

6. PROVING THEOREMS BY PATTERN RECOGNITION, I [1960(59)] .76

6.1 Introduction .76
6.2 A program that does 9 chapters of *Principia* in 9 minutes .83
6.3 The E_1A case solved with sequential tables .94
6.4 General remarks .96

7. OBSERVATIONS ON ATP .103

7.1 Mechanical mathematics and inferential analysis [1963(61)] .103
7.2 The mechanization of mechanical arguments [1963(62)a] .118
7.3 Formalization and automatic theorem—proving [1965(64)] .127

8. SOME DATA FOR ATP .139

8.1 On axioms of conditional set existence [1967(66)] .139
8.2 Natural hulls and set existence [1967(66)a] .143
8.3 A theorem on definitions of the Zermelo-Neumann ordinals [1967(66)b] .149

9. PROVING THEOREMS BY PATTERN RECOGNITION, II [1961(60)a] 159

9.1 A survey of the decision problem .159
9.2 The Skolem, case .167
9.3 The A_2E satisfiability case .174
9.4 The $A_1E_1A_1$ satisfiability case .176
9.5 A proof procedure for the predicate calculus .186
9.6 Remarks on mathematical disciplines .189

PART THREE. DECIDABILITY AND COMPLEXITY .193

10. GAMES, LOGIC AND COMPUTERS [1965a] .195

APPENDIX: NOTES ON A CLASS OF TILING PROBLEMS [1975(60)] .209

11. DOMINOES AND THE AEA CASE OF THE DECISION PROBLEM [1963(62)] 218

12. TOWARDS FEASIBLE SOLUTIONS OF THE TAUTOLOGY PROBLEM (with B.Dunhan) [1976(74)] 246

12.1 Computational complexity and Boolean validity 246
12.2 A brief overview with some general observations 248
12.3 Some basic properties of Boolean validity 252
12.4 Some calculations and classifications 257
12.5 Hard examples and negative results 263
12.6 A feasible decision procedure for biconditional expressions 266
12.7 Two partial methods and an indication of two generic methods 269

13. RANKED MATCHING AND HOSPITAL INTERNS (with D.A.Martin) [(1977)] 275
13.1 Preliminary 275
13.2 Deletion of useless names: Operations I and II 276
13.3 The canonical form T_1 of T 278
13.4 The student and hospital optimal assignments 280
13.5 Mixed assignments and a characterization of all stable assignments 284
13.6 The marriage problem 286

PART FOUR. TOPICS FROM THEORY TO PRACTICE 291

14. LOGICAL FRAGMENTS RELEVANT TO COMPUTER SCIENCE 293

14.1 Logic of many-sorted theories [1952(50)] 293
14.2 Ackermann's consistency proof [1962(53)] 304
14.3 Partial systems of number theory [1962(55)] 314
14.4 The calculus of partial predicates and its extension to set theory [1961(61)] 319
14.5 Model theory [1974(71)] 325

15. COMPUTERS AND MATHEMATICAL ACTIVITY 331

15.1 Remarks on machines, sets and the decision problem [1964(63)] 331
15.2 Logic and computers [1965] 344
15.3 Remarks on mathematics and computers [1970(67)] 349
15.4 On the long-range prospects of automatic theorem-proving [1970(68)a] 355

16. ON INFORMATION PROCESSING OF THE CHINESE LANGUAGE [1979] 361

THE LIST OF THE PUBLICATIONS OF THE AUTHOR 371

Note. The numbers in square brackets refer to the list of References at the end of the Introduction. The first number refers to the year when the material was published for the first time, and the second number, in paratheses, refers to the year when it was written. For instance, [1955(53)] *means that the item was written in* 1953 *and published initially in* 1955.

INTRODUCTION

Over the years I have thought and written about certain aspects of theoretical computer science, quite directly in connection with mathematical logic and broader conceptual issues. My greater and more continued concentration on this type of work was over the decade from 1953 to 1963, when computer science was in its infancy. Much has happened since that period, but computer science remains, in its fundamental theory, an immature discipline even today. Indeed, with the dominance of specialization and the pressure to publish, there is a tendency to lose sight both of larger issues and of guiding motivations more transparently revealed in simpler situations. In addition, the concern with longer range prospects and less obvious applications of mathematical logic in my early writings appears to remain relevant to a considerable extent for current research. There is of course also the general fact that familiarity with the initial stages of a subject often improves one's understanding of the field.

For all these reasons I am grateful to the late Professor Wu, Yunzeng for encouraging me to carry out the idea of collecting together those of my essays which are, in one way or another, directly or indirectly, related to computers, whose prevalence today is a socially important phenomenon. It may indeed be said that seemingly idle preoccupations with, e.g., formal precision, what is a 'small' number, the classification of (in particular, computable) sets and functions, the algorithmic components of pure mathematics, different concepts of infinity, etc. all begin to acquire more tangible significance through the surprising advance of computers.

The most surprising feature in the development of computers since the 1940s has been the radical and rapid improvements of the basic components: much smaller in size (more information for the same size), much faster, much more reliable, and much cheaper. This must be an exceptional phenomenon in the history of technology: usually not all factors can be maximized at the same time. This fact naturally induces an expansion of the range of application of computer-like devices to areas which would otherwise not be feasible (on account of cost or speed or size or reliability). Moreover, it has the consequence that more efforts are needed for software (to help the flexible use of computers) than hardware (to make computers). Even in the realm of theory, the unexpected development has its effects. For example, J. v. Neumann proposed a theory in 1952 to improve the reliability of computers ('to guard the guardian') on the basis of an estimate of the probability of a component (such as vacuum tubes at that time) malfunctioning at 0.5%, while the actual probability today is much smaller and the reliability is more than 10^7 times better than his estimate. (For a discussion of his idea, see my *Survey*, i.e. 1962(59) in the list of References at the end of this Introduction, chapter 5.6)

One aspect among the broad impacts of the computers is the creation of opportunities to utilize human talents which formerly had less chance to be exercised. One frequently encounters people doing significant research and other work with computers who would have been rather unsuccessful in more traditional disciplines. At the same time, there is a greater danger of pointless research in computer science both for the great practical importance of computers and for its interdisciplinary character. It often happens that work of neither theoretical nor practical interest is defended on two fronts: it is said to be of value for application to the mathematician and of theoretical value to the engineer. This seems to have been the case with a considerable portion of 'automata theory,' the 'resolution method' in theorem proving, 'fuzzy sets,' etc. Moreover, when more objective standards have not yet been formed, there is a tendency to exaggerate the future promises of insignificant beginnings. For many years, this has frequently happened in the field of artificial intelligence. It, therefore, seems necessary to cultivate a better scientific taste by being at home with some more mature discipline (such as mathematical logic or combinatorial analysis or electronics). In addition, a familiarity with some part of the brief history of computer science may also help to broaden one's perspective.

An advantage of computer science as a research area is the relatively slender accumulation of significant results so far. As a result less preparation is needed than in a mature subject to get to the frontier of research. This advantage is balanced by the disadvantage that much of the subject is not so well structured yet. Consequently, it is harder to distinguish more significant from less significant problems; indeed, discoveries in one context often turn out to be rediscoveries of what have long been familiar in other contexts. For example, various innovations in computer theory appear to repeat ideas formerly introduced for different purposes in mathematical logic. It must have been the frequency of this phenomenon that has led some leading computer scientists to urge beginners in computer science to acquire a greater familiarity with the relevant parts of logic.

If we compare computer science today with physics in its early stage, we see important differences as well as possible similarities. Physics deals directly with nature in its fundamental aspects while computer science deals with an exceptionally important class of man-made objects (viz., the computers) and uses more results from a variety of other disciplines. It does seem, however, that we can look forward toward a more substantive and more sophisticated theoretical computer science. In particular, one can anticipate making a sharper distinction in the future between theoretical and mathematical computer science, just as theoretical and mathematical physics are taken as two distinct (though related) areas. There is also a temptation to say that the importance of logic for theoretical computer science today is comparable to the importance of classical analysis for physics up till the last few decades (when, e.g., group theory became relevant too).

My interest in computer science has always been one-sided (as a logician and philosopher) and for many years my principal preoccupation has not been with computer science. Given this fact, my views are severely prejudiced and have serious limitations. In particular, I am not able to place the essays of this volume judiciously within the frame of the current state of computer science. I have no other alternative but to survey the essays of the volume from my highly restricted perspective. The only comfort is that, used with the proper reservation, a subjective account may be more suggestive and provocative than a more balanced one. To render more complete a statement of my perspectives I may mention that related considerations are included in my three other books (as listed in the references): *Survey* [1962(59), chapters 1 to 10], *MP* [1974(72), chapters 9 and 10] and *Popular lectures* [1981(78), chapers 3 and 6, as well as the three appendices]. I shall often refer to relevant parts of these books.

In broad terms there are included in my work relevant to computer science the following items: (1) reflections on the concepts of formalization and feasibility (in particular, feasible computation); (2) the first extensive programming development of idealized computers (including a proof of the theoretical dispensability of erasing); (3) extensive work in automated theorem proving (ATP); (4) a systematic treatment of sequential circuits (finite automata) revealing and leading to the study of an interesting type of monadic second–order theory; (5) the introduction and use of ′domino problems′ (or problems of colored tiles) which have since had applications in various directions not only in logic proper but also in computational complexity. Later in this introduction I shall try to give a more detailed summary of my work along these and other lines in chronological order. Unfortunately I have not kept up with recent literature and cannot, therefore, give anything like an adequate account of recent work related to my early ideas.

I would like to use this opportunity to present some vague idea of mine regarding the formidable NP problem. As is well known, Cook, in his 1971, first introduced a representation of nondeterministic computation by a formula in the propositional calculus and proposed the interesting question whether P (polynomial computation) is equal to NP (nondeterministic polynomial computation). The title of the paper appears to hark back to his early interest in ATP around 1965 at Harvard. As he stated in his oral presentation, his clever representation was inspired by the representation of Turing computation by the domino problems (and the AEA formulas): that is why there is a reference to my 1963(62) in the bibliography of the published version. This is an interesting example of arriving at a new discovery by combining two seemingly unrelated directions, as can be seen from the exclusive emphasis on the more obvious direction in Cook 1983, which contains an overview of the results not only on the NP problem but also in the broader area of complexity theory. (By the way, Cook′s survey presents, in my opinion, a relatively mature area of theoretical computer science in such a meticulous manner as to warrant the belief that computational complexity has become an established discipline matrix.) There appears to be also a more direct connection between complexity theory and the domino problems. For example, recently NP–completeness problems have been developed with bounded domino problems as the starting point in Lewis–Papadimitriou 1981 (chapter 7.6), and domino problems are said to be the best tool for studying computational complexity in Emde Boas 1983.

Most experts conjecture that P is not equal to NP. But there is really no persuasive argument to support this conjecture; we simply do not know enough to have a reasonable conjecture one way or the other. As is well known, many famous problems are NP complete so that each can be chosen according to one's preference to study the open question. Here I shall state my vague ideas in the context of the 'tautology problem' but I believe that similar ideas apply to other problems (on graphs, on equations, integer programming, etc.). The tautology problem asks whether there is a polynomial algorithm to decide the satisfiability of a propositional formula (oor Boolean expression) in conjunctive normal form (oor the validity of one in disjunctive normal form).

From 1973 to 1974 B. Dunham and I developed several fast partial decision procedures for the tautology problem [reported in 1976(74) which is included in this collection as chapter 12]. The hope was to develop a collection of such partial methods which taken together will decide all tautologies in polynomial time. In order to arrive at a solution of the tautology problem along such a direction, a clearer understanding of the partial methods seems necessary. Otherwise even if we had in fact an adequate collection of them, we would not be able to prove the adequacy. I have since that time come upon an open-ended research project which appears to promise cumulative progress toward a better understanding of the large central problem.

One idea is to try to determine more precisely the range of applicability of each partial method. More generally, we look for a characterization of the indefinite subclass of all propositional (oor Boolean) expressions which can be decided by one known method or another in polynomial time. Progress toward such a characterization will help us find counterexamples which fall outside of the subclass and suggest additional partial methods that will also decide them quickly. In addition, there is room to use computers to assist both in determining the characterization and in finding counterexamples. The solution of the four color problem suggests the idea of looking for quickly decidable categories of propositional expressions so that we can narrow down the range of expressions not obtainable by combining such categories. The least that may be attainable by such efforts is to separate out the harder to decide expressions and reduce the original problem to a more limited one. The most that can be hoped is to arrive at a good set of categories and prove by suitable reduction procedures that all propositional expressions are quickly decidable; such a solution would parallel the success with the four color problem (cf. an exposition in chapter 3.7 of my *Popular lectures*). An initial task could be the attempt to characterize the ranges of the partial methods described in 1976(74) (i.e., chapter 12).

Even this initial task does not appear easy at all. But I do feel vaguely that the task should be manageable and, if it is carried out in a moderately satisfactory manner, we shall be in a better position to look for further advances. One attractive feature of the project appears to be the promise of the likelihood of being able to make steady piecemeal progress by means of cooperative efforts. At any rate, I do not presently see where, if there is any, the insuperable obstacle along the envisaged path lies. I hope there will be attempts to render these vague suggestions more definite and thereby transform them into something more useful.

Let me now turn to a chronological summary of my diversified thoughts which may be relevant to computer science.

The evolution of the axiomatic method since Euclid's book on geometry eventually led to an intuitive concept of formal proofs and formal systems which requires that the proofs be mechanically checkable [see 1953(52)]. But the idea of formalization has a number of dimensions; these are discussed in 1955(53) (included in this collection as chapter 1). There are several essays which were written initially as topics in logic but are apparently useful in computer science: these are 1952(50), 1962(53), chapter 15 of *Survey*, 1961(61), and a part of 1974(71); they are included in this collection as chapter 14.

In 1953 I decided to look for some area of research which is more closely linked to the practical than my preoccupation with pure mathematical logic and the philosophy of knowledge. Theoretical issues centered around the emerging large computers appeared to me to be a natural choice on account of their intimate relations to logic, as well as a certain conceptual elegance that is often associated with any yet unexplored domain and seemed particularly striking in this area. For the next few years this interest was combined with a study of Wittgenstein's views on the philosophy of mathematics that, according to my interpretation, has much to do with the activity of *doing* mathematics, with special emphasis on 'surveyability,' 'perspicuity,' and feasibility (for proofs, which include computations as a special case).

At this time I was struck by the difference between primitive and general recursive functions. While the former are built up stage by stage, the latter are given all at once by a general condition of existence. I tried to find ways of classifying general recursive (or theoretically computable) functions which would give some sort of hierarchy according to their computational complexity. The inconclusive result is (1953), being published here for the first time as chapter 2. I understand that these considerations are relevant to current computer science.

My next work is 1957(54) (see *Survey,* chapter 6) which gives an elegant programming language for Turing machines and avoids the use of erasing. It was presented to a meeting of the Association for Computing Machinery in June 1954. In it is 'proved that a theoretically simple basic machine can be imagined and specified such that all partial recursive functions (and hence all solvable computation problems) can be computed by it and that only four basic types of instruction are employed for the programs: shift left one square, shift right one square, mark a blank space, conditional transfer. In particular, erasing is dispensable, one symbol for marking is sufficient, and one kind of transfer is enough.'

According to Minsky: 'The first formulation of Turing–machine theory in terms of computer–like models appears in the paper of Wang [1957], which contains results that would have been much more difficult to express in the older formalisms.' In addition, he speaks of 'the remarkable fact, first shown by Wang [1957], that for any Turing machine T there is an equivalent Turing machine T_N that *never changes a once–written symbol!*' (See Minsky 1967, p. 200 and p. 262). My programming formulation has been extended by Shepherdson—Sturgis 1963 to give languages which are even closer to the computer languages commonly in use. In addition, I have considered also universal

Turing machines in relation to self—reproduction and growth in 1957(57) (see *Survey*, chapter 7). A number of general observations are appended to 1957(54). For example, 'the question involved in the imitation of mind by machine or in the attempt to study the philosophy of mind by comparing mind with machine are surely fascinating but quite often we cannot even formulate the problems clearly, ... What has been discussed less frequently is the possibility of using machines to aid theoretical mathematical research on a large scale. One main contribution of mathematical logic is the setting up of a standard of rigor which is, at least by intention, in mechanical terms. ... The important point [in the use of computers] is that we are trading qualitative difficulty for quantitative complexity. On account of the great restriction on the mind's ability to handle quantitative complexities, we find it more necessary to rely on insight, ingenuity and vague intuition. Using machines, we find our ability in this respect increased tremendously and it is but natural to expect that we can then go a longer way even with less ingenuity.' (*Survey*, pp. 157–158).

On the relation between logic and computers (*Survey*, p. 154):

While mathematical logic had often been criticized for its uselessness, most professional logicians do not seem to have been overwhelmed by the extensive application of logic to the construction and use of computing machines in recent years. There is a strong feeling that the useful part of logic does not coincide or even overlap with the interesting part; or even a suspicion that what is interesting in logic is not useful, what is useful is not interesting. Yet it cannot be denied that there is a great deal of similarity between the interests and activities of logicians on the one side and designers and users of computers on the other. Both groups are interested in making thoughts articulate, in formalization and mechanization of more or less vague ideas. Certainly logicians are not more precise and accurate than the machine people who are being punished for their errors more directly and more vividly. Just as logicians speak of theorems and metatheorems, there are programs and metaprograms. Just as logicians distinguish between using and mentioning a word, automatic coding must observe the distinction between using an instruction and talking about it. Just as logicians contrast primitive propositions with derived rules of inference, there is the distinction between basic commands and subroutines. Shouldn't there be some deeper bond between logic and the development of computers?'

In autumn 1954 I went to England and gave six lectures at Oxford in spring 1955 'on formalizing mathematical concepts.' At that time I had an opportunity to read parts of Wittgenstein's manuscripts which were later published in 1956 as *Remarks on the foundations of mathematics*. In autumn 1956 I gave a seminar on this book and afterwards continued to reflect for several years on the notions of surveyability and feasibility in relation to the mathematical activity. Around 1957 I discussed the paradox of 'small' numbers: 1 is a small number, n + 1 is a small number if n is; therefore, by induction, every number is small. An extended consideration of this paradox is given in Dummett 1975(70) and a considerable amount of material has been published on this paradox since the appearance of his paper.

In this connection it is appropriate to say a few words about the relation between feasible computation and that in polynomial time. On account of the fundamental character of polynomials and the many nice closure properties of polynomal time, it is obviously of interest to investigate the realm of algorithms performable in polynomial time. At the same time, it is obviously false to dientify feasibility with executability in polynomial time. Indeed, we have here a special case of the paradox of small numbers: linear time (i.e., polynomial of degree 1) is feasible, and degree n + 1 is feasible if degree n is; therefore, by induction, any polynomial time is feasible. Put in this way, it appears clear that something is amiss in regarding polynomial time of high degrees as feasible. For example, a polynomial algorithm for linear programming has been given in Khachian 1979, but it is not accepted as generally feasible or indeed considered more efficient (for the majority of interesting cases) than alternative methods which do not always have polynomial bounds. (By the way, the relation between linear programming and the NP–complete problem of integer programming is instructive; I do not know whether, for other NP–complete problems, such as the tautology problem, the analogously related problems might not be of intrinsic or instrumental interest.) These observations also point to the elusive character of the notion of feasible (or 'small') computations.

A more accessible line of approach may be the consideration of feasibility, not in an absolute sense, but relative to given computer models and current states of technology. In this way one might look for a 'theory of relativity' for the concept of feasible computation. Indeed, by intention, feasibility is a relative concept. It is relative to the existing computer models of the day, in terms of cost, speed, size, reliability and structure (such as the intricacy of parallel computation, and the tolerance of local errors). Moreover, the feasibility of a given computational problem also depends on its importance, and consequently, also the ability of the people engaged in tackling it. Hence, the theoretical enterprise of elucidating the concept of feasibility enjoys an exceptional intimacy with practice; more than other theoretical pursuits, it directly faces the formidable task of unifying theory and practice. It is, therefore, not surprising that the development of a satisfactory theory of feasibility calls for a novel combination of experiences and abilities.

Related to the broader issue of feasibility are my paper 1961(60) (included in this collection as chapter 3) and the observation on 'anthropomorphism' (or 'strict finitism' which, however, appears to imply a less appropriate idea) in my 1958 (57). Since this latter paper is not included in this collection and since the material appears to be of current relevance, I should like to reproduce the section on anthropomorphism (or anthropologism?) here (from *Survey*, pp. 39—41).

'In mathematics, we constantly use the word "can" to refer to theoretical possibilities. If we are concerned with mathematics as a human activity, practical possibilities become more interesting. In his 1935, Bernays suggests studying such a use of "can" but says explicitly that he is not recommending that we do arithmetic with a restriction to "feasible" ("effectable") processes (pp. 61—62). Bernays mentions the fact that we pass without hesitation from k and j to k^j although nobody has given or can give the decimal expansion of, say, 67 to the power of 251^{729}. Yet intuitionists (and finitists, for that matter) do not question the meaningfulness and truth of the assertion that such an expansion exists. One may ask whether we have in this case truly intuitive evidence. "Isn't it rather the general method of analogy that is applied here, consisting in the extension to inaccessible numbers the relations which can be verified concretely for the accessible numbers? Indeed, the reason for applying this analogy is all the more strong since there is no precise limit between the numbers which are accessible and the ones which are not. One could introduce the notion of 'feasible' processes and restrict implicitly the range of significance of recursive definitions to feasible operations. To avoid contradictions, it would only be necessary to abstain from applying the law of excluded middle to the notion 'feasible'." In his *Remarks on the foundations of mathematics* (1956), Wittgenstein makes many cryptic observations (e.g., lines 23—25, p. 65; lines 16—22, p. 84; lines 5—9, p. 156) which become understandable if we keep in mind his preoccupation with the conception of mathematics as feasible activity.

'It is on this basis that we can distinguish a finitist proof from one which can actually be carried out and be kept in mind (what Wittgenstein calls "surveyable" or "perspicuous"). For example, from this point of view, a definition, even an explicit one, is not a "mere" abbreviation as it enables us to see a new aspect, to see an old expression as something different, and to grasp as a matter of fact a wider range of expressions. The "reduction" of numerical arithmetic to the predicate logic with identity or that of the decimal notation to the stroke notation entails, from this point of view, a great loss which consists in a considerable decrease of the range of numbers which we can actually handle.

'Mathematical induction codifies the analogy between accessible and inaccessible numbers. While the justification of the principle lies beyond what is concretely presentable, we are able, once we accept it, to bring a great deal of new things into the range of the surveyable. While we cannot survey the corresponding stroke numeral of every decimal numeral, we convince ourselves by induction that there exists a unique decimal or stroke notation for each positive integer. So also we get an indirect survey of all possible proofs of a system by an inductive consistency proof.

'As an actual calculating machine can only handle a restricted amount of data, we have to twist and turn the notations and techniques in order to increase the range of manageable calculations. If one views foundational studies as primarily concerned with the determination of the range of mathematics which we actually can do, then mathematical logic as is practised today could play at most a minor role and its dominance would be giving us a wrong impression of the problems of foundations. For example, if we had only the stroke notation, we could not manipulate with numbers much larger than 10, the decimal notation extends the range, and exponentiation extends it further still; in the use of calculating machines, what notation we use to represent numbers is an important problem. New definitions and new theorems interest the working mathematicians even though mathematical logic may claim that they were implicitly contained in the logical system to begin with. Anthropologism draws our attention to this distinction which is neglected by mathematical logic.

'From this approach, besides number, set, proof, notation also becomes officially an object of study. To describe a system, we have to include not only the basic rules but also the definitions and proofs since how much we can actually do with the system depends a great deal on what definitions and proofs are at our disposal. So also a proposition receives its meaning from a proof or a refutation of it because only afterwards can we place it at the right place in our understanding. Every proof changes our actual concept somewhat and may be said to give us a new concept. Or again, if we reflect on the human elements involved, it is doubtful that a contradiction can lead to a bridge collapsing.

'The comparison with machines must not give us a wrong impression. People actually engaged in the use and construction of calculating machines find the current automata studies not quite what they want, and there is demand to study, e.g., what the length of a calculation is, or how to develop something about machine operations that is similar to information theory in paying attention to quantitative details. However, these problems, or even the problems about an actual machine whose internal structures and tendency to make mistakes are not clear to us, are different from those for anthropologism. To deal with such machines, one might think of the application of statistics in gas dynamics, and, like von Neumann, talk about "probabilistic logics," but we would still get something similar to mathematical logic in so far as they all deal with something like the truth functions and their distributions. Anthropologism looks for a logic not of the static but of the developing, the becoming. Thus, since it is far beyond our present knowledge and understanding to treat fruitfully man as a machine, anthropologism suggests a behaviouristic or phenomenological, rather than a physiological, treatment of mathematical thinking. This seems to suggest a vague area of research quite different from mathematical logic, although there is no justification in believing that the different lines of research cannot enjoy coexistence.

'The intuitionistic logic might turn out to be applicable to anthropologism. But if one wishes to hold consistently to the position of anthropologism, he cannot accept the usual formulation of the intuitionistic calculus which allows for arbitrarily long formulae and arbitrarily long proofs.' [This completes the quotation on 'anthropologism.']

In the summer of 1956, I worked with A. W. Burks on the 'logic of automata' and wrote a long paper with him. The most interesting proof in the original paper turned out to contain a serious mistake. As a result, we had to revise it hastily, and the result was 1957(56) (included in *Survey* as chapter 9). Section 1 of the paper gives the following summary (*Survey*, p. 175):

'We are concerned in this paper with the use of logical systems and techniques in the analysis of the structure and behavior of automata.

'In Section 2 we discuss automata in general. A new kind of automaton is introduced, the growing automaton, of which Turing machines and self-duplicating automata are special cases. Thereafter we limit the discussion to fixed, deterministic automata and define their basic features. We give methods of analyzing these automata in terms of their states. Four kinds of state tables —— complete tables, admissibility trees, characterizing tables, and output tables —— are used for this purpose. These methods provide a decision procedure for determining whether or not two automaton junctions behave in the same way. Finally, a class of well–formed automaton nets is defined, and it is shown how to pass from nets to state tables and *vice versa*. A coded normal form for nets is given.

'In Section 3 we show how the information contained in the state tables can be expressed in matrix form. The (i, j) element of a transition matrix gives those inputs which cause state S_i to produce state S_j. Various theorems are proved about these matrices and a corresponding normal form (the decoded normal form or matrix form) for nets is introduced.

'In Section 4 we first show how to decompose a net into one or more subnets which contain cycles but which are not themselves interconnected cyclically. We then discuss the relation of cycles in nets to the use of truth functions and quantifiers for describing nets. We conclude by relating nerve nets to other automaton nets.'

It was in connection with these considerations that I got interested in adding the time element to the familiar representation of computer circuits by Boolean expressions. It seemed to me that an elegant extension of Boolean algebra can be made to take care of what is known as the 'delay' element in computer technology. It was at the beginning of 1959 that I decided to settle down and work out such an extension; the result was 1959(59) (included in *Survey* as chapter 10). While the formulation is intimately connected with the practical engineering task of circuit synthesis, it also suggests a monadic second order theory with a successor function which is of interest in logic. Indeed, the further developments of theories of this general type in, e.g., Büchi 1960 and Rabin 1969 all appear to be natural extensions of the problems on 'sequential Boolean equations.'

The content and motivation of the paper are briefly (*Survey*, pp. 269—270):

'There is an intrinsic ambiguity in the standard problems of analysis (given circuit, find formula) and synthesis (given formula, find circuit), because both the circuit and the formula (i.e., the condition to be satisfied) are to be specified in suitable symbolisms but there are no universally accepted symbolisms for these purposes. This is especially true of the initial languages expressing the conditions to be synthesised. Of course, the richer the language, the easier it is to express an intuitive requirement, but at the same time, the harder it is to obtain a systematic procedure for dealing with all conditions expressed in the language. A good starting language would seem to be that of sequential Boolean equations obtained by adding to the ordinary Boolean operations a sequential operator to take care of the time element.

'We shall study in this paper the general problem of solving such equations to get explicit representations of outputs as functions, or rather functionals, of time and the input functions. It turns out that in general such equations can have three different types of solution: deterministic ones, (effective) predictive ones, and noneffective ones. The first two types are effective and can be realized by circuits in one way or another. Algorithms for deciding solvability and exhibiting the solutions will be given for the different senses of solution. The last two types of solution do not seem to have been much studied in the literature.

'We extend ordinary Boolean algebra by introducing a distinction between input variables and output variables, and adding a time operator d so that intuitively, e.g., x means x_i and dx means x_{i+1}. More exactly, each equation is obtained by joining two terms with the equal sign =, and a term is either a constant 1 (on, true) or 0 (off, false), or an input variable i, j, etc., or an output variable x, y, etc., or obtained from given terms A, B by the Boolean operators', ., +, or by the time operator d, e.g., A', A.B, or AB, A+B, dA etc.'

Only in the summer of 1958 and 1959 did I actually work directly with computers: writing programs to prove theorems in first order logic. The results have been published in 1960(58) and 1960(59). From autumn 1959 on my attention shifted to more theoretical considerations. For the next decade I continued to think about and be invited to lecture on questions related to ATP (automated theorem proving). A review of my thoughts in this area is given in my 1984(82). which is included in this collection as chapter 5. I would like to add only a few supplements to that review.

In 1960(58) it is observed that the development of computers coincidentally vindicates an ideal of formal logic (*Survey*, pp. 257—258):

'The suspiciously aggressive term "mechanical mathematics" is not unattractive to a mathematical logician. It is a common complaint among mathematicians that logicians, when engaged in formalization, are largely concerned with pointless hairsplitting. It is sufficient to know that proofs can be formalized. Why should one take all the trouble to make exact how such formalizations are to be done, or even to carry out actual formalizations? Logicians are often hard put to it to give a very convincing justification of their occupation and preoccupation. One lame excuse which can be offered is that they are of such a temperament as to wish to tabulate all scores of all base ball players just to have a complete record in the archives. The machines, however, seem to supply, more or less after the event, one good reason for formalization. While many mathematicians have never learned the predicate calculus, it seems hardly possible for the machine to do much mathematics without first dealing with the underlying logic in some explicit manner. While the human being gets bored and confused with too much rigour and rigidity, the machine requires entirely explicit instructions.

'It seems as though that logicians had worked with the fiction of man as a persistent and unimaginative beast who can only follow rules blindly, and then the fiction found its incarnation in the machine. Hence, the striving for inhuman exactness is not pointless, senseless, but gets direction and justification.'

A summary of 1960(58) is the following (*Survey*, pp. 226—227):

'The writer wrote three programs last summer (1958) on an IBM 704. The first program provides a proof–decision procedure for the propositional calculus which prints out a proof or a disproof according as the given proposition is a theorem or not. It was found that the whole list of theorems (over 200) of the first five chapters of *Principia Mathematica* were proved within about 37 minutes, and 12/13 of the time is used for read–in and print–out, so that the actual proving time for over 200 theorems was less than 3 minutes.

'The second program instructs the machine to form itself propositions of the propositional calculus from basic symbols and select nontrivial theorems. The speed was such that about 14,000 propositions were formed and tested in 1 hour, storing on tape about 1000 theorems. The result was disappointing in so far as too few theorems were excluded as being trivial, because the principles of triviality actually included in the program were too crude.

'The third program was meant as part of a larger program for the whole predicate calculus with equality which the writer was unable to complete last summer due to lack of time. The predicate calculus with equality takes up the next 5 chapters of *Principia Mathematica* with a total of over 150 theorems. The third program as it stands can find and print out proofs for about 85% of these theorems in about an hour. The writer believes that slight modifications in the program will enable the machine to prove all these theorems within 80 minutes or so. The full program as envisaged will be needed only when we come to propositions of the predicate calculus which are much harder to prove or disprove than those in this part of *Principia Mathematica.*'

The other papers related to ATP are included in this collection as part two (chapters 5 to 9). Chatper 8 contains examples of rather simple and detailed mathematical arguments which may serve as useful examples for experimenting with ATP; in this regard several other papers in the list of reference are similar, viz., 1964a, 1966, 1966a, and 1966b. A helpful survey of ATP from 1958 to 1983 is the collection of papers *Automated theorem proving: after 25 years*, edited by W. W. Bledsoe and D. W. Loveland and published in 1984 as volume 29 of the AMS series on 'contemporary mathematics.' The paper by Chou, Shang–ching in this volume and a new paper by Wu himself (*J. sys. sci. & math. sci.*, vol. 4, no. 3, 1984, pp. 207—235) are helpful expositions of Wu, Wen–tsün's algorithm for ATP in geometry. (By the way, a point of conceptual interest is the status of the axioms of order in elementary geometry. As is well known, the axioms of order were implicitly assumed in Euclid's *Elements*. It was only in the 19th century that M. Pasch and others succeeded in uncovering this group of axioms. Rather surprisingly, the restriction of Wu's algorithm is to consequences of all except this group of axioms of elementary geometry. At the same time, ingenious applications of the algorithm appear to bypass the axioms of order in many cases. This phenomenon suggests the possibility of recovering, even from a mechanical procedure, some of our intuitive power to make leaps, as revealed by the fact that the human mind generally arrives at correct proofs even when making unconscious appeals to the axioms of order.)

Chapter 9, a reprint of 1961(60)a, is transitional in that it represents the shift of my attention from ATP to the more theoretical decision problem of first order logic. Several remarks on this paper are in order. This is the paper in which I introduced for the first time the class of problems of dominoes (or colored tiles) in connection with the search for a decision procedure for the AEA class of the formulas of first order logic. At the time (spring 1960) I expected the class to be decidable and conjectured, contrary to its resolution later, that the unrestricted domino problem is decidable. That was probably also the reason why I did not include my seminal proof of the result that the origin–constrained domino problem is unsolvable. Only a year later did I put it forth as the technical report 1961(60)b. This report is here included as an appendix to chapter 9, with which it belongs. Another point about this chapter is the brief observations on the inclusion of equality in 2.5. Like all others in the field, I believed at the time that the EA_2E case, as Gödel first asserted in 1933, remains decidable by easy extension when equality is included. It has recently been proved by W. Goldfarb (*Bulletin AMS*, vol. 10, 1984, pp. 113—116) that this (Gödel) case with equality is unsolvable and a reduction class.

I have briefly mentioned above the relations of the domino problems to the theory of computational complexity. In *Popular lectures*, I have tried to give a very incomplete list of the extensive literature on the domino problems and their unexpected applications (pp. 110—112). I shall not repeat the list here; nor can I make it more complete. Instead, I shall only mention its central place in the reduction problem (of first order logic). In autumn 1961, the AEA case was shown to be a reduction class by way of the domino problems and the proof was published in 1962(61). This result and further extensions were given a more leisurely exposition in 1963(62), which is included in this collection as chapter 11. Given the surprising fact that the seemingly very simple AEA case is a reduction class, all reduction classes in the prefix form (a natural principle of classification) follow as immediate corollaries. Indeed, the only other interesting prefix reduction classes are Suranyi's AAAE case and the Gödel case with equality. The former class can directly be seen to include the class of AEA formulas. The same is true of the latter class, once an axiom of infinity expressible by a formula in the class is available; indeed, it was Goldfarb's construction of such an axiom of infinity which had been the missing link for many years.

Of the remainder of this collection, I have already briefly commented on chapters 12 and 14. Chapter 13 is an incomplete paper based on casual discussions with D. A. Martin in 1977. It is included here as an example of problems which arise naturally in everyday life. The chapters 10, 15, and 16 are relatively nontechnical expositions which should be as easy to read as the initial four chapters on broad issues.

In the last two paragraphs of chapter 4 (originally written in 1966), I suggested as an alternative to the familiar idea of trying to obtain a grand system that yields all proofs, the idea of studying, for each interesting proof, all the systems in which it can be carried out. In particular, I said: 'Our intuition of the real numbers is not captured in any of the particular formal systems. A closer approximation could be obtained if we look instead for the class of formal proofs (and therewith the underlying formal systems) which all can represent a given intuitive proof. In this way, each intuitive proof would correspond to a class of formal proofs, and we can classify intuitive theorems according to the classes of formal systems in which they can be represented naturally.' I am under the impression that in the last decade or more a good deal of interesting work has been done which may be seen as in agreement with my loose suggestion.

With regard to the colored tiles (or dominoes), the combinatorial aspect has been carefully and extensively considered in a recent book: Branko Grünbaum and G. C. Shephard, *Tilings and Patterns*, 1987, W. H. Freeman and Company. In particular, the questions of aperiodic solutions and of the relations to other tiling ideas are discussed in chapter 11, entitled 'Wang tiles' (pp. 583—608). From their discussions, it appears that the several fruitful ideas for tilings are intimately connected. In this regard it is of interest to note that within this family of apparently 'pure' conceptions, what is called the 'Penrose tiling' has found applications in physics (see, for instance, the report by David R. Nelson on 'quasicrystals,' *Scientific American*, vol. 255, no. 2, August 1986, pp. 42—51).

References

P. Bernays

1935. Sur le platonisme, *L'enseignement math.*, vol. 34, pp. 52—69.

R. Büchi

1960. Weak second order theories and finite automata, *ZMLGM*, vol. 6, pp. 66—72.

S. A. Cook

1971. The complexity of theorem proving procedures, *Proc. 3rd ACM Symp. on Theory of Computing*, pp. 151—158.

1983. An overview of computational complexity, *Communications ACM*, vol. 26, pp. 401—408.

M. Dummett

1975(70). Wang's paradox, *Synthèse*, vol. 30, pp. 301—324.

P. v. Emde Boas

1983. Dominoes are forever, *First GTI workshop*, Paderborn, pp. 75—95.

D. Harel

1983. Recurring dominoes: making the highly undecidable highly understandable, *Proc. Int. Conf. Fund. Comp. Theory*, Burgholm, Sweden.

L. G. Khachian

1979. *Soviet math. doklady*, vol. 20, pp. 191—194. (For an exposition, see C. H. Papadimitriou and K. Steiglitz, *Combinatorial optimization*, 1982.)

H. Lewis and C. H. Papadimitriou

1981. *Elements of the theory of computation.*

M. Minsky

1967. *Computation: finite and infinite.*

J. v. Neumann

1952. *Probabilistic logics* (reprinted in his collected papers).

M. O. Rabin

1969. Decidability of second order theories and automata on infinite tree, *Transactions AMS*, vol. 141, pp. 1—35.

J. C. Shepherdson and H. E. Sturgis

1963. Computability of recursive functions, *Journal ACM*, vol. 10, pp. 217—255.

Hao Wang

1952(50). Logic of many-sorted theories, *JSL* (i.e., *Journal of symbolic logic*), vol. 17, pp. 105—116.

1953(52). Quelque notions d'axiomatique, *Revue philosophique de Louvain*, vol. 51, pp. 409—443. English version entitled 'The axiomatic method' is included in *Survey* [1962(59)] as chapter 1.

1955(53). On formalization, *Mind*, vol. 64, pp. 226—238. Reprinted in *Contemporary readings in logical theory*, edited by I. Copi and J. Gould, 1967, pp. 29—39; also included as the opening essay in their *Contemporary philosophical logic*, 1978, pp. 2—13.

(1953). The concept of computability. This essay was first written in 1953. It was then revised in 1954; but it has not been published before.

1962(53). Ackermann's consistency proof. These notes were written in 1953 and first published in *Survey* (pp. 362—375).

1955(54). On denumerable bases of formal systems. Invited hour lecture at the International Congress of Mathematicians, Amsterdam, 1954; published in *Mathematical interpretation of formal systems*, pp. 57—84.

1957(54). A variant to Turing's theory of computing machines, *Journal ACM* (i.e., of the Association for Computing Machinery), vol. 4, pp. 63—92. The paper was presented to the meeting of ACM in June, 1954.

1962(55). Partial systems of number theory. This material was written in 1955 and published for the first time in 1962(59), pp. 376—382.

1974(55). On formalizing mathematical concepts. Six essays delivered as the second series of John Locke Lectures at the University of Oxford in spring 1955; parts were published in revised form in 1974(72), chapters 1 and 2.

(1956). Elementary philosophy of mathematics. An uncompleted typescript of 450 pages written during 1955—56; only some fragments have been published.

1957(56). (With A. W. Burks). The logic of automata, *JACM*, vol. 4, pp. 193—218 and pp. 279—297. Reprinted in *Survey* as chapter 8.

1957(57). Universal Turing machines: an exercise in coding, *ZMLGM* (*i.e.*, *Zeitshrift für Mathematische*

Logik und Grundlagen der Mathematik), vol. 3, pp. 69—80. Reprinted in *Survey* as chapter 7.

1958(57) Eighty years of foundational studies, *Dialectica*, vol. 12, pp. 466—497. Reprinted in *Survey* as chapter 2.

1960(58). Toward mechanical mathematics, *IBM journal of research and development*, vol. 4, pp. 2—22. Reprinted in *Survey* as chapter 9; also reprinted in *The modelling of mind*, edited by K. Sayre and F. Crosson, 1963; Russian translation in *Problems of cybernetics*.

1959(59). Circuit synthesis by solving sequential Boolean equations, *ZMLGM*, vol. 5, pp. 216—239. Reprinted in *Survey* as chapter 10.

1962(59). *A survey of mathematical logic* (*Survey*), Science Press, 1962, 652 pp. + x; also distributed by North-Holland Publishing Company, 1963. Reprinted by Chelsea, New York, 1970 under the title *Logic, computers and sets.* The manuscript was completed and submitted in June 1959 (at Oxford).

1960(59). Proving theorems by pattern recognition, I, *Communications ACM*, vol. 3, pp. 220—234. Invited lecture at the ACM meeting in May 1960.

1961(60). Process and existence in mathematics, *Essays on the foundations of mathematics*, pp. 328–351. This was read to the Philosophical Club of Harvard University in spring 1960; a Russian translation came out in 1965.

1961(60)a. Proving theorems by pattern recognition, II, *Bell system technical journal*, vol. 40, pp. 1—41.

1961(60)b. An unsolvable problem on dominoes, The Computation Laboratory, Harvard University, *Report BL*–30, *I*, July 1961, 5 pp.

1975(60). Notes on a class of tiling problems, *Fundamenta mathematicae*, vol. 82, pp. 295—305.

1961(61). The calculus of partial predicates and its extension to set theory, I, *ZMLGM*, vol. 7, pp. 283—288. It was read to the Logic Society in England in spring 1961; the second part (extension to set theory) has not been written.

1963(61). Mechanical mathematics and inferential analysis, *Computer programming and formal systems*, edited by P. Braffort and H. Hirschberg, pp. 1—20. This is a revised version of an invited lecture at a seminar, spring 1961, sponsored by IBM in Holland.

1962(61). (With A. S. Kahr and E. F. Moore). Entscheidungsproblem reduced to the AEA case, *Proceedings of the National Academy of Science, U.S.A.*, vol. 68, pp. 528—532.

1963(62). Dominoes and the AEA case of the decision problem, *Mathematical theory of automata*, pp. 23—55. Invited lecture given April 1962 in New York.

1963(62)a. The mechanization of mathematical arguments, *Experimental arithmetic, high speed computing and mathematics*, pp. 31—40. This was an invited lecture at a meeting of the American Mathematical Society in 1962.

1963. Tag systems and lag systems, *Mathematische Annalen* (*MA*), vol. 152, pp. 65—74. This was included in *Popular lectures* [1981(78)] as appendix C5.

1963a. (With M. O. Rabin). Words in the history of a Turing machine with a fixed input, *Journal ACM*, vol. 10, pp. 526—527.

1964(63). Remarks on machines, sets and the decision problem, *Formal systems and recursive functions*, pp. 304—320. An invited lecture given at Oxford, England in summer 1963.

1964. Critique of logic for the computer sciences, *Communications ACM*. vol. 7, p. 218.

1964a. (with W. V. Quine). On ordinals, *Bulletin of American Mathematical Society*, vol. 70, pp. 297—298.

1965(64). Formalization and automated theorem proving, *Proceedings of the IFIP Congress 65*, pp. 51—58. This was an invited lecture to the Congress; Russian translation, *Problems of cybernetics*, vol. 7 (1970), pp. 180—193.

1965. Logic and computers, *American mathematical monthly*, vol. 72, pp. 135—140.

1965a. Games, logic and computers, *Scientific American*, vol. 213, no. 5 (November), pp. 98—106.

There is a Swedish translation in *Modern Datateknik*.

1965b. Note on rules of inference, *ZMLGM*, vol. 11, pp. 193—196.

1966. (With S. A. Cook). Characterizations of ordinal numbers in set theory, *MA*, vol. 164, pp. 1—25.

1966a. (With K. R. Brown). Finite set theory, number theory and axioms of limitation, ibid., pp. 26—29.

1966b. (With K. R. Brown). Short definitions of ordinals, *JSL*, vol. 31, pp. 409—414.

1971(66). Logic, computation and philosophy, *L'âge de la science*, vol. 3, pp. 101—115.

1967(66). On axioms of conditional set existence, *ZMLGM*, vol. 13, pp. 183—188.

1967(66)a. Natural hulls and set existence, ibid., pp. 175—182.

1967(66)b. A theorem on definitions of the Zermelo–Neumann ordinals, ibid., pp. 241—250.

1970(67). Remarks on mathematics and computers, *Theoretical approaches to nonnumerical problem solving*, pp. 152—160. An invited lecture given at Cleveland, Ohio in 1967.

1970(68). A survey of Skolem's work in logic, *Selected logical works of Th. Skolem*, pp. 17—52.

1970(68)a. On the long–range prospects of automated theorem–proving, *Symposium on automatic demonstration*, pp. 101—111. Invited lecture given at Versailles, France in December 1968.

1974(71). Metalogic, *Encyclopaedia Britannica*, vol. 11, pp. 1078—1086. All except the part on model theory is reprinted in 1974(72) as chapter 5.

1974(72). *From mathematics to philosophy*, Routledge and Kegan Paul, 413 pp. + xiv. Italian translation *Dalla mathematica alla filosofia*, Boringhieri, 1984, by Alberto Giacomelli.

1976(73). (With B. Dunham). A recipe for Chinese typewriters, IBM report RC4521, September 5, 1973. Chinese translation appeared in *Dousou bimonthly*, no. 14, March 1976, pp. 56—62.

1976(74). (With B. Dunham). Toward feasible solutions of the tautology problem, *Annals of mathematical logic*, vol. 10, pp. 117—154. (Originally issued as IBM report RC4924 on July 9, 1974).

(1977). (With D. A. Martin). Ranked matching and hospital interns. Some of the results are mentioned in 1981(78) under chapter 3.6.

1981(78). *Popular lectures on mathematical logic*, Science Press and van Nostrand Reinhold. 273 pp. + x. Chinese translation appeared about the same time, Science Press, 257 pp. + vii.

1979. On information processing of the Chinese language (in Chinese), *The state of the art report of computer technology*, no. 98 (June 1979), pp. 1—4.

1981(80). Specker's mathematical work from 1949 to 1979, *L'enseignement mathématique*, vol. 27, pp. 85—98.

1984(82) Computer theorem proving and artificial intelligence, *Automated theorem proving: after 25 years*, pp. 47—70. Lecture to accept the first Milestone Prize in automated theorem proving, awarded January 1983 at the annual meeting of the American Mathematical Society.

1984(83). The formal and the intuitive in the biological sciences, *Perspectives in biology and medicine*, vol. 27, pp. 525–542. Opening lecture at the Ninth International Congress of Thrombosis and Haemostasis, Sweden, on July 3, 1983.

PART ONE

BROAD ISSUES

1. ON FORMALIZATION *

1.1 Systematization

THE most striking results of formalization occur in logic and mathematics.

Here formalization provides at least one kind of systematization. We are led to believe that there is a fairly simple axiom system from which it is possible to derive almost all mathematical theorems and truths mechanically. This is at present merely a theoretical possibility, for no serious attempts seem to have been made to prove, for instance, all the theorems of an elementary textbook of calculus. Nevertheless, we seem to get a feeling of grandeur from the realization that a simple axiom system which we can quite easily memorize by heart embodies, in a sense, practically all the mathematical truths. It is not very hard to get to know the axiom system so well that people would say you understood the system. Unfortunately just to be able thus to understand the system neither gives you very deep insight into the nature of mathematics nor makes you a very good mathematician.

To say that physics uses the experimental method is not to say much about physics. To say that all theorems of mathematics can be proved from certain axioms by chains of syllogism (or *modus ponens*) is to say just as little about mathematics. Merely knowing the experimental method is not knowing the whole of physics; merely knowing an axiom system adequate for developing mathematics is not knowing the whole of mathematics.

There is another kind of systematization which is less superficial than learning the axiom system. It is an intuitive grasp of the whole field, a vivid picture of the whole structure in your mind such as a good chess player would have of the game of chess. This second kind of systematization is something that formalization (or at least formalization alone) would not provide us.

If we had never used logistic systems at all, the many interesting results about logistic systems (such as those of Skolem, Herbrand, and Gödel) would, of course, never have been expressed in the specific form in which they are now being expressed. But it is not certain that essentially the same results might not have been attained, though in other contexts and as the results about other things. Nevertheless, axiomatics or the axiomatic method has a strong appeal in that here we seem to be able to prove sweeping conclusions about whole fields. For many of us a significant theorem about a whole field appears more important than particular theorems in the field. In generating systems, formalization serves the function of enabling us to talk precisely about whole fields of learning.

---

* First published in *Mind*, vol. 64, pp 226—238. © Oxford Press, 1955, Reproduced by permission.

1.2 Communication

It is hard to say whether in general formalization renders a theory or a proof easier to understand.

Consider, for example, an oral sketch of a newly discovered proof, an abstract designed to communicate just the basic idea of the proof, an article presenting the proof to people working on related problems, a textbook formulation of the same, and a presentation of it after the manner of *Principia Mathematica*. The proof gets more and more thoroughly formalized as we go from an earlier version to a later. It is, however, questionable whether in general a more completely formalized version is clearer or serves better as a means of communication. Each step of it should be easier to follow since it involves no jumps. But even this is not certain, for there are many jumps which we are so used to making that we find it more natural to make the jumps than not to. Or alternatively, we may say that the step actually does not involve jumps and that our formal proof suggests that it does only because our formal system is defective as a map of our intuitive logic.

Who finds which proof easier to follow or who understands which proof in a shorter while depends pretty much on what background the man happens to have. In general, the better acquainted one is with the problem, the easier he finds the use of a more sketchy proof. But there is also a certain limit beyond which even the expert in the matter can no longer supply for himself the missing details. Moreover, there is always the possibility that the presentation would be much shorter if it were not so short. It seems safe, however, to say that a more thoroughly formalized proof is generally longer, provided that we do not appeal to abbreviations in its presentation and the less formalized version does not waste words.

We are all familiar with requests to explain a physical theory without using mathematics, to convey the basic idea of a proof without using symbols. Therefore, it would seem that in general the plain words or the less technical language provide a more efficient means of communication. Actually, however, we can easily think of examples which would indicate that this is not quite true.

To put thoughts on physics into mathematical symbols is one way of formalization. Through accumulation and tradition this way of formalization has also become a powerful way of communication: for those who understand the language, a short formula may express more precisely thought which could only be explained by many pages of ordinary words, and much less satisfactorily. Sometimes it becomes practically impossible to avoid the mathematical language in communicating with others. An elderly English political figure complains that none of the many eminent physicists with whom he has corresponded is courageous enough to pass any definite judgment on his proposed new theory of ether. Then he stresses the similarity between his theory and the concluding paragraph of a recent article by Dirac, and proceeds to discard as non–essential the accompanying mathematical passages in Dirac's article. It may be presumed that if he had also included comparable non–essential mathematical passages in his theory, he would have received more definite responses.

1.3 Clarity and consolidation

Does formalization help us to analyse and clarify concepts?

Often in formalizing ordinary concepts, we appear to have platitudes restated in pedantic obscurity; for instance, the mathematical definition of the continuity of a curve or the technical definition of the notion of effective computability. Moreover, the exact formalizations almost always distort our ordinary language at one place or another. For example, it has been pointed out that Russell's theory of descriptions does not apply to sentences such as "the whale is a mammal", and that sometimes in ordinary use the sentence "the king of France is bald" is neither taken as true nor taken as false.

In scientific investigations, we often recognize the advantage and even necessity of paying the price of considerable deviation from ordinary use of words in order to reach fairly precise terminology and notation. But, in what sense is, for instance, the technical notion of effective computability clearer than the corresponding common sense concept? Ordinarily, we would tend to say that the technical notion is *less* clear because it is more difficult to learn and a concept is clearer if and only if it is easier. We might speak of different kinds of clarity just as Mill speaks of different kinds of pleasure. Then we can also speak of a principle of preference: Only those who have experienced the feeling of clarity both of the ordinary notion and of the technical one are qualified to judge which is really clearer. And then, we hope, they will find the formalized notion clearer.

Perhaps we should also say that which definition of a term is clearer depends partly on the purposes we want the term to serve, and partly on our familiarity with the notions involved in each definition. The main advantage of the more articulate definition of a notion is, presumably, that it is sharper: for example, there are many cases where we can give a definite answer to the question whether certain given functions are effectively computable, only after we have made use of the technical notion of computability.

There are many cases where we could neither ask a univocal question nor obtain a univocal answer until we possessed the formalized notion. For example, we needed an exact definition of continuous curves before we could ask and answer the question whether there are space-filling continuous curves. And it was necessary first to formalize the notions of completeness and decidability before a negative answer could be given to the question whether number theory is complete or decidable.

Significant formalization of a concept involves analysis of the concept, not so much in the sense of analysis when we say that being a bachelor entails being unmarried, but more in the sense that an analysis of the problem of squaring the circle is provided by the proof of its unsolvability. When formalization is performed at such a level, it does serve to clarify and explicate concepts.

Another function of formalization is the clarification and consolidation of arguments or proofs. Sometimes we are not quite sure whether we have understood a certain given proof, sometimes we understand a proof once but fail to understand it again when reading it a few days later. Then there often comes the desire to work over the proof thoroughly, to make explicit all the implicit steps involved, and to write down the expanded result once and for all. With some people this desire to formalize and expand proofs may become a habit and a handicap to studying certain branches of mathematics. Yet occasional indulgence in this kind of thoroughness need not be a harmful thing.

In certain cases, there is no sharp line between formalizing and discovering a proof. There are many cases where essentially incomplete sketches, sometimes containing errors as well, get expanded and made into more exact proofs. Sometimes it is not until we have the thoroughly worked out proof on hand that we begin to perceive a connexion between it and the existing hint or sketch. Sometimes it seems hard to decide whether to consider the sketcher or the formalizer the true discoverer of the proof.

1.4 Rigour

In a sense, to formalize is to make rigorous.

There was Berkeley's attack on the mathematicians of his day entitled: " The analyst: or, a discourse addressed to an infidel mathematician. Wherein it is examined whether the object, principles, and inferences of the modern analysis are more distinctly conceived, or more evidently deduced, than religious mysteries and points of faith." There is the long story of how Lagrange, Cauchy, Weierstrass, and others strove to formalize exactly the basic notions of limits, continuity, derivatives, etc., providing thereby rigorous (though not necessarily reliable) foundations for mathematical analysis.

In the contemporary scene, we have logicians deploring how carelessly ordinary mathematicians use their words and symbols. Some logicians are puzzled that so many apparent confusions in mathematics do not lead more often to serious errors. On the other hand, mathematicians in turn complain about the inaccuracy of alleged proofs of mathematical theorems by physicists and engineers.

In the other direction, physicists consider that mathematicians are wasting their time when they worry about "foundational crisis"; mathematicians consider that logicians are indulging in learned hair-splitting when they devote pages and volumes to discussing the meanings of meaning or the use of quotation marks and brackets.

The right course is to be as rigorous and detailed as the occasion or the purpose requires. But this is more easily said than done. For example, certain authors seem to dwell tirelessly on the obvious, while skipping the crucial and more difficult steps.

The matter of distinguishing expressions from that which is expressed may serve to illustrate some of the questions about rigour. There were occasions when failure to be careful about the distinction actually hindered greatly the advance of logic. It is now customary in logic and philosophy to stress the difference, usually using quotation marks to separate, for example, the city Peking from the word "Peking". At present, even those who do not want to spend much time on using the quotation marks rigorously, often find it necessary to declare, for example, "quotation marks are omitted in most cases since we believe that no confusion will arise from this negligence". Every now and then, we run into certain articles in which the authors are so meticulous about using quotation marks that it becomes very difficult to read and understand what is being said.

One might even distinguish logicians into two groups depending on whether or not they always try to use quotation marks consistently and exactly. It may be a matter of temperament. Or it may also be a question of whether one happens to be either too lazy or too busy.

1.5 Approximation to intuition

To put thoughts in words or to describe a particular experience involves formalization of intuition. It has been contended that no finite number of propositions could describe exhaustively all that is involved in a particular experience. In other words, it is impossible to formalize without residue the complete intuition at the moment.

The matter of approximating intuition by formalization is clearer with regard to mathematics. For example, we know intuitively many things about integers. If we are asked to characterize our notion of integers, one way of answering is to say that integers form a group with respect to addition, they form an ordered set with regard to the ordinary relation of being greater than, and so on. The notions of group, ordered set, etc., are more exactly defined or more formalized than the notion of integers. Consequently, such answers tend to clarify somewhat our notion of integers, but they are usually inadequate because they fail to characterize unambiguously the integers.

We may compare the place of abstract structures such as group, field, ordered set, etc., in mathematics with the place of general concepts in ordinary life. They all can be considered as results of formalization or abstraction which serve as tools of thinking and research. As tools they help to economize our thought, as is often remarked. For example, not only integers, but transformations in space, etc., all form groups; anything that we prove about groups in general, of course, applies also to the special groups which may differ from one another in many respects. Similarly, there are many different chairs which can all be employed to support buttocks. In this way formalization, closely tied up with abstraction, produces useful tools.

On the other hand, it is often hard to characterize adequately our intuition through the use of formal structures. For example, it is not easy to describe exactly the colour, shape, etc., of a particular chair. Peano's axioms are thought to be capable of characterizing completely our notion of positive integers. Yet, as Russell observed long ago, Peano's axioms are satisfied by all progressions such as the odd positive integers, the negative integers. Russell thought that only by calling in a set theory could we make a univocal characterization. More recent advances in logic show that he was wrong even in believing this.

In fact, as we know, there are important results which indicate unmistakably that we can formalize without residue neither the fundamental intuitive notion of positive integers nor the basic notion of sets or classes.

Thus, there is Gödel's famous theorem according to which, for any fairly rich system, we can find some property expressible in the system such that we can prove for each of the integers 1, 2, . . . that it has the property, but we cannot prove the general statement that all positive integers have the property in question. In other words, although intuitively if P(1) (*i.e.*, 1 has the property P), P(2), P(3), . . . are all true, then it must be the case that all positive integers have the property P; yet in no fairly strong logistic system can we formalize adequately this intuition so as to guarantee the performability of such an inference for all the properties P expressible in the system. It also follows that no ordinary axiom system can preclude the interpretation that besides the ordinary 1, 2, . . . the set of positive integers also contains certain other queer things; there is no way to formalize in an ordinary logistic system our intuition that 1, 2, . . . are the only integers.

On the other hand, there is no axiom system in which we can get *all* the real numbers or the classes of positive integers. This follows easily from Cantor's famous argument for non-denumerability. Thus, given any axiom system, we can enumerate all the classes of positive integers which can be proved to exist in the system, either by applying Löwenheim's theorem or by reflecting on the fact that the theorems of existence in the system can be enumerated. Hence, if we define with Cantor a class K of positive integers such that for each *n*, *n* belongs to K if and only if *n* does not belong to the *n* th class in the enumeration, then the existence of K cannot be proved in the system. In other words, although in the system we can also speak of all the classes of positive integers, we cannot really formalize without residue the intuitive notion of "all" with regard to classes of positive integers; in each formalized axiom system, there is always some class of positive integers that is left out.

1.6 Application to philosophy

The application of mathematical logic to the treatment of philosophical problems may also be viewed as an attempt to formalize. Such applications often give the impression that a formidable technical book expresses in tiresome exactitude more or less commonplace ideas which could be conveyed more easily and more directly in a few sentences of plain language. Yet, undoubtedly, there are cases where the appeal to formalization is of more than pedantic interest. For instance, Heyting's formalization of the intuitionistic view of logic and mathematics helps quite a bit in conveying Brouwer's ideas to those people who have a radically different orientation. Another example is the gradual formalization of the notion of being a definite property, employed for defining sets in Zermelo's axiomatic treatment of set theory.

Perhaps we can compare many of the attempts to formalize with the use of an airplane to visit a friend living in the same town. Unless you simply love the airplane ride and want to use the visit as an excuse for having a good time in the air, the procedure would be quite pointless and extremely inconvenient. Or we may compare the matter with constructing or using a huge computer solely to calculate the result of multiplying seven by eleven. When the problems are so simple, even the task of translating them into a language which, so to speak, the machine can understand would already take longer than if we were to calculate the results by memory or with a pencil and a sheet of paper.

It is a practical problem to decide what means of transportation to use in making a certain particular trip, or to decide whether it is feasible to build a computer to handle a certain given type of question. As we know, there are many different factors which are ordinarily taken into consideration before making the decision. Similarly, it is also a practical problem to decide in each particular case whether it is profitable to apply mathematical logic in handling a definite kind of problem. The only difference is that the factors which have to be considered here are often more involved and less determinate.

Take the principle of verification. Various attempts at giving an exact definition of the notion of verifiability have failed. And systematic use of the logistic method has been recommended as the only way to a satisfactory solution. On the other hand, there is also the view that the important thing is a general attitude expressed vaguely in the rough principle of verification, rather than an exact definition of verifiability. Underlying this dispute, perhaps, are the varying attitudes toward the general desirability of crystallization of ideas.

This raises larger problems. Why should we want such crystallization in philosophy? What is the function and business of philosophy? Fortunately, general observations can be made without going into such hard questions.

1.7 Too many digits

After sketching an axiom system for his theory of probability, F. P. Ramsey goes on to say, "I have not worked out the mathematical logic of this in detail, because this would, I think, be rather like working out to seven places of decimals a result only valid to two". There are several disadvantages in working out a result to too many places. It uses up time which might be spent otherwise. It also makes the result harder to memorize or to include in future calculations, if anybody should want to make use of it. And pointless problems would arise regarding the last five places: do they exhibit any interesting pattern which would indicate the lawfulness of nature? Do they coincide with the five digits starting with the 101st in the decimal expansion of π? and so on.

How do we decide whether a result is valid only to two places? If the same experiment is repeated under different but, so far as we know, equally favourable circumstances, with results which agree satisfactorily only to the first two places, then we tend to conclude that the places after the second are not quite reliable. If most people refuse to calculate up to many places and a single person has an irresistible itch for reporting every result to at least seven places, it might be rather hard to decide whether his result is right.

The matter of constructing an exact theory of (say) probability contains an additional factor. Since ordinary language is not exact, new words are coined or ordinary words are given technical usage. In order to evaluate the theory, you have first to understand it. In order to understand it, you have first to learn a new language. Since it is usually impossible to explain clearly and exactly even the technical usages, a formal or exact theory can almost always be defended against charges that it does not conform to fact. As long as there is a sufficiently complicated system and a fairly big and energetic group of people who, for one reason or another, enjoy elaborating the system, we have a powerful school of learning, be it the theory of meaning, the sociology of knowledge, or the logic of induction. There is always the hope that further development of the theory will yield keys to old puzzles or fertilise the spirit of new invention. In any case, since there is mutual support between different parts of a given system, there is little danger that the discrepancy between one part and the facts should discredit the system. And of course if we are interested in the "foundations", there is no need to fear any immediate tests. The worst that can happen to such theories is not refutation but neglect.

1.8 Ideal language

Language is employed for expression and communication of thoughts. Failure in communication may either be caused by inadequate mastery of the language, or by internal deficiencies of the language: that is, if there is thought to be conveyed at all. Language is also sometimes used for talking nonsense. Here again, certain languages just seem to offer stronger temptations for doing so. And sometimes the language user is not careful enough, or he merely parrots others. In such cases he does not have thoughts or feelings to express, and there is, of course, no question of correct communication. A less serious disease is confused thinking, often involving internal inconsistency. This again is sometimes the fault of the language, such as the ambiguity of words and a misleading grammar.

The creation of an ideal language would yield a solution of these difficulties once for all. Such a language should be so rich, clear, and exact as to be sufficient both for expressing all thoughts and feelings with unmisunderstandable clarity, and for precluding nonsense. Given such a language, many problems now known as philosophical would be dissolved. Disagreement about what is to be taken as nonsense would lead to the construction of different ideal languages. There would be then the problem of understanding each other's ideal language.

An alternative to the ideal language is to handle each individual case separately and thoroughly. To explain at great length what we intend to say, to give concrete examples when possible, to invite questions and discussions. And to reflect carefully and ask what we really want to say, whether we do have something to say, whether we are not misled by false analogies or naive syntax.

The task of constructing a comprehensive ideal language is in many ways similar to that of finding a mechanical procedure to decide answers to all problems of mathematics. They are equally impossible. If and when these two tasks are clearly formulated, the impossibility can be proved definitely in both cases. In certain simple areas of logic and mathematics, we do possess decision procedures. Similarly in mathematical logic and theoretical physics we have more exact languages. But there is no mechanical method for finding decision procedures, and each significant mathematical problem calls for a special treatment. It is demonstrably impossible to reduce all mathematics to its decidable portion. It seems equally impossible to fit everything we say into the language of logic and physics. Moreover, these languages are more exact in their abstract setup than in their actual use. It is a familiar experience that mathematicians who know the language of mathematics very well often offer fallacious proofs.

The quest for an ideal language is probably futile. The problem of formalization is rather to construct suitable artificial languages to meet individual problems.

1.9 How artificial a language?

The contrast between natural and artificial languages suggests a sharp distinction. Russian is natural, while Esperanto is artificial. But is the language of the biologists or that of the philosophers natural or artificial? Is Mr. Woodger's proposed language for biology natural or artificial? Hilbert's language for the Euclidean geometry is more exact and artificial than that of Euclid's *Elements*. So far as the development of human scientific activities is concerned, the creation of the language of the classical mechanics or of the axiomatic set theory was rather natural.

We might speak of degrees of artificiality, as perhaps measured by the amount of deviation from the natural course. The Chinese language spoken today differs to a rather great extent from that used two thousand years ago, although the changes have been mostly natural. If we had attempted two thousand years ago to bring about the same changes in one year's time, we would have had to create at that time a language quite artificial. To introduce an artificial language is to make a revolution. Unless there are compelling natural needs, the resistance will be strong and the proposal will fail. On the other hand, when an artificial language meets existing urgent problems, it will soon get generally accepted and be no longer considered artificial. Hence, it may be more to the point if we compare artificial languages with Utopian projects.

Attempts to formalize the theory of probability are sometimes criticized on the ground that the efforts fail to make contact with the crucial and burning problems of physical science. One ready reply is that the situation is the same with many interesting investigations in branches of mathematics such as abstract algebra, set theory, and topology. One may argue, however, that more new ideas and methods are introduced through such studies than through the researches on foundations of probability theory. Or maybe there is more substance behind the new languages of algebra and set theory and results obtained there are not as easily discredited by slight shifts of emphasis or subtle mistakes in the original analysis.

Mrs. Joan Robinson somewhere remarks that economists are usually behind their time. An urgent practical problem often ceases to be urgent or practical long before the discovery of a theoretically satisfactory solution. Whether it is worthwhile to continue the search for the solution of a problem which is no longer urgent depends to a large extent on whether the particular problem is intimately connected with larger issues, whether it is sufficiently intriguing intellectually, and whether it is likely to recur in the near future. Similarly, the value of an artificial language has to be decided in accordance with its elegance and its usefulness either in its direct applications or as a model to be followed in future constructions. In a certain sense, an interesting artificial language must not be excessively artificial.

1.10 The paradoxes

Much time and space has been devoted to the discussion of the logical paradoxes or contradictions. Sometimes it is said that these paradoxes bring to light the self–contradictory character of our logical intuition. Indeed, as we know, the formalization of logic and set theory was largely motivated by a desire to avoid the paradoxes and yet obtain what we ordinarily want.

It has been suggested that we take the paradoxes too seriously, largely because of our preoccupation with formalization and our lack of flexibility.

What is proposed instead seems to be this. Suppose we find a contradiction by a seemingly plausible argument. Since we get a contradiction, we see that the argument is really not correct and indeed must be faulty. So let us remember never to use the argument again. And that is the end of the matter.

However, when we say that the argument looks plausible, we mean, among other things, that each step of the argument also looks plausible. It seems necessary not only to reject the whole argument as a unit but to pin down exactly which step or steps in the argument caused the trouble. Hence, there are the various attempts to reject one or another of the steps as unwarranted. But why can we not say that although each step is in itself all right, they must not be combined in the particular way that leads to the contradiction? Indeed, we may even use this possibility to justify the attitude of indifference, on the part of many working mathematicians, toward the paradoxes.

It is only when we come to constructing a formal system to embody our arguments that this procedure proves awkward. In a logistic system, we break up proofs and arguments into isolated steps so that if a step is valid at all, it is valid no matter where it occurs. In other words, certain combinations of shapes are taken as axioms so that they can be asserted as valid no matter where they occur; and certain (finite) sequences of combinations of shapes are taken as justified by the rules of inference so that any such sequence, wherever it occurs, is taken as determining valid steps. For instance, if we agree to take as an axiom, for two specific sets named a and b, the assertion "Either a belongs to b or a does not belong to b", we can no longer reject the same statement as an unwarranted step when it occurs in an argument that leads to a contradiction.

Two alternatives to the customary logistic method are: (1) not to attempt any exact characterization of all the valid arguments of any important branch of mathematics; (2) to list either all or samples of all the warranted and unwarranted whole specific arguments as inseparable units, instead of trying to break up all warranted arguments into a small number of basic atomic steps. The alternative (2) will either produce quite messy results or lead to something which is hardly distinguishable from a logistic system.

2. THE CONCEPT OF COMPUTABILITY *

2.1 Formalizing intuitive concepts

Many philosophical problems take the form: What is x? For example, what is time? What is truth? What is good? What is probability? What is a set? What is a number?

There are different kinds of answers to questions of this sort. Gauss and Wessel gave interesting and definite answers to the question: What are complex numbers? These answers are still preserved in textbooks, while speculative discourses by their contemporaries on the same topic have been quickly forgotten. The answers by Gauss and Wessel are usually not considered philosophy.

What Kant had to say on time is often taken as philosophy, and probably interesting philosophy. Frege's answer to the question "What are numbers?" is perhaps also philosophy. How about Dedekind's closely related theory on the same question? If we construe an axiomatic theory of sets (for example, Zermelo's) as an answer to the question "What are sets?" — is the answer also philosophy?

When Zermelo presented at first his theory of sets, he did not give a precise explanation of his notion of "definite property." Later on Skolem and others succeeded in making the notion more precise. Skolem's formalization of the notion is now generally accepted, just as the ε–δ–definition of continuous curves is now widely employed as a substitute for the intuitive notion of "traceable without lifting the pencil from the paper."

The notion of recursive or Turing computable function as a formalization of the intuitive notion of computable function is another example along the same line.

Such questions of formalization are sometimes called mathematics, sometimes called philosophy: a terminological dispute, important perhaps only for sociological reasons which need not concern us here.

Among systematic procedures, computation procedures are the most interesting mathematically.

2.2 The intuitive concept of computability

While in elementary schools and markets we are primarily concerned with individual problems of computation, mathematicians are mostly engaged in proving theorems or, in other words, deciding the truth and falsity of mathematical

---

* Not published previously.

propositions. The mathematician is, nonetheless, also quite interested in the problem of finding algorithms or computational routines, although particular questions of computation such as "what is 352 × 267" or "what is 27 + 76" are too simple to interest him.

An algorithm is a set of instructions which, when followed mechanically, will yield the answer to not only a single question, but any one of a whole class of related questions. For example, we have algorithms for addition and multiplication and for deciding whether a given number is a prime. As a result, we are inclined to think that we can carry out addition and multiplication simply by obeying a few easy rules and that machines can answer such questions too. Thus , while the schoolboy wishes to know the sum and product of particular numbers the mathematician is interested in the problem whether and how a particular class of questions can be answered by a general method. Since it is natural and customary to correlate a class of questions with a function (e.g., the class of questions of finding sums with the function of addition), the mathematician's problem becomes one of asking whether a given function is computable. A more general problem is to determine the totality of all computable functions. Each determination of the totality may be viewed as a proposed definition of the concept of computability. Moreover, it is quite satisfactory to confine ourselves to functions of (nonnegative) integers.

In order to determine the totality of computable functions, there are the two directions of beginning from outside and beginning from inside. Thus, knowing functions which are not computable can help; or, if we know sufficiently many computable functions, we can try to find out their common characteristics and then generalize.

First, let us find a function which is demonstrably noncomputable. The usual procedure is to assume given the totality of computable functions and then "diagonalize" to get a function that is not computable. We need not go into the familiar argument, as noncomputable functions got in this manner are clearly of no direct assistance to the task of determining the totality of computable functions.

We are left with the alternative of analysing given computable functions. Since addition and multiplication are the most familiar computable functions, it is natural to study first the rules governing the algorithms. These are embodied in the familiar recursive definitions:

$$\begin{cases} a + 0 = a, \\ a + b' = (a + b)'. \end{cases} \qquad \begin{cases} a0 = 0 \\ ab' = ab + a. \end{cases}$$

It is easy to convince ourselves that all functions defined by such recursions together with some other simple schemata are computable. In this way, we get the class of primitive recursive functions (compare below) and the conclusion that all primitive functions are computable.

There is a strong temptation to identify effective computability with primitive recursiveness, since practically all the effectively computable functions which are ever used in ordinary mathematics are primitive recursive. As we know, however,

Ackermann many years ago already found an effectively computable function which is not primitive recursive. Hence, there are more effectively computable functions than there are primitive recursive ones. The problem is to introduce and use a wider notion of recursiveness, and the answer is general recursive functions.

Instead of beginning with the recursive definitions for addition and multiplication, one can also analyse the process of human computation into simple steps and develop the idea that computation is essentially a mechanical procedure. This approach leads to definitions of computability by Post and Turing.

In 1934 Gödel introduced a definition of general recursive functions which he credited in part to an oral suggestion of Herbrand (see Gödel 1934 in bibliography listed at the end of the paper). Using a variant of this definition due to Kleene, Church proposed to identify the intuitive notion of effective computability with the notion of general recursiveness in Church 1936. A little later, Turing gave (in Turing 1936–37) an explication of the notion of effective computability in the form of a definition of computing machines (commonly known as Turing machines). It was afterwards established that a function is general recursive if and only if it is Turing computable. Since the questions which I wish to discuss concern Turing's thesis and Church's thesis in the same way and since they are more easily formulated in terms of the notion of recursiveness, I shall confine my discussions mostly to Church's thesis.

2.3 Computation by theoretical machines

I should, however, like to describe a variant of Turing's theory which is very simple and can be easily understood also by those who are familiar with ordinary digital computers but not accustomed to mathematical logic. (Compare my abstract in the *Bulletin of Am. Math. Soc.*, 1954.)

It can be proved that a theoretically simple basic machine can be imagined and specified such that all partial recursive functions (and hence all solvable computation problems) can be computed by it and that only four basic types of instruction are employed for the programs: shift left one space, shift right one space, mark a blank space, and conditional transfer. In particular, erasing is dispensable, one symbol of marking is sufficient, and one kind of transfer is enough. [For details, see my 1957(54).]

From this it follows that theoretically each computation can be performed by drawing up a program consisting of ←, →, *, and conditional transfer. Conditional transfer is clearly the most complex: it involves distinguishing a blank square from a marked square and responding accordingly. It seems rather comforting that in order to study the basic operations involved in computations, we can confine ourselves to these four types of operation.

2.4 General recursive functions

Of the two halves of Church's thesis, more weight is usually put on the half stating that all functions (of natural numbers) which are effectively calculable according to the intuitive notion are general recursive. Indeed, this is the half which has led to the many important negative results: there is no effectively calculable function (decision procedure) for such and such questions. The value of Church's thesis in proving these results is obvious: it achieves a great simplification in the form of all the functions which may be effectively calculable. If a sharp boundary is found so that a given indeterminate region falls completely within the interior of the boundary, then everything lying beyond the boundary lies also outside of the initial indeterminate region. Clearly the sharp boundary helps matters when we wish to establish that something is not in the initial region.

In this note I shall take for granted this part of Church's thesis but discuss the other half which is often considered less questionable and less important. I propose to distinguish two kinds of calculability. For lack of better words I shall speak of a priori calculability and effective calculability. In some ways the distinction resembles that between logical possibility and theoretical possibility. I shall argue that although it seems reasonable to identify a priori calculability with general recursiveness, effective calculability can best be identified with a more or less vague and indeterminate notion which is presumably narrower than general recursiveness but broader than primitive recursiveness.

To facilitate discussions, let me briefly review the usual definition of general recursive functions. I quote in rough form a version of Kleene 1943:

D.1 A function $\phi(x_1,...,x_n)$ is general recursive, if there exists a finite set E of equations involving the symbol = , the numerals, the variables, and one or more function symbols (f being the last one) such that for each set $m_1,...,m_n$ of numerals (representing the nonnegative integers $x_1,...,x_n$), an equation $f(m_1,...,m_n) = m$ is derivable from E for exactly one numeral m (representing the value of $\phi(x_1,...,x_n)$), by the following two rules:

R1: to substitute numerals for variables in a given equation;

R2: given a numerical equation $f(m_1,...,m_j) = k$ to replace a part $f(m_1,...,m_j)$ of the right member of an equation by k.

In the above definition, two assertions of existence are involved: the existence of a set E of equations containing = , variables, numerals and function symbols, and the existence for any given $m_1,...,m_n$, of a derived equation of E (by R1 and R2) of the form $f(m_1,...,m_n) = m$. If we assume a set E of equations given and ask whether it does provide a general recursive definition, then we have to face merely the following existential assertion: for any $m_1,...,m_n$, *there exists* a derived equation of E which has $f(m_1,...,m_n)$ as left member and a numeral as right member. The question arises how we are to arrive at the existential assertion.

Or, in arithmetized form, definition Dl amounts to this:

D2. A function h(n) is general recursive if there are primitive recursive functions f(n) and g(n,m) with the following properties:

(a) for every numeral $0^{(n)}$, there exists a numeral $0^{(m)}$ satisfying $g(0^{(n)}, 0^{(m)}) = 0$;

(b) for every $0^{(n)}$, $h(0^{(n)}) = f\{\mu_x[g(0^{(n)},x) = 0]\}$.

The phrase "there exists" in condition (a) has a clear and definite meaning. Thus, (n) (Em) [g(n,m) = 0] has a very simple arithmetic interpretation: for every numeral $0^{(n)}$, there exists a numeral $0^{(m)}$ such that $g(0^{(n)}, 0^{(m)}) = 0$ is verifiable (i.e., can be proved numerically). Nonetheless, definition D2 leaves open by what methods condition (a) is to be established.

Church (see Church 1936, p. 351, footnote 10; Kleene 1952, p. 319) anticipates this question and replies that all we have to do is to interpret the phrase "there exists" in a constructive sense or to make sure that the existence assertion is proved constructively. He then leaves the burden of determining the criterion of constructiveness to the reader. This seems too easy a way out. What happens appears roughly like this. Church proposes to tell us in exact terms what we mean vaguely by effectively calculable functions. In his explanation, he makes use of a difficult phrase "there exists" in a context of the kind which has caused constant trouble in attempts to explicate the notion of effectiveness. We ask Church to elaborate his uses of "there exists." He tells us that we need only construe the phrase in a constructive sense and supply our own criterion of constructiveness. But this is disappointing since we have always found it hard to say in exact terms what the constructive or effective sense of "there exists" is.

The great difficulty with this modified position is of course the problem of constructive proofs. No very satisfactory theory of constructive proofs is in existence, and it is apparently hard to obtain such a theory.

Definition I seems, however, to be faced with similar difficulties. [This apparently refers to a definition in an early version. I believe that it could be reconstructed from the considerations below in section 8 and is more closely related to the original definition of general recursive functions in Gödel 1934 than to D1 or D2.] The term "constructive ordinal" occurs explicitly in the definition, and the term "effective well-ordering" occurs implicitly, as it is used in the definition of ordinal recursive functions. There arise the questions: What is a *constructive* ordinal? What is an *effective* well-ordering? These questions appear to be more specific than the question of constructive proofs. But it is not clear that they are easier.

For example, I would consider the ordinal numbers represented in Ackermann 1951 as constructive ordinals. On the other hand, I do not feel sure that I would accept the whole "constructive second number class" of Church 1938 and Kleene 1938.

2.5 Constructive proofs

This reduces the problem of characterizing effectively computable functions to that of characterizing effective or constructive proofs.

Instead of asking what an effective proof is, one may ask, how is condition (a) to be established? First, not all conditions (a) can be proved in a single formal system (e.g., Z_μ in HB or Heyting's system), since the functions which are provably general recursive in this system can be enumerated and the diagonal argument can be applied to get a new general recursive function. Thus, for instance, it is not sufficient to identify effectively computable functions with those for which condition (a) can be established in Heyting's arithmetic (the system described, e.g., on p. 82 and p. 101 of Kleene 1952).

Second, if one looks at ordinary informal proofs, one can distinguish two classes. There are many proofs ("purely arithmetical" ones), where effective majorizing functions $\beta(m)$ can be extracted by the substitution method of HB. For example, by Ackermann 1940, if (m) (En) R (m,n), R primitive recursive, is provable in Z_μ, then a transfinite recursive function $\beta(m)$ of order ε_0 (compare below) can be extracted from the proof such that (m) (En) $[n < \beta(m) \& R(m,n)]$ holds. Similarly, one can also extract effective majorizing effective functions from proofs in systems of set theory which admit no impredicative sets. On the other hand, there are also set-theoretic proofs for which such effective bounds have not been extracted, e.g., in Zermelo's axiomatic set theory. Here, in fact, it is not even clear that the methods of proof are consistent.

The bounds for proofs in Z_μ can be obtained from Ackermann 1940 in the following manner. Thus, suppose (x)(Ey) Rxy (R primitive recursive) has been proved in Z_μ. Let g, e. j be respectively the number of ε-matrices, the number of ε-terms, and highest degree of terms in the proof. Let further ω, τ, ρ, λ be the functions as defined in Ackermann 1940. Then the bound $\beta(x)$ is given by:

$$\omega\{x + j,\ \lambda[\tau(j,e,g,1,\ \rho(g,e)),\ g + 1]\}.$$

We may wish to say that a general recursive function is effectively computable if the condition (a) in D2 of the form (x) (Ey) Rxy is proved constructively, and that a constructive proof of (x) (Ey) Rxy is one from which we can extract an effectively computable function $\beta(x)$ so that (x) (Ey) $(y < \beta(x) \& Rxy)$ holds. Clearly we get then into a circle which need not be harmful, provided we have an independent source to supply a sufficiently large class of computable functions to begin with.

We may also avoid the circle by leaving it to the light of nature to see whether any given proof is constructive or not.

In connection with constructive proofs, it may be of interest to consider the questions of formalizing constructive proofs in given formal systems. By Gödel's famous theorem, a formula of the form (x)Ax (A primitive recursive) which expresses the consistency of Z_μ is not provable in Z_μ. Yet from investigations of Gentzen and others, we know also that there is a simple primitive recursive well-ordering $<_e$ of order ε_0 such that we can prove in Z_μ and prove constructively: (1) A(0); (2) (n) $[n <_e m \supset A(n)] \supset A(m)$. The reason why in Z_μ we cannot infer (x)Ax from (1) and (2) is usually said to be the fact that we cannot prove in Z_μ the principle of transfinite induction with respect to $<_e$. The same fact can also be expressed by saying that we cannot prove in Z_μ that the well-ordering $<_e$ is actually a well-ordering, because to assert that $<_e$ is a well-ordering implies that for each formula A of Z_μ, we can infer (x)Ax from (1) and (2). Indeed, we can prove constructively, though not in Z_μ, that $<_e$ is a well-ordering.

This leads to the question whether, if we depend on the light of nature to judge the constructive character of proofs, it might not be possible to prove constructively all true propositions of the form (x)Ax (A primitive recursive) of the form (x)(Ey)Rxy (R primitive recursive) in a similar manner. In other words, whether it may not happen, for instance, that for each Rxy there is a simple well-ordering $<_\alpha$ such that we can prove constructively: (1) $<_\alpha$ is a well-ordering relation; (2) (Ey)Roy; (3) (m) (m $<_\alpha$ n $\supset$ (Ey)Rmy) $\supset$ Rny. If the answer is yes, then all true assertions of consistency and general recursiveness are provable constructively. Thus, in particular, it will follow that, given an arbitrary formal system, if it is consistent, then its consistency can be proved constructively, even though actually to discover the constructive consistency proof can be very difficult. It is not easy to see how such general questions of constructive demonstrability can be settled. Turing 1939 gives the impression of dealing with such problems, the treatment is hard to follow.

Incidentally, if all true formulae of the form (x)(Ey) Rxy are provable constructively, then of course all true formulae of the form (x)Ax are so provable too: we need only take (x) (Ey) (Ax&y = y) in place of (x)Ax. The converse is not true. By suitable use of the diagonal argument, we can prove that there is some true formula (x)(Ey)Rxy (R primitive recursive) which is not derivable from any (x)Ax (A primitive recursive) in the system Z_μ.

2.6 Effective methods

Kleene gives a more or less standard explanation of the notion of constructive proofs in the following words: "Therefore an intuitionistic proof of the proposition *there exists an n such that P(n)* must be *constructive* in the following (strict) sense. The proof actually exhibits an example of an n such that P(n), or at least indicates a method by which one could in principle find such an example." (Kleene 1952, p. 49.) Here, of course, the problem is shifted to knowing exactly what kind of *method* is acceptable. With regard to the case which concerns us, we do have a method of finding the corresponding equation for given n. The method consists in enumerating all the derived equations of E until we come upon one with f(n) as left member and a numeral as right member. Indeed, no matter how we have reached the existence assertion, this method is always applicable in the sense that we shall sooner or later arrive at the equation, since it does *exist* in the infinite sequence of derived equations. Either we find such methods acceptable, then the question whether the existence assertion is proved constructively makes no more difference. Or, what is more likely, we find the methods unacceptable because we want to have *effective* methods, and we feel that the method of enumerating derivative equations is not always an effective one for finding the desired equations. To decide between the two alternatives, we need a more exact explanation of the notion of effectiveness.

To put the matter in a different manner, there are two distinct ways of interpreting the notion of effectiveness. A method of finding the required equations is in one sense said to be effective if there is a mechanical procedure for writing out one by one the derived equations so that for each set of given argument n, an equation will eventually turn up which has Φ (f(n)) as left member and a numeral as right member (in other words, such an equation *exists* or occurs in the infinite sequence of derived equations of E). If we accept this sense of effectiveness, then it seems unimportant whether or not the existence of such derived equations is proved constructively. On the other hand, if, as is done in this note, a method of finding the desired equations is said to be effective only if we have, with regard to the mechanical procedure, a *pretty good idea* how soon the process will terminate for each set of given argument values $m_1,...,m_n$, then the question whether a set E of equations defines an effective method of calculating values of its corresponding function becomes almost a subjective matter or rather more correctly, a matter relative to our present state of knowledge. This is perhaps as it should be. The situation is somewhat similar to the distinction between logical possibility and theoretical possibility, i.e., something for which there is a blueprint based on currently accepted laws of nature. So far it seems theoretically impossible to travel to the moon, although, for all we know, the journey may become a theoretical possibility soon and be realized in another fifty years. The analogue of practical possibility is practical computability, i.e., a computation which could be carried out with the available computing machines.

Corresponding to the two different interpretations, a distinction between a priori computability and effective computability is being made.

When one asserts that all general recursive functions are effectively calculable, one has to add the proviso that the existence clause gets a constructive proof. If we wish to, this proviso can also be introduced in the definition (D1 or D2) of general recursive functions, so that one could speak without qualification, using the changed definition, that all general recursive functions are effectively calculable.

If one is not able or not willing to handle the problem of constructive proofs, there are two alternatives. One could simply drop the proviso and try to get along with the unqualified thesis that any function which is general recursive by D1 or D2 is effectively computable. Or, if such a course proves defective, one could propose some alternative course which avoids or at least postpones the problem of constructive proofs.

In what follows, I shall try to indicate some of the defects of the first procedure and discuss certain alternatives. We shall see that some such problem as the nature of contructive ordinals or constructive proofs will probably have to be faced at one stage or another. It may, however, be that different persons would prefer different problems even when they are similar.

2.7 Speed functions

The question of deciding the effective computability of a function can be reduced to one of knowing intuitively its speed functions: corresponding to each set of equations E which define an *a priori* calculable function f(n), there is some speed function $f_s(n)$ which, for each constant n_0, gives an upper bound to the number of steps needed for getting the value of $f(n_0)$ from E.

Obviously, each calculable function has many definitions and each definition has many corresponding speed functions. In most cases, we are concerned with what seem to be the "natural" definitions and the "natural" speed functions. I shall often speak of the speed function of a calculable function, as if there were a unique speed function.

To decide whether an *a priori* calculable function f(n) is effectively calculable, it suffices to estimate the values of its speed function $f_s(n)$. At first it might be thought that speed functions are simpler than their corresponding functions, and that an elegant classification of all general recursive functions can be obtained by a suitable grouping of their speed functions. A closer look, however, seems to reveal that in general the speed function of a given general recursive function is just about as complex as the function itself.

It is quite easy to establish that every primitive recursive function has a primitive recursive speed function. In each case, given the ordinary definition of a primitive recursive function, we can easily find its corresponding speed function.

It will be recalled that a function is primitive recursive if it can be defined by a sequence of applications of the following schemata.

(I) $\phi(x) = x'$,

(II) $\phi(x_1,\ldots,x_n) = c$,

(III) $\phi(x_1,\ldots,x_n) = x_i$,

(IV) $\phi(x_1,\ldots,x_n) = \theta[\chi_1(x_1,\ldots,x_n),\ldots,\chi_m(x_1,\ldots,x_n)]$, $\phi(0) = c$,

(Va) $\phi(y') = \chi[y,\phi(y)], \phi(0, x_1,\ldots,x_n) = \psi(x_1,\ldots,x_n)$,

(Vb) $\phi(y', x_1,\ldots,x_n) = \chi[y,\phi(y,x_1,\ldots,x_n)x_1,\ldots,x_n]$

To find the corresponding speed function of a primitive recursive function, it suffices to write out the corresponding speed function for each of the above schemata:

(I_s) $\phi_s(x) = 1$,

(II_s) $\phi_s(x_1,\ldots,x_n) = 1$,

(III_s) $\phi_s(x_1,\ldots,x_n) = 1$,

(IV_s) $\phi_s(x_1,\ldots,x_n) = (\chi_1)_s(x_1,\ldots,x_n) + \ldots + (\chi_m)_s(x_1,\ldots,x_n) + m$

$+ \theta_s[\chi_1(x_1,\ldots,x_n),\ldots,\chi_m(x_1,\ldots,x_n)] + 1$, $\phi_s(0) = 0$,

(Va_s) $\phi_s(y') = \phi_s(y) + 1 + \chi_s[y,\phi(y)] + 1$,

$\phi_s(0,x_1,\ldots,x_n) = \psi_s(x_1,\ldots,x_n) + 1$,

(Vb_s) $\phi_s(y',x_1,\ldots,x_n) = \phi_s$

$(y,x_1,\ldots,x_n) + 1 + \chi_s[y,\phi(y,x_1,\ldots,x_n), x_1,\ldots,x_n] + 1$

Since we feel we have a pretty good idea about the values of these speed functions for given argument values, we know that all primitive recursive functions are effectively computable. The "pretty good idea" can be obtained by examining the set of equations which define the function in question.

Similarly, the speed function of a general recursive function is again a general recursive function.

This can be proved by using the fact that every general recursive function can be defined by a sequence of schemata of the forms (I)—(V) and (VI) (see, e.g., Kleene 1952, p. 289):

$$\text{(VI)}\quad \phi(x_1,\ldots,x_n) = \mu_y[\rho(x_1,\ldots,x_n,y) = 0],$$

where the following holds:

$$\text{(VI–c)}\quad (x_1)\ldots(x_n(Ey))[\rho(x_1,\ldots,x_n,y) = 0].$$

It is known that (VI), subject to the condition (VI–c), is like (I)—(V), general recursive. For example, the function ϕ can be defined thus (Kleene 1943, p.45):

$$\text{(VI}'\text{)}\begin{cases}\sigma(0,x_1,\ldots,x_n,y) = y\\ \sigma(z',x_1,\ldots,x_n y) = \sigma[\mathrm{o}(x_1,\ldots,x_n,y'),x_1,\ldots,x_n,y'],\\ \phi(x_1,\ldots,x_n) = \sigma[\rho(x_1,\ldots,x_n,0),x_1,\ldots,x_n,0].\end{cases}$$

Here we have a typical case where, although we know, by (VI–c), that the process of computation will eventually terminate for any given argument values $x_1,\ldots x_n$, we have no idea how soon that will happen. The condition (VI–c) in general gives us no information as to how big a number y could satisfy the equation $\rho(x_1,\ldots,x_n,y) = 0$, for given $x_1,\ldots,x_n$. In other words, when the schema (VI) is involved in the definition of a certain function, the function is computable but not necessarily effectively computable.

This becomes even clearer, if we look at the speed function of (VI′) which is again general recursive:

$$\text{(VI}'_s\text{)}\quad \phi_s(x_1,\ldots,x_n) = 1 + \Sigma_{i=0}^{\mu_y[\rho(x_1,\ldots,x_n,y)=0]}[\rho_s(x_1,\cdots,x_n,i) + 3].$$

It follows that the speed function of a general recursive function is again general recursive: a conclusion which is, however, of little value so far as the question of classifying effectively computable functions is concerned.

2.8 Transfinite recursions

While it is accepted in this note that all effectively computable functions are general recursive, it now appears questionable that conversely all general recursive functions are effectively computable. In any case, it seems clear that knowing that a certain given function is general recursive does not entail knowing that it is effectively computable.

One can follow Church and identify effectively computable functions with those general recursive functions which can be proved to be recursive by constructive means. But this gives little clue to a classification of all effectively computable functions into some sort of hierarchy.

In order to reach a more informative characterization of these other functions, it seems most suitable to use the notion of transfinite recursion introduced in Ackermann 1940 and further developed by Péter (see, e.g., Péter 1951). As they have not made any general discussion of the matter, I venture to describe what I understand to be involved in their investigations.

As is well known, every ordinal number of Cantor's second number class is of the same cardinality (viz., $\aleph_0$) as the class of natural numbers. It is also known that theoretically we can rearrange the natural numbers to get a well–ordered set of any ordinal type of the second number class.

For example, if we put all even numbers after all odd numbers, we get a set whose ordinal number is $\omega 2$. Or again, we can rearrange natural numbers to get a set with the first ε–number as ordinal type in the following manner. Let A_i be the class of all positive integers divisible by 2^i but not by 2^{i+1} ($i = 1,2,3,\ldots$), $v_1(a_1,a_2)$ be $2^{a_1-1}(2a_2+1)$, and $v_{i+1}(a_1,\cdots,a_{i+2})$ be $2^{v_i(a_1,\ldots,a_{i+1})-1}(2a_{i+2}+1)$ $(a_1 > 0)$. Represent the ordinal 0 by the natural number 0, the final ordinal $n (n > 0)$ by the natural number $2n-1$, $\omega^i a_1 + \ldots + \omega a_i + a_{i+1}$ by the $[v_i(a_1,\ldots,a_{1+1})]$—th member of A_i. From this we can easily define a primitive recursive well–ordering relation that arranges the natural numbers in the ordinal type ε_0.

Let us assume that for every ordinal α of the second number class, whenever possible, an effective well–ordering of the class of natural numbers with ordinal type α is given and denoted by the sign $<_\alpha$. With no loss of generality, we assume also that the natural number 0 is always the earliest in each well–ordering. For every infinite ordinal α, a function is recursive of type α, if it is defined by a sequence of schemata which are either of type lower than α or of one of the forms (I)—(V) (given above for primitive recursive functions) and the following:

$$(\mathrm{VI}_\alpha)\begin{cases}\phi(0,x_1,\ldots,x_n) = \psi(x_1,\ldots,x_n)\\ \phi(y',x_1,\ldots,x_n) = \chi\{y,\phi[\theta(y'),x_1,\ldots,x_n],x_1,\ldots,x_n\end{cases}$$

where ψ, χ, θ are given functions such that for every y, $\theta(y') <_\alpha y'$.

It is easy to convince ourselves that all these are general recursive functions. Indeed, it is known that if we start from a given ordinal and proceed to smaller and smaller ordinals, we must come to a stop in a finite number of steps.

On the other hand, the situation here appears to be better than the situation with general recursive functions in the following respect. On the one hand, for certain numbers y, there may be infinitely many x, $x_\alpha < y$, and therefore the sequence consisting of the values of y, $\theta(y)$, $\theta(\theta(y))$,... could be longer than any preassigned finite length. On the other hand, the function θ is a given function so that we already have a pretty good idea how soon for each y, the sequence y, $\theta(y)$, $\theta(\theta(y))$,... would lead us to the value 0. In this way, we are led to these schemata and functions gradually from the primitive recursive functions. We have a hierarchy of functions which runs from primitive recursive functions to functions defined by transfinite recursions of types (say) ω^2, ω^ω, ε_0, ε_1, etc.

It seems reasonable to identify the totality of all these functions with the effectively computable functions.

The matter is actually, however, rather more complex. In the first place, the schema (VI_α) is not quite what we want. There is no reason to confine θ to a given function and thereby exclude "nested" occurrences of ϕ itself within its first argument. We shall return to this point later on. Meanwhile, we shall continue to use (IV_α) because (1) it is simpler, (2) it has been used by most people who discuss transfinite recursions, (3) most arguments based on (IV_α) can be modified quite directly to apply to the revised schema, to be given below, which admits "nested" occurrences of ϕ.

There are two further points which call for comment, viz., the relation $<_\alpha$ and the function θ. The well-ordering relation $<_\alpha$ is required to be effective. What is an effective well-ordering? The function θ is merely required to be a given function such that for every y, $\theta(y') <_\alpha y'$. It may be difficult to estimate the sequence $\theta(y)$, $\theta(\theta(y))$,... because θ is complex; or, θ may be very simple and yet because of the character of $<_\alpha$, it may still be difficult to estimate the sequence $\theta(y)$,$\theta(\theta(y))$,... .

In general, there is of course no reason to suppose that different relations $<_\alpha$ for the same ordinals α will yield the same class of recursive functions. Indeed, it has been established independently by N. A. Routledge and John Myhill in 1953 that given any general recursive function f, we can find a corresponding well-ordering $<_\omega$ (indeed, a primitive recursive well-ordering) such that f is recursive of type ω.

This result, surprising at first sight, is often taken as a blow to attempts to classify recursive functions by transfinite recursions. I am, however, inclined to disagree with this contention. In my opinion, rather than destroying the hope of classifying computable functions by transfinite recursions, the result suggests that the classification can be made in different ways.

For example, we may not wish to accept all primitive recursive well-ordering relations but only those which can be constructively proved to be well-ordering. Accordingly we can again classify computable functions and their equivalent recursive functions of order ω according to the proofs for the fact that the well-ordering relations used are indeed well-ordering relations. This alternative has the disadvantage that we have again to face the question of constructive proofs.

Another way is to use simple well-ordering relations $<_\alpha$ fixed for each α throughout all investigations so that recursive function of a type always means recursive of type α relative to a particular well-ordering $<_\alpha$. For instance, in Peter 1951, $<_\omega$ is the natural ordering, and $2^a(2b + 1) - 1 <_{\omega^{k+1}} 2^c(2d + 1) - 1$ if and only if either $a < c$ or $a = c$ but $b <_{\omega^k} d$. Using these well-ordering relations, Péter has, by the diagonal argument, established that for every k, there is a recursive function of type ω^{k+1} which is not of type ω^k. In her proofs, she introduces no restriction on the given function θ apart from the condition that $\theta(y') <_{\omega^k} y'$, although she has to permit "nested" occurrences in the definitions of the functions introduced by transfinite recursion.

The possibility of enumerating all functions of type ω relative to the particular well–ordering $<_{\omega^k}$ but with no restriction on θ seems to depend essentially on the fact that for the particular ordinals ω^k and the particular well–orderings $<_{\omega^k}$, all the given functions θ satisfying the condition $\theta(y') <_{\omega^k} y'$ can be characterized and enumerated in a simple way. This is no longer obviously true when we come to functions of type ω^ω.

Thus, for instance, let us assume the well–ordering relations $<_{\omega^\omega}$, $<_{\omega^{\omega^\omega}}$, etc. as given in Ackermann 1940. It is not unlikely that all general recursive functions can be shown to be of order ω^ω already, if we make no restrictions on the given functions θ for which only $\theta(y') <_{\omega^\omega} y'$. In other words, if we wish to classify computable functions by ordinal numbers, as we go to recursive functions of higher ordinal types, we have to fix not only the well–ordering relations we use, but also in some simple manner the functions θ to be employed in the schema of transfinite recursion.

This is quite reasonable and unobjectionable if we remember that the purpose of classifying computable functions is after all to have a clearer picture of the different types of computable functions. Any device which can make us understand clearly a slice of the body of computable functions is welcome. Indeed, quite often piecemeal devices could be more informative than whole sweeps.

It may be asked whether Péter's result about functions of ordinal type ω^k can be extended to arbitrary fixed well–orderings $<_{\omega^k}$. It can also be asked whether ordinal recursive functions of order ω are coextensions with the primitive recursive functions.

Now let us say a few words on the more correct form of transfinite recursion. The general idea is merely this: to evaluate $\phi(y')$ we can only make use of values $\phi(x)$ for $x <_\alpha y'$. From this, there is surely nothing to prevent nested occurrences of ϕ in the definition. Indeed, many useful functions (cf. Péter 1951) require such nested occurrences in their definitions. Using just one parameter, we can so revise the schema (VI_α):

$$(VI_\alpha{}^*)\begin{cases}\phi(x,0) = \alpha(x),\\ \phi(x,y') = \beta\{x,y,\phi[x,\Gamma(x,y')]\}\end{cases}$$

where $(x)[\Gamma(x,y') <_\alpha y']$, Γ may contain given functions and also itself, subject only to the condition that for each part of the form $\phi[x, \Gamma_1(x,y')]$ in Γ, we have always $(x)[\Gamma_1(x,y') <_\alpha y']$.

2.9 The indeterminate domain of computable functions

In short, the domain of general recursive functions is too determinate to correspond exactly to the indeterminate domain of computable functions. To preserve the indeterminateness, we can either follow Church in adding the proviso of constructivity in proving existence. Or else, we can identify computable functions with recursive functions for each of which there exists some constructive ordinal α, some simple well-ordering $<_\alpha$ and some simple schema Γ such that the function is of order α according to the schema (IV_α*). This latter alternative can lead us to informative classifications of computable functions. But as it stands, it is more a program of research than a definition. Thus, we are faced with the questions of explaining *constructive* ordinal, *simple* well-ordering, *simple* schema Γ. In each case, the more suitable course is perhaps not to look for general theories of constructive ordinal or simplicity, but to judge individual cases as they come by. No matter which of the two alternatives we may adopt, we are led to a position rather different from the straight identification of effective with general recursive functions.

It would be of interest to decide the effects of taking this position in place of the identification of general recursiveness and effectiveness. According to this position, every effectively computable function is general recursive, while conversely it is not known that every general recursive function is effectively computable. The boundary of the region of general recursive functions forms a sort of upper or external bound to the totality of effectively computable functions.

Hence, the adoption of this position does not affect the validity of results which depend on the thesis that all effectively computable functions are general recursive but not on the converse of the thesis. Thus, the interesting results (on the unsolvable problems) of Church, Kleene, Markov, Post, Turing, and others (e.g. the theorems on p. 301 and p. 383 of Kleene 1952) remain unaffected. In other words, all impossibility results or negative results still hold. Indeed, when we want to prove conclusions about all effectively computable functions, it is often desirable or even necessary to speak of the broader but less indeterminate totality of all general recursive functions.

There appear to be cases where we tend to apply the notion of general recursiveness but not the notion of effectiveness. These cases provide evidence in support of positions such as that suggested here which question the unqualified identification of general recursiveness with effectiveness.

One interesting example is Péter's remark to the following effect. Although we can decide whether each given number has a certain given property and although it is assumed that there is some number which has the property, we cannot always effectively seek out from all natural numbers the smallest one which has the property. (See Péter 1951, bottom of p. 9.) In his review, Robinson observes that Péter's remark contradicts her later identification of effectiveness with general recursiveness, and seems to imply that therefore Péter's remark is wrong. In my opinion, however, one could as well accept Péter's remark as right and consequently reject the identification. This seems to me a case in support of the kind of position formulated here.

It is known that if both a class C (of natural numbers) and its complement can be enumerated by general recursive functions (say by f_1 and f_2 respectively) then we can find a general recursive function f(m) such that for each n, f(n) is 0 or 1 according as n belongs to C or not (first noticed in Turing 1936—37). Roughly, since each n belongs to either C or its complement, each n is a value of either f_1 or f_2. Hence if we check $f_1(0)$, $f_2(0)$, $f_1(1)$, $f_2(1)$,... successively, we shall sooner or later hit either an $f_1(i)$ or an $f_2(j)$ which has the value n; in the first case, put f(n) = 0, in the second case, put f(n) = 1. In this way, we get the desired general recursive function f(m).

The function thus defined can, we feel, hardly be called effective. It is true that by computing the values of f_1 and f_2, we can, for each n, sooner or later decide whether f(n) is 0 or 1, but we have absolutely no idea how long it will take us to get the answer.

The discrepancy disappears if we either use the indeterminate totality of transfinite recursive functions or add Church's proviso of constructivity. For instance, when the proviso is added, the example of enumerating the class C and its complement can be reformulated thus. If C and its complement can be enumerated by f_1 and f_2 which can be constructively proved to be general recursive, then there is a function f which can be constructively proved to be general recursive, etc. Similarly, Péter's remark could easily be reconciled with the qualified identification.

Incidentally, we might, with regard to ordinals, wish to use a zig–zag definition of constructive ordinals and ordinal recursive functions:

Step(1_a): we start with constructive ordinals, say those $< \varepsilon_0$.

Step(1_b): we define ordinal recursive functions of order $< \varepsilon_0$, e.g., as in Kreisel 1952.

Step (n_a): we define constructive ordinals by some suitable definite method using the functions defined in steps (m_b) for $m < n$.

Step (n_b): using the ordinals of step (n_a) we define new ordinal recursive functions.

If we choose the right method of defining new ordinals from given recursive functions, we may be able to get a self–expanding process. For instance, at step (n_a) we might wish to define constructive ordinals by the method of Church 1938 and Kleene 1938 using (instead of general recursive functions) the functions defined in steps (m_b) for $m < n$. It is known that in the Church–Kleene definition of constructive second number class one can get just as many constructive ordinals using only primitive recursive functions as one can using general recursive functions. Therefore, if we use the Church–Kleene method in step (n_a), there will be no self–expanding process.

There are at least two objections to the positions suggested in this note. In the first place, the totality is not clearly defined. Although we often speak of the totality of all ordinals of Cantor's second number class, very little is known about the members of this totality. Those members which have been studied fairly carefully and given names constitute a very small portion of the whole. Moreover, the matter seems very difficult: the theory of the second number class which we possess today is practically the same as that first invented by Cantor. We have a general theory and some knowledge about a group of lower ordinals, but we know practically nothing about the peculiarities of special higher ordinals of the class.

In the second place, it appears impossible to specify any clear totality that would serve our purpose. Since it seems hopeless to deal with the absolutistic totality of all ordinals of the second number class, we have to be content with as large a portion of it as we can clearly envisage. However, as we know, any such totality is denumerable and we could, by the familiar diagonal methods, get new effectively calculable functions which are different from all those determined by the given totality of ordinals.

I feel vaguely that these two objections cancel each other. In the very nature of the matter, neither the class of effectively computable functions nor the totality of all (say) constructive ordinals of the second number class is or should be (at least to our knowledge) a sharply delineated whole. If we leave the boundary somewhat indeterminate, there is no danger of automatic expansion by the diagonal arguments. The fact that if we assume both a sharp boundary and a constructive approach, we would get into difficulty seems to me to prove that no sharp boundary should be drawn. Others seem to have been driven to the different conclusion that, therefore, nothing short of all general recursive functions (a sort of transcendental totality) could catch all the effectively computable functions.

It is true that every time we think of a given α of the second number class and consider the totality of all recursive functions of type, we can also think of a new effectively computable function which lies beyond. This, however, does not justify the big jump to general recursive functions which include not only all these, but also those about whose speed functions we know nothing positive at all (except that they are *defined* everywhere).

This matter may be compared with what Brouwer says about denumerable ordinals (Brouwer 1912, p. 91):

Let us consider the concept: 'denumerably infinite ordinal numbers.' From the fact that this concept has a clear and well-defined meaning for both formalist and intuitionist, the former infers the right to create the 'set of all denumerable infinite ordinal number,' the power of which he calls aleph-one, a right not recognized by the intuitionist. [etc.]

References

Wilhelm Ackermann

1950. "Zur Widerspruchsfreiheit der Zahlentheorie." Mathematische Annalen, Vol. 117, pp. 162—194.

1951. "Konstruktiver Aufbau eines Abschnitts der zweiten Cantorschen Zahlenklasse." Mathematische Zeitschrift, vol. 53.

L. E. J. Brouwer

1912. "Intuitionism and formalism." Trans. in Bulletin of American Mathematical Society (November, 1913).

Alonzo Church

1936. "An unsolvable problem of elementary number theory." American journal of mathematics, vol. 58, pp. 345—363.

1938. "The constructive second number class." Bulletin of American Mathematical Society, vol. 44, pp. 224—232.

Kurt Gödel

1934. On undecidable propositions of formal mathematical systems. Princeton, N. J., 30 pp.

D. Hilbert and P. Bernays (HB)

1939. Grundlagen der Mathematik, vol. II.

S. C. Kleene

1943. "Recursive predicates and quantifiers." Transactions of the American Mathematical Society, vol. 53, pp. 41—73.

1952. Introduction to metamatheamtics. Amsterdam.

G. Kreisel

1952. "On the interpretation of non-finitist proofs, Part II." Journal of Symbolic Logic, vol. 17, pp. 43—58.

Rózsa Péter

1951. Rekursive Funktionen. Budapest, 206 pp. Compare also review of the work by R. M. Robinson, Journal of Symbolic Logic, vol. 16 (1951), especially last paragraph of p. 281.

A. M. Turing

1936—37. "On computable numbers." Proceedings of the London Mathematical Society, vol. 42, pp. 230—265.

1939. "Systems of logic based on ordinals." Ibid., vol. 45, pp. 161—228.

Added note (Nov. 1, 1984). Recently I have found in my papers this manuscript and some correspondence on it. One letter dated November 1, 1953 comments on a revised version of the original paper, entitled 'Recursiveness and calculability.' As far as I can recall, the initial paper was submitted to the *British journal of philosophy of science* in the summer of 1953 and afterwards received a lengthy referee's report which brought in various extraneous issues. I wrote a revision trying to take into consideration whatever points in the report which appeared relevant. The letter of November 1, 1953 managed, however, to introduce a lot of other ramifications. The version here appears to be a further revision made in the early part of 1954, which, however, has remained in a somewhat unfinished state. I must have decided to drop the matter instead of trying to satisfy the referee.

Professor Wu, Yunzeng, after seeing the manuscript, urged me to include it in this collection. I have decided to let it stand as it was written in 1954 and made no attempt to make any reference to work after that date. Rereading the paper reminds me of my preoccupation with ordinals (of Cantor's second number class) all through the 1950s, as well as the frustrations of watching the technicians turning every interesting suggestion into repulsive complexities which bury the initial more attractive simple idea under mechanical but inpenetrable elaborations. It is hoped that the computer scientists, with their more concrete concerns, might find something suggestive in this discursive old paper. In any case, that appears to be Professor Wu's view.

3. PROCESS AND EXISTENCE IN MATHEMATICS *

1. In learning elementary geometry, we are asked to prove the equality of the base angles of an isosceles triangle. We draw a rough diagram of an isosceles triangle and observe it. The happy idea of constructing a new line from the top vertex to the base enables us to notice relations between the parts of the new diagram, thereby proving the conclusion. Or, alternatively, we can do this by observing the possibility of a rigid motion in space that interchanges the two base vertices.

We are asked to find the sum of the first 10000 positive integers, and hit on the device of rearranging the numbers to look like:

$$\begin{array}{cccc} 1 & 2 & \cdots & 5000 \\ 10000 & 9999 & \cdots & 5001 \end{array}$$

We notice that each of the 5000 columns adds up to 10001.

When the service of a mathematician is requested by an engineer or a physicist, he reformulates the problem into a more idealized form, striking out all the factual details he judges to be irrelevant. This reformulation may require the joint efforts of a mathematician and a practitioner of the source subject, sometimes combined in one person. The new problem is more abstract and retains only a skeleton of the original problem. It is more perspicuous, at least to the properly trained mind which is often able to juggle it to get a method of solution either by standard techniques or by inventing new mathematics. Sometimes the application of the method to the specific problem may be tedious and, for example, calculating machines may have to be used to supply an actual solution.

2. In each case, there are interplays of schematic representations (diagrams, graphs, arrays of characters such as numerals, variables, schematic letters, logical and mathematical constants) and mental experimentations ("Gedankenexperimente"). We are interested in schemata or diagrams rather than pictures or portraits, because we are concerned not with all the factual details about them, but rather their skeletons and structures the "formal facts" about them, the forms and patterns revealed by them. They are aids to our imagination in the process of reasoning, and, as such, essential to mathematics. This does not mean that we always have to draw the diagrams on paper or blackboards, nor that mathematics is a manipulation of symbols. It is not the physical production of the diagrams that distinguishes the mathematical activity, but the possibility of using them to assist our mental experimentations in the search for desired necessary connections.

The mind participates actively in seeing, e.g., an array of numbers as paired off suitably to create a new uniformity. Thus this "seeing as" enables us to take in at a

* First published in *Essays on the Foundations of Mathematics*, edited by Y. Bar-Hillel et al., pp 328—351. © Magnus Press, 1961. Reproduced by permission.

glance the 5000 pairs of numbers which all have the same sum 10001. In this respect, the dots are not "mere abbreviations" either, because they, or something else like them, are indispensable for grasping the array of numbers at one go; they embody the formal fact that we see the 5000 pairs as a whole string with a definite beginning, a definite end, and a definite way of continuation. In doing this calculation, one is likely to make (mental) experiments such as trying to look for suggestions from summing up a small number of integers. But calculation is not itself an experiment, since once the path is found, certainty intervenes.

3. To prove that for every prime p, there is a greater prime, the crucial construction is, of course, the function $p! + 1$. Here it is not natural to describe this function as obtained by the act of "seeing as." In general, the types of construction are varied and heterogeneous.

In searching for a solution, the activity is directed to a definite goal. One is easily led to ask how the mental experiments are chained together. The technical problem about methods of discovering solutions ("how to solve it") is not one for the philosophy of mathematics, although it is of pedagogic interest and central for the mechanical simulation of the mathematical activity. The nature of inferring and the compulsion of the logical must, once the inference is made, is indeed the concern of philosophers. We accept, as a matter of fact, a sequence of symbols as an application of a certain rule, e.g., the modus ponens. Here we may easily get into the slippery ground of truth by convention, synthetic a priori, self–evidence. But an underlying foundation is the sociological fact that it is so accepted.

4. That Beethoven continued to compose good music after he had gone deaf is important for the study of the activity of composing music. Similarly, blind mathematicans are a phenomenon which should shed some light on the nature of the mathematical activity. It is very striking that most of us would find it very difficult, if not impossible, to multiply three 7–digit numbers in our head. For one thing it is not easy to retain the question without the assistance of paper and pencil. If a child asks his father who is blind to help him to do such a sum, he would probably ask the child to serve as his pencil and paper to record the question and the intermediate results. If such assistance is denied a blind mathematician who wishes to do complicated numerical calculations, he would have to train himself to be a calculating prodigy.

That pencil and paper is indispensable to complicated calculations is certainly an important fact about the calculating activity. Most of us do not memorize a large number of telephone numbers but we remember, or rather know, different methods of finding them out. We do not learn the multiplication table to 100 times 100 but only to 9 times 9, or 12 times 12. In more advanced mathematical activities, most of the things which a mathematician knows have not come to him through a deliberate effort to memorize. Interlinks not only increase the size of things remembered but also their duration, their quality. Certain things are kept simultaneously in the head, and these enable one to spin out a great deal of things in sequence. The spinning power of a head with structured memories and dispositions determines the power to experiment mentally and the ability to do mathematics. When one says that mathematics is an activity of the pure intellect, it cannot be to deny that sense perceptions and memory form an integral part of it, but rather that an excellent eyesight or a good memory is not a distinguishing characteristic

of better mathematical capabilities.
5. Some problem–solving is prompted by practical needs, others by analogy with existing problems. Not all mathematical activity is problemsolving. Esthetic needs and the desire to systematize and smooth out things lead to the development and improvement of mathematical theories. It is among such results that the reduction of mathematics to logic comes in.

And it is along such a path that one is led to the librarian's definition of pure mathematics as the class of all conditional propositions in which all constants are logical constants.

"All *A* are *B*, all *B* are *C*; therefore, all *A* are *C*" is a diagram and traditional logic is a sort of mathematics, as ticktacktoe is a sort of board game. One may feel that, being so crude and inefficient, it hardly deserves the fair name of mathematics. However, continuity with mathematical logic seems to lend some colour to it.

Traditional logic is more a hindrance than a help to right reasoning that is quite adequately taken care of by our natural power. This is seen from the fact that the more purely rational an activity is, the less is it needed. Mathematics is least in need of it while election politics, judging from Susan Stebbing's studies, needs it most.

Mathematical logic has to a considerable extent suffered the same kind of misfortune. Perhaps certain considerations in modern topology, which involve more definitions than arguments, are an exception. Logic is primarily interested in the analysis of a proof into as many distinct steps as possible, and not, like mathematics, in efficient methods of reasoning which can produce remote consequences in one swoop or unravel an involved entanglement. When, e.g., an elementary branch of it gets practical applications in making machines, it does this only, so to say, accidentally and against its own will. It is by leaving behind the basic concerns of logic and pursuing the subject as a simple sort of mathematics that the application is made.
6. The breaking up of a proof into a large number of small steps is desirable in so far as the set of all possible different small steps is in general less complex than the set of a smaller number of different bigger steps. This seems obvious, since the union of all small steps which make up one big step is simpler than the big step which contains the simple steps (possibly with some repetitions) plus a special mode of combination, and some small steps generally occur in a number of different big steps. There is, however, no equally obvious reason why such simplification should be desirable for the mathematical activity. In fact, since we are quite at home with the bigger steps, one is inclined to think that by multiplying the pieces in each proof, the breaking up only serves to slow us down and make it harder for us to take a proof in.
7. Few mathematicians have taken the trouble to learn the theory of quantifiers and they are none the worse for their ignorance. It sounds idle to rejoice over the accomplishment that when a logician has analyzed and reformulated a proof, even a machine can check it for correctness. Nobody, not even a logician, checks an elaborate mathematical proof in this manner, and so far machines have not been used to check proofs.

Twenty years ago it must have appeared that if man finds such a way of checking proofs tedious, machines would not do it any better either in speed or in accuracy. The appearance of large machines and the rapidity with which their speed and reliability have been improved, is one of the unexpected occurrences in history which yield consequences which are hard to predict.

There is, however, a distinct possibility that in this connection a basic application of logic will be found that is based on the essence rather than the accidents of logic: viz., to handle inferences as efficiently as calculations. For example, some preliminary work has already enabled a common machine to prove all theorems of quantification theory with equality in *Principia* (9 chapters) in less than nine minutes.

8. The application of the analysis of inference by mathematical logic to a project of mechanizing mathematics depends on a quite prosaic but yet fundamental sense of the reducibility of mathematics to logic: every proof could be formalized in logic as establishing purely by logic a conditional proposition that the theorem follows from the relevant mathematical axioms. This is not surprising since no claim is made about proving in logic the axioms as well. As a result, there is also no need to stretch the range of logic to include extraneous things such as sets or classes.

Grammar is of little help to learning one's native language or cultivating elegant writing. And we do not worry about the theory of sound waves when learning to speak. Phonetics is a little more relevant, although few can afford tuitions from Professor Higgins. If mathematical logic were a little less pure, perhaps it could assist a mathematician to learn some alien branch of mathematics. In its present aloof form, however, a training in mathematical logic is neither necessary nor likely to speed up the pursuing of other branches of mathematics.

On the other hand, if a machine is to do mathematics, it is necessaty that methods of logic be explicitly included. This provides incentive for doing more detailed formal work on the decision problem and proof procedures for logic.

9. Moreover, considerations about the practical feasibility of alternative procedures are pushed to the forefront. This supplements the basic concern that a mathematical argument should be perspicuous, surveyable, or capable of being taken in. These two aspects of the problem of efficiency are not identical. For example, a less efficient proof procedure is generally easier to describe, and the argument for proving its adequacy is generally easier to grasp. On the other hand, the two aspects combine to account for and give direction to much of our mathematical activity. To stress the requirements that procedures be feasible and that proofs be surveyable, one might coin the label "praximism."

If a machine produces a proof of Fermat's conjecture with one million lines, we still have the somewhat easier task of making the proof perspicuous. This would be a situation where we could say, in a clear sense, that a proof exists but nobody has understood it. Somebody would undoubtedly prefer to say that there is no proof yet, just as he would say that a machine cannot calculate, cannot prove, because there must be a final contact which lights up the whole thing and only a man can establish this contact by taking in the whole process that makes up the calculation or the proof.

When interesting mathematical questions can be settled by machines, our chief concern will be shifted to the methods and their representation in mechanizable languages, And we do not expect to obtain any program with 10^6 lines of coding. We synthesize and abbreviate as we make progress, in order to press more and more into a brain as a bounded finite machine. With the increased power of mechanized methods, an economy in storage is achieved by substituting general methods for particular arguments.

In a different direction, the project of mechanical mathematics calls our attention to the problem of formalizing methods of finding proofs. Here we have another hitherto largely neglected domain which is susceptible of a treatment by the methods of mathematical logic. Such problems are on a different level from the study of the psychology of mathematical invention. We may be able to simulate the external circumstances under which Poincaré's subconscious functions marvelously. But it seems preposterous to suppose we are capable of endowing a machine with a subconscious comparable to Poincaré's.

10. The more sensational reduction of mathematics to logic is the thesis that definitions of mathematical concepts can be found in logic such that mathematical theorems can be transformed unconditionally into theorems in logic. This is plausible only if "logic" is understood in a very broad sense to include set theory as a part.

The term "set theory" is less familiar than the term "logic", but then, at the same time, more unambiguous too. Since set theory is itself a branch of mathematics, the question is that of reducing other branches of mathematics to this particular one. In this sense, the matter is initially a domestic affair of mathematics. The concern of philosophers has come about partly as a result of the historical accident that Frege and Russell, rightly or wrongly, connect them with philosophy, and that at least one of them is such a good propagandist. None the less, the persistence of such interest surely cannot be discarded simply by deploring the poverty of philosophy. After all, even if set theory is but another branch of mathematics, the claim that all other branches are reducible to it makes it a proper concern of philosophers.

11. The most interesting case is number theory. If we are concerned only with numerical formulae containing addition and multiplication, it appears possible to find theorems of logic which correspond to them rather naturally. On the other hand, if we are concerned with general laws of arithmetic as well, the reduction is only possible when we take set theory rather than logic proper.

It is puzzling that Kant called "$7 + 5 = 12$" synthetic a priori and that Frege believes himself to have refuted this by his reduction of arithmetic to logic. One way to make the two viewpoints plausible seems to be the following. In order that an equation be analytic, the two sides must have the same sense, not just the same denotation. One is tempted to say that "$7 + 5$" and "12" have different senses, although the same denotation. Hence, it is synthetic and a priori, the necessity part not being questioned here. But there is a natural way of reducing "$7 + 5 = 12$" to a theorem of logic. Suppose we use the abbreviations:

$$(E!_1x)\,Gx \text{ for } (Ex_1)(y)\,[Gx_1 \,\&\, (Gy \supset y = x_1)],$$

$$(E!_2x)\,Gx \text{ for } (Ex_1)(Ex_2)(y)\,[x_1 \neq x_2 \,\&\, Gx_1 \,\&\, Gx_2 \,\&\, (Gy \supset (y = x_1 \vee y = x_2))],$$

....

Then the corresponding theorem of logic is:

$$(*)\ [(E!_7x)\,Gx \,\&\, (E!_5x)\,Hx \,\&\, (u) \sim (Gu \,\&\, Hu)] \supset (E!_{12}x)(Gx \vee Hx).$$

Since it is natural to regard all theorems of logic, i.e., the theory of quantifiers with equality, as analytic, Frege seems to have shown that "7 + 5 = 12" is analytic.

12. There are a number of difficulties in this explanation. The negation of something like (*) does not give us what we want if we are interested in proving, e.g., "$7 + 6 \neq 12$". The obstacle arises because the letters G, H serve as free variables so that we have to quantify them to get the correct negation. We certainly do not wish to say that "7 + 5 = 12" is analytic but "$7 + 6 \neq 12$" is synthetic a priori. Moreover, there is no way to get around the need of existence assumptions in one form or another. If there are not enough entities in the universe of discourse, the antecedent of (*), for instance, could be always false, and we can change the consequent and prove, e.g., 12 = 13. We are led back to the reduction of arithmetic to set theory, and there is an obvious choice between saying that arithmetic has been shown to be analytic (Frege) and saying that logic (more correctly, set theory) has been shown to be synthetic (Russell at one time).

Although the numerals 5, 7, 12 occur in (*) as subscripts, there is no direct circularity in the reduction, because we can expand (*) and avoid the use of numerals by employing sufficiently many distinct variables. A striking feature of the reduction is that short propositions are reduced to long ones. As a result, it would be very clumsy if one were to do arithmetic in such a notation, and we are quickly forced to introduce abbreviations. This is rightly considered an inessential complication for the simple reason that the reduction is not meant to introduce a new technique of calculation but only yields incidentally an informal result about calculations, that one could do arithmetic in the complicated symbolism too. This depends on the reduction plus the information that one can do arithmetic in the customary notation.

A more basic difficulty of the reduction is the accompanying increase in the conceptual complexity. If we attempt to give a proof of (*) in the expanded form, we find ourselves counting the distinct variables, and going through, in addition to operations with logic, exactly the same kind of moves as in elementary calculations. We are able to see that (*) is a theorem of logic only because we are able to see that a corresponding arithmetic proposition is true, not the other way round. By tacking "frills" on an arithmetic proof of "7 + 5 = 12," we get a proof of (*) in logic. "A definition of christening in a particular church is no longer a definition of christening."

13. There are different ways of defining arithmetic concepts in set theory. If we imagine a determinate situation with one specific formal system of set theory, one of arithmetic, and one specific set of linking definitions, then there is a theorem in the primitive notation of set theory that corresponds to the arithmetic theorem "1000 + 2000 = 3000" by dint of the linking definitions. The formula would be forbiddingly long. Does it mean the same thing as the original formula of arithmetic? When one who is not aware of the definitions is faced with the long formula, he might be at a loss to see any clear connections between the two formulae. He may be sufficiently familiar with set theory to understand the long formula and still not recognize its relation to the short one. Or even if he knows the definitions and is asked to simplify the long formula according to them, chances are he will make errors and arrive at some incorrect result. We are inclined to think that such considerations are irrelevant as far as the intended meaning of the formulae is concerned. But if a man fails to see the equivalence of the two formulae even after hours of hard labour, can we still say that the two formulae mean the same thing to him?

This is an artificial question because nobody is expected to write out or work with the long formula in order to do arithmetic calculations. We have a short argument to show that there must be such a formula, and that nearly exhausts the meaning of the hypothetical assertion that we could work directly with it too. When it is a matter of doing mathematics, we naturally fall back on the best available technique we have. If we had only the long version at first, then we would as a matter of fact not be able to do much calculating until we hit on some systematic way of changing it into a short version. We may spend many hours to read a long formal proof, but when we understand it, we do not give each line the same status, but work out an easily memorizable structure which may include known theorems, lemmas, subcases, reminders that certain successions of steps are of certain familiar forms. We do not have to keep all details of the structure in mind at the same time. The proof may be a mile long, we can still plant posts as we go along and not worry about parts changing when we are not looking at them. As soon as we are convinced that some parts do give us a subtheorem which is the only contribution which those parts can make toward proving the final theorem, we need retain only the subtheorem in our head.

14. Through the linking definitions, the theorems of set theory can be divided into two classes: those corresponding to theorems of arithmetic and those which do not. Theorems of both classes are, one is inclined to think, in the system all along; the linking definitions do not change their meaning but merely provide a different way of looking at those in the first class. Most of us have seen pictures which appear to be a mess at first, but reveal, e.g., a human face upon closer scrutiny. The physical object that is the picture is not affected by the different impressions which we get from it. The picture, however, means different things before and after we discern a face. This, one feels, is also the situation when linking definitions enable us to see certain formulae of set theory as disguised arithmetic formulae. If one is afraid that next time he will forget how he can discern a face, he may, as a reminder, trace certain parts of the picture by a red pencil. As a result, everybody can immediately see a face, although the configurations in the picture remain the same. Does it make an essential difference whether the stress is made by a red pencil or just seen in our mind's eye?

Does a proof change the meaning of a hitherto unproved mathematical proposition? Does a new proof of a mathematical theorem change its meaning? Think of the proposition as a station in a formal stysem. The country is there, but we do not know whether there is any road which leads to the station. Presently we find one road, then we find another. But the country is the same, the station is the same. Both of us understand the proposition that there are infinitely many prime numbers. You know a proof of it but I do not. Does it have the same meaning for both of us? It is not yet known whether there are infinitely many pairs of primes n and $n + 2$ ("twin primes"). Will a proof of the proposition change its meaning? The proof will reveal new connections and provide reminders which enable every member of the mathematical community to see the proposition as true. Does the increase of knowledge affect the meaning of a proposition or is the relation between knowledge and meaning only an external one resembling the relation between the weight of an elephant and our knowledge of it?

The elephant exists independently of our knowledge but in what sense does a proof exist independently of all knowledge? Once a proof is found, it can be codified and put at its proper place within a textbook, but where did it reside previously? Moreover, to call several pages of printed marks a proof presupposes a good deal of the sociological circumstances which make them a proof. For instance, they are sufficient to recreate in a few people the gradual process which culminates finally in seeing that the concluding proposition of the several pages must be true. We are reluctant to deny that every possible proof in a formal system exists even before we have singled it out and digested it by constructions, mental or with red pencils. Under suitable conditions of size and endurance, a machine can eventually grind it out. In this sense, the undigested proof has existed all along, even though the digested proof has to be invented. Is, however, an undigested proof a proof? To say that it is a proof because it is, though undigested, digestable, leads to the question of distinguishing digestable in principle from digestable as a matter of fact. Even if a miracle reveals that there is a way of seeing the geographical contours on Venus as a proof of Fermat's conjecture, how do we know we shall ever be able to find suitable perspectives to make such an undigested "proof" perspicuous? It seems like a dogma to say that every undigested proof will eventually be digested. If one does not wish to assert so much, then it is hard to provide, without circularity, a sense of "digestable" according to which every undigested proof is digestable.

15. I think I know how to add and multiply. But it would be easy to find complicated problems which I cannot do within two hours. For instance multiplying 78 by 78, 78 times. With some effort, we can also find computation problems which I cannot do, at any rate by the ordinary technique, within a month, or within my lifetime. In what sense do I know how to add and multiply? Not just in the sense that I can handle small numbers, because I feel I can deal with large numbers too. Or perhaps, if I live long enough, say by keeping myself fit like a great athelete, I shall be able to complete even the most complicated additions and multiplications? But then surely I cannot do them with the ordinary technique for there would not be enough chalk, would not be sufficiently large blackboards.

These considerations strike one as utterly irrelevant. When I say I can do addition and multiplication, I do not mean to preclude the possibility that practical difficulties may prevent me from carrying out certain complicated calculations. I feel I can do them, shalll we say, in principle. One is generally not expected to do artificially elaborate calculations. If it were the case that nobody is interested in multiplications of less than 300 numbers each with more than 10 digits, then one might say that nobody can multiply unless he is assisted by a machine.

The words "can," "decidable," etc. mean different things in pure mathematics and applied mathematics, in actual mathematical activities and in the discussions of mathematical logicians. A man says that the further expansion of π is a further expansion of mathematics and that the question changes its status when it becomes decidable. Since what the millionth place of the decimal expansion of π is, is a theoretically decidable question, the man seems to be inconsistent in saying that a ground for the decision has yet to be invented. This is so only if we think of decidable in the logician's sense. In the sense of actually doing mathematics, the question is not yet decidable because it is to be expected that some ingenious general argument is required to supply the required digit and prove to the satisfaction of mathematicians that it is indeed the desired one. And it strikes one like dogmatism to assert categorically that such an argument will be found. It is true that finitists and intuitionists do not worry about such questions because once a problem is decidable in the theory, they lose all interest in it. This, however, does not mean one cannot interest oneself in feasibility as a concept worthy of philosophical considerations.

Confusions arise when two men choose the two different senses and refuse to recognize that there is also the other sense. Perhaps a phenomenologist is one who permits both senses and distinguishes them from each other. At any rate, it seems convenient to make use of both senses, at least until we have more successfully unified them.

16. There is a great gap between what can be done in principle and what can be done in practice. Often we are interested in broadening the range of the latter. That is why such techniques as the use of the Arabic notation, logarithmic tables, computing machines are important. Are they only of practical importance or are they also of theoretical interest? Shall we say that theoretical and practical significances merge in such fundamental improvements in the technology of mathematics?

It is not always easy to draw the line between the theoretical and the practical. Numbers of the form $2^{2^n} + 1$ are called Fermat's numbers because Fermat conjectured that all such numbers are prime. It has been proved since Fermat's time that, for $n = 5$, 6, 7, 8, Fermat's numbers are composite. A proof for each case was a nontrivial piece of mathematics, even though, with patience, these questions could be settled simply by the ordinary methods of calcualtion. One might say that the proofs provide us new techniques for deciding problems which could otherwise be solved by uninspired laborious computation.

In mathematics the introduction of new techniques is important and definitions do serve to introduce new techniques. It is therefore misleading to speak of them as "mere abbreviations." Even if, after a proof of a theorem in number theory has been discovered, it is possible to eliminate defined terms and translate the proof into the primitive notation of set theory, the translated proof would not have been discovered by one who worked exclusively with the primitive notation of set theory. Nor could the translated proof be understood correctly even if one is aware of the definitions. If set theory alone is given but the linking definitions with arithmetic are still missing, then we do not yet have arithmetic in full force because we would not and could not, as a matter of fact, do the arithmetic proofs and calculations in set theory. If both set theory and the linking definitions are given, we continue to do arithmetic as before only with the awareness that there is a sense in which our proofs and calculations could be translated into set theory. But doing arithmetic is still different from doing set theory. We do not change our manner of doing arithmetic. That is the sense in which arithmetic has not been reduced to set theory, and, indeed, is not reducible to set theory.

Do we reduce mathematics to abstract set theory or do we get set theory out of mathematics by padding? In analysis, we find certain real numbers such as π and e of special significance. Somehow we are led to the search for a general theory of real numbers. Since we want the theory to be general, we postulate many more real numbers in order to make the surface smooth. When we find that real numbers, natural numbers and many other things can all be treated as sets, we are induced to search for a general theory of sets. Then we add many more other sets in order to make the surface appear smooth. "If tables, chairs, cupboards, etc., are swathed in enough paper, certainly they will look spherical in the end." In this process, we lose sight of the distinctions between interesting and uninteresting sets, useful and useless real numbers. In order to recover the distinctions once more, we have to take off the padding. Could we perhaps describe this reverse process as reducing (e.g., "Mrs. E is on a diet") abstract set theory to mathematics?

If we think in terms of true propositions about natural numbers, then set theory is also reducible to arithmetic at least in the sense that, given any consistent formal system for set theory, a translation can be found such that all theorems turn into true arithemtic propositions. The same is true of any other branch of mathematics on account of the possibility of an arithmetic representation of formal systems. Hence, we can also say that all mathematics is reducible to arithmetic, but in a sense quite different from, for instance, what was known as the arithmetization of analysis. Arithmetization of logic involves a change of subject from talk about classes, etc., to talk about how we talk.

17. When we ask, what is a number? what is the number one? we seem to be after an answer as to what numbers *really* are. If numbers are neither subjective nor outside of us in space, what could they be? And then it is gratifying to get the answer that they are really certain classes. One is relieved to have thus unmasked numbers. What does the unmasking accomplish? Frege's definition of number seems to resemble rather closely our unanalyzed concept of number so that we are sometimes inclined to take it as providing a true analysis of our intentions. But what more?

Apparently there is some belief that the reduction puts mathematics on a more trustworthy basis. Otherwise, the paradoxes would not have induced Frege to say that the foundation of arithmetic wobbles. This is, as we now know, unjustified. We understand arithmetic better than set theory, one evidence being the highly informative consistency proofs of arithmetic. The foundation of arithmetic is more trustworthy than that of set theory: what would be of greater interest is rather to found set theory on arithmetic, or an extension of arithmetic to infinite ordinals.

There are different ways of defining numbers in terms of classes. Each of them leads to and from the undefined concept of number, and they are seen to be equivalent not through the interconnection between themselves but by way of the channels connecting them to the naked concept of number. Perhaps this indicates a certain priority of numbers to their corresponding classes?

Another advantage of identifying numbers with a suitable class is said to be "recommended by the fact that it leaves no doubt as to the existence-theorem." "Postulating" a limit to fill the gap for each Dedekind cut is said to have advantages which are the same as those "of theft over honest toil," while the course of honest toil is to identify the limit with the class of ratios in the lower section of the cut. It is in a sense true that the latter course "requires no new assumptions, but enables us to proceed deductively from the original apparatus of logic." This is so, however, only because in the original apparatus of logic we have already made assumptions of the same kind. If the existence of the postulated limit is called in question, the existence of its corresponding class is equally doubtful. There is no reason to suppose that numbers evaporate but classes are rocks.

The reduction to set theory gives "the precise statement of what philosophers meant in asserting that mathematics is a priori." This is neither an informative statement nor a true one.

"In speaking of arithmetic (algebra, analysis) as a part of logic, I mean to imply that I consider the number concept entirely independent of the notion of space and time, that I consider it an immediate result from the laws of thought." (Dedekind) It seems, however, clear that, instead of resolving the foundational difficulties in the separate branches, the reduction to set theory merely jumbles all difficulties together and adds a few new ones.

It is said that the axioms of arithmetic admit diverse interpretations while the reduction eliminates such ambiguities. True, the concept of set is involved in the axiom of induction and the intended interpretation of the concept of set assures the intended interpretation of the axioms of arithmetic. But arithmetic presupposes only inductive sets which are a particular type of set. Moreover, we should not confuse the possibility of incorrect interpretations with the impossibility of correct interpretations. It is possible both to interpret the axioms of arithmetic correctly and to interpret the axioms of set theory incorrectly. Moreover, interpreting the axioms of set theory involves greater conceptual difficulties.

18. Surely one cannot deny that Frege's definition has the great virtue of taking care of applications? But the application of number to empirical material forms no part of either logic or set theory or arithmetic. This is undoubtedly the case if we perform a multiplication just in accordance with the rules of calculation or argue formally by observing the rules of logic. There may be some doubt if we consider the proposition "Paris has 4 million inhabitants" as an application of the number 4 million, and the proposition "two rabbits plus two rabbits yield four rabbits" as an application of the mathematical proposition "$2 + 2 = 4$."

Such applications can appear neither in arithmetic nor in set theory for the simple reason that words such as "Paris," "rabbits," "inhabitants" do not occur in the vocabularies of these fields and the settheoretical definition of numbers offers no help. If it is meant that the definition enables us to apply numbers within the framework of a wider language, then it is not clear why the same does not apply without the definition. Suppose we are to infer the proposition "she has two virtues" from the proposition "her only virtues are beauty and wit." It is apparently thought that the inference can only be made by using Frege's definition of the number 2, because otherwise the class of her virtues cannot be shown to have the number 2. If, however, the full richness of ordinary discourse is permitted, we can surely make the inference without appeal to Frege's definition.

In any case, why should such applications be taken as the proper business of set theory or of arithmetic? Mathematics and its applications are two things which can coveniently be studied separately. If the desire is to have a general language which includes both mathematics and other things, the link between numbers can just as well be provided by axioms which assert for example that a class has $n + 1$ members if and only if it is gotten from a class with n members by adding a new member. In other words, if we adopt the course of taking numbers as undefined, we can still, if we wish, add axioms to do the job of Frege's definitions. The effects are the same except that mathematics and its application are divided at a more natural boundary.

19. It is remarkable that the Russell–Zermelo contradiction led Frege to doubt whether arithmetic can possibly be given a reliable foundation at all. Actually the contradictions in no way make it necessary to modify the definitions of number in terms of sets, only the project of formalizing a general theory of sets is affected. We are, therefore, faced with the task of designing a consistent and adequate formal system of classes. We were at first struck by the fact that natural numbers, real numbers, and many other things can all be gotten out of sets. Then we found that contradictions can be gotten out of sets too. We are now to design a calculus which includes as much of the other things as possible but not the contradictions.

If we do not think in terms of a system, why can we not treat a proof of contradiction as just another piece of mathematics which could be judged interesting or uninteresting more or less in the same manner as other mathematical proofs? True, the conclusion, being a contradiction, cannot be significant in the same way as an ordinary theorem is. The proof establishes either more or less than usual. It either shows "the unreliability of our basic logical intuition," or reveals some confusion on the part of the owner of the proof. Dividing both sides of the correct equation $3 \times 0 = 2 \times 0$ or $3(2 - 2) = 2(2 - 2)$ by 0, we easily get the contradiction: $3 = 2$. Such a discovery does not excite us because it is well–established that the restriction $c \neq 0$ is essential in inferring $a = b$ from $ac = bc$. Why can one not discard the contradictions in set theory as easily? The reason on the surface is the lack of any comparably simple and natural restriction which would do the job. This is, it is sometimes said, an indication of the more basic fact that our concept of class is not sufficiently clear.

20. Formal systems are to suit the actual proofs in living mathematics, not the other way about. If a formal system adequate for analysis yields a contradiction, we say that we no longer trust the formal system. How would this affect the many mathematical results in analysis, accumulated through the centuries? It is hard to speculate on the basis of such an indeterminate hypothesis. We may, however, remind ourselves that practically no significant mathematical theorems or proofs have been given up because of the contradictions of set theory which have, according to some people, discredited the fundamental methods of argumentation in set theory.

The emphasis on a consistent adherence to the rules we use in mathematical reasoning has generated a sharper distinction between confusion and contradiction. To treat an infinitesimal sometimes as zero sometimes as a positive quantity in the same proof is a confusing and inconsistent procedure, but does not yet yield an explicit contradiction. The criticism of a proof using infinitesimals is ambiguity and not that a contradiction follows.

Why should contradictions worry anybody? Imagine a mathematician pleased with the discovery of a group of new theorems publishes a book. A rival studies the proofs and comes along with the challenge, "Using your kind of argument, I can prove even contradictions." Could it then be replied "Well, how nice, my methods have interesting applications of which I was not aware, let me add another chapter entitled 'Further Applications of the Above Methods'"? Even though contradictions are often interesting and new methods are often recommended by the interesting theorems which they enable us to prove, nobody, unless his purpose were to experiment with contradictions, has recommended a method on the ground that it is powerful enough to yield contradictions. The usual reaction to the discovery of a contradiction is to analyze the moves involved in the derivation and pronounce some of the moves unwarranted. The repercussions of a contradiction include the rejection of all proofs which involve similar moves. In this sense, contradictions are contagious. Proofs which were otherwise considered healthy are put into concentrated isolation on account of their contact with contradictions.

21. It is customary to use formal systems as a tool for separating desirable from undesirable arguments. Formal systems are constructed under the guiding principle that when an argument is found to be faulty, all arguments of the same *kind* are to be excluded. This gives the impression of being less arbitrary in our exclusion of certain arguments because we are rejecting not only one particular argument but all arguments of its kind. Given any argument, there is, however, inevitably an element of arbitrariness in any attempt to determine the kind to which it belongs, that is, to account for the troubles. Indeed, there are so many different ways by which we can determine the underlying category of an argument. We do not even have to use formal systems for this particular purpose.

Suppose given a group of theorems and a formal system in which proofs for these theorems can be carried out, and a contradiction is discovered in the formal system. The system is thereby discredited. What about those theorems which were originally discovered with no regard to this formal system? True, there is now a uniform method of proving all these theorems *in the formal system* because there is a generally accepted principle that a contradiction implies everything. We may yet distinguish proofs of the system which go through contradictions from those which do not. Every proposition of the system has a proof of the first kind but not necessarily one of the second kind.

Does the inconsistency of a formal system destroy the value of those proofs which do not go through contradictions? The first question is, of course, what values we did attach to the proofs to begin with. Were we originally interested in the proofs on account of their beauty, or the truth of the conclusions they establish, or the utility? Proofs in an inconsistent system or a system not known to be consistent can often have heuristic value: for instance, there are theorems of number theory which were at first proved by methods of analysis and later received more elementary proofs.

22. It is known that we can derive the differential and integral calculus in some system of set theory which is not known to be consistent. Suppose the system is found to be inconsistent. It follows that we can derive all sorts of false and absurd consequences in this system, some of them having to do with the differential and integral calculus.

Since the calculus can be applied in constructing bridges, we may be able to prove that a pillar whose diameter is three feet long is strong enough although actually we need a pillar whose diameter is seven feet long. Hence, it might be argued, bridges may collapse because of the inconsistency of the particular system in which we can develop the calculus.

Actually no such thing can happen. For one thing, those who construct and develop axiomatic foundations of the calculus are usually not the same people as those who apply the calculus in the construction of bridges. It is not impossible that, by accident, the same person may be engaged in both kinds of activity. Even then, he is not going to do his calculations by going all the way back to his favorite axiomatic set theory. Moreover, even if he does take the trouble to justify his calculations, after it is done, by citing explicitly the axioms and theorems of the set theory, he is still in no danger of getting the wrong result because he does not use all the complicated apparatus that is available in the system but makes only such turns as could also be justified in consistent systems.

It is not necessary to formalize mathematics nor to prove consistency of formal systems if the problem is that bridges shall not collapse unexpectedly. There are many things which are more pertinent in so far as bridges are concerned.

23. So far as the present state of mathematics is concerned, speculations on inconsistent systems are rather idle. No formal system which is widely used today is under very serious suspicion of inconsistency. The importance of set–theoretical contradictions has been greatly exaggerated among some quarters. When the non–Euclidian geometries were discovered and found to be unintuitive, it was natural to look for consistency proofs by modelling considerations. And then it was a short step before one asked for the basis on which the model itself is founded. When Kronecker thought of classical analysis as a game with words, it was again natural that he did raise the question whether such a game was even consistent. But the more modern search for consistency proofs is differently motivated and has a more serious purpose than avoiding contradictions: it seeks for a better understanding of the concepts and methods.

"The superstitious fear and awe of mathematicians in face of the contradiction." But Frege was a logician and Cantor was a mathematician. Cantor was not a bit worried about the contradictions. In fact, he said: "What Burali–Forti has produced is thoroughly foolish. If you go back to his articles in Circolo Matematico, you will remark that he has not even understood properly the concept of a well–ordered set." Admittedly Cantor's well–known definition of the term "set" is difficult, yet it cannot be denied that the definition does exclude, through the mildly "genetic" element, the familiar detivation of contradictions.

24. The explanation of mathematical existence in terms of consistency appears to be an evasive twist: since we cannot give a suitable positive characterization of all mathematical objects, let us say that in mathematics, all that is not impossible is real. On the one hand, constructibility seems to leave out some desirable mathematical objects and face us with the question of explaining the existence of a construction. On the other hand, a Platonic world of ideas, unlike material things in space and time which form the basis of the physical sciences, seems to have very little explaining power in mathematics.

The classical definition of the existential quantifier in terms of the universal quantifier has the flavor of identifying existence with consistency, while our experience with the physical world suggests that although the actual is not impossible, the possible does not always exist. While in physics there is a natural distinction between things and laws, laws and constructions seem to be all–pervading in mathematics. Radical phenomenalism is idle and futile as far as the foundations of empirical knowledge are concerned, but even there the basic distinction is hesitantly preserved in the dubious entities called sense–data. Yet mathematical objects are primarily connections, relations, and structures.

25. In doing mathematics, it might even increase some people's power of penetration to think of mathematics as a study of the natural history of numbers and classes. As a philosophical position, such a view would lead too quickly to mysticism and make an articulate philosophy of mathematics well nigh impossible, except perhaps as a sort of metaphysical poetry.

If, e.g., numerals are treated as proper names there is no point to ask then whether positive integers exist, sinve otherwise numerals would not be proper names. The question of existence has to be directed to the satisfiability of a property, a relation, a condition, a theory: is there some object or some set of objects with a suitable structure that satisfies a given condition? There exist non–Euclidean spaces since axioms of non–Euclidean geometries have models in the Euclidean. There exist complex numbers since axioms for them can be satisfied by pairs of real numbers. Each particular complex number, e.g., i, has a derived existence as a constituent of the whole structure of complex numbers, satisfying certain relations to other complex numbers.

It is familiar that such modelling considerations generally come to an end with positive integers and the continuum: there is in any case a sort of circularity in the explanation of existence by consistency and consistency by satisfiability. We need some basic stuff to begin with: in what sense do they exist?

It seems reasonable to suppose that if a theory is consistent, it must have some interpretation. It may be very difficult to fabricate a model, but how can a theory be consistent and yet sutisfied by no model whatsoever? The fundamental theorem of logic gives a sharper answer for theories formulated as formal systems within the framework of logic, i.e., the theory of quantifiers: any such theory, if consistent, has a relatively simple model in the theory of positive integers, simple in the sense that rather low level predicates in the arithmetic hierarchy would suffice.

26. Hence, we may feel that the basic question is the sense in which positive integers exist. More exactly, we are concerned with the existence of a structure or a relation that would satisfy the axioms of arithmetic; the individual positive integers would enjoy a derived existence in such a structure.

It appears at first sight that the proof–theoretical consistency proofs of the axioms of arithmetic provides a (modified) finitist solution to this question, and that the translation into the intuitionistic system of arithmetic gives an intuitionistic solution of the problem. If this were indeed so, we could at least concentrate on what Hilbert calls the combinatorial hard–core of mathematical thinking or what Brouwer calls the basic intuition of two–in–one. There are, however, a number of difficulties accompanying the incompletability of the axioms of arithmetic.

The arithmetic translations of theorems in the usual systems of set theory are often no longer theorems of the usual systems of arithmetic. As a result, a consistency proof of the axioms of arithmetic does not settle the consistency question of classical analysis or of set theory. Even in the consistency proof of arithmetic, there appears to be an indeterminacy in the notion of finitist proofs.

Moreover, there is a choice between different axiom systems of arithmetic not only in the simple sense that alternative equivalent formulations of, say, the Euclidean geometry are familiar, but in the deeper sense that extensions of the usual set of arithmetic axioms seem to be just as natural, e.g., the addition of transfinite induction up to the first epsilon number. This tends to indicate that there is something absolute in the concept of number and we only gradually approximate it through mental experimentations. Or at least, we have no full control over our intentions and mental constructions which, once in existence, tend to live a life of their own.

In a different direction, the existence of consistent systems which have no standard models (e.g., are omega—inconsistent) points to a certain discrepancy between existence and consistency. The usual axioms require that certain sets or numbers exist but remain mum on what things to exclude. On account of this, we can add unnatural numbers to the natural numbers without violating the axioms, and, indeed, consistently add new axioms to require that there must be unnatural numbers too. One might argue with reason that although these unnatural numbers are required by the axioms of a consistent system, they should not be said to exist. Such a position would foil the unqualified identification of consistency with existence.

27. There is a temptation to cut through the foundational problems by using the nonconstructive rule of induction (the omega-rule) and similar semantic concepts to characterize all true propositions in arithmetic, classical analysis, and set theory. In this way, of course, e.g., unnatural numbers are excluded by the basic principles. However, there is not much explaining left to be done, since what is to be explained is simply taken for granted. With it, more is accepted which is a projection by analogy of the finite into the infinite. We can never go through infinitely many steps in a calculation or use infinitely many premises in a proof unless we have somehow succeeded in summarizing the infinitely many with a finite schema in an informative way. Both mathematical induction and transfinite induction are principles by which we make inferences after we have found by mental experimentations two suitable premises which summarize together the infinitely many premises needed. A very essential purpose of the mathematical activity is to devise methods by which infinity can be handled by a finite intellect. The postulation of an infinite intellect has little positive content except perhaps that it would make the whole mathematical activity unnecessary.

It might seem puzzling that, e.g., the Peano axioms, in particular an alternative explicit formulation with only a finite number of axioms, should contain so many surprises. The essential thing is of course the possibility of iterated applications of the same old rules over and over again in an unbounded number of combinations. This is also why proving the consistency of such a system is no easy matter.

With regard to the nature of the continuum, there are conceptual difficulties of an order different from those confronting the positive integers. This has largely to do with the use of impredicative definitions in the customary formal systems for classical analysis. One indication of the difference is the fact that no comparable informative consistency proof is available for any formal system of the classical analysis that is as natural as the usual system for arithmetic. It is fair to say that on the basis of our present knowledge, we have full confidence in being able to devise only consistency proofs for predicative systems.

28. Several alternative courses for dealing with the continuum have been suggested.

It seems desirable to develop further and study more formally Brouwer's theory, also expounded by Weyl, with the distinction of effective and free choice sequences.

There is the course of restricting sets, in particular sets of positive integers, to some less nonconstructive totality which enjoys agreeable closure properties. For example, hyperarithmetic set theory, or some yet to be determined domain of predicative set theory.

A third course is to use what Bernays calls the quasi–combinatorial principle to justify the impredicative formulation of classical analysis. This depends on a natural but uncontrolled generalization of a situation with finite sets to infinite sets. A set of positive integers either contains 1 or not, either contains 2 or not, etc., hence, there must be $2^{\aleph_0}$ possible sets which includes all the number sets definable in any set theory. While this supplies a sort of inaccessible model for the continuum, it does not yield a consistency proof in the proof–theoretical sense. The most important mathematical problem on the continuum appears to be a proof–theoretical consistency proof for some familiar formal system adequate to the impredicative formulation of the classical analysis.

For this purpose, as well as the purpose of proving the consistency of predicative systems, the central problem is to rearrange by mental experimentations positive integers to form suitable well–ordered sequences and to see that the arrangements are indeed well–ordered. In this way, one is justified in applying the principle of transfinite induction relative to such rearrangements. Such well–orderings are the hard facts about mathematical structures. It seems that as long as mathematical thoughts are expressed in language and symbolism, there is no compelling reason to go beyond the second number class.

29. There are other viewpoints which are somewhat too one–sided so far as the foundations of mathematics are concerned. Algebraists tend to favor an abstract point of view. It used to be said that group theory contains the essence of all mathematics. Since, however, groups are of so many diverse types and so much more can be added to groups to give other structures, group theory does not supply the correct emphases on the basic concepts and methods of mathematics.

Sometimes it is thought that foundations of mathematics can be obtained on the basis of physical things. Given the physical things, we can think of sets of them, sets of sets of them, and so on. If one recognizes no ideal constructions at all, then there are no empty sets, and a unit set is identical with its only member. Moreover, there are no sets but only sums so that, e.g., $\left\{\{x, \{y, \{x, y\}\}\}\right\}$, $\left\{\{x, \{x, y\}\}, \{y, \{x, y,\}\}\right\}$ are both the same as $\{x, y\}$. In either case, unless we assume there are infinitely many physical things or permit infinite repetitions of the process of forming new sets, we cannot arrive at the positive integers. Such approaches deal with infinity only as an afterthought. They are, therefore, unsuitable for the study of mathematics the essence of which is infinity. Moreover, it is difficult to give a meaning to the supposition that there are infinitely many physical things.

4. LOGIC, COMPUTATION AND PHILOSOPHY *

4.1 Logic and logical positivism

The great attraction of the deductive method is that it serves to «divide and conquer». By breaking a proof into minute steps, difficulty gives way to complexity the comprehension of which usually requires a lower order of intellectual capacity. In the words of Descartes, «For whenever single facts have been immediately deduced the one from the other, they have been already reduced, if the inference was evident, to a true intuition. But if we infer any single thing from various and disconnected facts, often our intellectual capacity is not so great as to embrace them all in a single intuition; in which case our mind should be content with the certitude attaching to this operation. It is in precisely similar fashion that though we cannot with one single gaze distinguish all the links of a lengthy chain, yet if we have seen the connection of each with its neighbour, we shall be entitled to say that we have seen how the first is connected with the last.»

Spinoza proved in the geometrical order not only Descartes' *Principia* but his own major work, *Ethics*. Universal mathematics was for Descartes and Spinoza a sort of bigger geometry which proposes to comprehend all knowledge including principles of ethics. At first sight it all seems very reassuring: Our youthful dreams of a completely solid basis of knowledge and action seem to be realized in their systems. How often do we wish we knew for certain what we should do? If we had an ultimate goal and knew how much every piece of action contributed to that goal, we sould be able to live and behave resolutely. Spinoza now tells us, the highest good is the knowledge of the union existing between the mind and the whole of nature. He wishes to direct all sciences to one end so that we may attain this supreme perfection.

History of philosophy and science since Spinoza's time has taught us to distinguish scientific knowledge from moral principles, and mathematics from empirical sciences. Most of us are not fortunate enough to have our life and values organized entirely around one ultimate end. Still less are we able to decide courses which would contribute most to our ends, such as the vague concepts of happiness and the common good. We are nonetheless more hesitant to relinquish the hope of organizing human knowledge in an all-inclusive framework with a solid unquestionable basis.

In this century, we find a hybrid of Descartes' deductivism and Hume's empiricism that promises a bigger and neater science with all empirical knowledge founded on the rational basis of my sense-data here now and empirical induction, using the a priori

* First published in *L'âge de la science*, vol. 3, pp 101—115. Editions Bordas, 1971. Reproduced by permission of the author.

framework of logic and mathematics. This oversimplified view of knowledge of some logical positivists is open to many serious criticisms and has undergone numerous refinements and watering–downs. The most basic drawback is perhaps its utter irrelevance to life and the exact sciences. It not only implies a rather limited philosophy of life, but, more seriously, it fails to provide an adequate account of the natural sciences and mathematics. This is especially damaging since the founders of this school got most of their inspirations from their views of physics and mathematics. Ironically, the only directions along which logical positivism can claim to have produced some beneficial effects are those most unexpected at first, viz., the social sciences, linguistics, and the implications of computers.

In order to rebuild philosophy, or, less ambitiously, a philosophy of knowledge, on the ruins of logical positivism, it seems natural to reexamine first the nature of mathematics and then the conceptual foundations of the exact sciences. In each case, it is essential to avoid what might be called ≪disembodied generalities.≫

It seems likely that logical positivism requires more than one successor in order to take account adequately of the complexity of actual knowledge. Mathematics is a stumbling block because it is most closely tied up with its own language, so that it is hardest for a non–specialist to have a correct overview while specialists are rarely concerned with the varied ways in which mathematics is related to the other sciences. In what follows an attempt is made to sketch a few points relevant to arriving at an overview of mathematics. Such outlines are inevitably weak, since a genuine argument could conceivably be obtained only if such outlines are carried out.

4.2 What is mathematics?

The most impressive features of mathematics are its certainty, its abstractness and precision, its broad range of applications, and its dry beauty. The precision and certainty is to a large extent due to the abstractness which also in part explains the wide applicability. But the close connection to the physical world is an essential feature which separates mathematics from mere games with symbols. Mathematics coincides with all that is the exact in science.

According to Kant, mathematics is determined by the form of our pure intuition so that it is impossible to imagine anything violating mathematics. If we agree that the physical world, including our brains, is a brute fact, this view can be said to imply that the external world, including the physiological structure of our mind, determines mathematics. The discovery of non–Euclidean geometries need not be regarded as refuting Kant's doctrine, since we can construe them as superstructures on the Euclidean or even weaker foundations. A more serious objection is that Kant's theory does not provide enough elucidation of the principles by which these and other superstructures are to be set up.

As we all know, Shaw was accustomed to exaggerations. He defended himself by arguing that the shock value is the best way to call attention to new ideas. In a similar spirit, we may hope to clarify our vague thoughts by examining a few one-sided views of mathematics.

4.2.1 Mathematics is the class of logically valid propositions ≪p implies q≫. Thus, given any theorem q, we can write the conjunction of the axioms employed as p, and ≪p implies q≫ is a theorem in elementary logic. In this somewhat trivial sense, all mathematics is reducible to elementary logic. This really says nothing about mathematics proper, since one would like to assert p and q unconditionally. This evades the whole question why certain p, e.g. the Peano axioms, is accepted as a mathematical truth. A less clear and less clearly inadequate view would be to permit a broader domain of logic and take, e.g., propositions such as ≪For all x and y, if x and y have no common members, x has 7 members y has 5 members, then $x \bigcup y$ has 12 members.≫ Then one has to define numbers in logic, and so on. Such a view is akin to the next one.

4.2.2 Mathematics is axiomatic set theory. In a definite sense, all mathematics can be derived from axiomatic set theory. To be definite, we can adhere to a standard system commonly referred to as ZF. This is the counterpart of Frege's and Russell's reduction of mathematics to logic and paradoxically also of Poincaré's 1900 remark on the arithmetization of mathematics (≪numbers and their sets≫). This is what most impressed the logical positivists, leading to, among other things, an emphasis on axiomatization and formalization. There are several objections to this identification. As we shall discuss later, there are many difficulties in the foundations of set theory. This view leaves unexplained why, of all the possible consequences of set theory, we select only those which happen to be our mathematics today, and why certain mathematical concepts and results are more interesting than others. It does not help to give us an intuitive grasp of mathematics such as that possessed by a powerful mathematician. By burying, e.g., the individuality of natural numbers, it seeks to explain the more basic and the clearer by the more obscure. It is a little analogous to asserting that all physical objects, such as tables, chairs, etc., are spherical if we swathe them with enough stuff. There is a side issue of logicism which continues to be upheld in some quarters in the face of definitive evidence against it. In at least one important case, this mysterious state of affairs is based on a mistaken identification between Frege's logical theory of sets (extensions of predicates) with Cantor's mathematical theory of sets. The argument goes like this. Since Frege's theory looks like logic and mathematics can be reduced to Cantor's theory; therefore, by the identification, mathematics is reducible to logic.

4.2.3 Mathematics is the study of abstract structures. This appears to be the view of Bourbaki. To the extent that a sequence of books has been written to substantiate this view, it deserves a careful consideration. A conscious attempt to divorce mathematics from applications is not altogether healthy. The inadequacy of this outlook is revealed not only by the omission of various central results of a more combinatorial sort, but especially by the lack of intrinsic justification in the selection of structures which happen to be important for reasons quite external to this approach. Constructive contents of mathematical results are not brought out. There is also a basic inconsistency insofar as lipservice is paid to an axiomatic set theory as the foundations, while serious foundational researches are frowned upon. It would conform more to the general spirit if number, set, function were treated in a more intuitive manner. That would at least be more faithful to the actual practice of working mathematicians today.

4.2.4 Mathematics is to speed up calculations. Here calculations are not confined to numerical ones. Algebraic manipulations and juggling with logical expressions (e.g., in switching theory) are also included. A somewhat broader view would be to say that every serious piece of mathematics must have some algorithmic content. A different, though related, position would be to say that all mathematics is to assist science, to assist us to understand and control nature. These views seem to make it impossible to explain, e.g., why we often prefer more elegant proofs with higher bounds and why we take great delight in impossibility results. One could argue that there is in addition the human element in mathematical activities so that it is essential, even for applications, that the situation should be perspicuous. Thus, we can better grasp an elegant proof and, indirectly, are enabled to look for more efficient algorithms; and impossibility results tell us the limitations of given methods, helping the search for positive results in the long run. This kind of argument is, however, typical of philosophers stretching a position to try to fit in unwanted facts.

So much for oversimplifications.

If we review quickly the history of mathematics, we would find quite a few surprises. What appears particularly attractive is there is room for a serious and fruitful synthesis of mathematics, for work in the philosophy of mathematics which would help the progress of mathematics itself by making the subject more appealing and by fighting against excessive specialization. Let us, however, postpone such discussions till the last section.

4.3 Logic and computation

Familiar connections between logic and computers are the possibility of representing circuits by Boolean functions and the close resemblance between formal systems and programming languages. As early as 1656, Leibniz dreamed of a universal scientific language in his first published work; and the search for a universal language for computers is a rather central concern today. In an oblique way the performance of arithmetic by computer circuits may be said to accomplish a reduction of arithmetic to logic in a particular down-to-earth manner.

A more basic link between logic and computers is the common interest in explicit procedures. The long evolution of attempts to formalize mathematical proofs finally led, with the help of mathematical logic, to mechanizability as the ultimate external criterion of a successful formalization. A common complaint among mathematicians is that logicians, when engaged in formalization, are largely concerned with pointless hairsplitting. Computers seem to supply, more or less after the event, one good reason for studying formalization. While many mathematicians have never learnt the predicate calculus, it seems hardly possible for the machine to do much mathematics without being able to deal with the underlying logic in some fairly explicit manner. While the human being gets bored and confused with too much rigour and rigidity, the computer requires entirely explicit instructions.

There are some pleasant surprises in initial attempts to mechanize mathematical arguments. But it is clear that extensive work is needed before computers can influence mathematical research in an overall manner. For example, it is hard for the computer to hit on $x! + 1$ and prove there are infinitely many primes. This type of work should in due course affect the methods of research in logic and mathematics, as well as improve drastically the ability of computers to do all sorts of more sophisticated intellectual tasks. A particularly instructive direction would be to compare mechanizability with teachability.

While the discussions on whether machines can think tend to degenerate into terminological debates, it cannot be denied that sophisticated computers provide a useful framework for helping to think more fruitfully on the nature of mind, both philosophically and scientifically. Only one must resist the temptation to jump to conclusions. Many exciting problems in this area are simply too difficult. The present state of the study of ≪artificial intelligence≫ has been compared to alchemy both in its negative and its positive aspects. The great present challenge is to select and formulate fruitful problems.

The idealized models of computers developed by Turing and others are of great philosophical and mathematical significance (see next section), and form the beginning of a serious theory of computation. An area of current research interest is to refine the theory so that we can take care of not only theoretical possibilities but also practical feasibility. For example, different models which are equivalent in theoretical power can be distinguished because some can compute faster than others. In this respect, the intuitively obvious conclusion that multiplication is more complex than addition is, as a mathematical theorem, at present a much sought–after result. More specifically, it is desired to show that there is no potentially infinite machine on which we can multiply any two n–digit numbers (for all n) in linear time (i.e., in kn operations, k a constant). On the other hand A. L. Toom and S. A. Cook have recently shown that, for any ε, there exists some multitape Turing machine on which, for every n, two n–digit numbers can be multiplied in $n^{1+\varepsilon}$ operations. This area of computational complexity is very much a virgin land at present.

Another apparently promising area of research is to develop an elegant theory of programming languages.

4.4 Relatively undecidable propositions and absolutely unsolvable problems

The non–Euclidean geometries show that the parallel postulate is independent of the other postulates. Or, in other words, it is undecidable in the axiom system consisting of the remaining axioms. The impossibility of squaring a circle by ruler and compass gives an infinite class of problems (one for each circle) which cannot be solved by a given method of construction.

These two interesting results of the 19th century have stronger analogues in the 20th century. On the one hand, we have Gödel's general result that any fairly rich system is incomplete; on the other hand, we have a group of results that certain infinite classes of problems are unsolvable by *any* mechanical method.

It should be emphasized that these two types of result, while obtainable by related methods, are conceptually quite different. What is undecidable is in each case a single proposition relative to a given axiom system, what is unsolvable is in each case an infinite class of problems (a ≪mass≫ problem) for which there exists in an absolute sense no general method that can settle every member of the class. For example, there exists no general method to decide whether an arbitrary Turing machine will eventually stop, there exists no general method by which, given any proposition, we can tell whether it is a consequence of the Peano axioms.

The fundamental import of Gödel's incompleteness result is to exhibit exactly the limitations to the possibilities of formalizing our intuitive concept of natural numbers and sets of them. The undecidable propositions are tailor–made to fit each given system. They are not the common and garden variety. People have tried unsuccessfully to show Fermat's or Goldbach's conjecture undecidable in the usual system. This situation is similar to the difference between Cantor's proof of the existence of transcendental numbers and the proofs that e and π are transcendental. The former is philosophically more significant while the latter is mathematically more substantial.

Another important result on the limitations of formalization is the Löwenheim–Skolem result on set theory. In particular, this result shows that we cannot even have an adequate axiomatization of sets of integers. This general situation has led not only to Skolem's nonstandard model for arithmetic but also P. J. Cohen's clever proof that the continuum hypothesis is undecidable in the usual axiom system of set theory. We shall discuss this result in the next section, in connection with the foundations of set theory.

The absolute notion of unsolvable problems in terms of Turing computability or other equivalent notions is of special importance in that with this concept one has for the first time succeeded in giving an absolute definition of a basic epistemological idea, i.e., independent of any formal system. It is not necessary to distinguish different levels, and the diagonal argument does not lead outside the defined notion. Somehow we have succeeded in looking at computability from outside. This is quite different from definability and provability. So far, we have only been able to define them relative to a given formal system, and for each individual system it is clear that the one obtained is not the desired one. One of the most fascinating problems in logic and foundations is to characterize absolute notions of provability and definability.

The exact concept of computability and the accompanying phenomenon of unsolvable problems have wide applications in different branches of mathematics. An especially appealing feature is that we have often problems which are very easy to state but very hard to settle. Moerover, one is able to tackle many of the problems by native wits (especially combinatorial skills). Furthermore, several problems originated initially from other branches of mathematics. We give a brief list of some of the settled and unsettled problems.

4.4.1 Word problem for groups. Proved unsolvable by P. S. Novikov (1955). Related more general results have since been obtained by G. Higman.

4.4.2 The 4–dimensional homeomorphy problem. Shown to be unsolvable by A. A. Markov (1958).

4.4.3 Various partial results on Hilbert's tenth problem; i.e., unsolvability of various broader classes of problems.

4.4.4 The tiling (or ≪domino≫) problems and applications to logic.

4.4.5 G. S. Tsentin and D. Scott (1955) have shown unsolvable the word problem of the following simple system.

ac— ca, ad— da, bc— cb, bd— db, adac— abace, eca— ae, edb— be.

4.4.6 The word problem is solvable if we replace e by c in $adac \Rightarrow abace$ of the above system.

Unsettled problems.

4.4.7 The 3–dimensional homeomorphy problem.
4.4.8 Hilbert's tenth problem (Diophantine equations).
4.4.9 The concatenation analogue of Hilbert's tenth problem.
4.4.10 Word problem on $\{0, 1\}$ for the tag system:

$$0\_\,\_ \rightarrow 00$$
$$1\_\,\_ \rightarrow 1101.$$

With regard to Hilbert's tenth problem, one may note that it is equivalent to the decision problem for quantifier–free number theory with addition and multiplication. Moreover, it is of interest to remark that the problem of enumerating finite sequences is of central importance.

—Added in proof: Recently I. V. Mateyasivich has proved that Hilbert's tenth problem is indeed unsolvable, and that all recursively enumerable sets are Diophantine. The proof was reported at Novosibirsk on February 9, 1970 and a paper has just been published.

4.5 Foundations of set theory

Wittgenstein complains of the fuss mathematicians made over the contradictions. This is rather inaccurate since Frege and Russell are logicians while Cantor is a mathematician. Cantor's view is not that bridges will not collapse because of an inconsistent system of set theory, but rather that contradictions arise only because of an inadequate understanding of the concept of set. Admittedly it remains debatable whether Cantor's theory of sets might not after all contain contradictions. It is unquestionable that the extensive discussion of contradictions has little direct relevance to Cantor's theory.

According to Cantor, by a "set" we shall understand any result M of collecting together definite, distinct objects (the "elements" of M) of our intuition or of our thought (1895, *Gesammelte Abhandlungen*, p. 282). In his letter to Dedekind (1899, *ibid.*, p. 443), Cantor is more explicit and distinguishes sets from inconsistent multiplicities. In particular, that of all sets and that of all ordinals are inconsistent multiplicities. Hence, the Burali–Forti paradox of 1897, known to Cantor two years earlier, does not apply.

A familiar interpretation of Cantor's idea, first explicitly formulated by D. Mirimanoff (1917, *L'enseignement Math.*), is the maximum iterative concept of set which is determined by the power set operation P relative to a basic domain D_0 of things and the totality Ω of ordinal numbers. For example,

$$D_0 = \Lambda$$
$$D_{n+1} = P(D_n)$$
$$U = D_\omega = \bigcup D_n$$

determines a model of set theory with finite sets only. Using all ordinals Ω in place of ω, one obtains not only a model for the current axioms of set theory but even, it is believed, a characterization of Cantor's original concept. Since the choice of the basic domain D_0 is extraneous to set theory proper, the problem reduces to explaining the power set operation P and the totality Ω of ordinals.

Not only is the maximum iterative concept familiar, it is claimed to be natural on the ground that it is generally accepted and that one learns rather quickly to apply it. On this claim of naturalness it is possible to expend endless philosophical discussions. For the present purpose, we simply accept the choice as a postulate to exclude what we regard as peripheral controversies.

Ordinals are central and we have five principles of generating ordinals.

*P*1. A new ordinal is generated by the addition of a unit to an ordinal which has already been formed.

*P*2. There is an infinite ordinal ω.

*P*3. Given any ordinal and an arbitrary well–ordering of all earlier ordinals, we get another ordinal. If we speak of any definite succession of ordinals, we would have to impose at least a restriction on cardinality.

*P*4. Given any cardinality, we can consider the totality of all ordinals of that cardinality and introduce an ordinal of a next higher cardinality.

*P*5. Given any set of ordinals, we can introduce new ordinals by replacement.

The first principle is concerned with the potentially (countably) infinite, and a thorough absolute theory is available in terms of computability, finite methods, algorithms, syntax, two–oneness, the unity of contraries, or the intuition of the effective.

*P*2 and *P*3 should be combined either to give only countable ordinals or, better, to give definable ordinals in some mildly constructive sense. What is involved is the basic difficulty of surveying all countable ordinals. It is not unreasonable to suppose that all definable sets and ordinals are countable. In any case, we face here the question of definability.

*P*4 introduces us directly to the position of ≪platonism≫ and embodies an analogue of the power set operation. It is also the natural place to introduce the continuum hypothesis (Cantor, op. cit., p. 192, 1883).

Ideally the impredicative should be a limit of the predicative. There is no philosophical reason why the impredicative and the uncountable should be tied together. We think of them together only because, as a result of our inability to survey either all the countable or all the predicative, we can only arrive at the uncountable from the impredicative and we do not have a good understanding of non–standard countable models of the impredicative set theory.

*P*5 gets us beyond and is in itself less basic. Since, however, we do not have a good understanding of the range and variety (the width) of the countable, we are tempted to increase the width by using bigger ordinals introduced through *P*5. One is, for example, led to speculate on stronger axioms of infinity in order to obtain more sets of integers.

Cantor was pleased with his power of creating a lot of new numbers. The late philosopher (a critical realist) W. P. Montague tried to imitate Cantor by introducing the rainbow series. He thought that by taking logarithm base 2 of aleph zero and repeating, he could introduce cardinals between the finite and the infinite.

Cantor did not use an axiomatic approach. There is, in fact, no obvious reason why his intuitive theory should be axiomatizable. In fact, it is clear from the Skolem paradox that Cantor's theory is not axiomatizable. Hence, philosophically, the axiomatic approach is not satisfying except perhaps on the ground that so far we have not been able to get very far from any but the axiomatic approach, which, at any rate, is useful as an auxiliary tool, if indeed not the only humanly possible tool.

Cantor thought he had a proof of the continuum hypothesis long before the development of axiomatic set theory. And there does not appear to be any intrinsic reason why a better understanding of the concepts of sets and wellorderings of integers could not yield a decision of the continuum hypothesis independently of particular axiom systems.

Even Zermelo's axiom system left open what a definite property is. While the exact formulation by Skolem sharpened the system and indeed for the first time made it into a genuine axiom system, the vaguer formulation is closer to our intuitive requirement.

Once the axiomatic approach is adhered to, problems such as the continuum hypothesis change their nature drastically. It was expected that a solution of the problem relative to the current axiom system would inevitably illuminate the original problem. Cohen's solution has contradicted this expectation. Instead of yielding any strong hint as to how to solve the original problem, it reveals how incomplete the axiom system is. It almost seems to be man's fate that on the most fundamental questions the best results can only be negative ones.

In his paper «Some remarks on the axiomatic founding of set theory» of 1922, Skolem diagnosed the inadequacy of axiomatic set theory (on account of Löwenheim's Theorem) and predicted rather accurately Cohen's result. Skolem suggests it would be interesting and probably difficult to add a new set of integers to the set theory and prove relative consistency. In this connection, he adds a footnote 9.

9. Since Zermelo's axioms do not uniquely determine the domain B, it is very improbable that all cardinality problems are decidable by means of these axioms. For example, it is quite probable that what is called the continuum problem, namely the question whether $2^{\aleph_0}$ is greater than $\aleph^1$, is not solvable at all on this basis; nothing need be decided about it. The situation may be exactly the same as in the following case: an unspecified commutative field is given, and we ask whether it contains an element x such that $x^2 = 2$. This is ust not determined, since the domain is not unique.

Later Skolem repeated this remark in connection with a review of Hilbert's paper «On the infinite» (1925).

In 1938, Gödel announced a proof of relative consistency. This in itself did not say much about the truth and falsity of the continuum hypothesis. But the very special character of the proof enhances Skolem's conjecture that the continuum hypothesis may be independent of the usual axioms.

Finally in 1963, Cohen proved the independence, and in addition, brought out spectacularly how incomplete the axioms of set theory are.

It is natural to ask whether extending the axioms might change the picture. One precise way of getting a richer axiom system is by assuming measurable cardinals. Very recently it has been proved that the addition does not affect the undecidability of the continuum hypothesis.

There are two other ways of attacking this question (apart from finding specific richer set theories):

To prove a more general theorem: any axiom system if consistent must leave the question undecided. Here, of course, we cannot quite achieve the same generality as the incompleteness of arithmetic. Since C. H. is a single statement, we can always refute the claim by adding C. H. or its negation. Hence, we must give some delimitation of the type of axioms we are permitted to add-- e.g., strong axioms of infinity. But then we have to give some general characterization of axioms of infinity.

We may give up the axiomatic approach altogether by returning to Zermelo's indefinite concept of definite properties or by introducing a maximum principle: no greater universe satisfies the axioms (this is a statement of a different sort and transcends the usual situation that any theory with a model of certain size always has bigger models. We are supposed to envision *all* sets).

Apart from the difficulty in accepting and interpreting such a principle, there is a basic difficulty about deriving consequences from it, in particular on the truth or falsity of C. H.

While Cantor was mainly interested in the ontic question of the existence of sets and ordinals, Brouwer was primarily concerned with the epistemic question of provability. This has its obvious parallel in the history of philosophy. The concern of the French semi–intuitionists with definability was something in between.

It is of interest to see how these questions may be related. From the epistemic approach, one might argue that we should stop worrying about meaningless questions such as the continuum hypothesis. From the ontic approach, someone has conjectured that all number–theoretic questions become decidable provided we get enough strong axioms of infinity.

There is a familiar comparison between the continuum hypothesis and Fermat's conjecture, the latter being true if undecidable. This fact does not, however, contradict Post's suggestion that Fermat's conjecture might be undecidable in some absolute sense. If, however, one gives any proof of such absolute unprovability, one would get into trouble because then Fermat's conjecture would be proved, seeing that we cannot reasonably exclude the ability to produce numerical counterexamples. In fact, Brouwer accepts the double negation of the law of excluded middle which is interpreted as saying that we can never prove something absolutely unprovable.

Many feel that to speak of improving our understanding of the concept of set implies an absolutist conception of set, or that all statements about sets have a definite truth value. This does not seem necessary. As a historical fact, we have come to accept axioms of choice, regularity, and replacement and feel they conform to our intuitive concept. We feel that we understand the concept of set better as a result.

4.6 What is mathematics? (continued)

Foundational studies in this century have been very fruitful in several ways. The analyses by Brouwer of basic concepts of mathematics are, for some people anyhow, philosophically satisfying. And the study of proof theory has yielded many fundamental results, especially negative ones, on decidability and solvability. On the whole, there remains, however, the impression that foundational problems are somewhat divorced from the main stream of mathematics and the natural sciences. Whether this is as it should be seems to me a highly debatable point.

The principal source of detachment of mathematics from mathematical logic is that logic jumps more quickly to the more general situation. This implies a neglect of mathematics as a human activity, in particular, of the importance of notation and symbolism, and a neglect of the more detailed relations of mathematics to applications. It is philosophically attractive to study in one sweep all sets, but in mathematics we are primarily interested in only a very small range of sets. In a deeper sense, what is more basic is not the concept of set but rather the existing body of mathematics. For example, the distinction between linear and nonlinear problems, the invention of logarithms, the different ways of enumerating finite sequences, the nature of complex numbers and their functions, or the manipulation with infinities by physicists (such as Dirac's delta function and the intrusion of infinities in quantum electromagnetic theory) all seem to fall outside the range of problems which interest specialists in foundational studies. Rightly or wrongly, one wishes for a type of foundational studies which would have deeper and more beneficial effects on pedagogy and research in mathematics and the sciences.

As a first step, one might envisage an ≪abstract history≫ of mathematics that is less concerned with historical details than conceptual landmarks. This might lead to a resolution of the dilemma between too much fragmentation and too quick a transfer to the most general.

4.6.1 Concrete arithmetic began with practical problems. The idealization of the indefinite expandability of the sequence of numbers and the shift from individual numbers to general theorems about all numbers give rise to the theory of numbers. Only around 1888, was Dedekind able to formulate the so-called Peano axioms by analyzing the very concept of number.

4.6.2 The solution of equations together with the use of literal symbols such as letters for unknowns marks the beginning of algebra (≪transposition and removal≫). Only in 1591 (Viète) were letters used for known quantities as well (variables and parameters).

4.6.3 Geometry deals with spatial forms and geometrical quantities such as length and volume. The number of a set is an abstraction of that which is invariant under any changes whatsoever of the properties and mutual relations of the objects in the set (e.g., color, weight, size, distance), provided only the identity of each object is not disturbed (by splitting or merging). Similarly a geometrical figure or body is an abstraction of an actual body viewed purely with regard to its spatial form, leaving out all its other properties. Rather surprisingly, such an abstract study led not only to pure geometry but also to the first extensive example of the deductive method and axiomatic systems. There was even a geometrical algebra in Greece.

4.6.4 Measurement of length and volume is a union of arithmetic and geometry, applying units to calculate a number. This, just as the solution of equations, is a natural way of leading to fractions and even irrational numbers. The desire to have an absolutely accurate or rather indefinitely improvable measurement leads to the general concept of real numbers. Algebra led to negative numbers and complex numbers. But a better understanding of complex numbers was only reached through their geometrical representations.

4.6.5 By the way, in terms of speeding up computations, the invention of logarithms (Napier, 1614) was a great advance.

4.6.6 In an indeterminate equation, say $3y - 2x = 1$, we may view x and y as unknowns or as variables so that the given equation expresses the interdependence of these two variables. The general concept of function or interdependence is the subject matter of analysis. Using the Cartesian coordinates, we get a connection of algebra and geometry, with function playing the central role. In this sense, analytic geometry may be said to be the simplest branch of analysis. It is implicitly assumed that we deal with at least all real numbers.

4.6.7 If we add in addition the concept of change or motion, and study a broader class of functions, we arrive at the calculus. The original source was from geometry and mechanics (tangent and velocity, area and distance.) Theories of differential and integral equations search for functions rather than numbers as solutions. They are natural both from applications and from an intrinsic combination of the calculus with the algebraic problem of solving equations. In the same spirit, functional analysis is not unlike the change from algebra to analysis, the interest being no longer confined to finding individual functions but rather the general interdependence of functions.

4.6.8 It is not easy to understand why functions of complex variables turned out to be so elegant and useful. But it certainly was a gratifying phenomenon that an extension serves to clarify many facts in the original domain. Incidentally, if we require the axioms of fields be satisfied, extensions of complex numbers are not possible. For quaternions, e.g., multiplication is not commutative.

4.6.9 The lively development of the theory of probability has been connected with statistical mechanics, and its foundations are a fascinating but elusive subject.

4.6.10 In algebra, Galois theory not only gives a conclusive treatment of the solution of equations but opens up a more abstract study of abstract structures dealing with operations on arbitrary elements rather than just numbers.

4.6.11 The greatest changes in geometry have been the discovery of non-Euclidean geometries and Riemann's general idea on the possibility of many different «spaces» and their geometries. Figures are generalized to arbitrary sets of points.

4.6.12 The development in functions of a real variable touches on various philosophical problems such as the definition of real numbers and the meaning of «measure».

In this century, the development of logic, the emergence of computing machinery, and the prospect of new applications such as in the biological sciences and in linguistics all tend to emphasize what might be called «discrete mathematics», even though continuous mathematics is well-entrenched and as lively as ever.

The fond hope of a meaningful synthesis of all mathematics and its applications belongs perhaps only to one who is ignorant and quixotic. But then it is a heart-warming dream.

One of the very basic problems is that we still do not have any definitive theory of what a real number or what a set of integers is. Perhaps we can never have a definitive theory. It seems quite unknown how this fundamental unclarity affects the rest of mathematics and the novel applications of mathematics in physics.

Relative to different concepts of set and proof, one could reconstrue most of mathematics in several different ways. What is not clear is, whether these different formulations are just essentially equivalent manners of describing the same grand structure or there exists a natural framework in which everything becomes more transparent.

I would like to suggest a more schematic approach to the formalization of mathematics. It is a striking fact that in diverse systems of different strength we can prove counterparts of all ordinary theorems on real numbers. This, suggests that no proof in any formal system formalizes faithfully the true mathematical result. Our intuition of the real numbers is not captured in any of the particular formal systems. A closer approximation could be obtained if we look instead for the class of formal proofs (and therewith the underlying formal systems) which all can represent a given intuitive proof. In this way, each intuitive proof would correspond to a class of formal proofs, and we can classify intuitive theorems according to the classes of formal systems in which they can be represented naturally. This approach would avoid the futile dispute as to which formal system is the correct one. Also it would reveal more fully the intuitive content of ordinary informal mathematical proofs[1].

---

(1) The section on set theory was included in a talk presented at the joint meeting of the American Philosophical Association and the Association for Symbolic Logic in December, 1965. The whole paper was presented with omissions at The Rockefeller University in January 1966, and without omissions (in two sessions) as the «Class of 1927 Lecture» at The Rensselaer Polytechnic Institute in March 1966.

PART TWO

AUTOMATED THEOREM PROVING (ATP)

5. COMPUTER THEOREM PROVING AND ARTIFICIAL INTELLIGENCE*

It gives me a completely unexpected pleasure to be chosen as the first recipient of the milestone prize for ATP, sponsored by the International Joint Conference on Artificial Intelligence. (See appendix.) I have worked in a diversity of fields; I am correspondingly limited in my capacity to appreciate, or express my appreciation of, a large range of efforts in each of these fields; and I tend to shun positions of power. Undoubtedly to a considerable extent as a result of these innocent shortcomings, honors have a way of passing me by. I have indeed slowly grown used to this. Hence, the present reward has surprised me.

I had hoped to arrive at an understanding of current work in the field in order to relate it to my early aspirations and expectations. For this purpose, I had asked Professor Woody Bledsoe for help, who has kindly sent me some papers and references. I soon realized that with my present preoccupation with philosophy I shall not be able to bring myself to concentrate enough to secure a judicious evaluation of these writings which at first sight appear largely alien to my own way of thought. Indeed several years ago Joshua Lederberg thoughtfully sent me a copy of Dr. D. Lenat's large dissertation (*AM: an artificial intelligence approach to discovery in mathematics as heuristic search*) which I found thoroughly unwieldy and could not see how one might further build on such a baffling foundation. This previous experience strengthens my reluctance to plunge into a study of current work. Hence, I have decided to limit myself to a summary of my own views with some comments only on one line of current research with which I happen to have some familiarity. Actually I have just noticed, in the announcement of the prize, an explicit statement that "the recipients will present one-hour lectures on their work." Hence, restricting to work familiar to me is quite proper. It is possible that certain aspects of my thought have been bypassed so far and they may be of use to future research.

Around the beginning of 1953 I became dissatisfied with philosophy (as seen at Harvard) and, for other reasons, I also wanted to do something somewhat more obviously useful. Computers struck me as conceptually elegant and closely related to my training. The first idea was to see in computers a home for the obsessive formal precision of (the older parts of) mathematical logic which mathematicians tend to find irrelevant and worse, pedestrian and perhaps a hindrance to creativity. With little conception of the industrial world, I accidentally got an appointment as 'reserch engineer' at Burroughs. Unfortunately I was not permitted to use the only local computer and even

* First published in *Contemporary Mathematics*, (1984), pp 49—70. Reproduced by permission of the American Mathematical Society.

discouraged from taking a course for electronics technicians. I was reduced to speculations which led to a programming version of Turing machines and some idle thoughts on 'theorem-proving machines.' The material was written up only quite a bit later (probably in 1955) and published in January 1957 (see reference 1, §6 is devoted to theorem-proving).

The section begins with the observation that logicians and computer scientists share a common concern. 'Both groups are in terested in making thoughts articulate, in formalization and mechanization of more or less vague ideas.' Decidable theories are mentioned in passing. But the main emphasis is on 'the possibility of using machines to aid theoretical mathematical research on a large scale.' A few modest research goals are mentioned: the independence of axioms in the propositional calculus; 'we often wish to test whether an alleged proof is correct'; this is related to the matter of filling in gaps when 'the alleged proof is only presented in sketch'; to confirm or disprove our hunches;' to disentangle a few exceedingly confusing steps' in a (proposed) proof. Herbrand's theorem on cut free proofs is mentioned as possibly helpful. 'We can instruct a proving machine to select and print out theorems which are short but require long proofs (the 'deep' theorems).' And a general observation: 'The important point is that we are trading qualitative difficulty for quantitative complexity.'

It was not until the summer of 1958 that I began to work actually with a computer (an IBM 704, using SAP, the Share Assembly Programming Language), thanks to Bradford Dunham's arrangement. It was an enjoyable long summer (owing to the fact that Oxford begins school only late in October) when I learned programming, designed algorithms, and wrote three surprisingly successful programs. 'The first program provides a proof-decision procedure for the propositional calculus' and proves all such theorems (about 220) in *Principia mathematica* (briefly *PM*) in less than three (or five?) minutes. 'The second program instructs the machine to form itself propositions of the propositional calculus from basic symbols and select nontrivial theorems.' 'Too few theorems were excluded as being trivial, because the principles of triviality actually included in the program were too crude.' The third program dealt with the predicate calculus with equality and was able to prove 85% of the over 150 theorems in *PM* in a few minutes. This was improved in the summer of 1959 (see below) to prove all these over 150 theorems. The improved program embodies a proof-decision procedure for the 'AE predicate calculus with equality' '(i.e., those propositions which can be transformed into a form in which no existential quantifier governs any universal quantifier) which includes the monadic predicate calculus as a subdomain.'

What is amazing is the discovery that the long list of all the theorems of *PM* in the undecidable predicate calculus (a total of ten long chapters) falls under the exceedingly restrictive domain of the AE predicate calculus. This seems to suggest that of the vast body of mathematical truths, most of the discovered theorems are among the easier to prove even in mechanical terms. Such a suggestion would be encouraging for the project of computer theorem proving. A more definite general implication of the situation of *PM* with regard to the special case of the predicate calculus is the desirability of selecting suitable subdomains of mathematical disciplines which are, like the AE calculus, quickly decidable and yet contain difficult theorems in the human sense. The recent work of Wu (see below) in elementary and differential geometry would seem to be an excellent illustration of such a possibility.

The work of the summer was written up and submitted to the *IBM Journal* in December 1958. But for some curious reason it appeared only in January 1960. The paper (see reference 2) begins with a comparison between calculation and proof, arguing that the differences present no serious obstacles to fruitful work in mechanical mathematics. In particular, 'much of our basic strategies in searching for proofs is mechanizable but not realized to be so because we had had little reason to be articulate on such matters until large, fast machines became available. We are in fact faced with a challenge to devise methods of buying originality with plodding, now that we are in possession of slaves which are such persistent plodders.' 'It is, therefore, thought that the general domain of algorithmic analysis can now begin to be enriched by the inclusion of inferential analysis as a younger companion to the fairly well-established but still rapidly developing leg of numerical analysis.'

The paper concludes with a long list of remarks such as the relation between quantifiers and functions, quantifier-free number theory, the central place of induction in number theory, the *Entscheidungsproblem*, 'formalizing and checking outlines of proofs, say, from textbooks to detailed formulations more rigorous than *PM*, from technical papers to textbooks, or from abstracts to technical papers,' etc. (By the way, I now feel that the reverse process of condensing is of interest both for its own sake and for instructiveness to the original project.) 'It seems as though logicians had worked with the fiction of man as a persistent and unimaginative beast who can follow rules blindly, and then the fiction found its incarnation in the machine. Hence, the striving for inhuman exactness is not pointless, senseless, but gets direction and justification.'

At that time, several reports were in circulation by Newell-Shaw-Simon which deal with the propositional calculus in an inefficient manner and contain claims against 'automatic' procedures for producing proofs. Their claims were refuted conclusively by my work. Hence, I included some discussion of their work. 'There is no need to kill a chicken with a butcher's knife. Yet the net impression is that Newell-Shaw-Simon failed even to kill the chicken with their butcher's knife.' A larger issue is the tendency at that time to exclude what are now called 'expert systems' from artificial intelligence. Relative to the project of theorem proving, I put forward at that time a plea for 'expert systems':

'Even though one could illustrate how much more effective partial strategies can be if we had only a very dreadful general algorithm, it would appear desirable to postpone such considerations till we encounter a more realistic case where there is no general algorithm or no efficient general algorithm, e.g., in the whole predicate calculus or in number theory. As the interest is presumably in seeing how well a particular procedure can enable us to prove theorems on a machine, it would seem preferable to spend more effort on choosing the more efficient methods rather than on enunciating more or less familiar generalities. And it is felt that an emphasis on mathematical logic is unavoidable because it is just as essential in this area as numerical analysis is for solving large sets of simultaneous numerical equations.'

In the summer of 1959 I went to Murray Hill to continue with the work of the preceding summer, and before long a program adequate to handling the predicate calculus with equality in *PM* was completed. It contains about 3,200 instructions. All the over 350 theorems of this part of *PM* are proved in 8.4 minutes with an output of about 110 pages of 60 lines each. The result was written up before the end of 1959, published in April 1960 (see reference 3) and then presented in May at the ACM Conference on Symbol Manipulation. Most of the other papers at the conference were devoted to introducing programming languages. I recall that several of the speakers illustrated the power of their languages by giving a succinct presentation of 'the Wang algorithm' viz. the simple program for the propositional calculus described in reference 2. With a special emphasis on the role of logic (which I now consider an unnecessary limitation), I suggested that 'computers ought to produce in the long run some fundamental change in the nature of all mathematical activity.' The recent extensive use of computers in the study of finite groups and in the solution of the four color problem would seem to confirm this expectation and point to an effective general methodological principle of paying special attention to and taking advantage of potentially algorithmic dimensions of problems in pure mathematics.

Toward the end of 1959, I was struck by the interesting patterns of the Herbrand expansions of formulas in the predicate calculus. My thought at that time was that if we can handle the interlinks within each pattern in an effective way, we would not only decide many subdomains of the predicate calculus but also supply a generally efficient procedure to search for proofs as well. At that time Paul Bernays and Kurt Schütte happened to be at the Princeton Institute. Schütte impressed on me the importance of the challenging and longoutstanding problem of whether the apparently simple AEA case is decidable. The attraction of such a clean problem lured me away from the computer enterprise which seemed to suggest indecisively a large number of different directions to continue.

While working on this theoretical problem, since I enjoyed discussions with my colleagues who had little training in logic, I came to introduce a class of tiling problems (the 'domino problems') that captures the heart of the logical problem. I was surprised to discover that the 'origin-constrained domino problem' can simulate Turing machines and is therefore undecidable. These results were written up in reference 4 but unfortunately I neglected to include the easy result on the origin-constrained problem, which was made into a technical report only a year later (in the summer of 1961). Soon after that, with the assistance of my student A. S. Kahr, I was able to prove, using the domino problems, that the AEA case is undecidable, contrary to my previous conjecture.

The paper 4 also includes a few concrete examples. An analysis is given of how $x \neq x + 1$ and $x + y = y + x$ would be proved in quantifier-free number theory from the Peano axioms. A research paper by J. Hintikka is reformulated into a relatively simple example of deciding a formula in the predicate calculus. This last example was further elaborated in reference 6 in which the vague concept of 'stock-in-trade systems' was introduced and two examples from number theory (the infinitude of primes and the irrationality of $\sqrt{2}$)

were mentioned. It was only in the 1965 IFIP address (reference 8) that I offer more detailed discussions of four examples from number theory and three examples from the predicate calculus. I am not aware that anybody has actually solved any of these examples on a computer, either by working out my suggestions or by other means. I believe that these examples remain useful today because particularity often sharpens the challenge to provoke sustained thinking. Some general points underlying the comments on these examples are made more explicit in reference 9, which also goes into more general speculations such as the following.

'We are invited to deal with mathematical activity in a systematic way.... one does expect and look for pleasant surprises in this requirement of a novel combination of psychology, logic, mathematics and technology.' There is the matter of contrasting the formal and the intuitive: for example, Poincaré compares Weierstrass and Riemann, Hadamard finds Hermite's discoveries more mysterious than those of Poincaré. The four stages of an intellectual discovery are said to be: (1) preparation; (2) incubation, (3) illumination; (4) verification. To mechanize the first and the last 'stages appear formidable enough, but incubation leading to illumination would seem in principle a different kind of process from the operations of existing computers.'

A chapter of reference 10 is devoted to a general discussion of 'minds and machines,' with a long section on the relation to Gödel's incompleteness theorems, as well as a section contributed by Gödel. An account of some aspects of computer theorem proving is included in reference 11.

I should like to take this opportunity to publicize an idea of using some simple mathematics to bridge the gap between computer proofs and human proofs. I have in mind proofs which include more formal details than usual which are more likely candidates for comparatively easier computerization. In particular, around 1965 I wrote a number of set-theoretical exercises, some by myself and some with my students: see references 12 to 17. In all these cases I am confident that I can without a good deal of effort turn the fairly detailed outlines into completely formal proofs. My suggestion is to use these papers as initial data to help the designing of algorithms for discovery and verification of sketches of proofs.

For some time I have assumed that it is desirable to have accomplished mathematicians involved in the project of using computers in their special areas. This has certainly taken place in several regions as auxiliaries to broader projects of settling certain open problems (say on finite groups and map coloring). A less frequent phenomenon is to design efficient algorithms wherever possible and then examine their ranges of application. It used to be the case that pure mathematicians tended to shun computers as something ugly and vulgar. In recent years, with the rapid spreading of computers, more and more younger mathematicians are growing up with computers. Hence, a basic obstacle to their use has been removed and the remaining division is less sharp: some would only use computers as a hobby while others would more or less consciously think of taking advantage of computers to extend their own capacity for doing more and better mathematics.

Along the line of searching for algorithms in branches of mathematics, I happen to have some familiarity with the work of Wu Wen-tsün, an accomplished geometer. (At one time Wu taught secondary schools and, for lack of contact with modern mathematics, became an exceptional expert of elementary geometry. This misguided training has since not only helped him in his research in advanced mathematics but also turned out to be very useful for his recent project of mechanizing certain parts of geometry.) In 1977 Wu discovered a feasible method which can prove mechanically most of the theorems in elementary geometry involving only axioms of incidence, congruence and parallelism but not axioms of order and continuity (in Hilbert's famous axiomatization). The method was later extended to elementary differential geometry. These are reported in 1978 and 1979 (references 18 and 19).

Among the theorems proved on an HP computer (9835A and then 1000) are the gou-gu or Pythagorean theorem, some trigonometric identities, the Simson-line theorem, the Pappus theorem, the Pascal theorem, and the Feuerbach theorem. Among the 'new theorems' discovered and proved are: (1) the 'anticenter theorem and the anticenter line theorem'; (2) the 'Pappus-point theorem'; (3) the 'Pascal-comic theorem.' In differential geometry the results proved include Dupin's theorem on triply orthogonal families of surfaces, the affine analogue of the Bertrand curve-pair theorems, and the Bäckland theorems on the transformation of surfaces. For more details, please consult Wu's recent paper, reference 20.

In 1978 Douglas McIlroy and I experimented with Wu's algorithm for elementary geometry and found that without proper caution the requirement for storage tended to be large. For example, a crucial step is to test whether a polynomial g vanishes modulo given irreducible polynomials $p_1, \cdots, p_n$. If we work with each coefficient separately rather than dealing with the whole expansion of g, the size of calculation is reduced considerably. McIlroy has since continued to toy with matters surrounding the algorithm and accumulated an amount of interesting data.

I should now like to turn to a few general remarks on the larger topic of artificial intelligence (briefly AI) as well as on its relation to theorem proving.

The area of AI is rather indefinite. In a broad sense it includes 'expert systems' (or 'knowledge engineering') which, according to Michie (p. 195) 'transfer human expertise in given domains into effective machine forms, so as to enable computing systems to perform convincingly as advisory consultants. Expert systems development is becoming cost effective.' Michie goes on to list more than ten examples. One might argue about the cost effectiveness but at least we get here a tangible criterion of success. Most of the 'basic' or 'theoretical' work in AI does not admit any such transparent standards of evaluation: this is undoubtedly one of the several reasons why the field is so controversial. But let me delay over expert systems a bit.

DENDRAL takes the pattern generated by subjecting an unknown organic chemical to a mass spectrometer, and infers the molecular structure. This program is commonly regarded as a success. Indeed I was full of enthusiasm about it when I first heard of the project in the 1960s. But the outcome fell short of my high expectations. One problem is that experienced chemists are often unable to state explicitly the rules they know how to use when confronting actual particular situations. An unquestionable authority told me that the direct outcome of DENDRAL was not worth all the investment put into it but that when one took into consideration its more general influence in starting the trend to develop expert systems, the evaluation was different. The more interesting meta-DENDRAL project has apparently been left in abeyance for complex sociopsychological reasons.

I can no longer be sure whether my programs for proving theorems in the predicate calculus qualify as an expert system. They did use some 'expert' knowledge of logic. And I was told that people did use my programs for other purposes. In this case it is clear that they were cost effective since it involved only a few months' labor of one person plus less than ten hours on a 704 machine. At any rate, my getting a prize from an international AI organization shows that I am finally being accepted into the AI community, twenty years after I was commended as a 'cybernetist' in the Soviet Union. This example illustrates why I get confused over the range of AI which, like any other subject, undoubtedly evolves with time.

Critics of AI quote the less interesting part of my work only in order to berate the unprofessional job of Newell-Shaw-Simon (e.g., Dreyfus, first edition, p.8 and Weizenbaum, p.166). (By the way, Dreyfus reduces all the 220 theorems to only the 53 selected by Newell-Shaw-Simon.) They miss the central point that although the predicate calculus is undecidable, yet all the theorems of *PM* in this area were so easily proved. Enthusiasts for AI simply leave me alone. The professional writer McCorduck at least makes a bow to theorem proving: 'and I rationalize such neglect by telling myself that theorems would only scare away the nonspecialist reader this book is intended for'. This incidentally leads to some other factors which render the field of AI controversial. It incites popular interest yet, unlike physics, it is a more mixed field in which natural scientists and social scientists (not to mention philosophers) with quite different standards meet. More, since it is near big technology, it is close to industrial and government money. As a result of all these factors, public relations tend to play a larger role than in more mature disciplines. And that tends to put some people off even though they find the intellectual core of many problems in the area appealing and challenging. Exaggerated and irresponsible claims and predictions, instead of being chastised, appear to be a central ingredient of the glory of many of the 'giants' in this field.

The controversies over AI are a mixed bag. At one end there are relatively solid accomplishments which may be evaluated in terms of 'cost effectiveness' and cleancut 'milestones.' There is a middle region which contains results which are not particularly decisive in themselves but seem to promise more exciting future developments. But in most cases the promises are fairly subjective and precarious. When we move to 'basic' AI which in many ways is the most exciting for the widest audience, we seem to be arguing over conclusions for or against which our universal ignorance of crucial aspects offers so little evidence that individual psychology over ethical, esthetic, political, and philosophical matters generally appears to play a dominant role. The fascinating question seems to shift from who is right (science?) to the motley of sources of the vehement disagreements (socio-psychology of convictions). Let me try to sort out some of the larger issues as I darkly see them.

Is the brain a computer? By now it seems generally agreed that the brain is certainly not a computer of the kind we are familiar with today. This point was argued, for example, by John von Neumann, who used to say that the brain is a professional job (which the computer isn't), in *The computer and the brain*, 1958. There remains the position that whatever the brain can do, some computer of the current type can do it as well, some day. This seems implausible and in any case so indefinite that we don't know how to begin assembling evidence for such a position. Recent proponents of mechanism take rather different forms. For example, according to Hofstadter (p. 579), the AI thesis is: 'As the intelligence of machines evolves, its underlying mechanisms will gradually converge to the mechanisms underlying human intelligence.' I presume that 'the evolution of machines' will require quite a bit of human intervention. I can't help wondering that if we postulate the possibility of such convergence, why not include also a divergence? Since human intelligence is not perfect, if the machine is believed to be on the way of converging to the complex human intelligence, surely it can be expected to acquire other superiorities (apart from the obvious one of speed in certain situations). There is no indication how long the evolution will take.

Webb undertakes to argue that Gödel's incompleteness theorems are for rather than against mechanism. (By the way, he also argues the stronger thesis that these theorems are for rather than against Hilbert's program. Here he is certainly wrong since Hilbert's requirement of 'finitist viewpoint' is sufficiently definite to admit a conclusive refutation of his program by Gödel's results.) Webb is certainly only arguing for a matter of principle. 'But the ultimate test would be to simulate Hilbert's metatheorem about Thales' theorem, undertaken to analyze his implicit commitment to spatial intuition in plane geometry. Existing machines are presumably light years away from results like this, much less the philosophical motivation for them, but I see no obstacle of principle' (p. 111). Is he merely saying that nobody has mathematically refuted mechanism (or, for that matter, the existence of God')?

The more familiar battleground over larger claims of AI involves the issues whether machines can handle natural language, acquire commonsense understanding, and especially situated understanding. In this area we of course easily come upon perennial problems in philosophy and the methods of psychology. Before getting into this more nebulous area, it may be better to discuss a somewhat neutral and modest thesis: Every 'precisely described' human behavior (in particular, any algorithm) can be simulated (realized) by a suitably programmed computer. (In the 1950s I heard something like this attributed to John von Neumann who believed that not all human behavior can be precisely described.) There is a slight ambiguity in this thesis which relates to the by now familiar distinction between theoretical and feasible computability.

We can precisely describe a procedure for finding a nonlosing strategy for playing chess or deciding theorems in elementary geometry. But, as we know, these computations are not feasible. It could be argued that carrying out such computations is not normal human behavior. But, for example, in deciding tautologies in the propositional calculus or dealing with cases of the travelling salesman problem, we do naturally try the standard simple algorithms which are in the general case not feasible.

One would have thought at first that this type of behavior is exactly what the computer can profitably simulate and get beyond our limited capacity in managing complexity. The point is that there are many different levels of precise description which, even with the computer's increased speed and storage, are often inseparable from commonsense understanding or sophisticated reasoning, if the descriptions are to lead to feasible computation. For example, if we increase a thousandfold the matrix for pattern recognition, the computers could do a much better job, except for the fact that the procedures are no longer feasible for the computers.

There is a tendency to view theorem proving as of little central interest to the large goals of AI. The idea is that it is too pure and clean to run into the tougher problems. But I disagree. For example, the surprising human ability to make connections may be illustrated by the discovery of hundreds of problems equivalent to P = NP. I mentioned earlier the stage of incubation leading to illumination which is typically applicable to mathematical discoveries. Poincaré reports on how dreams played a crucial role in the process of some of his discoveries. If we use the distinction of conscious, preconscious and unconscious, it is said that phenomenology places its main emphasis on the line between the conscious and the preconscious, while psychoanalysis is preoccupied with the line between the unconscious and the preconscious. As we know, dreams are one of the main tools for getting at the unconscious. Hence, the experience of Poincaré points to the involvement of the unconscious in mathematical discoveries. That is certainly not something 'pure and clean' nor irrelevant to the larger contrast of minds and machines.

I recently observed to a colleague, 'How are we to introduce the unconscious into the machines?' After a pause, he answered. 'I would say that machines are entirely unconscious.' In other words, we haven't yet made machines 'conscious' or able to simulate our conscious behaviors, it seems very remote to get at the unconscious and then endow the machines with a reservoir of the unconscious. This sort of consideration may be what has led to the talk of evolution, child development, and the 'Society of Minds.' Minsky speaks of trying 'to combine methods from developmental, dynamic, and cognitive psychological theories with ideas from AI and computational theories. Freud and Piaget play important roles. In this theory, mental abilities, both "intellectual" and "affective" (and we ultimately reject the distinction) emerge from interactions between "agents" organized into local, quasi-political hierarchies.' (p. 447). This is certainly an ambitious and awe-inspiring project which would seem to be remote from existing computers. I am much in sympathy with the project, but more as one in philosophy rather than one in AI. Perhaps the idea is to simulate the development of mankind and the universe but to speed up evolution and individual as well as societal developments to such a high degree that the mental states of living human beings all get mirrored into a gigantic computer program? In one sense, since the project is interesting in itself, a good theory is welcome more or less independently of its connections with AI. The problem is rather whether a preoccupation with AI helps or hinders the development of a good theory of such a dimension.

Minsky and Seymour Papert draw on Freud and Piaget but say nothing about the sources of their political theory. Dreyfus rather appeals to Heiddeger, to Wittgenstein, to Merleau-Ponty, and, more recently, also to John Dewey. Dreyfus sees a convergence of Minsky's earlier 'frame theory' to Husserl's phenomenology which has, for Dreyfus, been superseded by the work of the four philosophers just mentioned (p. 36). In fact, Dreyfus explicitly follows Heidegger in his view of the history of western philosophy and it is refreshing to see him relate such a view to a critique of AI. While I find the debates stimulating for my philosophical pursuit, I am unable to render clear to myself direct connections between these interesting observations and my mundane reflections on AI.

One problem I find tangibly enticing is how one chooses an appropriate research project. Of course we are often ignorant of the real causes behind our choices and we all make more or less serious mistakes in our choosing. (Closely related to the individual's choice, there is a problem of the society's choice.) For instance, given the present state of AI, what sort of young people would be making a good choice to enter the field? That means, among other things, they would be happier or do better work or both, than in some other field. 'Happier' and 'better work' are tough and elusive concepts. Yet I feel that even using such considerations as a guide would be more realistic than trying to find out whether either the AI enthusiasts or their opponents are more in the right. For one thing, they seem to argue over a motley of diverse issues. More, it is particularly difficult to decide on remote possibilities and impossibilities, even harder to have them guide one's action and short-term goals. -- Certainly the problem for a popularizer is quite different from that for a prospective research worker, and will not be discussed here.

Once Freud argued against mysticism in the following words (letter of 1917 to Groddeck):

'Let us grant to nature her infinite variety which rises from the inanimate to the organically animated, from the just physically alive to the spiritual. No doubt the Unconscious is the right mediator between the physical and the mental .... But just because we have recognized this at last, is that any reason for refusing to see anything else? ... the monistic tendency to disparage all the beautiful differences of nature in favour of tempting unity. But does this help to eliminate the differences?'

That the unconscious should play such a key role certainly throws a long shadow over the attempts to understand better the relation between mind and body. It does in a way tempt people to some form of mysticism. But the larger question raised is the dilemma between the cleanliness of universality and the richness of particularity. It is so difficult to take full advantage of the proper particularity while striving successfully for universality. Applied to problems in AI, for instance, somebody's proving a theorem or recognizing a pattern is a particular event which generally resists the universalizing process of a 'precise description' of what is involved in it.

A particular act of recognition or reasoning depends on one's genes, the history of one's mind and body, some essential relation to desire, even race, class, sex and many other factors. Of most of these factors we have at best an imperfect knowledge and understanding. In AI we can only take into consideration some small residues of these factors which happen to be noticeable by perception or introspection. This is not to deny that we have often unexpected surprises in technology for one reason or another. But the courses of advance are hard to predict and certainly do not coincide with the history of human development. For example, I once thought of getting a gadget which would do the work of a dog. I was surprised to get a box which sounds the alarm when bodily temperature is suddenly introduced. It is surprising that the computer can play a good game of chess. Yet it can't yet drive a car well in traffic. But for the human intelligence situated in a technological society the latter task is easier than the former.

In short, I am not able to find the right mix of participation in and distancing from the AI project to offer any clear and distinct ideas on the loudest current controversies in this area. To rationalize my own failure in this regard, I venture to give my impression that the strong voices come usually from committed believers who begin with couclusions, perhaps reached through incommunicable private sources. Hence, it is easier to be more decisive since merely assembling arguments to support predetermined positions suffers from less distractions. Those who wish to examine the available evidence with an open mind, as is to be expected, are at a disadvantage.

References

Hao Wang.

1. "A Variant to Turing's Theory of Computing Machines," *Journal of the Association for Computing Machinery*, vol. 4 (1957), pp.63—92; reprinted in reference 5 below.
2. "Toward Mechanical Mathematics," *IBM Journal*, vol. 4 (1960), pp.2—22; reprinted in reference 5 below.
3. "Proving Theorems by Pattern Recognition," Part I, *Communications of the Association for Computing Machinery*, vol. 3 (1960); pp. 220—234.
4. "Proving Theorems by Pattern Recognition," Part II, *Bell System Technical Journal*, vol. 40 (1961), pp. 1—41 These two parts also appeared as Bell Technical Monograph 3745.
5. *A Survey of Mathematical Logic*, Science Press, Peking, 1962, 652 pp. + x; also distributed by North-Holland Publishing Company, Amsterdam, 1963. Reprinted by Chelsea, New York, 1970 under the title Logic, computers and sets.
6. "Mechanical Mathematics and Inferential Analysis," *Seminar on the Relationship Between Nonnumerical Programming and the Theory of Formal Systems*, P. Braffort and D. Hirschberg (eds.) (1963), pp. 1—20.
7. "The Mechanization of Mathematical Arguments," *Proceedings of Symposia in Applied Mathematics*, vol.15, American Mathematical Society (1963), *Experimental Arithmetic, High Speed Computing and Mathematics*, pp. 31—40.
8. "Formalization and Automatic Theorem Proving", *Proceedings of IFIP Congress 65*, 1965, Washington, D.C., pp. 51—58. Russian translation in *Problems of Cybernetics*, vol. 7 (1970), pp. 180—193.
9. "On the Long-range Prospects of Automatic Theorem-Proving," *Symposium on automatic demonstration*, Springer-Verlag, 1970, pp. 101-111.
10. *From Mathematics to Philosophy*, Routledge & Kegan Paul, 1974, 431 pp. + xiv.
11. *Popular Lectures on Mathematical Logic*, Science Press, Beijing; Van Nostrand Reinhold Company,

New York, 1981.

12. (With S. A. Cook) "Characterizations of Ordinal Numbers in Set Theory, "Mathematische Annalen, vol. 164 (1966), pp. 1—25.
13. (With K. R. Brown) "Finite Set Theory, Number Theory and Axioms of Limitation," *ibid.*, pp. 26—29.
14. (With K. R. Brown) "Short Definitions of Ordinals," *Journal of Symbolic Logic*, vol. 31 (1966), pp. 409—414.
15. "On Axioms of Conditional Set Existence," *Zeitschr. f. Math. Logik und Grundlagen d. Math.*, vol. 13 (1967), pp. 183—188.
16. "Natural Hulls and Set Existence," *ibid.*, pp. 175—182.
17. "A Theorem on Definitions of the Zermelo-Neumann Ordinals," *ibid.*, pp. 241—250.

Wu Wen-tsün.

18. "On the Decision Problem and the Mechanization of Theorem Proving in Elementary Geometry," *Scientia Sinica*, vol. 21 (1978), pp. 159—172.
19. "Mechanical Theorem Proving in Elementary Differential Geometry," *Scientia Sinica, Mathematics Supplement*, I (1979), pp. 94—102.
20. "Mechanical Theorem Proving in Elementary Geometry and Differential Geometry," *Proc. 1980 Beijing DD-symposium*, vol. 2 (1982), pp. 1073—1092.

A. M. Turing.

21. "Computing Machinery and Intelligence," *Mind*, Vol. 59 (1950), pp. 433—460.

Joseph Weizenbaum.

22. *Computer Power and Human Reason: From judgment to Calculation*, W. H. Freeman and Co., 1976.

Hubert L. Dreyfus.

23. *What Computers Can't Do: The Limits of Artificial Reason*, revised edition, 1979 (original edition 1972).

Douglas R. Hofstadter.

24. *Gödel, Escher, Bach: An Eternal Golden Braid*, 1979.

Judson C. Webb.

25. *Mechanism, Mentalism, and Metamathematics*, 1980.

Donald Michie.

26. *Machine Intelligence and Related Topics: An Information Scientist's Weekend Book*, 1982.

Pamela McCorduck.

27. *Machines Who Think*, 1979.

Marvin Minsky.

28. 'The society theory of thinking,' *Artificial Intelligence: An MIT Perspective*, Vol. 1, 1979, pp. 421—450.

APPENDIX: CITATION FOR HAO WANG AS WINNER OF "MILESTONE" AWARD IN AUTOMATIC THEOREM-PROVING

The first "milestone" prize for research in automatic theorem-proving is hereby awarded to Professor Hao Wang of Rockefeller University for his fundamental contributions to the founding of the field. Among these, the following may be listed:

1. He emphasized that what was at issue was the development of a new intellectual endeavor (which he proposed to call "inferential analysis") which would lean on mathematical logic much as numerical analysis leans on mathematical analysis.
2. He insisted on the fundamental role of predicate calculus and of the "cut-free" formalisms of Herbrand and Gentzen.
3. He implemented a proof-procedure which efficiently proved all of the over 350 theorems of Russell and Whitehead's "Principia Mathematica" which are part of the predicate calculus with equality.
4. He was the first to emphasize the importance of algorithms which "eliminate in advance useless terms" in a Herbrand expansion.
5. He provided a well-thought out list of theorems of the predicate calculus which could serve as challenge problems for helping to judge the effectiveness of new theorem-proving programs.

Articles by Hao Wang on automatic theorem-proving

The list consists of items 2 through 4 and 6 through 9 in the references given above and is, therefore, omitted.

That concludes the citation.

In the text of the paper I have supplemented the citation with a few additional remarks. (a) My work can be viewed as an early example of 'expert systems' in contrast with the related work of Newell-Shaw-Simon. (b) I have stressed that in order to use computers in a mathematical discipline, it is essential to go beyond the predicate calculus and put emphasis on principles special to the discipline (e. g., induction in number theory). (c) It is extremely helpful that the computerization be undertaken by accomplished mathematicians in the particular discipline to be computerized.

In the announcement of prizes for advancements in automatic theorem proving, it is stated:

'A committee of mathematicians and computer scientists has been formed to supervise all aspects of the prizes. The current members of this ATP Committee are: Woody Bledsoe (Chairman), Robert Boyer, Martin Davis, Bill Eaton, Daniel Gorenstein, Paul Halmos, Ken Kunen, Dan Mauldin, John McCarthy, Hugh Montgomery, Jack Schwartz, Michael Starbird, Ken Stolarsky and Francois Treves.'

6. PROVING THEOREMS BY PATTERN RECOGNITION, I*

6.1 Introduction

Certain preliminary results on doing mathematics by machines ("mechanical mathematics") were reported in an earlier paper [20]. The writer suggested developing inferential analysis as a branch of applied logic and as a sister discipline of numerical analysis. This analogy rests on the basic distinction of pure existence proofs, elegant procedures which in theory always terminate, and efficient procedures which are more complex to describe but can more feasibly be carried out in practice. In contrast with pure logic, the chief emphasis of inferential analysis is on the efficiency of algorithms, which is usually attained by paying a great deal more attention to the detailed structures of the problems and their solutions, to take advantage of possible systematic short cuts. The possibilities of much more elaborate calculations by machines provide an incentive to studying a group of rather minute questions which were formerly regarded as of small theoretical interest. When the range of actual human computation was narrow, there seemed little point in obtaining faster procedures which were still far beyond what was feasible. Furthermore, on account of the versatility of machines, it now appears that as more progress is made, strategies in the search for proofs, or what are often called heuristic methods, will also gradually become part of the subject matter of inferential analysis. An analogous situation in numerical analysis would be, for example, to make the machine choose to apply different tricks such as taking the Fourier transform to obtain a solution of some differential equation.

The present paper is devoted to a report on further results by machines and an outline of a fairly concrete plan for carrying the work to more difficult regions. A fundamentally new feature beyond the previous paper is a suggestion to replace essentially exhaustive methods by a study of the patterns according to which extensions involved in the search for a proof (or disproof) are continued. The writer feels that the use of pattern recognition, which is in the cases relevant here quite directly mechanizable, will greatly extend the range of theorems provable by machines.

As is to be expected, the actual realization of the plan requires a large amount of detailed work in coding and its more immediate preparations. The machine program P completed so far on an IBM 704 contains only a ground-work for developing the method of pattern recognition. It already is rather impressive insofar as ordinary logic is concerned but has yet a long way to go before truly significant mathematical theorems can be proved. For example, the program P has to be extended in several basic directions before a proof can be obtained for the theorem that the square root of 2

* First published in *Communications of the ACM*.vol. 3, pp 220—234.

is not a rational number. On the other hand, theorems in the logical calculus can be proved very quickly by P. There are in *Principia Mathematica* altogether over 350 theorems strictly in the domain of logic, viz., the predicate calculus with equality, falling in 9 chapters (1 to 13, since there are no 6, 7, 8, and since 12 contains no theorems). The totality of these is proved with detailed proofs printed out by the program P in about 8.4 minutes. To prove these theorems, only about half—and the easier half—of P is needed. The other half of P can prove and disprove considerably harder statements and provides at the same time groundworks for handling all inferential statements. This program P will be deseribed in section 2.

Since the central method to be discussed is primarily concerned with the predicate calculus, its wider significance may be appreciated better, if we review briefly certain familiar facts about the relation of the predicate calculus to mathematics in general.

Thus, it is well known among logicians that if we add equality and the quantifiers "for all x," "for some y" to the propositional connectives "and," "if", "or," "not," etc., we obtain the predicate calculus in which every usual mathematical discipline can be formulated so that each theorem T in the latter becomes one in the former when the mathematical axioms A applied are added as premises. That is to say, if T is the theorem in the mathematical discipline, "if A, then T" is one of logic. This, rather than the constructions of Frege and Dedekind, is the significant sense in which mathematics is reducible to logic. From this fact it is clear that in order to prove mathematical theorems by machines a major step is to deal with theorems of the predicate calculus.

There is a natural uneasy feeling that this cannot be a feasible way of handling mathematics since we expect the methods to be largely dictated by the peculiar mathematical content of each individual branch, which presumably gets partly lost when the disciplines are thus uniformly incorporated into the predicate calculus by formalization and abstraction. This is quite true, and indeed we have to add special methods for each special mathematical discipline. But the point often neglected is that an adequate treatment of the predicate calculus is of dominating importance and that for each discipline the basic additional special methods required are relatively homogeneous. For number theory, the essential new feature is largely concentrated in mathematical induction as a method of proof and as one of definition; for set theory, in the axiom of comprehension, i.e., the axiom specifying all the conditions which define sets. Hence, there is the problem of choosing the formula to make induction on, or the condition for defining a set. While it seems doubtful that there is any uniform efficient mechanical method for making such selections, there are often quite feasible partial methods. For example, for making such selections in number theory the obvious uninspired method of trying a conclusion and its subformulae as the induction formula should suffice in many cases.

Thus, it would seem that, once a feasible way of doing logic is given, fairly simple additional methods could carry us quite some way into special mathematical disciplines. Moreover, the method of pattern recognition is basically number-theoretic, and as such recovers a considerable amount of the mathematical content of each branch of mathematics. This is so because, in order to establish, e. g., a conclusion (x) (Ey) Rxy, it aims at choosing a simple correct function f such that (x) Rxfx. And it seems not unreasonable to contend that a good deal of originality in mathematics consists precisely in the ability to find such functions.

The proposed method for doing logic always begins from scratch for each theorem. This is quite different from the type of proof we encounter in Euclid, where it is essential that later theorems are proved with the help of earlier ones. While this problem of selecting relevant earlier theorems to apply appears unimportant in the domain of logic as dealt with by the method to be described, it has to be faced at some stage, and the writer does not have a ready general solution of it. Two remarks, however, seem to be relevant. In the first place, because of the greater speed of calculations by machines, it is natural to expect that it is often faster to provean easy old theorem anew rather than look it up. Hence, we may neglect easy theorems and record only hard ones, perhaps as new axioms. In this way we arrive at a conception of expanding axiom systems which include difficult new theorems as additional axioms. Here it is irrelevant that the new axioms are not independent, since the goal is to prove other new theorems as quickly as possible. When we use such expanding axiom systems, we arrive at a compromise between pedantry and ignorance. This is not much different from the practice of a good mathematician who remembers only a number of important theorems and works out simple consequences as he is in need of them. In the second place, although it is of interest to extend the range of problems which the machine can do without human intervention, it is fair to expect that when we arrive at the stage of having machines try to prove theorems which we cannot prove, we shall not hesitateto feed the machine all the useful suggestions we can think of. Eventually machines are to be an aid to mathematical research and not a substitute for it; there is no point of running a handicap race by refusing to lend the machine a hand to complement its shortcomings. In fact, once the general framework is available, one would expect that, compared with a mere expert of the general techniques, a mathematician working on a particular problem will more likely succeed in using the framework with additional hunches appended to get a proof of the desired theorem by machine.

Another question is that the axiom of induction and the axiom of comprehension both have infinitely many instances. Or, in the usual formulation of number theory and set theory, each contains infinitely many axioms beyond the predicate calculus. Hence, if we ask whether, e. g., a statement T is a theorem of number theory, we are actually asking whether it is a logical consequence of the infinitely many axioms. If there are only finitely many axioms, we can write their conjunction A, and ask simply whether $A \rightarrow T$ is a theorem of the predicate calculus. This trick is denied us when the axioms are infinite in number. It, therefore, seems desirable to use only finitely many axioms when possible, and indeed there are standard methods for reducing usual sets of axioms to finite sets (see, e. g., [18] for one such formulation of number theory). The matter is, however, not very clear since we have to make selections from the axioms anyhow and the finite set only gives an enumeration of the infinite set, introducing meanwhile complexities through another avenue. Finite sets of axioms are however, undoubtedly useful for many purposes of mechanization, e. g., when one comes to classifying theorems according to their logical forms.

Since most of us learned Euclid and number theory without worrying about the predicate calculus, it might seem that the natural course is to bypass logic and go directly to mathematics. The writer is opposed to such an approach if the aim is to prove more and harder theorems rather than to study the psycholgy and history of mathematical thinking. Obviously what is natural for man need not be natural for the machine. More specifically, if logic is not treated in an explicit and systematic manner, constant additions of adhoc new devices make the progress toward less trivial theorems slower and slower, as well as more and more confusing. As a result, one may, e.g., even mistake the introduction of familiar logical principles for genuinely giant steps. Devising a vast machinery specifically designed to obtain a few easy theorems is wasteful. The writer feels that results obtained from different approaches ought to be measured against the generality and economy of the machinery behind them, and that prelimainary steps should be capable of supporting large superstructures yet to be erected. It is the writer's conviction that the alternative approach of treating logic only by the way would score very poorly by both criteria.

This is, however, not to deny that some of the problems encountered in dealing directly with mathematics will still have to be faced by the present approach. It is merely contended that the alternative approach does not take advantage of the possibility of "divide and conquer." As a result, what could be handled simply with the help of known techniques is mixed up with the less easily manageable further details, so that an intrinsically complex problem is made even more complex than necessary. The present attempt is concerned less with obtaining partial results which immediately excite man's undisciplined imagination, but rather more with setting up a framework capable of yielding rich results in the long run. There is a third approach which concentrates on coding known decision procedures for isolated areas such as elementary geometry, or arithmetic with only multiplication. Since these areas do not include very many interesting theorems and do not form organic parts of proof procedures for more interesting areas, the writer feels they are not of central importance to the program of proving theorems by machines. At a later stage they may serve as useful auxiliary devices to assist the more basic techniques. It cannot be denied, however, that this type of problem has the advantage that only more restricted theoretical considerations are needed for their mechanical implementation.

In the previous paper [20], the writer has suggested an Herbrand–Gentzen type proof procedure which is also an efficient decision procedure in the realms of the propositional calculus and the AE predicate calculus (i. e., those formulae which can be transformed to ones with prefix $(x_1)\cdots(x_m)(Ey_1)\cdots(Ey_n)$). In the more general case, there is the well–known unbounded search procedure illustrated in the following simple example.

Example (1).

$$(x)(Ey)(Gyy \;\&\; Gxx) \supset (Ex)(z)(Gzx \;\&\; Gzz);$$

or, alternatively,

$$(Ex)(y)(z)[(Gyy \;\&\; Gxx) \supset (Gzx \;\&\; Gzz)].$$

According to Herbrand's theorem to be described in Part II, (1) is a theorem if and only if there exsts some N such that $S_1 v \cdots v S_N$ is a truth–functional tautology, where the S_i's are:

S_1:$(x, \ y, \ z,) = (1, \ 2, \ 3)$: (G22 & G11) ⊃ (G31 & G33)
S_2:$(x, \ y, \ z) = (2, \ 4, \ 5)$: (G44 & G22) ⊃ (G52 & G55)
S_3:$(x, \ y, \ z) = (3, \ 6, \ 7)$: (G66 & G33) ⊃ (G73 & G77)
S_4:$(x, \ y, \ z) = (4, \ 8, \ 9)$: (G88 & G44) ⊃ (G94 & G99)

Since (1) is not a theorem, there can exist no tautologous disjunction $S_1 v \cdots vS_n$, or briefly, D_n. If we are to test successively the disjunctions D_1, D_2, etc., we can never reach an answer. One can undoubtedly use some special argument to show that (1) is not a theorem, but then there is the question of formalizing the argument and generalizing it to apply to some wide range of cases. As it happens, (1) falls under a simple decidable class, viz., the E_1A case of all formulae beginning with a prefix $(Ex)(y_1) \cdots (y_n)$, and it has been shown that for each formula A in the class, one can find some N, such that either D_N is tautologous or A is not a theorem. Hence, it may seem that simply adding the method of calculating N to the unbounded search procedure would already provide a decision procedure for the class in question. This is, however, only a theoretical possibility and hardly feasible ever on machines. For example, according to Ackermann's evaluation for N (see [3], p. 265) the value of N for the simple example (1) is no less than $2^{48} - 1$. Of course, the bounds for more complex formulae in the class, and formulae in more complex decidable classes, are much higher according to the traditional decision procedures. This situation led the writer to envisage in [20] the prospect of not using more decision procedures but trying to simplify directly the brute force search procedure of proof as soon as we get beyond the AE predicate calculus.

More recently, steps along such a direction have been taken by Gilmore, Davis and Putnam. Gilmore has written a program using essentially the brute force method and tested a small sample of examples [8]. Davis and Putnam have in [4] devised efficient techniques for testing whether a given disjunction $S_1 v \cdots vS_N$ is tautologous. An efficient test for truth-functional tautologies in general, proposed earlier by Dunham-Fridshal-Sward, has also been coded and run on a machine [7].

The writer feels that Gilmore's result is basically negative, i.e., it shows that without fundamental improvements the brute force method will not do. Perhaps the two most interesting examples, a nontheorem (2) and a theorem (3), which his program fails to decide, are fairly simple and can indeed be decided by the method of pattern recognition quite easily, as will be shown in Part II. His examples are (drawn from [3], p. 262):

Example (2).

$(Ex)(Ey)(z)\{[Gxz \equiv Gzy) \ \& \ (Gzy \equiv Gzz) \ \& \ (Gxy \equiv Gyx)] \supset (Gxy \equiv Gxz)\}.$

Example (3).

$(Ex)(Ey)(z)\{[Gxy \supset (Gyz \ \& \ Gzz)] \ \& \ [(Gxy \ \& \ Hxy) \supset (Hxz \ \& \ Hzz)]\}.$

Davis and Putnam have indicated that by their improved method of testing for tautologies, a treatment of (3) becomes feasible. Since their method is concerned only with the last stage, viz., that of testing each disjunction, it can of course do nothing with nontheorems such as (2). Moreover, since it provides no device for deleting useless terms among S_1, S_2, etc., it is not likely to be of use even when a formula is indeed a theorem but the smallest n for which $S_1 v \cdots vS_n$ is tautologous is large. For example, with regard to (3), $S_1 v \cdots vS_{25}$ is the earliest tautology; in 21 minutes on an IBM 704, only $S_1 v \cdots vS_7$ has been handled by Gilmore's program. Although the particular example appears to be mechanically manageable by the method of Davis and Putnam, one would expect that expressions can easily become too long to handle by this method.

The writer now feels that a more basic step is to eliminate in advance useless terms among S_1, S_2, etc., or, alternatively, instead of actually constructing and testing the disjunctions; examine in advance, for each given problem, all the possible courses along which counterexamples to S_1, $S_{1v}S_2$, etc. may be continued. Using the second alternative, we obtain at the same time a disproving procedure for most cases. The detailed techniques for achieving these-goals are here called the method of proving theorems (and disproving nontheorems) by pattern recognition, or, more specifically, the method of sequential tables. When applied to Example (1), the method gives the desired answer in the following manner. When we substitute numbers for the variables, each elementary part Gyy, Gxx, Gzx, Gzz gives way to infinitely many new elementary parts which occur in S_1, S_2, etc. We now. ask whether we can so assign truth values (true or false) to the infinitely many elementary parts that S_1 S_2, etc. all become false. If that is impossible, (1) is a theorem, otherwise we get a counterexample and (1) is not a theorem. If we look at the matrix of (1):

$$(Gyy \ \& \ Gxx) \supset (Gzx \ \& \ Gzz),$$

we see that it is false only when:

Gxx	Gyy	Gzx	Gzz
t	t	f	t
t	t	t	f
t	t	f	f

Since, as happens in this case, different variables are always replaced by different numbers, each of the above rows can make each of S_1, S_2, etc. false if we imagine that the variables are replaced by their corresponding numbers. The problem is whether we can select simultaneously one row for each S_i which, taken together, will not conflict with one another. For example, although each row can falsify S_1, and even both S_1 and S_2, S_3 becomes true when G33 gets the value f. Hence, to falsify S_1, S_2, and S_3, we must take the first row for S_1:

G11	G22	G31	G33
t	t	f	t

Then any row can falsify S_2 and S_3. But in order to falsify also S_5, G55 has to be t, so that again we can use only the first row for S_2:

G22	G44	G52	G55
t	t	f	t

Similarly, in order to falsify S_7 and S_3 one has to use the first row for S_3. It is clear that by always using the first row we can simultaneously falsify all S_i's since the constraints imposed on later S_i's by earlier S_i's are uniform. Hence, we conclude that (1) is not a theorem. In fact, as will be discussed in section 3, all we have to do is to cross out every falsifying row in which Gyy or Gzz gets a value which Gxx does not get in any row. After repeated application of this operation, either no row is left and then the original statement is a theorem, or else some row is left and then a countermodel is possible.

This last method, called the method of sequential tables, seems to be a new feature that goes beyond the general method of pattern cognition.

As will be shown in Part II, the method can be generalized and rigorously justified for a number of broad classes. The type of considerations involved in such a method should be clear, however, from the above example.

The basic ideas of the general method of pattern recognition, though not the special addition of the method of sequential tables directed at efficiency, go back to Herbrand [10] and, in a less general form, also to Skolem [16]. By this method, Herbrand was able to give in a uniform way a treatment of most solvable cases of the decision problem (for logic) known at his time, and to discover two interesting new cases, viz., a generalization of the E_1A case and the disjunctive predicate calculus dealing with formulae with a matrix that is a disjunction of elementary parts and their negations. Church gave along a similar line a more exact treatment of these same cases plus two cases by Skolem but minus Herbrand's generalization of the E_1A case [2, 3]. The chief additional case at first obtained by the more usual sort of technique, shortly after Herbrand's treatment, has recently been handled by Klaua [12] with this general method. Dreben has pursued the matter further and announced in general terms a number of results [6]. We understand Dreben is writing a monograph on the subject.

The writer believes that in several direction's the important implications of the method has not yet been fully exploited in the works just cited. First, the method can be used to give decision procedures for well–known unsolved cases. The writer has found a partial solution of the decision problem for the class of formulae with the prefix (Ex)(y)(Ez) (this open problem is mentioned, e.g., by Church, [2, p. 271], and by Ackermann, [1, p. 85]; it seems to go back to the early thirties). The solution will be given in Part II. This case is of special interest since it is a natural class and includes simple examples which are nontheorems but have no finite countermodels. A well–known example due to Schütte [1, p. 83] is:

Example (4).

$$(Ex)(y)(Ez)\{Gxy \supset [Gxxv(Gyz \ \& \sim Gxz)]\}.$$

The negation of this ia an axiom of infinity, i.e. a statement satisfiable only in an infinite domain. A related but more familiar form of axioms of infinity is the conjunction of:

$$\text{(i)}\ (x) \sim Gxx; \quad \text{(ii)}\ (x)(Ey)Gxy; \quad \text{(iii)}\ (Gxy \ \& \ Gyz) \supset Gxz.$$

Secondly, the method and ideas of pattern recognition can be extended to give some quasi–decision procedure for the whole predicate calculus. By this is meant a procedure which in theory always gives a proof if the given formula is indeed a theorem, and which in "most" cases gives also a counterexample if the given formula is not a theorem, so that the mudecidable formulae become, one might say, points of singularity. In general, it is no longer a question whether a nontheorem has finite countermodels but whether it has either finite or simple infinite countermodels. For example, if a nontheorem has no recursive countermodels, it is to be expected that a natural quasidecision procedure will not be able to refute it. It is possible to design different quasi–decision procedures which have different ranges of application. The way to get such procedures is roughly to apply the consideration of patterns to all formulae or to all in a reduction class, i.e., a class such as all formulae with the prefix

(Ex)(Ey)(Ez)(w), such that there is an effective method by which each formula can be transformed into an equivalent one in the class. There are many ramifications in carrying out the matter in detail.

Thirdly, as is only natural, not sufficient attention has been paid to the question of efficiency, or the difficulties in actually applying the procedures by man or by machines In particular, the method of sequential tables is an example of the possible ways to improve efficiency. A related minor point is that the more difficult decision procedures are usually not illustrated by examples.

In view of these three explorable areas, the writer feels that there is a good deal of interesting theoretical work which is yet to be done. This is one of the reasons why it seems to the writer difficult to make definite estimates and predictions as to how fast and how far theorem proving can be mechanized. At present, it appears that there are a succession of rather difficult but by no menas humanly impossible steps yet to be taken which do not embody any known limitations. Man will have to devise methods or methods for devising methods, but the machine will use the methods to do things which man cannot feasibly do. There is nothing paradoxical in this. Even today long multiplications and other calculations provide ample examples. Since we should, as a fundamental methodological principle, expect no miracles, the fact that there is much yet to be done and that we have a fairly good idea of the sort of thing to be done seems a very good indication that we are not after a will–o'–the–wisp.

After a longer paper had been nearly completed, the writer learned of the restriction on the length of the paper As a result, the paper is divided into two parts. Nearly all detailed theoretical considerations are given in Part II, which will be made a memorandum at the Bell Laboratories and presumably published eventually.

In this first part, section 2 gives a general description of the completed machine program mentioned above. Sketches are given of results on elementary domains such as the restricted AE predicate calculus (similar to Qp in [20, p. 10], except for a method of eliminating functors) and the AE calculus, as well as general preliminary steps useful for extension of the program to the whole predicate calculus. In particular, devices will be stated which are useful for the systematic simplification of formulae so that many additional formulae are reduced to members in classes known to be decidable. It should be emphasized that in the program actually completed procedures using the method of pattern recognition in the specific sense explained above have not been included, although the program is oriented toward a systematic preparation for the treatment of such procedures.

Section 3 gives a solution of the simple E_1A case as an illustration of the general method of pattern recognition, and, more specifically, for the method of sequential tables. In Part II, all the main decidable cases, the new case (Ex)(y)(Ez), as well as quasi–decision procedures, will be considered, all along a line similar to that followed by the method in section 3.

Finally, section 4 contains a number of general remarks.

6.2 A program that does 9 chapters of principia in 9 minutes

The running program P on an IBM 704 accepts any sequent S in the predicate calculus (with equality), in particular, any sequent expressing that a theorem follows from certain axioms in some special mathematical discipline; reduces it to a finite set of atomic sequents (in other words, gives its quantifier–free matrix in a conjunctive normal form); and compiles an economic quantifier tree (see below) for S which can be used directly or as a basis for selecting a most favorable prefix (i. e., a quantifier string that does not violate the relations of dominance in the tree). When no negative variables [20, p. 9] occur in the set of atomic sequents, or briefly, the matrix, of S, the program P can decide always whether S is a theorem, and give a proof or a counterexample. The program P can often, though not always, do the same for S, when no positive variables (U–variables) are governed by negative variabies (E–variables) in the matrix of S, i.e., when S is reducible to the AE–form. It is fairly easy to extend the program P to include a procedure for dealing with all AE cases. We have not done this so far because of a rather paradoxical situation. On the one hand, the restricted AE method which is included in P already suffices to decide a clear majority of the examples encountered in books on logic, and there are few actual examples which are undecidable by the restricted AE method but decidable by the full AE method. On the other hand, considered in the abstract, even the full AE method can deal with only a very restricted class of sequents which are of interest to us.

Hence, on the one hand, for the simple purpose of illustrating the surprising ease with which machines can be employed to prove and disprove common examples in logic, the theoretically narrow range of the restricted AE method is more remarkable than some more extended method. On the other hand, when we wish to use examples obtained by formalizing the statement of mathematical theorems as consequences of certain axioms, our needs will quickly go beyond even the full AE method. That is why considerable thought has been given to the question of reducing a given sequent S to the simplest possible form as a uniform basis for the treatment of many diverse cases. Most pieces in the program P are designed in such a way that they can be efficiently useful in handling all more complicated cases. Thus, the elimination of logic connectives, the reduction of each problem to as many simpler subproblems as possible, the construction of the simplest quantifier trees, and the relatively fast comparison routine for deciding atomic sequents: all these are designed as a common part in further extensions.

Familiarity with [20] should be helpful, though not necessary, for understanding the following more detailed description of P. Even though P is essentially an extension of the program III as described in [20], and we shall try to avoid repetitions, there are a number of differences in those parts which are dealt with in both programs. Some of these differences should be mentioned in advance to prevent misunderstandings. While negative variables (E–variables) are replaced by numbers in [20], it has been decided to use in P the more natural course of replacing initial positive variables (U–variables) by numbers. The reduction to the miniscope form envisaged in [20] has in part been abandoned; in P, what is taken as a better procedure is used instead. Functors with explicit arguments attached to them are not used in P; rather the governing relations among the letters which replace the variables are tabulated separately.

We proceed to give a more detailed description of the program P. The program is written entirely in the language of SAP except that the subroutines of reading and writing tapes are from the Bell Monitoring System. This impurity could, if one wishes, be gotten rid of by changing just a few instructions. The whole symbolic deck of the program contains about 3200 cards. About 13,000 words of the core storage are assigned for use by the program, although a lot of these words are only reserved spaces for handling more complex problems. For the problems actually run so far, it should be easy to fit everything into a machine with 8000 words. Auxiliary storages are not needed except that, as a convenience, tapes are used to avoid going through on-line input-output equipments.

At present, there are two somewhat irksome restrictions on the program. It can only deal with a problem expressible with no more than 72 characters (i.e., one card long). This is quite adequate for handling common theorems of logic, but insufficient when we wish to apply the program to, e.g., number theory or elementary geometry It is highly desirable to remove this restriction, which is the sort of thing that has been taken care of in several systems for symbol manipulation. Since we have no immediate plan for using such systems, we only envisage a modification that will accept, say, a problem 10 cards long.

A less fundamental restriction is the use of a sort of Polish notation, partly to speed up operations, partly to reduce the length of the sequent stating a given problem, and partly necessitated by the fact that machine printers do not have the familiar logic symbols. This has the consequence that it is not so easy to read the outputs. A translation routine could be added to bring the outputs (and, if one wishes, also the inputs) into a form which resembles some ordinary notation more closely. To assist exposition, we shall, in what follows, neglect this notation feature, and speak always as if everything had been done in a more familiar notation. (For a "dictionary," see [20, p. 6].)

A readily quotable, albeit misleading, indication of the power of the program P is the fact that it disposed of nine chapters of *Principia* in about 8.4 minutes, with an output of about 110 pages of 60 lines each, containing full proofs of all the theorems (over 350). This is misleading for two reasons. On the one hand, proving these theorems does not require the full strength of the program which can do considerably more things that are basically different from this particular task. Hence, this does not give a fair summary of what the program is capable of doing. On the other hand, while many college and graduate students, especially in philosophy, find it not so easy to prove these theorems in their homeworks and examinations, the methods we use are a bit easier than the usual methods and of the type that is specially suitable for machines. As a result, the theorems in *Principia* are far easier to prove than expected, and it is not very remarkable that they can be proved in a reasonably short time. The actual time required came, however, as a bit of a surprise At the very outset, the writer guessed 20 hours on an IBM 704 as the probable time required to prove these (over 350) theorems. In [20], the theorems of the propositional calculus (over 200) were proved in about 37 minutes with the on-line printer, and it was estimated that the computing time was only about 3 minutes; the majority of the theorems with quantifiers (over 150) were also proved than, and it was conjectured that about 80 minutes would be needed to prove the lot. The final result with the program P is that the 200 strong theorems in the propositional calculus took about 5 minutes, while the 150 strong theorems with quantifiers took less than 4 minutes. (The writer has not been able to determine how much of the time was spent on input and output operations.) In every case, it seems that the machine did better than expected. While this fact presumably means very little, it was natural that one felt encouraged by it.

On second thought, there is an uneasy feeling that the efforts to improve the restricted AE method were a bit wasted. As the time when [20] was written, it was already quite clear that even with rather slight changes, the program available then would yield proofs of the desired theorems in a few hours. What is the point of spending many week's efforts to bring the time down to a few minutes? Would it not have been better if the efforts had been spent on studying more difficult cases? As a matter of fact, however, in the process of trying to handle the restricted case more efficiently, one also got a clearer view of more general questions. Moreover, while the difference between a few minutes and a few hours is not very important, the difference between a few hours and several hundred hours may prove to be decisive; and it is of interest to have a definite example of the degree of increase in speed which one can expect from improved methods.

A possible objection to the present approach to the problems of coding is that a good deal of time has been spent in those parts of the procedures which are relatively easy to carry out even by hand. The more sensible alternative approach would seem to be a concentration of efforts on testing those parts where there is a serious doubt whether machines can feasibly succeed at all. To this objection, the answer can only be that the gradual approach adopted here is meant as the beginning of a long range scientific project rather than a quick test whether crude standard methods can already produce amazing results. In fact, it is fair to say that we have learned enough to see a healthy situation with regard to the question of proving theorems by machine: the prospects are encouraging, but one has no right to expect fast miracles. This being so, the gradual approach has the advantage that we can more easily test a large number of sample problems because little work is needed in preparing them by hand.

The master control of the program P has 45 instructions. Subroutines are heavily employed. When a problem on a single card is accepted, the program first takes the following preliminary steps:

S1. If the first 6 characters of the card all are blanks, the program P interprets this as an indication that no more data cards are to be accepted. The machine stops or goes to some other job (e. g., run or compile the next program in waiting). Otherwise, P searches for the arrow sign → (actually the sign / could be used to serve the same purpose).

S2. If → does not occur, the input is treated as ordinary prose. It is printed out without comment and P proceeds to receive the next data card. If the arrow sign does occur, P searches for quantifiers. By the way, standard BCD representation of characters is used in the cores, except that the BCD representations of zero and blank are interchanged.

S3. If quantifiers do not occur (the equal sign may occur), proceed more or less in the same way as in the treatment of the quantifier–free case in [20]. Otherwise, we have the principal case.

S4. When the input problem contains quantifiers, the following preliminary simplifications are made. (i) All free variables are replaced by numbers, distinct numbers for distinct variables. (ii) Vacuous quantifiers, i. e., quantifiers whose variables do not occur in their scopes, are deleted. (iii) Different quantifiers are to get distinct variables; for example, if (x) occurs twice, one of its occurrences is replaced by (z), z being a new variable. This last step of modification is specially useful when occurrences of a same quantifier are eliminated more than once at different stages.

S5. After the above preliminary simplifications, each problem is reduced to as many subproblems as possible in the following manner. (i) Eliminate in the usual manner every truth–functional connective which is not governed by any quantifiers. (ii) Drop every initial positive quantifier (i. e., universal in the consequent or existential in the antecedent that is not in the scope of any other quantifier) and treat its variable as free, i. e., replace all its occurrences by those of a new number. (i) and (ii) are repeated for as long as possible. As a final result of this step, each problem is reduced to a finite set of subproblems such that the problem is a theorem if and only if all the subproblems are.

To illustrate the steps S1 to S5, we give in an ordinary notation a proof obtained by the program P:

$11*53/\rightarrow (x)(y)(Gx \supset Hy) \equiv ((Ex)Gx \supset (y)Hy)$		
$/\rightarrow (x)(y)(Gx \supset Hy) \equiv ((Ez)Gz \supset (w)Hw)$		(1)
$1/(x)(y)(Gx \supset Hy) \rightarrow (Ez)Gz \supset (w)Hw$		(2)
$2/G1, (x)(y)(Gx \supset Hy) \rightarrow H2$		(3)
$1/(Ez)Gz \supset (w)Hw \rightarrow (x)(y)(Gx \supset Hy)$		(4)
$4/(w)Hw \rightarrow G3 \supset H4$		(5)
$5/G3,(w)Hw \rightarrow H4$		(6)
$4/\rightarrow G3 \supset H4,(Ez)Gz$		(7)
$7/G3 \rightarrow H4,(Ez)Gz$		(8)
$8/G1 \rightarrow H2,(Ez)Gz$		(1)
$1/G1 \rightarrow H2,G1$	SVA	(2)
	PQED	
$6/G1, (w)Hw \rightarrow H2$		(1)
$1/G1, H2 \rightarrow H2$	SVA	(2)
	PQED	
$3/G1, (x)(y)(Gx \supset Hy) \rightarrow H2$		(1)
$1/G1, (y)(Gx \supset Hy) \rightarrow H2$		(2)
$2/G1, Gx \supset Hy \rightarrow H2$		(3)
$3/G1 \rightarrow H2, G1$	SVA	(5)
$3/G1, H2 \rightarrow H2$	SVA	(4)
	QED	

In the above example, all quantifiers are made distinct in (1). By (i) of S5, (1) is reduced to (2) and (4). By (i) and (ii) of S5, (2) is reduced to (3), which can be reduced no further by (i) or (ii) of S5. By (i) of S5 and then (ii) of S5, (4) is reduced to (5) and (7). Finally, by (i) of S5, (5) and (7) are respectively reduced to (6) and (8). Hence, the original problem 11∗53 is reduced to the 3 subproblems (3), (6), (8). It is possible to show the following:

T2.1. *The original problem is a theorem if and only if all its subproblems* (*in above sense*) *are.*

Hence, if any subproblem is refuted, then the original problem is also. If no subproblem is refuted, but some subproblem is undecidable by some restricted method, then the original problem is undecidable by the same restricted method.

Hence, the remaining problem is to study each subproblem (in the above sense). In theory, this reduction to subproblems is rather wasteful, since it could be automatically taken care of by tackling the whole problem directly in a uniform manner similar to the way in which each subproblem is tackled. In practice, however, it is clearly desirable to isolate separate problems whenever possible. It is to be noted that further reductions are of a different nature because the subproblems would be interconnected through variables attached to some common quantifiers. This will soon become clear.

Now, e.g., we may study (3), (6), (8) in the above example each as a separate problem in itself. Each is stored away temporarily until we have obtained the last, (8) in the example. Then (8) is taken as a new first line and treated. Afterwards, (6) and (3) are called back and handled similarly. Note also that the numbers in each subproblem begin from 1 both at the end of each line and inside the body of the proofs. Each of (3) (6), (8) happens to be a theorem, so we conclude at the end that 11*53 is itself a theorem.

We now explain how each subproblem is to be treated. In order to do this, we have to describe first how quantifiers in general are handled, as well as how different comparison procedures are performed on atomic sequents.

Variables can be replaced by 3 kinds of symbol according to the status of their quantifiers.

(i) Free variables and initial U–variables: by 1, 2, 3, ⋯, 9 (numbers).

(ii) All E–variables: by s, t, u, v, w, x, y, z. (In fact, unchanged.)

(iii) U–variables governed by E–variables: k, l, m, n, o (functors).

With U–variables governed by E–variables and E–variables governed by such U–variables, a record is kept separately of the letters which govern them.

The present method avoids the necessity of reducing a formula first to a prenex normal form, as well as an unpleasant feature about ≡ For example, if we have a formula (Ex)Gx ≡ (y)Hy, then we should get 4 quantifiers when ≡ is eliminated. The situation may be seen from the reduction of (1) to (2) and (4) in 11*53. Now, since it is necessary, for certain purposes, to have distinct variables for distinct quantifiers, we may feel we have to double the number of variables in such cases. Since, however, the two quantifiers resulting from one quantifier always have different signs, the above convention about the replacement of variables automatically assures us that the two new quantifiers get different symblos. Thus, in 11*53, the variable x in (4), being an initial U–variable, is replaced by the number 3 in (5), while the variable x in (2), being an E–variable, will remain unchanged after the quantifier (x) is dropped, as is seen in (2) in the last part of the whole proof.

In determining what variables or functors govern a given quantifier Q, we use a somewhat more economic criterion. Instead of recording all quantifiers which contain Q in their scopes, we use all the variables (and functors) which are free in the scope of Q and distinct from the variable of Q. This requires a theoretical justification that can be stated (true only under restrictions):

T2.2. *We can separate out Q and its scope from those quantifiers whose variables do not occur in the scope of Q.*

Another device is employed to simplify the governing relations among variables and functors when one subproblem is reduced to a finite set of atomic sequents (a matrix). Two symbols, each a variable or a functor, are connected if there is an elementary part in the matrix which contains both symbols or contains one of the two symbols as well as a variable or functor connected to the other. Then a variable or functor is really governed by another if its quantifier was originally governed by the latter and they are connected. A subroutine EFCTR serves to reduce the governing relations in this way. This will be justified in Part II by:

T2.3. *If two symbols, each a functor or a variable, are not connected in the final matrix, we can always so transform the original sequent as to separate the two quantifiers which give way to them.*

We make use of several kinds of comparison procedure in deciding an atomic sequent. Given an atomic sequent, we first compare the antecedent with the consequent as if no quantifiers occur, i.e., whether a same atomic formula occurs on both sides, or, if = occurs, whether a selfidentity occurs in the consequent or substituting equals for equals would yield an atomic formula on both sides. This is COMP, a procedure described in [20]. If the answer is yes, then the atomic sequent is a thoerem, and we put a VA on it and print it out. We do not have to worry about it anymore.

If the answer is no, we generally go to a different comparison routine COMQ, which permits us to make substitutions on the variables: each variable can be replaced by any number, as well as by any functor not governed by the variable, or by another variable. If in this way, we can obtain a result valid by the previous criterion, we put tentatively the label SVA on and store the sequent away. This comparison routine is quite complex because we require that the same variable get the same substituent not only in each atomic sequent but in all the atomic sequents which come from one given subproblem. The presence of = makes this part doubly complex.

If every atomic sequent from a subproblem gets SVA by compatible substitutions, the subproblem is proved. If at one stage, an atomic sequent fails to get SVA with any substitution compatible with earlier substitutions, we shall test no more atomic sequents by substitution until we have, if possible, simplified the governing relations of variables and functors with regard to the whole set of atomic sequents obtained from the orginal subproblem.

The substitutions are made in a sensible way in the sense that we do not try all possible substitutions but try only the most likely ones (compare [20, p. 11]). This is an important factor in making P more efficient than the earlier program.

At a later stage there is also a negative comparison test called NTEST in which each atomic sequent is tested separately by substitution, possible conflicts with substitutions for other atomic sequents being neglected. When this is not possible and the atomic sequent contains no functors, the atomic sequent, and therewith the original subproblem, is refuted. This step again requires a theoretical justification.

Let us now give some examples (at the right) from the outputs of P and then summarize the main steps.

These should be a fair sample of the shorter results among the problems beyond *Principia* which have been handled by the program P. We now give a summary of the steps needed in solving each subproblem and illustrate them by the above examples. When a problem has only one subproblem the subproblem is of course the problem itself.

S6. Eliminate quantifiers and truth–functional connectives whenever possible, i.e., whenever a sequent under consideration is not an atomic sequent. By the way, before all subproblems were obtained, atomic sequents were not dealt with, e. g., (5) in 19*10.

S7. If the sequent is atomic, try to decide it by COMP. Put on VA if it is valid and continue with next sequent. If it is not valid, put on NO and finish a subproblem (hence, also the problem) which contains no quantifiers. This is the case with the subproblem (8) in 19*10.

S8. If the subproblem contains quantifiers, go to COMQ. If this makes it valid, i. e., there are acceptable substitutions to make the sequent valid, put on SVA and store it away. If this is the last atomic sequent of a subproblem, then we have proved it. We put the line out together with all earlier SVA sequents which have been stored away. For example, this is the case with 14*4 and 15*16.

S9. If this cannot make the atomic sequent valid, we store it away and record the fact. We then continue with the problem but test no more atomic sequents beyond COMP. When all the atomic sequents are obtained, we use EFCTR to simplify the governing relations between the functions and variables. There are four possibilities given under S10, S11, S12, S13.

S10. If there is no governing relations in the result, i. e., the result is in the AE form; then either this was so all along, or this is so only because certain functors could be eliminated (i. e., not really governed and can therefore be replaced by new numbers). In the first case, test whether there is only one undecided atomic sequent or only one number occurs in the undecided atomic sequents. In either case, the restricted AE method is sufficient, and

14*4 / $(x)((Hx \,\&\, Hy) \supset Gx),\ p,\ (x)Hx \rightarrow Gy$		
/ $(x)((Hx \,\&\, H1) \supset Gx),\ p,\ (z)Hz \rightarrow G1$		(1)
1 / $(Hx \,\&\, H1) \supset Gx,\ p,\ (z)Hz \rightarrow G1$		(2)
2 / $Gx,\ p,\ (z)Hz \rightarrow G1$		(3)
2 / $p,\ (z)Hz \rightarrow G1, Hx \,\&\, H1$		(5)
5 / $p,\ Hz \rightarrow G1, Hx \,\&\, H1$		(6)
6 / $p,\ H1 \rightarrow G1,\ H1$	SVA	(8)
6 / $p,\ H1 \rightarrow G1,\ H1$	SVA	(7)
3 / $G1,\ p,\ H1 \rightarrow G1$	SVA	(4)
	QED	

14*15 / $(Ex)(y)Gxy \rightarrow (x)(Ey)Gxy$		
/ $(y)G1y \rightarrow (Ew)G2w$		(1)
1 / $G1y \rightarrow (Ew)G2w$		(2)
2 / $G1y \rightarrow G2w$	SNO	(3)
	NOT VALID	

14*6 / $\rightarrow (Ex)(y)(z)((\sim Gxw \supset Gwy) \supset (Gzw \supset Gwz))$		
/ $\rightarrow (y)(z)((\sim Gx1 \supset G1y) \supset (Gz1 \supset G1z))$		(1)
1 / $\rightarrow (y)(z)((\sim Gx1 \supset G1y) \supset (Gz1 \supset G1z))$		(2)
2 / $\rightarrow (z)((\sim Gx1 \supset G1k) \supset (Gz1 \supset G1z))$		(3)
3 / $\rightarrow (\sim Gx1 \supset G1k) \supset (Gm1 \supset G1m)$		(4)
4 / $\sim Gx1 \supset G1k \rightarrow Gm1 \supset G1m$		(5)
5 / $G1k \rightarrow Gm1 \supset G1m$		(6)
5 / $\rightarrow Gm1 \supset G1m,\ \sim Gx1$		(8)
8 / $Gm1 \rightarrow G1m,\ \sim Gx1$		(9)
6 / $G31,\ G12 \rightarrow G13$	NO	(7)

NOT VALID

14∗16 / (x)Gxu → (Ew)(Gyw & Gzw)
/ (x)Gx1 → (Ew)(G2w & G3w) (1)
1 / Gx1 → (Ew)(G2w & G3w) (2)
2 / Gx1 → G2w & G3w (3)
3 / Gx1 → G2w F (4)
4 / Gx1 → G3w F (5)

NONE

15∗16 / (x)x = x, (Ey)(Ez)y ≠ z
→ (Ex)(Ey)(Ez)(x ≠ y & y ≠ z)
/ (x)x = x, 1 ≠ 2 → (Ew)(Ev)(Eu)(w ≠ v & v ≠ u) (1)
1 / (x)x = x → (Ew)(Ev)(Eu)(w ≠ v & v ≠ u), 1 = 2 (2)
2 / x = x → (Ew)(Ev)(Eu)(w ≠ v & v ≠ u), 1 = 2 (3)
3 / x = x → w ≠ v & v ≠ u, 1 = 2 (4)
4 / x = x → w ≠ v, 1 = 2 (5)
4 / x = x → v ≠ u, 1 = 2 (7)
7 / 2 = 1, x = x → 1 = 2 SVA (8)
5 / 1 = 2, x = x → 1 = 2 SVA (6)

QED

19∗10 / → ((Ey)Guy & ∼Guu) & (Guv ⊃ (Gvw ⊃ Guw))
/→ ((Ey)G1y & ∼G11) & (G12 ⊃ (G23 ⊃ G13)) (1)
1 / → (Ey)G1y & ∼ G11 (2)
2 / → (Ey)G1y (3)
2 / → ∼ G11 (4)
4 / G11 → (5)
1 / → G12 ⊃ (G23 ⊃ G13) (6)
6 / G12 → G23 ⊃ G13 (7)
7 / G23, G12 → G13 (8)
8 / G23, G12 → G13 NO (1)

NOT VALID

19∗13 / → (Ex)(y)(Ez)((∼ GxyvGxx)v(Gzx & ∼ Gzy))
/ → (Ex)(y)(Ez)((∼GxyvGxx)v(Gzx & ∼ Gzy)) (1)
1 / → (y)(Ez)((∼GxyvGxx)v(Gzx & ∼ Gzy)) (2)
2 / → (Ez)((∼GxkvGxx)v(Gzx & ∼ Gzk)) (3)
3 / → (∼ GxkvGxx)v(Gzx & ∼ Gzk)) (4)
4 / → ∼ GxkvGxx, Gzx & ∼Gzk (5)
5 / → ∼ Gxk, Gxx, Gzx & ∼ Gzk (6)
6 / Gxk → Gxx, Gzx & ∼ Gzk (7)
7 / Gxk → Gxx, ∼ Gzk (9)
7 / Gxk → Gxx, Gzx FNO (8)
9 / Gzk, Gxk → Gxx FNO (10)

k by x
z by k

we have sufficient data to conclude that the subproblem is not a theorem. This is the case with 14∗15. Otherwise, we go to S14. This is the case with 14∗16.

S11. If the result contains no more functors after elimination, we know that the subproblem can be treated by the full AE method. In general, it is, however, necessary to first transform the matrix by a procedure similar to the reduction to a miniscope form (see[20]). Such a procedure is not included in the program P. Instead, we use the original matrix with functors replaced by numbers and go directly to S14 for a negative test only. 14*6 is an example.

S12. If the result is not in AE form and no eliminations have been made, go to S14 directly. This is the case with 19*13.

S13. If the result is not in the AE form but some eliminations have been done, repeat S8 and S9 except that upon failure the program goes to S14.

S14. Make a negative test. If some atomic formula with no functors cannot be made valid even by NTEST, append NO to it, and refute the whole problem. This is the case with 14*6.

S15. Otherwise, the question is undecided. In this case, restore functors if eliminations have been made. And then, put F after each atomic sequent. If an atomic sequent could not be made valid by NTEST but contains functors, add also NO after F. The two cases are seen in the last lines of 14*16 and 19*13.

S16. Finally, print out the best possible governing relations among the functors and variables. In the case of 14*16, there is none. This means the problem can be settled by the full AE method. In the case of 19*13, we have an irreducible string (Ex)(y)(Ez) of quantifiers.

This completes the summary of the program P. It is clear that at the end we are ready to add more powerful methods which are usually classified according to the string of quantifiers. The governing relations we list in general give quantifier trees ("trees" in an intuitive sense). Thus, if x governs k, y governs m, m governs z, and we treat k, m as variables for the moment, then we are free to use several different strings as long as the governing relations are preserved, e. g., (Ex)(k)(Ey)(m)(z), (Ex)(Ey)(k)(m)(z), (Ey)(Ex)(k)(m)(z), etc. That is why quantifier trees give us in general a better reduction of a given problem.

It should be emphasized that although the methods of the program P are essentially confined to a subdomain of the AE method, it can solve problems which are not solvable by the ordinary full AE method. Quite a number of problems can be decided by the auxiliary procedures introduced along the way. So far we have only been able to give a small sample of shorter problems. Now we list below a number of further examples which have been definitely

---

LIST I. *Theorems Proved by P*

14*7 / (Ez)Hxz ⊃ (z)Gxz,(z)(Gzz ⊃ Hzy) → Hxy = (z)Gxz
15*3 / → (Ex)(Ey)[(x = u & y = v) ⊃ (Gu ⊃ Gv)]
15*4 / → (Ew)(Ex)(y)(z){(Gvy & Gwz) ⊃
  [(Gwy & Gxy)v(Gvz & Gxz)]}
15*6 / → (Eu)(v)(Ew)(x){[((Gux ≡ Gxw) ≡ Gwx) ≡ Gxu] &
  [((Gvx ≡ Gxw) ≡ Gwx) ≡ Gxv]}
15*7 / → (Eu)(v)(Ew)(x){(Gux ≡ Gvx) ⊃ [((Gux ≡ Gxw) ≡
  Gwx) ≡ Gxv]}

15∗9 / → (Eu)(Ev)(z){[(Gxu ⊃ Gzx) ⊃ Gxx] ⊃ (Gxx & Guv)}
15∗18 / → (Ex)(y)(Ez){Gyy ⊃ [Gxxv((GxzvGyz) & GzxvGzy))]}
19∗2 / → ~ (Ex)(y)(Gyx ≡ ~ Gyy)

proved or disproved by the program P. We shall not list problems for which the program P has given no complete solutions. (All the examples are drawn mainly from [3] and [1].)

It seems fair to say that the program P can decide some rather complex propositions. With a more advanced program, one can naturally expect mechanical proofs of more elaborate theorems of logic. Since only quite simple logical principles are employed in actual methematics, it seems reasonable to expect that mechines will often turn out proofs rather different from those obtainable by man. In fact some fairly complex but quite useful logical principles might be suggested by mechanical proofs of even familiar mathematical theorems.

We discuss now briefly the possible full AE methods. Among the examples given earlier, 14∗16 is not solvable by the restricted AE method but solvable by a full AE method. One method is this. In general, whenever all functors can be eliminated, we simply take the conjunction of all the undecided atomic sequents and make all possible substitutions of numbers for variables, and test the disjunction of all the instances. In the case of 14∗16, we have:

x = 1, w = 1: G11 → G21; G11 → G31

and

(x, w) = (1, 2),(2, 1), (2, 2), (1, 3), (3, 1), (2, 3), (3, 2), (3, 3)

This is like Qr [20,p.12], except that the elimination of functors extends the range beyond Qr.

It is, however, clear that this is not efficient and we can improve the method by using pattern recognition. In cases, however, like 14∗16, which can be seen to be of the AE form at the beginning, we may proceed simply as follows. From the original sequent, we see that there are 3 initial positive quantifiers and 2 negative quantifiers. Hence, we need to consider just:

G11, G21, G31 → G21 & G31, G22 & G32, G23 & G33

which is easily seen to be a tautology. (Compare Qq, [20,p.12].)

In the original version of the program P, step S11 contained also an erroneous method of proving a subproblem directly after the elimination of all functors. As a

---

LIST II. *Nontheorems Disproved by the Program* P

14∗12 / (y){[Jx = (Jy ⊃ Gy)] & [Gx ≡ (Jy ⊃ Hy)] &
[Hx ≡ ((Jy ⊃ Gy) ⊃ Hy)]} → Hz & Gz & Jz

15∗5 / → (Ex)(y)(Ez)[(GxyvGxz) & (∼Gxyv ∼ Gxz)]
15∗10 /→ (Ey)(z)(x ≠ zvy ≠ z)
15∗13 / → (Ex)(Ey)(Gxy ⊃ p) ≡ (x)(y)(Gxy ⊃ p)
19∗5 / → (Ey)(x)[Gxy ≡ (z) ∼ (Gxz & Gzx)]
19∗12 / → (Ey)(z)[(Gxy & ∼ Gxx) & (Gzx ⊃ Gzy)]
19∗14 / → (Ey)Gxy & [Gxy ⊃ (z)(Gxz ⊃ y = z)] & [Gyx ⊃ (Gzx ⊃ y = z)] & (Ex)(y) ∼ Gyx
19∗17 / → (Ey)Gxy & (Ex)(y) ∼ Gyx & [Gyx ⊃ (Gzx ⊃ y = z)]

result, some nontheorems were asserted to be theorems in the print–out. This fact was noticed by John McCarthy, and has led to the revised S11.

6.3 The E_1A case solved with sequential tables

The E_1A case is simple becuse, as can be seen from example (1) in the introduction, we need only worry about those elementary parts each of which contains only occurrences of a same variable, and then only the relations between the U–variables and the single E–variable require considerations. A simple subcase is explained quite thorughly in [3, p. 259].

In general, let us consider:

(3.1) $(Ex)(y_1)\cdots(y_n)Mx\cdots y_n$,M *containing* N *predicates* $G_1,\cdots,G_N$.

To form S_1, S_2, etc.,we need only replace $(x, y_1, \cdots, y_n)$ by $(1, 2, \cdots, n+1)$, $(2, n+2, \cdots, 2n+1)$, etc. The number of possible elementary parts in M depends of course on the number of places of the predicates $G_1, \cdots, G_N$. If, e. g., they are all dyadic, then there are 3^2N possible elementary parts G_ixx, G_ixy, G_iyx, G_iyy, G_ixz, G_izx, G G_iyz, G_izy, G_izz $(i = 1, \cdots, N)$. In the present case, the number of places of each predicate is immaterial since we have to consider only $G_ix\cdots x$, $G_iy\cdots y$, $G_iz\cdots z$, and each predicate behaves like a monadic one.

Thus, for example (1) in the introduction, we need consider only the following table T of all possible assignments of t and f to Gxx, Gyy, Gzz which would make the matrix of (1) false:

Gxx	Gyy	Gzz
t	t	t
t	t	f

Since S_1, S_2, S_3, etc., are obtained by substituting (1, 2, 3), (2, 4, 5), (3, 6, 7), etc., for (x, y, z), we get a tree structure:

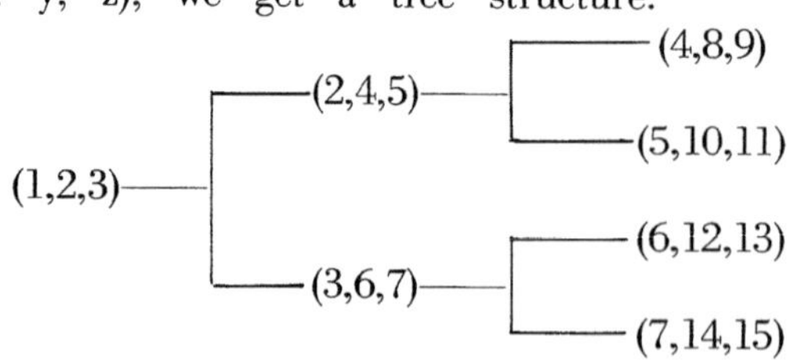

In order that a row Q can falsify S_1, i. e., M123, it is necessary that there is a row R which falsifies S_2, i. e., M245, and a row S which falsifies S_3, i.e., M367. For this purpose, it is only necessary that Gyy in Q is the same as Gxx in R, and that Gzz in Q is the same as Gxx in S. Since Gxx can take at most two values, t and f, if a table T contains two rows, one with Gxx taking t as value, one with Gxx taking f as value, we can always find a countermodel because, for each row falsifying S_1, we can always find two more rows which together with it falsify simultaneously S_1, S_2, S_3; and the same is true for any S_i and, therefore, for any D_i. If in every row of the table T, Gxx always takes one value, say t (or f), then it is necessary and sufficient to have one row in which both Gyy and Gzz take the same value, viz., t (or f). Hence, it is very easy to decide whether a statement (Ex)(y)(z)M, with a single predicate is a theorem, since, by the fundamental theorem of logic, it is a theorem if and only if there is some k such that D_k is a tautology.

In general, if a statement $(Ex)(y_1)\cdots(y_n)M$ contains a single predicate, the criterion is the same, viz.,

(3.2) *It is not a theorem if and only if either (i) its table T contains two rows in which $Gx\cdots x$ get different values, or (ii) it contains one row in which $Gx\cdots x$, $Gy_1\cdots y_1\cdots$, $Gy_n\cdots y_n$ all get the same value.*

If now there are N predicates $G_1, \cdots G_N$ in the martrix of $(Ex)(y_1)\cdots(y_n)Mx\cdots y_n$, then the matter is a little more complex, because we have to consider a table T with (n + 1) N columns:

$$G_1x\cdots x\cdots G_Nx\cdots xG_1y_1\cdots y_1\cdots G_Ny_1\cdots$$
$$y_1\cdots G_1y_n\cdots y_n\cdots G_Ny_n\cdots y_n$$

In this case, $G_1x\cdots x, \cdots, G_Nx\cdots x$ together have 2^N possible sets of values, if every set occurs in some row of T, then the original formula is of course not a theorem. In general, use the following "sequential method".

Examine each row R of T and determine for each i(i = 1, ⋯ , n), whether there is a row R_i such that the values which $G_1x\cdots x, \cdots, G_Nx\cdots x$ take in R_i are respectively the same as the values which $G_1y_i\cdots y_i, \cdots, G_Ny_i\cdots y_i$ take in R. If there is one such R_i for each i, retain R, otherwise, cross out R. Each time a row is crossed out, the same process is repeated with the reduced table until either the table is empty or the table is not empty but no further reduction is possible. Using this procedure, it is easy to prove the following theorem:

(3.3) *The formula $(Ex)(y_1)\cdots(y_n)M$ is a theorem if and only if its reduced truth table is empty.*

This method seems considerably more efficient than existing alternatives in the literature which are usually based on a determination of some constant K such that the given formula is a theorem if and only if D_k is tautologous.

When all the predicates are dyadic, Ackermann gives the bound J in terms of validity in a domain of J members as a sufficient condition of general validity. If we recall that nK + 1 numbers occur in D_k, we can calculate that his bound for K is no better than:

$$n > 1, \quad K = \frac{n^{3k} - 1}{(n-1)}, \quad \text{where} \quad k = 2^{Nn^2},$$

when

$$n = 1, \quad K = 3(2^N) - 1.$$

Church does not give the general bound, but calculates that [2, p. 213]; [3, pp. 260, 261]:

$$n = 1, \quad K = 2^N,$$

when

$$n = 2, \quad K = 2^k - 1, \quad \text{where} \quad k = 2^N.$$

It appears that if one extends Church's argument, by using a tree with n branches at each node and of height $2^N - 1$, the general bound would be (compare Herbrand [10, p. 46]):

$$n > 1, \quad K = \frac{n^k - 1}{n - 1}, \quad k = 2^N$$

In particular, when $N = 1$, $K = n + 1$.

It seems quite clear that the sequential method is faster than testing the Herbrand disjunctions D_1, D_2, etc. Take a simple example with n = 2, N = 4 and a table T with 16 rows such that in R_1, Gx, Hx, Px, Qx gets tttt or 0000 = 0, but Gy, Hy, Py, Qy, as well as Gz, Hz, Pz, Qz all get tttf or 0001 = 1, and similarly in every R_i, Gx, Hx, Px, Qx, get the truth values corresponding to the binary notation i -- 1, where the y, z parts both get the values corresponding to i. In such a case, it is easily seen by the sequential method that the formula is not a theorem since the reduced table is the same as the original table, yet by the alternative methods, we have to test D_k with $K = 8^{65536} - 1$ by one method and $K = 65535$ by the other. This example incidentally illustrates that just speeding up the method of testing each D_j is not sufficient to handle many interesting formulae.

Incidentally, there is a striking similarity between the type of argument involved in these decision procedures and the method of "sequential tables" developed in [19]. The similarity suggests the question of a more abstract mathematical treatment of more basic underlying principles which govern such sequential methods. The writer, however, has no inkling as to whether results will be obtained on this question and, if so, how interesting they will be.

6.4 General remarks

One is naturally curious to know how far we are from machine proofs of truly significant mathematical theorems in different domains. The writer cannot see sufficiently far and clearly into the future to make any responsible predictions, except that the simplicity of all theorems of *Principia* in the predicate calculus came as a great surprise, suggesting the opinion that one could be too conservative in estimating the potentialities of machines in theorem proving. It seems that several types of objectives are likely to be achieved with just a few more months of programming efforts along the present approach which precludes ad hoc measures designed specially for a few immediate specific problems. Among these are proving a large portion of theorems in Landau's booklet [14] on the number systems, proving a fair number of theorems in high school algebra and geometry, formalizing fairly interesting theorems in set theory. In the last category, it seems likely that the machine will soon be able, e. g., to do the tedious but less inspired part of the work needed to establish the main conclusion of [11], viz., to derive the contradiction from the few axioms chosen in advance by man. Those who have worked on this type of problem would appreciate that such assistance is not to be despised. A considerably more difficult and remote task would be to formalize Specker's derivation of a contradiction [17] in Quine's "New Foundations" plus the axiom of choice, which is a more complex system. However, in view of the apparent artificiality of the formal system concerned, the advantage of man over the machine is greatly reduced when results on such systems are to be established.

On the whole, it seems reasonable to think that machines will more quickly excel in areas where man's intuition is not so strong. Hence, the author is now in- clined to feel that difficult theorems in analysis and set theory will more easily be proved by machines than those in number theory. For example, the writer feels that among possible targets for the next year or two the irrationality of $\sqrt{2}$ and the unique factorization theorem may tax the ingenuities of machines and their programs equally heavily as theorems in set theory and analysis, such as the Heine–Borel theorem and the Bernstein theorem, which man finds considerably harder to understand. This may appear rather nonsensical in view of the great conceptual difficulties we have with the continuum and higher infinities. In the writer's opinion, however, the decisive factor is rather the fact that we are capable of making much more varied and extended moves when general theorems about natural numbers are being considered; this is likely to make it harder for machines to catch up with us. On the other hand, it is well known that machines are good at dealing with essentially combinatory problems, which are, however, not the chief concern of number theory.

When imagination is given free rein, Fermat's and Goldbach's conjectures, the Riemann hypothesis, the four–color problem, the consistency of impredicative analysis, the continuum hypothesis, and other famous overwhelmers all come to mind. Of these celebrated problems, it seems fair to concede that at present we have no idea how machines might assist in arriving at a settlement of any. It is of course possible that machines may get hold of things which have eluded man for decades or centuries, since, after all, viewed in the context of Plato's realm of ideas, man's path must have been pretty narrow. Since, however, machines will, we hope, never become the Master and they have no higher masters than man in the horizon, it would be very surprising indeed if they should quickly surpass man in areas which demand the highest human creative genius and prove a whole lot of theorems which the best mathematicians have strived and failed to establish for very long. At any rate, it is a happier thought that machines will increase the power of mathematicians rather than eliminate them, and there is no evidence at present that the latter alternative will ever materialize.

The writer, as an amateur programmer and as one who has given little thought to designing monitoring systems for symbol manipulation, has little to say on the theory of programming. One obvious suggestion is that the writer's program P as it exists now should be used as a guinea pig for testing symbol manipulation systems: see how much easier it is to rewrite the program in each system and determine how much slower the new program runs. A careful study of the program P may also suggest to the experts to add or modify certain devices in their systems in order to meet the natural demands by neutral programs which involve a good deal of symbol manipulation.

(After the above paragraph had been written, the writer saw John McCarthy's "The Wang algorithm for the propositional calculus programmed in LISP," Artificial Intelligence Project, MIT, Symbol Manipulating Language, Memo 14. This deals roughly with program I of [20]. "It took about two hours to write the program and it ran on the fourth try." Even making allowance for the fact that this part of the program P is relatively simple and specially suited to LISP, the coding time required is still amazingly short. The running time is also much better than the writer had expected. Apparently, for the same problems, the LISP program takes no more than 10 times longer than the original SAP program; and presumably this can be further improved. Moreover, it is stated in the memo that the algorithm has also been useful in suggesting a general concept of ambiguous functions to be used in computations.)

Like many people embarked on ambitious computing projects, the writer used to think how nice it would be to have at disposal a STRETCH computer or something beyond. Recently, however, the writer has come to feel a bit differently. Of course a larger machine can handle more difficult problems and one can afford to use more crude methods which are hopeless on some slower machine. But so often the difference in speed between different methods is so great that even an increase by 100 is quite inadequate to compensate for the deficiencies in a more crude method. How far can 100 go when people indulge in exponentiations? For example, if one were to use the so-called British Museum algorithm or even the less wasteful brute force search method mentioned above, a simple formula, say with a conjunction of two disjunctions of elementary parts as its quantifier-free matrix, would have for $S_1 \vee \cdots \vee S_{100}$ a formula with 2^{100} long disjunctions to test.

If, however, one is to seek more efficient methods, it often happens that we can only begin with simpler cases and proceed to more complex cases. Then it is likely that a good deal of effort is needed before one can do full justice to a large machine. For example, the writer has made no special efforts to economize storage and feels sorry to have been able to use only less than half of the 32,000 words on the particular IBM 704. Similarly, the writer is sorry to have found no natural problems for the program which actually require a long running time that is justified by the intrinsic interest of the problems. The building of more powerful machines and the designing of more efficient methods to use them both require time and human efforts. A big lagging behind in either direction causes waste. At least financially, a long lag in the using direction is the worse of two evils. When the best existing method does demand the full capacity of the largest existing machines, we have the happy situation of a harmonious coordination. This is truly nice only when there is no immediate prospect for improving the method. Under those circumstances, the problems solved must be quite interesting, because otherwise we should say that the method is not yet good enough for use on machines, or that the problems are not yet suitable for machine treatment. It seems that the machines, when cursed for being too slow or too small, may often with justification demand in turn that the user do some more thinking.

We have a feeling that there are things which machines can do and things they cannot do, things for which they are especially adapted and things for which they are not quite suitable. The familiar ambiguities of the word "can" seem to give at least two different conclusions. In one sense, it is silly to make machines do what they cannot do. In another sense, this is precisely the exciting thing Compared with numerical calculations, proving theorems seems like forcing machines to do what they cannot do. Compared with proving theorems, mechanical translation, for example, seems even more remote from the natural aptitudes of machines.

Quite justifiably, mechanical translation has a wider appeal than mechanical mathematics. For one thing, more people use languages than mathematics. However, so far as the near future of the two projects is concerned, it would seem fair to say that mechanical mathematics has a brighter prospect. While inferential analysis can draw from a body of profound exact results in theoretical logic, a mathematical treatment of language is very much something yet to come and seems to involve fundamental intrinsic difficulties, particularly in dealing with meaning and ambiguities. Any substantial progress in mechanical translation would be impressive in so far as machines would be doing something for which they are apparently not suitable. It is, however, highly unlikely that machines can give better translations than expert human translators. On the other hand, machine proof of theorems is quite likely to yield impressive results in the more absolute sense: Viz., the doing of things which lie beyond unassisted human capabilities.

It seems undeniable that computers have changed somewhat the face of applied mathematics. The writer feels that the cross–fertilization of logic and computers ought to produce in the long run some fundamental change in the nature of all mathematical activity. Such a development will not only make pure mathematicians take computers more seriously but provide proud applications for the parts of logic which are logicians' primary concern. Modern logic is intrinsically interesting, yet it is customary among mathematicians to think of it as a bit irrelevant. As logic matures, the relevance is being felt, e. g., in the general study of effective procedures and the analysis of existence proofs. Advances in mechanical mathematics will spread the influence of logic further and bring to the forefront detailed works on logic in the narrower sense of dealing with inferences in the first place.

The relevance of logic is perhaps rather gratifying to experts in advanced programming which seems to attract capable people who enjoy thinking but dislike extensive implicit presuppositions. The energy, created by their unfulfilled desire for solid theories unhampered by such presuppositions, finds an outlet in logic, and with the increasing relevance of logic they can now spend time to learn the new trade with a clear conscience. On the other hand, thinking on many logic problems can be assisted when the more exacting demands of machines and programming are kept in mind.

As far as we know, machines can only do significant things by means of algorithms. But in the human mathematical activity one also speaks of intuition, insight, and heuristic methods, which do not seem to possess easy exact definitions. When these terms are applied to machines, the unity of contraries sounds distinctly paradoxical. In particular, the term "heuristic method" has gained some currency among sophisticated users of machines, since Polya revived the term in a rather orthodox context.

In ancient times Archimedes distinguished a method of proof (the method of exhaustion in this case) from a method of discovery (the heuristic method) in finding out the area or volume of a configuration. His heuristic method includes considerations of the law of the lever and the center of gravity, and also of intuition such as that cylinders, spheres, and cones are made up of parallel circular discs. (An extensive discussion of the matter is given in T. L. Heath, *The Methods of Archimedes*, 1912.) In modern times, Euler was perhaps the best-known mathematician who told of how his theorems were often first discovered by empirical and formalistic experimentations.

In these examples, there appears to be an element of inexactness in the methods so that they cannot be rigorously formulated, cannot be taught in words alone but only by awakening certain latent understanding implicit in the pupil's mind. For the same reason heuristic methods are usually not taught explicitly, and while, e. g., Polya's attempt to teach them is very likely a good pedagogic idea, one would expect that pupils in the same class would achieve very different results which are determined largely by the ability of each pupil. There is nothing important in this trite prediction except the equally obvious conclusion that a machine in such a class can hardly be expected to do as well as the average pupil. This would seem to indicate that there are rather serious difficulties in teaching machines heuristic methods. It is very doubtful that any completely mechanized method, available so far, can properly be called a heuristic one. Much more elaborate instructions are necessary in order to produce the impression that the machine is actively participating as a creative agent in the search for a mathematical proof.

Sometimes it is suggested the heuristic methods are just the methods which a man would find natural to use. This is not very helpful. For example, it is not hard to contend that the method of doing propositional calculus described in [20] is quite natural, but by no stretch of imagination can it be called a heuristic method.

It can be a useful thing to fill an old bottle with new wine. But if this is being done when specific mechanical methods are said to be heuristic, it would perhaps be less misleading if the constituents of the new wine are explained a little more exactly. One might think it silly to make heavy weather on what is just a trivial terminological matter. The truth is, however, that such expansive obscurity can arouse useless enthusiasm in some lighthearted quarters and generate harmful suspicion among the more seriously-minded scientists.

If we leave aside at the present rather primitive stage the emotion-laden term "heuristic method," there are a few more prosaic distinctions which can be made among the algorithms for proving theorems and disproving nontheorems. A partial algorithm may be able to give yes and no answers only in some part of a domain, e. g., a partial decision procedure such as the monadic within the predicate calculus, or only yes as answer, e.g., the proof procedure for the predicate calculus. Among partial decision procedures, there are those for which we know in advance the ranges of application, i.e., there are some simpler effective methods by which we can test whether any given problem is decidabie by the method. There are also those with ranges of application which are undecidable in advance or simply undecidable. All the procedures which are commonly studied have, however, one thing in common, viz., when the method does give a decisive answer, the answer should be correct. In actual research, however, we also use alternative methods which tend to give us an answer more quickly, and often although not always correctly. Then we can use more elaborate methods to verify whether the tentative answers are correct. Such methods, when mechanizable, would resemble the methods of approximation in numerical analysis and have, indeed, some flavor of the heuristic methods.

On the whole, we need interlocked hierarchies of methods, and the mysterious elements in the creative activity seem likely to be replaced by a complex web of clearly understood, definite and deterministic algorithms, rather than random elements or obscure machine programs. If man gets results by intuitive methods we cannot easily formalize, it does not mean that by introducing uncontrollable elements, a machine will more likely behave like man. If we do not understand how certain turns are made at crucial junctures, casting a die each time to guide the machine will scarcely ever produce the desired final effect. With machines, large masses of well-organized minute details seem to be the only sure way to make the correct surprises emerge.

References

1. W. Ackermann. *Solvable Cases of the Decision Problem.* 114 pp., 1954, Amsterdam.
2. Alonzo Church. Special cases of the decision problem, *Revue philosophique de Louvain 49* (1951), 203—221; a correction, ibid., *50*(1952), 270—272.
3. Alonzo Church. *Introduction to Mathematical Logic, I.* 376 pp., 1956, Princeton.
4. M. Davis and H Putnam. A computational proof procedure. AFOSR, 1959 (submitted to *Journal of Association for Computing Machinery*).
5. B. Dreben. On the completeness of quantification theory. *Proc. Nat. Acad. Sci, USA, 38* (1952), 1047—1052.
6. B. Dreben. Systematic treatment of the decision problem. Summaries of talks at the Summer Institute of Symbolic Logic, p. 363, 1957, Cornell.
7. B. Dunham, R. Fridshal, and G. L. Sward. A nonheuristic program for proving elementary logical theorems (abstract). *Comm. ACM, 2* (1959), 19—20
8. P. C. Gilmore. A proof method for quantification theory: its justification and realization. *IBM J. Res. Develop. 4* (1960), 28—35.
9. J. Herbrand. *Recherches sur la Théorie de la Démonstration.* 128 pp., 1930, Warsaw.
10. J. Herbrand. Sur le problème fondamental de la logique mathématique, *Compt, rend. Soc. Sci. Lettres Varsovie,* Classe III, *24*(1931), 12—56.
11. K. J. .J Hintikka. Vicious circle principle and the paradoxes. *J. Symbol. Logic, 22*(1957), 245—249.
12. L. Kalmar. Über die Erfüllbarkeit derjenigen Zählausdrücke, welche in der Normalform zwei benachbarte Allzeichen enthalten. *Math. Ann. 108*(1933), 466—484.
13. Dieter Klaua. Systematische Behandlung der lösbaren Fälle des Entscheidungsproblems für den Prädikatenkalkül der ersten Stufe. *Zeit. math. Logik Grundl. Math. 1* (1955), 264—270.
14. E. Landau. *Grundlagen der Analysis,* 1930, Leipzig.
15. W. V. Quine. *Methods of Logic.* 1950 and 1958, New York.
16. T. Skolem. Über die mathematische Logik. *Norsk Mat. Tidsskrift, 10* (1928), 125—142.
17. E. Specker. The axiom of choice in Quine's new foundations for mathematical logic. *Proc. Nat. Acad. Sci. USA, 39* (1953), 972—975.

18. HAO WANG. A theory of constructive types *Methodos 1* (1949), 374—384.

19. HAO WANG. Circuit synthesis by solving sequential Boolean equations. *Zeit. math. Logik Grundl. Mathe. 5* (1959), 291—322.

20. HAO WANG. Toward mechanical mathematics. *IBM J. Res. Develop. 4* (1960), 2—22.

7. OBSERVATIONS ON ATP*

7.1 Mechanical mathematics and inferential analysis

1. General Speculations

If we compare calculating with proving, four differences strike the eye: (1) Calculations deal with numbers; proofs, with propositions. (2) Rules of calculation are generally more exact than rules of proof. (3) Procedures of calculation are usually terminating (decidable, recursive) or can be made so by fairly well-developed methods of approximation. Procedures of proof, however, are often nonterminating (undecidable or nonrecursive, though recursively enumerable), indeed incomplete in the case of number theory or set theory, and we do not have a clear conception of approximate methods in theorem-proving. (4) We possess efficient calculating procedures, while with proofs it frequently happens that even in a decidable theory, the decision method is not practically feasible. Although shortcuts are the exception in calculations, they seem to be the rule with proofs in so far as intuition, insight, experience, and other vague and not easily imitable principles are applied. Since the proof procedures are so complex or lengthy, we simply cannot manage unless we somehow discover peculiar connections in each particular case.

Undoubtedly, it is such differences that have discouraged responsible scientists from embarking on the enterprise of mechanizing significant portions of the activity of mathematical research. The writer, however, feels that the nature and the dimension of the difficulties have been misrepresented through uncontrolled speculation and exaggerated because of a lack of appreciation of the combined capabilities of mathematical logic and calculating machines.

Of the four differences, the first is taken care of either by quoting Gödel representations of expressions or by recalling the familiar fact that alphabetic information can be handled on numerical (digital) machines. The second difference has largely been removed by the achievements of mathematical logic in formalization during the past eighty years or so. Item (3) is not a difference that is essential to the task of proving theorems by machine. The immediate concern is not so much theoretical possibility as practical feasibility. Quite often a particular question in an undecidable domain is settled more easily than one in a decidable region, even mechanically. We do not and cannot set out to settle all questions of a given domain, decidable or not, when,

---

* First published in *Computer Programming and Formal Systems*, edited by P. Brafford and H. Hirschberg, pp 1—20. © North–Holland Publishing Company, 1963. Reproduced by permission.

as is usually the case, the domain includes infinitely many particular questions. In addition, it is not widely realized how large the decidable subdomains of an undecidable domain (e.g., the predicate calculus) are. Moreover, even in an undecidable area, the question of finding a proof for a proposition known to be a theorem, or formalizing a sketch into a detailed proof, is decidable theoretically. The state of affairs arising from the Gödel incompleteness is even less relevant to the sort of work envisaged here. The purpose here is at most to prove mathematical theorems of the usual kind, e.g., as exemplified by treatises on number theory, yet not a single "garden-variety" theorem of number theory has been found unprovable in the current axiom system of number theory. The concept of approximate proofs, though undeniably of a kind other than approximations in numerical calculations, is not incapable of more exact formulation in terms of, say, sketches of and gradual improvements toward a correct proof.

The difference (4) is perhaps the most fundamental. It is, however, easy to exaggerate the degree of complexity which is necessary, partly because abstract estimates are hardly realistic, partly because so far little attention has been paid to the question of choosing more efficient alternative procedures. The problem of introducing intuition and experience into machines is a bit slippery. Suffice it to say for the moment, however, that we have not realized that much of our basic strategies in searching for proofs is mechanizable, because we had little reason to be articulate on such matters until large, fast machines became available. We are in fact faced with a challenge to devise methods of buying originality with plodding, now that we are in possession of servants which are such persistent plodders. In the more advanced areas of mathematics, we are not likely to succeed in making the machine imitate the man entirely. Instead of being discouraged by this, however, one should view it as a forceful reason for experimenting with mechanical mathematics. The human inability to command precisely any great mass of details sets an intrinsic limitation on the kind of thing that is done in mathematics and the manner in which it is done. The superiority of machines in this respect indicates that machines, while following the broad outline of paths drawn up by people, might yield surprising new results by making many new turns which man is not accustomed to taking.

The attempt to mechanize, as much as possible, mathematical thinking opens up a large area of research. In my opinion, the theoretical core of this area is a new branch of applied logic which may be called inferential analysis, characterized by an emphasis on the explicitless and practical feasibility of methods of inference. This discipline enjoys a measure of autonomy not shared by numerical analysis which, for example, does not deal with logical operations on the choice and arrangement of numerical methods. It is believed that the development of mechanical mathematics will influence pedagogical and research methods in mathematics, as well as affect certain epistemological questions of a specifically mathematical coloring.

The governing general principle is: what can be formulated exactly can be put on a machine, subject to the practical limitations on the manageable dimension and complexity. Repetitions are a good indication of the suitability of a mechanical treatment. Thus, for example, much of the activity of teaching mathematics is tedious and requires patience. If no interaction between pupil and teacher were necessary, televisions, or sometimes just gramophones, would be sufficient to replace teachers. As it is, these ready-made conveniences are only used as partial substitutes but, even so, teaching has already begun to enjoy to a certain extent the advantages of mass production. However, interesting problems of mechanical mathematics arise only when we come to tasks which call for an active agent to give answers, advices, and criticisms, the simplest being the correction of exercises and examination papers. Psychologically, the pupil has many reasons for preferring a patient machine teacher when the problem is, as in the majority of situations, a matter of drill rather than inspiration. The result may be that human teachers will employ mechanical devices as teaching assistants.

In a similar fashion, since in mathematical research there is also a great deal of mechanizable tedious work, mechanical devices may be used to aid individual mathematicians. In this connection, in view of the fact that specific mathematical discoveries are made essentially once and for all, there are less of exact repetitions, but more of the problem of using mechanical devices flexibly by, for example, designing and combining programs on general purpose computers. In order to use machines either to aid research or to aid teaching, the results, methods, and spirit of formalization in mathematical logic are to play an essential role.

The advance of mechanical mathematics may also affect some of our concepts in the philosophy of mathematics. We get, not only in theory, but even in practice, an objective criterion of mathematical rigor in mechanical terms. The range of feasible mathematical methods will be extended so that the theoretically feasible and the practically feasible begin to converge, and we have a more realistic guidance to the improvement of feasibility. As we understand more fully the range of mechanical mathematics, we get a clearer view of the relation between complexity and conceptual difficulty in mathematics, since we would probably wish to say that mechanizable pieces, even when highly complex, are conceptually easy. When alternative proofs are fully mechanizable, we obtain also a quantitative measure of the simplicity of mathematical proofs, to supplement our vaguer but richer intuitive concept of simplicity. With the increasing power to formalize and mechanize, we are freed from tedious details and can more effectively survey the content and conceptual core of a mathematical proof.

2. The Central Role of Logic

In theory all mathematical arguments can be formalized in elementary logic (quantification theory, predicate calculus). If we add equality and the quantifiers "for all x" and "for some y" to the propositional connectives "and", "if", "or", "not", etc., we obtain the predicate calculus, in which, as logicians will know, every usual mathematical discipline can be so formulated that each theorem T in the latter becomes one in the former when the relevant mathematical axioms A are added as premises. That is to say, if T is the theorem in the mathematical discipline, then "if A, then T" is a theorem of logic. From this fact it is clear that in order to prove mathematical theorems by machines a major step is to deal with theorems of the predicate calculus.

One may question the advantage of thus handling mathematics, on the ground that the peculiar mathematical content of each individual branch is lost when the disciplines are thus uniformly incorporated into the predicate calculus by formalization and abstraction. Now it is indeed true that we must add special methods for each special mathematical discipline. An adequate treatment of the predicate calculus is, however, of dominant importance, and for each discipline the basic additional special methods required are fairly uniform. For number theory, the essential new feature is largely concentrated in mathematical induction as a method of proof and of definition; for set theory, in the axiom of comprehension, i.e., the axiom specifying all the conditions which define sets. So there is the problem of choosing the formula to make induction on, or of choosing the condition for defining a set. While it seems doubtful that there is any uniform efficient mechanical method for making such selections, there are often quite feasible partial methods. For example, for making such selections in number theory the obvious uninspired method of trying the desired conclusion of one or another of its clauses as the induction formula should suffice in many cases. It would seem that, once a feasible way of doing logic is given, fairly simple additional methods could carry us quite some way into special mathematical disciplines.

Since most of us learned Euclid and number theory without worrying about the predicate calculus, it might seem that the natural course is to bypass logic and go directly to mathematics. But in fact such an approach is ill-advised, so long as the aim is to prove more and harder theorems rather than merely to re-enact the history of mathematical thinking. What is natural for people need not be natural for the machine. If logic is not treated in an explicit and systematic way, constant subsequent additions of *ad hoc* devices keep slowing our progress toward interesting theorems, while multiplying the sources of possible confusion. In general, a vast machinery specifically designed to obtain a few easy theorems is wasteful; results obtained from whatever approaches should be measured against the generality and economy of the machinery used. Foundations, furthermore, should be scaled to large future superstructures. It is our conviction that to treat logic only by the way would score very poorly by both criteria.

3. Some Possible Directions for Further Exploration

Results so far are too rudimentary to provide us with any decisive conclusions as to the dimension of the long-range effects of the pursuit of mechanical mathematics. Nevertheless, I shall venture a few comments drawn from my own restricted experience. (a) I have examined the theoretically undecidable domain of the predicate calculus and managed to make an IBM 704 prove all theorems (over 350) of Principia mathematica in this domain in less than 9 minutes; this suggests that we usually do not use the full power of strong mathematical methods and should not be prevented from trying to handle an area on account of pessimistic abstract estimates of the more difficult cases in the region. (b) Care in the theoretical design of the procedures is essential and a certain amount of sophistication in mathematical logic is indispensable, because most existing methods are not immediately applicable on machines. (c) In particular, one often has to reformulate available methods or even invent fundamentally new ones; sometimes theoretically insignificant improvements could increase the speed or reduce the necessary storage by several orders of magnitude, for example, a device to try out certain "preferred" substitutions first. (d) Long-range planning and efforts to make results cumulative are necessary; *ad hoc* measures and desire for quick sensation should be avoided because otherwise the limit of diminishing return will be reached too soon; the correct course would increase reward per unit of work more and more quickly with greater and greater efforts. (e) While more can be done with larger machines, the design and choice of methods is, at least at the present stage, more crucial because we are far from having made full use of an IBM 704 or 7090 yet. (f) Distrust luck and do not, for example, use obscure methods with the hope that something wonderful might happen since we do not know what will happen; the chances of undesirable consequences are much bigger.

At the present stage, mechanical mathematics seems to be one of the areas in information processing which promise the highest reward for each unit of labor. Only accidental circumstances such as the lack of alliance of potential contributors in administration, programming, and logic have so far sabotaged more rapid developments. The laziest solution of this practical difficulty is for one to attack problems in isolation and hope that the pieces will miraculously fit together in due course. This is not the most satisfactory solution but perhaps the most feasible, given all the facts of competition, sales exaggeration, desire for liberty and independence. There are at least three groups of preliminary work necessary for genuine advances in the long run: a good common (idealized programming) language for crystallization, communication, and accumulation of results; a decent library of subroutines for simple algebraic and logical manipulations of symbols, as well as for simple basic proof and decision procedures; and a fairly sophisticated logical analysis of a number of specific mathematical proofs with a view to bringing out the details which have to be taken care of in order to thoroughly formalize and mechanize them.

It is of course not excluded that one would often run into blind alleys. But I am confident major wastes can be avoided through careful planning and alert flexibility. With these provisos in mind, I now proceed to list a few possible directions which, in my opinion, are worthy of at least some preliminary exploration.

That proof procedures for elementary logic can be mechanized is familiar. In practice, however, were we slavishly to follow these procedures without further refinements, we should encounter a prohibitively expansive element. It is desirable to study underlying properties of such expansions in order to increase efficiency. In this way we are led to a closer study of reduction procedures and of decision procedures for special domains, as well as of proof procedures of more complex sorts. Such deeper considerations of elementary logic also provide us with a systematic approach to axiomatic theories viewed as applied predicate calculus. The insights thus obtained can complement our direct treatment of specific mathematical disciplines.

For the sake of a more concrete goal to guide the choice of theoretical questions, we may set ourselves the aim of programming machines to formalize and "discover" proofs in quantifier-free number theory and axiomatic set theory. These areas are chosen both because they are so central and because it seems desirable to isolate the two basically difficult mathematical concepts: functions and quantifiers. It is possible that the quantifier-free theory of positive integers, including arbitrary simple recursive definitions, can be handled mechanically with relative ease, and yield fairly interesting results. It is clear from works in the literature that this restricted domain of number theory is rather rich in content. It goes beyond logic in an essential way because of the availability of (quantifier-free) mathematical induction. On the other hand, in axiomatic set theory, the explicit use of functions can be postponed for quite a long time. Moreover, here certain general concepts often prove difficult; perhaps machines will more quickly excel in areas where people's intuitions are not strong. A case in point would be Quine's axiomatic system "New Foundations," which was obtained by relaxing certain syntactical restrictions of the theory of types.

While the ulterior aim is to use machines to aid mathematical research with the assistance of logic, machines can also be used to aid our theoretical research in logic at the present stage. Computers can be put to good use in the quantity production of concrete examples, which we constantly need as a means of clarifying our concepts and so expediting general theoretical results.

Already in the limited experience in the mechanizing of logical procedures, the machine outputs have from time to time brought out features of the procedures which one had not thought out clearly in advance. Such experiences have sufficed to persuade us that we would do well to experiment with computing machines even if it were only for purposes of theoretical logic.

Some other possible directions are: (1) Experiment with redoing school and college mathematics by machines; instruct a machine to compete with the average student by using its patience to compensate its lack of intuition; partial decision procedures in algebra, trigonometry, analytic geometry, the calculus; prove theorems in elementary geometry and algebra with extensive use of methods dealing with the underlying logic. (2) Try to combine numerical and inferential methods so that principles can be introduced for the machine to choose particular methods to apply according to the nature of the given problems; this aims at delegating to the machine as much as possible of the work which now requires a mathematical analyst. (R.W. Hamming is much interested in work along this direction.) (3) In fields like algebraic topology where often definitions are long but proofs are short, it is not unlikely that mechanized logical methods will prove to be of practical use in helping to sort out logical consequences of new concepts. (4) Fairly simple mathematical researches involving combinatorial considerations such as questions of completeness, independence, deducibility in the various systems of the propositional calculus can presumably be helped radically by a few suitably devised machine programs. (5) Use this type of work as data to guide us in the design of more realistic idealized programming languages.

With regard to the formulation of programming languages, it seems desirable not to pursue the task in isolation and then look for applications afterwards. One must not let the initial investment in a programming language control the choice of problems to be programmed, but frequent revisions of a fixed language require a prohibitive amount of energy and work which can easily prevent one from meeting new demands with an open mind.

A good compromise between rigidity and complete lack of organization would seem to be the isolation of necessary devices such as the designing of MACRO instructions at every stage, as is called for by specific but typical occasions. In this way, a body of quite well-or-ganized data would gradually emerge as more programs are written. Attention to the accumulation of good MACRO instructions also brings into the somewhat routine task of programming a theoretically more interesting element of formulating exactly concepts which are intuitively familiar.

4. Case Studies and Stock-of-Trade Systems

To analyze in detail specific mathematical proofs is clearly a useful preliminary step toward the mechanization of types of arguments. One might attempt to work out a few examples of such case studies drawn from number theory, geometry, and axiomatic set theory.

In number theory, one might compare quantifier and free variable proofs of the infinitude of primes and of the fundamental theorem of arithmetic, putting emphasis on recursive functions and mathematical induction. In geometry and axiomatic set theory, one might consider mildly difficult theorems which are proved with quantifiers but without functions. In each case, two types of problems can be conveniently separated: deriving very elementary properties such as the commutativity of addition from the basic axioms on the one hand, and organizing such elementary properties to obtain a basis for further derivations on the other. For the human being, the first type of problem is rather artificial and contraintuitive. For example, the very early theorems in elementary geometry are more abstruse than the simple exercises about parallels, triangles, etc. A good organization of elementary properties plus an exact formulation of familiar methods of trying to find a proof would presumably yield in each discipline something similar to the principles obtained from what is often called the "heuristic approach". It is here proposed that such organizations be called stock-of-trade systems. However, despite terminological disputes, everybody probably agrees as to roughly what sort of thing is to be done, and the more relevant question is how good a result one gets. It is with regard to this last point that a patient study of special cases with ample use also of the stocks in trade of mathematical logic appears indispensible. For instance, even a formalization of Euclid's proof of the infinitude of primes contains a few surprises, and there are quite a number of theoretically interesting questions connected with the problem of proving the irrationality of 1/2 with no appeal to quantifiers.

We consider here only an example in axiomatic set theory derived from a paper of Hintikka [13].

If a theorem is proved in a system, even one with only a finite set of axioms, it would seem that one major problem is to select the axioms needed for the proof. It stands to reason to expect that it would be easier for the machine to begin with the selected axioms. In so doing, we may lose some alternative proof which uses other axioms but that is something which we do not have to worry about yet. Moreover, it appears easier to select and prove intermediate lemmas and break up the whole proof into parts. In both cases, if we do not have the selection to begin with, it is not easy to decide whether it is advantageous to take all to begin with, or to add routines to select. In the long run, one would expect to use the latter alternative. But when the methods of selecting subproblems and branching out are as cumbersome as some existing crude attempts appear to be, it is not necessarily more efficient to use the selection alternative.

The example to be considered is of special interest because it lies in an area which has not been developed nearly as much as old subjects such as number theory. Consequently, we can draw very little from a cumulative intuition, and our advantages over machines are not great. Moreover, this area has been pursued with a considerable emphasis on formal arguments.

Let Hxy and $\exists zGxyz$ be short for:

$$\exists z(z \neq x \wedge z \neq y \wedge Fzy \wedge Fyz).$$

(1) $u \neq v.$

(2) $y \neq a \supset (Fya \equiv Hay).$

(3) $y \neq b \supset (Fyb \equiv \neg Hby).$

(4) $y \neq c \supset (Fyc \equiv (y=a \vee y=b)).$

(5) $y \neq d \supset (Fyd \equiv y=c).$

The assertion is that the conjunction of (1)—(5) is contradictory. More exactly, this says that the following formula is a theorem of the predicate calculus.

$$\text{(I)} \quad \neg\exists u\exists v\exists a\exists b\exists c\exists d\forall y\forall z\exists w\exists x$$

$$\begin{aligned}
&\{u \neq v \\
&\wedge [y \neq a \supset ((Fya \wedge Gayw) \vee (\neg Fya \wedge \neg Gayz))] \\
&\wedge [y \neq b \supset ((Fyb \wedge \neg Gbyz) \vee (\neg Fyb \wedge Gbyx))] \\
&\wedge [y \neq c \supset (Fyc \equiv (y=a \vee y=b))] \\
&\wedge [y \neq d \supset (Fyd \equiv y=c)]\}.
\end{aligned}$$

If the system does not include $=$, then we have to treat $a=b$ as an abbreviation for

$$\forall x(Fxa \equiv Fxb),$$

and add the axiom:

$$a=b \supset \forall y(Fay \equiv Fby).$$

This incidentally illustrates the fact that for mechanical mathematics it is in practice desirable to include $=$ to begin with. In that case, the formula is in one of the familiar decidable cases since it contains only two consecutive $\exists$'s (for validity). In terms of satisfiability, the part without the initial $\neg$ has no model and can be decided by the $\exists\forall_2\exists$ satisfiability case (see [23]).

On the whole, it seems easier to make machines do some of the formalizing work which logicians sometimes have to do. This may be viewed as an application of the principle "Charity begins at home." Some malicious soul might use this as evidence for his favorite view that logic is trivial, and he will be wrong for too many reasons which it is tiresome to elaborate.

In general, what is needed for mechanization is not just axiomatic systems with emphasis on economy and elegance but rather "stock-of-trade systems" and formalizations which are exact and yet remain as close to good common expositions as possible.

5. Some Theoretical Difficulties

In order to mechanize proof procedures of the predicate calculus, it seems natural to use Herbrand's Theorem. This has been suggested and carried out to varying degrees of completion by different people (see [21], [22], [10], [16], [6]). The crucial part contains the generation from a given formula of a sequence of propositional or Boolean conditions, and the testing of them. It is clear, both by theoretical estimates and from results obtained so far, that (i) doing the expansion and the testing both by a brute force approach is not feasible; (ii) even greatly speeding up the testing part is not adequate to dealing with fairly interesting cases because often we have to generate a large number of Boolean conditions.

Hence, a central theoretical problem is to find ways of selecting only useful terms from each sequence of Boolean conditions. This problem has been explored in a preliminary manner in [22] and [23]. One element is to develop decision procedures for subdomains of the predicate calculus. Another element is to use miniscope forms instead of prenex forms. A third element is to develop semidecision procedures whose range of application we do not know exactly in advance ([23], p.30).

The decision procedures appear not to include the formulas which are of the most interest to us. More specifically, the decision procedures mostly deal with formulas in the prenex form, and when we derive a theorem from a few axioms, even though the theorem and the axioms are separately of simple forms, putting the implication (of the theorem by the axioms) in a prenex form quickly gets us outside the decidable subdomains. This suggests that we should try to extend the decision procedures to truth functions of formulas in the prenex form. Property C in Herbrand's dissertation [11] (see below) seems relevant to this question.

The semidecision procedure of [23] is not developed far enough in that the conditions under which we are to terminate the procedure are not specified explicitly. For example, if we encounter a periodic situation (a torus), we can naturally stop; but since the initial columns occupy a special place, we permit also cases while the initial columns and the others have two periods which can be fitted together. Closer examination is necessary in order to lay down broader stopping conditions.

The miniscope form defined in [22] is different from the one developed in Herbrand's dissertation because it permits the breaking up of a quantifier into several. While this permits more extensive reductions, it makes an elegant general treatment difficult. Hence, it seems desirable to return to Herbrand's treatment which is again connected intimately with his Property C and Property B.

Both for these reasons and for the additional reason that Herbrand's dissertation contains a wealth of relevant material which has been largely overlooked hitherto in the literature, we shall reproduce here in part lecture notes given at Oxford in the Michaelmas term of 1960 on Herbrand's dissertation, especially on the Properties B and C. His Property A also appears interesting and has been revived in Ackermann's book ([1], p.93), but will not be discussed here because we do not understand fully its implications for mechanization.

6. Herbrand's Dissertation

6.1 Herbrand's System H. The primitive logical constants are $\neg$, $\vee$, $(+v)$ (or $\forall v$), $(-v)$ (or $\exists v$), with $\supset$, $\vee$, $\equiv$ defined in the usual manner.

To avoid confusion, we shall use $p,q,r,...Fx$, Gxy, ... as atomic formulas and X, Yx,... as arbitrary formulas which may or may not be atomic. By a "tautology" we shall mean a formula that is always true according to the customary interpretations of truth-functional (Boolean) connectives, abstracting from an analysis of parts which contain quantifiers.

The system H contains six rules ([11], pp.31—32).

RT. Rule of tautology. Every quantifier-free tautology is a theorem. (For example, $p \supset p$, although not $\forall xFx \supset \forall xFx$, falls under this.)

RI. Rules of inversion. Given a theorem X of H, we get another theorem of H if we replace within X a part which is of one of the following forms by its dual:

$$\begin{array}{ll} \neg\neg Y & Y \\ \neg(\pm v)Yv & (\mp v)\neg Yv \\ (\pm v)(Yv \vee Z) & (\pm v)Yv \vee Z \end{array}$$

Z not containing v.

RU. Rule of universal generalization. $Xxx \rightarrow +yXyy$ ("$\rightarrow$" for infer).

RE. Rule of existential generalization. $Xxx \rightarrow -yXxy$.

RC. Rule of contraction. $X \vee X \rightarrow X$.

RD. Rule of detachment (cut, modus ponens). $X, X \supset Y \rightarrow Y$.

The difference between RU and RE can be brought out by:

$$x = x \rightarrow +y(y = y)$$
$$x = x \rightarrow -y(x = y)$$
$$x = x \not\rightarrow +y(x = y).$$

The first important result is a direct proof of the following ([11], p.36).

6.2. Theorem 1. *Every tautology is a theorem of H; in other words, if we substitute quantifier expressions for parts in RT, we again get theorems of H.*

6.2.1. $X \vee \ldots \vee X \rightarrow X$.

6.3. Positive and Negative Parts. It is familiar that every formula can be brought into a prenex normal form with a matrix in the conjunctive (or disjunctive) normal form:

$$(\pm v_1)\ldots(\pm v_n)\ Xv_1\ldots v_n$$

such that X is in, say, a conjunctive normal form.

If we wish to determine whether a quantifier turns into $+$ or $-$, or whether an atomic proposition (an occurrence of it) gets negated or not in such a normal form, we do not have to carry out the transformation but may "calculate" directly by using the notion of positive and negative parts.

6.3.1. Signs of occurrences of propositional parts ([11], p.21 and p.35).

(a) The sign of the occurrence of X in X is $+$.

(b) The sign changes from Y to $\neg Y$; i.e., if an occurrence of $\neg Y$ in X has one sign, the same occurrence of Y in X has the other sign.

(c) The sign does not change from $X \vee Y$ to X or Y.

(d) The sign does not change from $(\pm v)\, Xv$ to Xv.

6.3.2. A positive occurrence of $+v$ or $-v$ remains $+v$ or $-v$ in a prenex form; a negative occurrence of $+v$ or $-v$ becomes $-v$ or $+v$.

6.3.3. When a formula X without quantifiers is transformed into a conjunctive normal form Y, the sign of each occurrence of each atomic formula is preserved; in particular, a negative or a positive occurrence of e.g., a letter p turns respectively into one or more occurrences of p preceded by $\neg$ or not.

6.3.4. If p has only positive occurrences in $X(p)$, then

$$\vdash_H (p \supset q) \supset (X(p) \supset X(q));$$

if p has only negative occurrences in $X(p)$, then

$$\vdash_H (p \supset q) \supset (X(q) \supset X(p)).$$

Proof by induction on the number of logical constants in X. ([11], p.36.)

6.3.5. General and restricted variables. A quantified variable is general if it is $+$ and begins a positive part, or $-$ and begins a negative part; in the other two cases it is called a restricted variable. ([11], p.35.)

6.3.6. Miniscope form. A formula is in the miniscope form if the quantifiers cannot be driven inwards any further by the rules of inversion RI.

Each formula has a unique miniscope form.

6.4. Champs (sections, cycles). In a given formula, we may replace each general variable v by an indexing function (a Skolem function, an Herbrand function) with the restricted variables governing v in the formula as arguments. Since ordinary functions (the descriptive functions) add no basic complications, we shall exclude them. Functions for ungoverned general variables are (indexing) constants, they are always contained in C_2 as defined below.

For an arbitrary set S of indexing functions, we can define the associated champs as follows. The champ of order 1, C_1, contains only one fixed object a_1, which may be taken as the number 1. Given the champs $C_1,\ldots,C_k$, C_{k+1} consists of all and only the indexing constants and the values of the indexing functions in S with arguments drawn from $C_1,\ldots, C_k$. We shall assume that the values of different functional expressions are always different. Thus, given a set of indexing functions, we can always determine the number n_k of the members of the union of $C_1,\ldots,C_k$.

7. Property *B* and Property *C*

Given an arbitrary formula, we replace each general variable by a function of all the restricted variables which govern it. Then we can define the champs and a function n_k which gives the number of numbers occurring in $C_1,\ldots,C_k$.

7.1. The reduced form of order k of a formula X. Let $N=n_k$.

(1) An atomic formula is its own reduced form.

(2) If the reduced forms of Y and Z are Y^* and Z^*, those of $\neg Y$, $Y \vee Z$ are $\neg Y^*$, $Y^* \vee Z^*$.

(3) If $+x$ or $-x$ is a general variable and the reduced form of Yx is Y^*x, that of $+xYx$ or $-xYx$ is $Y^*x(y_1,\ldots,y_m)$, where $y_1,\ldots,y_m$ are the restricted variables which govern x.

(4) If $-x$ is restricted, the reduced form of $-xYx$ is:

$$Y^*1 \vee \ldots \wedge Y^*N.$$

If $+x$ is restricted, that of $+xYx$ is:

$$Y^*1 \wedge \ldots \wedge Y^*N.$$

7.2. A proposition X has Property C of order k if its reduced form of order k is a tautology. It has Property B of order k if its miniscope form has Property C of order k.

A very important theorem in Herbrand's dissertation ([11], pp.101—105) is:

Theorem 2. *If a proposition has Property C (or B) of order k, then every proposition derivable from it by the rules of inversion RI has Property C (or B) of order k.*

The first rule for negation exchanging Z and $\neg\neg Z$ obviously makes no difference.

The rules of exchanging $(\pm v) \neg Zv$ and $\neg(\mp v) Zv$, $(\pm v)$ $(Yv \vee Z)$ and $(\pm v)$ $Yv \vee Z$ do not affect the reduced forms when $(\pm v)$ is a general variable, since a general variable remains general by the transformation.

When $(\pm v)$ is a restricted variable, the rules of negation do not affect Property C because of the definition of $\wedge$ in terms of $\vee$, from which we have:

$$(\neg Y1 \wedge \ldots \wedge \neg YN) \equiv \neg(Y1 \vee \ldots \vee YN)$$
$$(\neg Y1 \vee \ldots \vee \neg YN) \equiv \neg(Y1 \wedge \ldots \wedge YN).$$

The complex cases are with the rules of disjunction when $(\pm v)$ is a restricted variable.

Case (a). The exchange of $(-v)$ Yv $\vee$ Z and $(-v)$ $(Yv \vee Z)$.

Suppose U and V are the same except that an occurrence of $(-v)$ $Yv \vee Z$ in U is replaced by $(-v)$ $(Yv \vee Z)$ in V. Let U^* and V^* be the reduced forms of U and V of order k. We wish to show that U^* is a tautology if and only if V^* is.

The difference between U^* and V^* is that, for every general variable y in Z, we use $y(x_1,\ldots,x_n)$ to get U^*, $y(x_1,\ldots,x_n,v)$ to get V^*. The number n_k in the two cases are, say, N_1 and N_2, $N_1 < N_2$.

(a1). If V^* is a tautology, then U^* is one. If V^* is a tautology, then, in particular, if $y(x_1,\ldots,x_n,1),\ldots,y(x_1,\ldots,x_n,N_2)$ are all identified, the result is again a tautology, since we are thereby merely restricting the choice of truth values of certain atomic propositions. Hence, if we delete repetitions, we can derive U^* from V^*. Hence, U^* is also a tautology.

(a2). If some assignment falsifies V^*, then there is also some assignment which falsifies U^*. We choose for each falsifying assignment of V^* a falsifying assignment of a suitably chosen U^{**} which can be shown to be implied by U^*.

Since $x_1,\ldots,x_n$ are the (only) free variables in $(-v)\ Yv \vee Z$ and $(-v)(Yv \vee Z)$, the two parts may multiply their occurrences in U^* and V^* according to the values given to $x_1,\ldots,x_n$ from $\{1,\ldots, N_1\}$ and $\{1,\ldots,N_2\}$ respectively. Now we define U^{**} from U^* by choosing one number i from $\{1,\ldots,N_2\}$ for each fixed set of values of $x_1,\ldots,x_n$ from $\{1,\ldots,N_2\}$, and replacing $y(x_1,\ldots,x_n)$ in U^* by $y(x_1,\ldots,x_n,i)$. It is essential that U^{**} is modified to be over $\{1,\ldots,N_2\}$ rather than $\{1,\ldots,N_1\}$.

First, we make the choice of i for U^{**} for a fixed falsifying assignment T of V^* and a fixed choice of $(x_1,\ldots,x_n)$ from $\{1,\ldots,N_2\}$. The corresponding parts in U^{**} and V^* are:

$$Y^*1 \vee \ldots \vee Y^*N_2 \vee Z^*i \tag{1}$$

$$(Y^*1 \vee Z^*1) \vee \ldots \vee (Y^*N_2 \vee Z^*N_2). \tag{2}$$

If in the assignment T, (2) gets the value true, then either Y^*j is true for some j, and we can take any number, say 1 as i, or else, Z^*j is true for some j, and we choose the smallest such j to be i. If (2) gets the value false, then i can be any number, say 1. It is clear that (1) and (2) get the same truth value by the choice, and we can similarly choose an i for each set of values $x_1,\ldots,x_n$ from $\{1,\ldots,N_2\}$. In this way, it follows that the assignment T falsifying V^* also falsifies U^{**} which is like V^* except for containing parts like (1) in place of parts like (2).

Now we have to show that U^* implies U^{**}. Just the replacement of $y(x_1,\ldots,x_n)$ by $y(x_1,\ldots,x_n,i)$ makes no difference since we merely give the former a new name. But the fact that U^* is relative to $\{1,\ldots,N_1\}$, but U^{**} is relative to $\{1,\ldots,N_2\}$, $N_1 < N_2$, means there are more clauses in U^{**} than in U^*. If we look at (4) of 7.1, we see that a positive occurrence of $(-x)\,Wx$ is replaced by a disjunction, a negative occurrence of $(+x)\,Wx$ is replaced by a conjunction. Hence, using 6.3.4, we can get $U^* \supset U^{**}$ from:

$$(W1 \vee \ldots \vee WN_1) \supset (W1 \vee \ldots \vee WN_2)$$
$$(W1 \wedge \ldots \wedge WN_2) \supset (W1 \wedge \ldots \wedge WN_1)$$

Therefore, a falsifying assignment of V^* yields one for U^*, i.e., V^* is a tautology if U^* is.

By (a1) and (a2), we have proved Theorem 2 for Case (a).

Case (b). The exchange of $(+v)\ Yv \vee Z$ and $(+v)(Yv \vee Z)$.

The proof is similar to Case (a) except that in choosing i for U^{**}, we use:

$$(Y^*1 \wedge \ldots \wedge Y^*N_2) \vee Z^*i \tag{1*}$$

$$(Y^*1 \vee Z1) \wedge \ldots \wedge (Y^*N_2 \vee Z^*N_2). \tag{2*}$$

If (2*) gets true in the assignment T, then take any number, say 1, as i. If (2*) gets false, then $(Y^*j \vee Zj)$ must be false for some j, take the smallest such j as i.

This completes the proof of Theorem 2.

It is clear that, from this theorem, we can use a prenex form or the miniscope form and retain the same order k of Property C (oor Property B).

8. Herbrand's Theorem

The fundamental theorem proved by Herbrand ([11], pp.112—113) is somewhat different from what is commonly known as Herbrand's Theorem:

Herbrand's fundamental theorem

(i) *If X has Property B of order k, then* $\vdash_H X$; *one can effectively find the proof of X from the number k.*

(ii) *Given a proof in H of a proposition X, we can effectively find a number k such that X has Property B of order k.*

By Theorem 2, if X has Property B of order k, a prenex form of X has Property B of order k and can be proved in H with the help of 6.2.1, in a manner which is familiar nowadays. Hence, by RI, we get a proof of X in H. And (i) is proved.

To prove (ii), we make induction on the number of steps in the proof.

An axiom must be a case of RT and has Property B or C of order 1.

By Theorem 2, RI does not affect the order of Property B or C. RU does not change the reduced form.

If Xxx has Property C of order k with $n_k = N_1$, then $(-y)Xxy$ has Property C of order k or $k+1$. We may assume Xxx in the prenex form (by Theorem 2), and then the reduced form of $(-y)Xxy$ of order $k+1$ must be a disjunction $X^*j1 \vee \ldots \vee X^*jN_2$ which includes all the disjuncts of the reduced form of Xxx of order k, since $N_1 < N_2$. For example, $(-z)(Gzx \vee \neg Gvx)$ is of order 1 since its reduced form $G12 \vee \neg G12$ is a tautology; $(-y)(-z)(Gzx \vee \neg Gvy)$ is of order 2 since its reduced form $(G12 \vee \neg G11) \vee (G22 \vee \neg G11) \vee (G12 \vee \neg G12) \vee (G22 \vee \neg G12)$ is a tautology and y does take the fixed value 2 of x. Hence, RE preserves Property B or C with the possible increase of the order by 1.

RC does not change the order of Property B because, given a falsifying assignment of the reduced form of X, we get one for that of $X \vee X$ by the same assignment of truth values to the atomic formulas.

RD preserves the order of Property B. First, if X and Z have Property B of order k, then $X \wedge Z$ has the same order. Thus, take the reduced form $X_1 \vee \ldots \vee X_j$ of X and $Y_1 \vee \ldots \vee Y_t$ or Y. Then the reduced form of $X \wedge Z$ clearly contains all disjuncts $X_a Y_b$, $1 \leq a \leq j$, $1 \leq b \leq t$, and is therefore a tautology.

Hence, if X and $X \supset Y$ are of order k, so is also $X \wedge (X \supset Y)$. Let X be $(\pm y_1 \pm \cdots \pm y_p)\ My_1 \cdots y_p x_1 \cdots x_n$, and Y be $(\pm z_1 \pm \cdots \pm z_q)\ Nz_1 \ldots z_q x_1 \ldots x_n$. To avoid conflicts of variables, we replace $y_1, \ldots, y_p$ in X by $u_1, \ldots, u_p$ to get X_1, and consider $X \wedge (X_1 \supset Y)$. To get this into a prenex form, we can pair off u_i with y_i, one of which must be positive while the other is negative. We put them pairwise at the beginning, with the negative preceding the positive in each case. Let the negative one of the pair y_i and u_i be v_i and the other be w_i, then we have:

$$(-v_1 + w_1 - \ldots - v_p + w_p \pm z_1 \pm \ldots \pm z_q)$$
$$[My_1 \ldots y_p x_1 \ldots x_n \wedge (Mu_1 \ldots u_p x_1 \ldots x_n \supset Nz_1 \ldots z_q x_1 \ldots x_n)].$$

Now we form a champ of order k for this formula, and identify suitable elements in it to get one for Y. Thus all elements $w_{i+1}(v_1, \ldots, v_i)$, with arbitrary $v_1, \ldots, v_i$, are identified with v_{i+1}; all elements $z_i(v_1,\ldots,v_p,z_{i1},\ldots,z_{ij})$ with a fixed $z_i(a_1,\ldots,a_p,z_{i1},\ldots,z_{ij})$.

In this way, we get a champ for Y of order p since the only indexing functions are those for the negative quantifiers among $z_1,\ldots,z_q$ plus the indexing constants for the free variables $x_1,\ldots,x_n$.

By hypothesis, $X \wedge (X \supset Y)$ has Property B of order p and has, therefore, a disjunction:

$$[X_1 \wedge (X_1 \supset Y_1)] \vee \ldots \vee [X_N \wedge (X_N \supset Y_N)]$$

which is a tautology. After the identifications, we get a new tautology in which certain disjuncts are repeated and can be deleted. Suppose the formula is the result after the deletions, then we can derive by the propositional calculus:

$$Y_1 \vee \ldots \vee Y_N.$$

But then this tautology is essentially the reduced form of order k of the formula Y.

This completes the exposition of Herbrand's fundamental theorem.

Recently Peter Andrews and Burton Dreben discovered in Herbrand's thesis (see their forthcoming paper "Some difficulties in Herbrand's Theorem") a subtle mistake which is preserved in the above exposition in the sketched proof of Case (b) on p.17. We have to replace the two occurrences of "order k" in the statement of Theorem 2, on p.15 above, by "some finite order." The fundamental Theorem remains correct except that the detailed calculations of the order of Property B become more complex. Dreben is planning a thorough discussion of these questions. (October, 1962).

References

[1] ACKERMANN, W., *Solvable Cases of the Decision Problem.* Amsterdam, 1954.

[2] Goodstein, R.L., *Recursive Number Theory.* Amsterdam, 1957.

[3] Péter, R., *Rekursive Funktionen.* 2nd edition, Budapest, 1957.

[4] Quine, W.V., *Mathematical Logic.* Revised Edition, Cambridge, Mass., 1950.

[5] Surányi, J., *Reduktionstheorie des Entscheidungsproblems.* Budapest, 1959.

[6] Davis, M. and H. Putnam, "A Computing Procedure for Quantification Theory." *Journal ACM*, vol.7 (1960), pp.201—215.

[7] Dunham, B., R. Fridshal, G.L. Sward, "A Nonheuristic Program for Proving Elementary Logical Theorems." *Proceedings I.C.I.F.*, Paris, 1959 (pub. 1960), p.284.

[8] —, J. H. North, *Exploratory Mathematics by Machine* (to be published in a symposium volume at Perdue University).

[9] Gelernter, H., "Realization of a Geometry Theorem Proving Machine." *Proceedings I.C.I.F.*, Paris, 1959 (pub.1960).

[10] Gilmore, P.C., "A Proof Method for Quantification Theory: Its Justification and Realization." *IBM Journal*, vol.4 (1960), pp.28——35.

[11] Herbrand, J., *Recherches sur la Théorie de la Démonstration.* Warsaw, 1930.

[12] —, "Sur le Probléme Fondamental de la Logique Mathématique." *CR*, Warsaw, No.24 (1931).

[13] Hintikka, K. J. J., "Vicious Circle Principle and the Paradoxes." *Journal of Symbolic Logic*, vol.22 (1957), pp.245—248.

[14] Minsky, M., "Steps Toward Artificial Intelligence." *Proceedings I.R.E.*, vol.49 (1961), pp.8——30.

[15] Newell, A., J.C. Shaw, H.A. Simon, "Empirical Explorations of the Logical Theory Machine: A Case

Study in Heuristics," *Proceedings W.J.C.C.*, (1957), pp.218——230.

[16] Prawitz, D., H. Prawitz, N. Voghera, "A Mechanical Proof Procedure and its Realization in an Electronic Computer." *Journal ACM*, vol.7 (1960), pp.102—128.

[17] —, "An Improved Proof Procedure." *Theoria*, vol.26 (1960), pp.102—139.

[18] Robinson, A., "On the Mechanization of the Theory of Equations." *Bulletins of the Research Council of Israel*, vol.9F, No.2 (Nov. 1960), pp.47—70.

[19] Shepherdson, J., *The Principle of Induction in Free Variable Systems of Number Theory*. (Lecture at the Polish Academy, Spring, 1961, to be published).

[20] Skolem, Th., "Begründung der elementaren Arithmetik." Kristiania, 1923, 38 pp.

[21] Wang, H., "A Variant to Turing's Theory of Computing Machines." *Journal ACM*, vol.4 (1957), pp.63—92.

[22] —, "Toward Mechanical Mathematics," *IBM Journal*, vol.4 (1960), pp.2—22.

[23] —, "Proving Theorems by Pattern Recognition." Part I, *Communications ACM*, vol.3 (1960), pp.220—234; Part II, *Bell System Technical Journal*, vol.40 (1961), pp.1—41.

7.2 The mechanization of mathematical arguments *

1. Interaction of logic and computers

The attempt at theorem-proving by machine owes its chief attraction to the prospect of delegating more and more of the tedious part of mathematical research to machines, while reserving the conceptual innovations and designs of plots to people. Unthinking scepticism against this sort of enterprise may perhaps be alleviated a little if we think of developing mechanical assistants or collaborators for mathematicians. Once a stock of mechanical methods is available and when it is a question of finding significant new results, we shall certainly not wish to have the machines run a handicap race but will readily feed them with all suggestions we can think of.

The pursuit of mechanizing mathematical arguments has quite immediate consequences in computer and logical research.

1.1. The demands in this area of extensive manipulations with nonnumerical data in a systematic manner suggest new principles of organizing computers and programming languages.

1.2. The requirements of mechanization introduce a new criterion of classifying mathematical arguments.

1.3. It provides an incentive to formalize more thoroughly concepts, theorems, and rules of inference, with the guidance of a more objective mechanical norm.

1.4. It yields an external standard of applicability on machines for evaluating mathematical results having to do with decision and proof procedures; one does not desert a useful domain just because at one stage the work gets messy and offends the good sense of mathematicians.

1.5. Concern with machines leads to a tendency to be entirely literal and explicit which is helpful for certain types of research; for example, this factor is helpful in obtaining some new results on the decision problem and the reduction problem such as

* First published in *Proceedings of Symposia in Applied Mathematics*, (1963), Vol. 15, pp 31—40. Reproduced by permission of the American Mathematical Society.

those reported in [15].

A more immediate example is a new view of the project of a systematic development of mathematics as pursued by Frege, Peano, Whitehead-Russell, Quine, Bourbaki, Rosser. In all these cases, one is forced to a choice of two evils either stop at the very early stage or sacrifice rigor. After a development of rudimentary set theory and arithmetic Rosser contented himself with a brie sketch and a pious hope [23, p.517] that number theory and analysis can be obtained. Bourbaki, who uses a different approach with special emphasis on abstract structures, explicitly stresses the importance of "abuses of language" for mathematicians and appears to think of them like the boxer's gloves, to borrow a simile from Hilbert. The prospect of genuine mechanization recommends to us a whole-hearted pursuit of the project with consistent efforts to analyze and rende harmless abuses of language. Even though it is probably natural and more efficient for people to misuse language in mathematics, it is certainly not too much to believe that the patience of machines is sufficient to compensate the loss in efficiency caused by being denied the ability to misuse language, i.e., to vary the meanings in different contexts without explicit declarations of intention.

A major theoretical undertaking being contemplated is to redo the Bourbaki books more exactly in the system H of [28, p.237].

There is nothing to prevent us from mechanizing abbreviations and derived rules of inference. It is somewhat harder to get rid of voluntary misuses of language. An even harder task is to formalize the abilities needed to fill in the gaps which are normally left to the intelligent reader. All these three aspects are most conveniently considered in connection with the undertaking just stated.

In terms of writing machine programs to do specific works, the lack of a generally accepted programming language makes accumulation and communication rather difficult. In addition, the specific methods underlying the particular programs are also not easy to compare.

An example of results obtained to date is a program which embodies a fairly general procedure along familiar lines that turn out to produce proofs in a few minutes for all the over 350 theorems in the predicate calculus with equality actually included in the book *Principia Mathematica*. The most interesting lesson from these results is perhaps that even in a fairly rich domain, the theorems actually proved are mostly ones which call on only a very small portion of the available resources of the domain.

Usually a good deal of theoretical work ought to precede the writing of a particular program. There are, however, several methods which appear to be ripe for the preparation of machine programs. (For longer discussions of these, compare later sections of this paper.)

1.6. Decision procedure for the EA_2E case.

1.7. Free variable systems for algebra and rudimentary number theory.

1.8. Reduction procedure of the decision problem to games with dominoes.

1.9. Modification of the semidecision procedure of Friedman.

1.10. Modification of the semidecision procedure through AEA formulas.

1.11. Dunham's fast test of Herbrand expansions.

2. Types of application

There are simple mathematical results which depend essentially on certain combinatorial considerations well within the range of the capacity of existing machines. An example is the various questions of independence in complete and partial systems of the propositional calculus. In such areas the possibility of using machines as an aid is quite obvious.

For more serious questions in combinatorial analysis and number theory, one can often experiment with special cases and employ the results as a heuristic aid to the study of more general statements. Questions of this sort have been discussed extensively by Ulam.

What we wish to discuss here is rather a different type of possibility which may be said to be nonnumerical in so far as mathematical propositions and inferences are dealt with directly. In this connection the knowledge about formalization accumulated in mathematical logic is especially relevant. Thus, for example, we have decidable regions of logic and mathematics, and unbounded proof procedures which can be mechanized.

The major problem is, however, that the decidable domains are usually not broad enough to include sufficiently interesting mathematical domains, while the proof procedures are too inefficient to yield mechanically proofs of significant theorems even by large computers. This is not very surprising since mathematical logic has been developed with an interest more in theoretical mechanizability than in practical performability. It is to a certain extent for the same reason that mathematical logic has not been useful in increasing the mathematician's ability to discover new theorems.

This defect is not intrinsic to the pursuit of formalization but is a result of the additional motive, natural to every mathematician, of achieving theoretical elegance in the proofs of decidability or completeness, as well as economy in the basic apparatus. It is familiar that such elegance and economy is often bought at the price of slower speed and more cumbersome operations. For example, Turing machines are theoretically much more elegant than the commercial machines currently in use. A mathematician, in trying to prove a new theorem, does not go back to the logician's axioms.

Hence, what we have to pursue is formalization with special attention to practical feasibility. This requires a kind of work that is like a mixture of mathematics, physics, and engineering. It is like mathematics in so far as we strive for decision or proof procedures which are certain and general. But the additional constraint of feasibility means that there is an external criterion which often contradicts the aesthetic sense of a mathematician. In fact, one would have to work quite hard to get results which are by the current mathematical standard nothing new. Since, at least at the present stage, this type of work is not easier than more orthodox research, one is troubled by the lack of incentive to venture into a field of research which is not immediately rewarding.

An intermediate type of work is developed in R.W. Hamming's recent book on numerical analysis in which methods are given for selecting calculation methods to apply as called for by the problem on hand.

3. Decision procedures

We may wish to adapt known decision procedures to the task of proving theorems. One example that appears to be of some general interest is the result from the 1930's that there is a general procedure by which, given any formula of the form

$$Ex\cdots EyAuAvEz\cdots EwMx\cdots yuvz\cdots w,$$

we can decide whether the formula has any models at all.

From this result it follows that if in a formal discipline without function symbols we have a theorem B which can be deduced from the axioms $A_1, \cdots, A_n$, then we can decide whether

(1) $$(A_1 \wedge \cdots \wedge A_n) \rightarrow B$$

is true, provided each of $A_1, \cdots, A_n$ contains no more than two (consecutive) universal quantifiers and B contains no more than two (consecutive) existential quantifiers. This is so because (1) is a theorem of the predicate calculus if and only if its negation

(2) $$A_1 \wedge \cdots \wedge A_n \wedge \neg B$$

has no model at all. Since $AxGx \wedge AxHx$ is equivalent to $Ax(Gx \vee Hx)$, the prenex normal form of (2) falls under the $E\cdots EAAE\cdots E$ case if $A_1, \cdots, A_n$, $\neg B$ fall under the case.

In set theory, there appear to be fairly interesting theorems with proofs falling under the form (1). Thus, among the permissible axioms is the axiom of extensionality. As to the axioms of set existence, we may begin with an axiom for the empty set $EyAx(Fxy \equiv x \neq x)$, F for the membership relation. We obtain a collection of sets and axioms of the required form: given two sets in the collection, their pair set is in the collection; given a set in the collection, its power set and sum set are in the collection; in general, given a set in the collection, any subset with a defining condition containing only one quantifier belongs to the collection. In fact, any axiom with a one-quantifier defining condition belongs to the collection. This can be employed to experiment with possible selections from axioms of the form $EyAx(Fxy \equiv -x-)$. It is essential that no axiom of infinity can be of the required form. One application is to a paper by Hintikka [30, p.37] and it is likely that there are other similar applications.

The equality sign is included. If we wish to apply the decision procedure to geometry, it is desirable that we extend the procedure to a many-sorted theory, to take care of points, lines, angles, etc. In fact, for certain purposes, it is better to extend the predicate calculus to include ordered pairs and n-tuples, as well as certain fundamental operations on predicates such as taking the intersection, etc., perhaps also the Hilbert selection symbol.

The particular case is not useful for number theory because functions are not permitted, and in order to represent $g(x)=y$ by Gxy, we need

$$AxAyAz[(Gxy \wedge Gxz) \supset y=z]$$

as an axiom, which is no longer of the given form.

4. Free variable systems

While it seems advisable to study axiomatic set theory at first without reference to function, number theory and algebra may be studied initially with quantifiers excluded. An underlying region is the study of algebraic manipulations, in particular, the question when two algebraic expressions are reducible to each other. A large part of the secondary school education on mathematics is to teach this stock of trade which is useful in algebra, analysis, and physics. This calls for a basic language with a set of organized rules which are natural to anybody with a reasonable mathematical education.

With a moderately good solution of this problem, it is a small step to set up a system to deal with, e.g., trigonometric identities. One can use a free variable system (propositional calculus, equality, perhaps two sorts of variables) and supply a decision procedure by a reduction to some normal form.

To deal with number theory is more serious. Following Skolem, we can develop a free variable system that allows proofs by induction, as well as arbitrary primitive recursive definitions of functions and predicates. This, however, does not quite give us what we want, because it takes a long time to get to the simple theorems which ordinarily we do not bother to prove. Instead, we normally begin with a body of familiar theorems and strategies. Hence, we are faced with the problem of setting up a stock-of-trade system. As a start, we may try to analyze theorems such as there are infinitely many primes, the fundamental theorem of arithmetic, and 2 is not a rational number. This would involve an unsatisfactory empirical experimentation more like Peano than Dedekind. Perhaps Grassmann's work and Peano's work can be of some assistance here.

5. Proof verifiers, proof formalizers, proof discoverers

In order to restrict one's task at the beginning, one might wish to consider first the question of writing a program that will verify that a proposed proof is indeed a proof. This seemingly attractive idea is somewhat confused. If we are concerned with completely formal proofs, the project is of little theoretical interest and would be something like a FORTRAN editor with the additional disadvantage that the verifier has little application. If a verifier checks just things like punctuation, passing the test does not assure us that the result is a proof. If the test is to give the assurance, then since practically no proofs are completely formalized, in order to have a reasonable range of application, the verifier must be able to fill in gaps and even decide whether certain gaps can be filled in.

For a proof to be formal does not mean that each step must be a very small one recognized in standard axiomatic systems devised by logicians. It is perfectly all right that bigger steps are taken, provided we somehow possess a list or a procedure of testing all the permissible steps. Since we permit a large variety of moves, for a checking to be feasible, we have to have an organized list of permissible steps so that we do not have to check off slavishly each item on the list one by one.

As a result, there is no way to distinguish a reasonably interesting proof verifier from a proof formalizer that proposes to supply a formal proof starting from a rather detailed sketch. In either case, the more serious problem is not so much the writing of a program as the devising of the organized body of permissible steps and strategies for checking them.

In this connection we ought to introduce a criterion for judging work in this area. Roughly, the work is not determined just by a selection of the things which a program can do but, more important, by how the things are done. In other words, a good piece of work automatically advances our knowledge and produces something new that can be embodied in our accumulation of scientific information. This in particular excludes strictly ad hoc devices introduced just to produce a small number of desired results.

Viewed in this light, a proof formalizer is an unavoidable integral part of a proof discoverer, and the difference can only be one of degrees in complexity and significance.

6. Mathematical problems as games of dominoes

The comparison of mathematics with games has often produced spirited controversies in philosophical discussions. It is, however, possible to say that mathematics can be viewed as games with symbols for the purpose of mechanization without having to defend the emotion-laden thesis that mathematics is just a game. It has been shown, for example, that the problem of proving a statement by certain axioms can be reinterpreted as one of telling whether a Turing machine will stop, which in turn may be treated as a question of whether one can win a suitable game with dominoes.

We assume there are infinitely many square plates (the dominoes) of the same size (say, all of the unit area) with edges colored, one color on each edge but different edges may have the same color. The type of domino is determined by the colors on its edges and we are not permitted to rotate or reflect any domino. There are infinitely many pieces of every type. The game is simply to take a finite set of types and try to cover up the whole first quadrant of the infinite plane with dominoes of these types so that all corners fall on the lattice points and any two adjoining edges have the same color.

6.1. A (finite) set of domino types is said to be solvable if and only if there is some way of covering the whole first quadrant by the dominoes of these types.

The following simply stated problem which is two years old remains open:

6.2. The (unrestricted) domino problem. To find an algorithm to decide, for any given (finite) set of domino types, whether it is solvable.

On the other hand, it is not hard to show that if we require a fixed type to occur at the origin, the (origin-constrained) domino problem is unsolvable because to every Turing machine we can find a set of domino types such that the former stops if and only if the latter has no solution. Hence, to each mathematical problem we can find a question of solving a corresponding finite set of domino types.

It turns out that the result can be extended to the diagonal-constrained domino problem and then yield a reduction of the general decision problem to that for the class of all formulas of the form $AxEuAyMxuy$ or $AxAyMxx'y$, x' being short for $x+1$ (see [15]). This reduction to the AEA case makes it appear plausible that we can devise more powerful decision procedures with the help of our better intuitive grasp of the simpler structure. In fact, we can get a semidecision procedure (see [30, pp.25—32]).

7. Semidecision procedures

The reduction procedure depends on the expansion of a formula F to a sequence S_F of truth-functional formulas and replaces the predicates in F by predicates connecting the formulas in S_F. Moreover, we can modify the reduction procedure so as to preserve finite models; every formula in the reduction class has a recursive model if it has any model at all.

The semidecision procedure is such that it always terminates if the original formula F either has no model (i.e., its negation is a theorem) or has finite models. In fact, it is easy to show that an *AEA* formula has no model (finite model) if and only if the corresponding semidecision procedure has no solution (a periodic solution). Moreover, the procedure can also decide a number of cases when the formula has only infinite models. Thus, when we recognize that apart from the first few rows or columns the rest has a periodic solution, we get a formula with only infinite solutions. It is also possible that we can find conditions which enable us to recognize formulas with only more complex infinite models. To accomplish this, we may wish to experiment with examples on a machine program to look for new recognizable patterns of recursion.

When we deal with the general case, we find that often, for a given F, too many terms of S_F are needed before we can even recognize that F has no model at all. It, therefoer, seems desirable to look for some relevance algorithm such as some generalization of the considerations on [30, p.18]. Since, however, it is not easy to find a good one for the general case, it is natural to deal with this problem relative to some normal form. And the semidecision procedure may be viewed as a relevance algorithm using the *AEA* normal form of each formula F. This is more than a proof procedure since it does not only give an answer when F has no model. It is not a decision procedure because we cannot tell in advance the exact range to which it applies. But it does include all classes known to be decidable in the sense that any formula falling under one of the known decidable classes can be decided by this procedure. The belief is that, in addition, the procedure is more efficient than the existing proof and decision procedures for the predicate calculus and its subdomains.

Recently it came to the author's attention that Joyce Friedman has, since 1956, introduced in her unpublished report [10] a very nice semidecision procedure which applies directly to formulas in the Skolem normal form: for formulas belonging to most of the classical decidable cases, the procedure yields a decision procedure, and for all formulas, the procedure in general yields a reduction to a somewhat simpler form.

It seems desirable to make several modifications on the procedure: (1) to permit formulas in prenex form and miniscope form; (2) to use Herbrand's Property B; (3) to avoid "saturation" or, in other words, expansions for the "don't care" cases; (4) to include equality.

With these modifications, and combined with Dunham's fast test of Herbrand expansions, the procedure should be very powerful and efficient.

8. Methods of testing Herbrand expansions

Let us assume the Herbrand expansion H written in the disjunctive normal form. Compare every pair of clauses in H, whenever one contains some proposition letter p and the other contains $\neg p$, circle both p and $\neg p$. For example, if one clause is $\neg p \wedge q \wedge r$ and another clause is $p \vee \neg r \wedge s$, circle r and $\neg r_1 p$ and $\neg p_1$. After we have introduced all possible circles, we can delete every clause which contains a literal without a circle. Then we erase the circles and repeat the process until finally we arrive at a disjunction in which no clause contains any literal without a circle. Call the resulting disjunction minus the circles H^*. In his lecture [5], Dunham introduces the above procedure and proves:

Dunham's theorem. *If H is a tautology, so is H^*.*

This is clearly useful in eliminating irrelevant clauses. Dunham further discusses how we can efficiently carry out the above procedure on a computer. In particular, he envisages very long disjunctions and gives efficient ways of record-keeping and the use of magnetic tapes. A forward-looking feature is introduced to take advantage of the characteristics of the process to date in deciding what to do next, and also, perhaps in a nonvalidity determination. He stresses the *ordering* of the steps in the method of solution in order that we can anticipate storage needs and use the tapes without loss of time.

Details of these results will be fully described in the paper [8].

9. Direct characterization of recursive predicates: a digression

The consideration of models of AEA formulas suggests a type of recursive definition which uses predicates rather than functions. These definitions should be useful for studying models of formulas in the predicate calculus without function symbols. Although the relation to theorem-proving is rather indirect, we include it here on the ground that it is desirable to give a mathematical result in a paper for a mathematical symposium.

Definition. A sequential proposition is a proposition obtainable in the following manner. There are the individual constant 0 and infinitely many variables x, y, z, etc., and infinitely many predicates F, G, etc., including $=$. A term is either 0 or a variable, or obtained from a given term t by applying the successor operation, viz., t'. An atomic proposition is a predicate followed by a suitable number of terms. Applying truth functions to given propositions we get again propositons. Consequences of a set of sequential propositions are obtained by substitution of numerals $0^{1,\cdots,1}$ for variables, equals for equals, and truth-functional inferences.

Theorem. *A sequential proposition defines a set of recursive predicates if and only if:* (i) *it has a unique model relative to the natural interpretation of* 0 *and the successor function, or* (ii) *for every set of numerals $n_1,\cdots,n_j$ and every predicate P in the* *proposition, either* $p(n_1,\cdots,n_j)$ *or* $\neg P(n_1,\cdots,n_j)$ *is a consequence, but never both.*

This is related to 9.7 on p.316 of [27]. Some essential steps in the proof were suggested by Specker. We merely sketch the proof.

If a sequential proposition has a unique model, then we can try out, from small numerals to large, all possible truth values of the predicates involved followed by numerals. Since each predicate, unlike a function, has only two possible values (true or false) for each given set of arguments, we can, by the infinity lemma. narrow down the value of $P(n_1,\cdots,n_j)$ to the correct one after a finite number of trials, where $n_1,\cdots,n_j$ are given numerals. Hence, the predicates are recursive, and we can derive a unique truth value for $P(n_1,\cdots,n_j)$ in each instance.

Conversely, we can also simulate every general recursive definition by a sequential proposition as follows, using $Ax_1\cdots x_ny$ for $f(x_1,\cdots,x_n)=y$.

9.1. $Axy \equiv y=x'$.

9.2. $Ax_1\cdots x_nz \equiv z=c$.

9.3. $Ax_1\cdots x_nz \equiv z=x_i$.

9.4. Composition. $(Ax_1\cdots x_ny_1 \wedge \cdots \wedge A_mx_1\cdots x_ny_m \wedge By_1\cdots y_mu) \supset (Cx_1\cdots x_nv \equiv u = v)$.

Observe that C defines a function when $A_1,\cdots,A_m$, B define functions.

9.5. Primitive recursion.

$$[Bx_1\cdots x_nu \supset (Ax_1\cdots x_n0v \equiv u=v)] \wedge [(Ax_1\cdots x_nyv \wedge Cx_1\cdots x_nyvw) \supset (Ax_1\cdots x_ny'z \equiv z=w)].$$

9.6. Use $Ax_1\cdots x_nu$ to represent $u_yR(x_1,\cdots,x_n,y)=u$.

$$[Bx_1\cdots x_n0 \equiv Rx_1\cdots x_n0] \wedge [Bx_1\cdots x_ny' \equiv (Bx_1\cdots x_ny \wedge Rx_1\cdots x_ny')] \wedge [Ax_1\cdots x_n0 \equiv Bx_1\cdots x_n0] \wedge [Ax_1\cdots x_ny' \equiv (\neg Bx_1\cdots x_ny \wedge Bx_1\cdots x_ny')].$$

References

1. W. Ackermann, *Solvable cases of the decision problem*, North-Holland, Amsterdam, 1954.

2. N. Bourbaki, *Éléments de mathématique*, Hermann, Paris.

3. J.R. Büchi, *Turing machines and the Entscheidung sproblem*, Notices Amer. Math. Soc., **8**(1961), 354.

4. M. Davis and H. Putnam, *A computing procedure for quanification theory*, J. Assoc. Comput. Mach., **7**(1960), 201—215.

5. B. Dunham, *Theorem testing by computer*, Lecture at Harvard Computation Laboratory, February 15, 1962.

6. B. Dunham, R. Fridshal and G. L. Sward, *A nonheuristic program for proving elementary logical theorems*, Proc. Internat. Conf. on Information Processing, UNESCO, Paris, 1959 (published in 1960), p.284.

7. B. Dunham, R. Fridshal and J.H. North, *Exploratory mathomatics by machine*, Proc. Sympos, Decision and Information Processes, Macmillan, New York, 1961.

8. B. Dunham and J. H. North, *Theorem testing by computer*, Sympos. Math. Theory Automata, Brooklyn Polytechnic Institute, Brooklyn, N.Y., April, 1962.

9. G. Frege, *Grundgesetze der Arithmetik*, Jena, 1893 and 1903.

10. J. Friedman, *A semi-decision procedure for the functional calculus*, J. Assoc. Comput. Mach.,

10(1963), 1—24.

11. H. Gelernter, *Realization of a geometry theorem proving machine*, Proc. Internat. Conf. on Information Processing, UNESCO, Paris, 1959 (published in 1960).
12. P. C. Gilmore, *A proof method for quantification theory: its justification and realization*, IBM J. **4**(1960), 28—35.
13. H. Grassmann, *Lehrbuch der Arithmetik*, Berlin, 1861.
14. J. Herbrand, *Recherches sur la théorie de la démonstration*, Warsaw, 1930.
15. A. S. Kahr, Edward F. Moore and Hao Wang, *Entscheidungsproblem reduced to the AEA case*, Proc. Nat. Acad. Sci. U.S.A., **48**(1962), 365—377.
16. M. Minsky, *Steps toward artificial intelligence*, Proc. I.R.E., **49**(1961), 8—30.
17. A. Newell, J. C. Shaw and H. A. Simon, *Empirical explorations of the logical theory machine: a case study in heuristics*, Proc. Western Joint Computer Conference, IRE, New York, pp.218——230, 1957.
18. G. Peano, *Formulaire de mathématiques*, Turin, 1894—1908.
19. D. Prawitz, H. Prawitz and N. Voghera, *A mechanical proof procedure and its realization in an electronic computer*, J. Assoc. Comput. Mach., **7**(1960), 102—128.
20. D. Prawitz, *An improved proof procedure*, Theoria **26**(1960), 102—139.
21. W. V. Quine, *Mathematical logic*, Harvard Univ. Press, Cambridge, Mass., 1951.
22. A. Robinson, *On the mechanization of the theory of equations*, Bull. Res. Council Israel (9F) No.2(1960), 47—70.
23. J. B. Rosser, *Logic for mathematicians*, McGraw-Hill, New York, 1953.
24. Th. Skolem, *Begründung der elementaren Arithmetik durch die rekurrierende Denkweise ohne Anwendung scheinbarer Veränderlichen mit unendlichem Ausdehnungsbereich*, 38 pp., Skr. Videnskapsselskapet i Kristiania, I. Mat. -Naturv. Klasse No.6, 1923.
25. J. Suranyi, *Reduktionstheorie des Entscheidungsproblems*, Ungarischen Akademie, Budapest, 1959.
26. S. M. Ulam, *A collection of mathematical problems*, Interscience, New York, 1960.
27. H. Wang, *Circuit synthesis by solving sequential Boolean equations*, Z. Math. Logik Grundlagen Math. **5**(1959), 291—322.
28. ———, *Ordinal numbers and predicative set theory*, Z. Math. Logik Grundlagen **5**(1959), 216—239.
29. ————, *Toward mechanical mathematics*, IBM Journal **4**(1960), 2—22.
30. ————, *Proving theorems by pattern recognition*. I, Comm. ACM **3**(1960), 220—234; II, Bell System Tech. J., **40**(1961), 1—41.
31. A. N. Whitehead and B. Russell, *Principia mathematica*, Cambridge Univ. Press, Cambridge, 1910—1913.

7.3 Formalization and automatic theorem-proving*

General survey

The main purpose of this paper is to give some new examples of mechanizing proofs in number theory and quantification theory, which, it is believed, suggest a few additional steps to be taken to advance the area of mechanical mathematics. Before

---

* First published in ***Proceedings of the IFIP Congress***, pp 51—58. Spartan Books, 1965. Reproduced by permission of the author.

plunging into such technical matters, a somewhat incomplete summary of existing results is given to conform to the wishes of the organizers of the meetings.

There are different ways of using computers to assist the proving of theorems. The highly interesting work of Lehmer[23] in number theory consists essentially in reducing (by the human being) a general theorem to a number of complex numerical instances which are verified by machines. New theorems have been proved in this way which is certainly a clever application of machines. But this is not the sort of application envisaged here, because the computer is making indisputably only numerical calculations. It is not that logicians will insist on beginning exclusively with the Peano axioms. In fact, the writer also envisages an organization of the stock of trade in mathematics, as well as frequent interventions by people. What is different is the desire to extend the well-known power of computers in numerical calculations to the range of making logical inferences.

Among those who attempt to make this extension, two basic trends are discernible which might be labeled psychocentric and logocentric. The distinction is in principle not a sharp one. In practice, the psychocentric approach appears to have a deceptive glamour which generally leads to disappointments. At any rate, it would seem better to evaluate each individual piece of work on its own merit rather than indulge in debates of a political flavor.

The well-known work of Newell, Shaw, Simon[25,26] has had important by-products in the study of programming languages but, as far as theorem-proving is concerned, appears to be too ill-defined to permit any solid superstructure to be built on such foundations. Moreover, although it is likely that thinking on questions of mechanical simulation can be suggestive to a psychologist, it is not easy to see how the results of actually carrying out these ideas on computers can contribute in any substantial way to the progress of the field of psychology.

The geometry-proving scheme of Gelernter and his colleagues[17] offers results which were puzzling at first. But the careful analysis by Gilmore seems to show that much less had been accomplished than one was at first led to believe: "The programming that the authors have done is entirely conventional in the sense that well-defined algorithms have been realized on a computer" (Gilmore[19], p.26).

Slagle[34] has a successful program to have computers do integration problems in the calculus. It would be of interest to have a more systematic study of the whole problem of integration. The program by Bobrow[2] to have computers solve story problems in algebra is carefully worked out and explained. It is interesting that a few simple devices enable him to handle a moderately wide range of English sentences. However, the emphasis is on matters not central to the type of problem considered in this paper.

In the more logocentric approach, the main emphasis thus far is on quantification theory. Outside quantification theory, several people have worked on the Sturm-Tarski decision procedure for elementary algebra, in particular, Collins[5]. A. Robinson proposes, in a somewhat different direction, a system for the theory of equations[29,30]. Brown has developed an efficient practical system for algebraic manipulations[3].

The first serious attempt to do quantification theory on computers seems to date from the summer of 1958 as reported in Ref.36, with quite pleasantly surprising results. A lot of related work has been done since then, both on proof procedures* and on decision procedures for interesting subdomains**. It is encouraging to observe a slow but steady progress toward more and more efficient procedures.

The basic theoretical result for both approaches goes back to the work of Skolem and Herbrand, commonly known as Herbrand's Theorem. In order, however, to make the theoretically possible expansions practically feasible, a novel species of question arises which demands considerations of a rather high order of complexity.

In the area of decidable subdomains of quantification theory, Friedman[16] not only gives an efficient program for deciding whether a formula of the form

$$(x_1)\ldots(x_m)\,(Ey_1)\,(Ey_2)\,(z_1)\ldots(z_n)\;M$$

is a theorem, but also treats systematically the question of amplification (*don't care* clauses). Desirable immediate extensions are the inclusion of equality, and the removal of the restriction to dyadic (and monadic) predicates.

Among the proof procedures studied on machines, the most powerful methods thus far completed would seem to be Chinlund's et al.,[4] and Robinson's[31]. A more efficient procedure is given in Robinson[32]; similar ideas were independently reported in Aanderaa[1]. Neither method has been realized on a computer, although Aanderaa seems to specify more details to guide the programmer. It is desirable to add equality to both methods.

None of the programs completed so far have succeeded in dealing with the examples ExQ1, ExQ2, ExQ3 (in order of increasing difficulty) discussed below. It is hoped that these examples will be accepted as test cases for new programs.

A more general question is whether the method of Herbrand expansion is basically sufficient. It may be necessary to introduce special strategies which have no direct connections with testing Herbrand expansions. For example, the strategy of eliminating quantifiers by substituting Fu for $(Ex)\,(x=u \wedge Fx)$ or $(x)\,(x=u \supset Fx)$ is considered below.

In the case of number theory, little has been done to mechanize the arguments. Very likely, many people have tried and found the task too formidable. The section covering examples from number theory, below, will be devoted to a preliminary exploration of some of the possibilities. It appears that a suitable goal for the near future would be to mechanize the proofs in Skolem[33].

---

* See Refs.4,6—8, 18, 27, 28, 31, 32, 40.

** See Refs.14—16, 37.

The writer has speculated extensively elsewhere* on the future of automatic theorem-proving and will refrain from repetitions. Four short items seem worth stressing.

Although formalization is only part of the battle, since, for mechanization, we need in addition a method of choosing the next step, it is highly desirable to do more work on formalization, with a view to developing a repertoire of precise tricks. An attractive feature of this line of work is the possibility of avoiding computers. Hence, we have an area where one does not have to *get one's hands dirty*. This helps the division of labor.

It is unrealistic to expect practical methods which are fully "automatic" from the beginning, since the goal is to extend human ability by delegating more tedious steps to machines**.

It might be instructive to contrast mechanizability with teachability.

The fitting together of diverse methods into a more widely applicable structure presents many serious difficulties. For example, Boolean operations are cheap, but if we inject such operations at many different stages, it is essential that at each stage not too many such operations are required.

Examples from number theory

Given a formula of number theory to be proved, we assume it false and take a least counter-example LC, which is represented by an *ambiguous constant*. In general, an existence statement (Ex)Fx yields an ambiguous constant which is ambiguous but has the property F. Universal quantifiers are dropped and free variables are used. We shall use a, b, c, x, y, z, etc., for variables; m, n, k, x_m, y_m, etc., for ambiguous constants. The basic strategy is to derive properties about these ambiguous constants in the hope of finding contradictory ones. Variables range only over positive integers (excluding 0).

To make such an approach feasible, it is necessary to use a stock-of-trade file SF and a list TF of simple truth-functional transformations.

The four examples to follow are merely meant as illustrations of the sort of thing one may do. The proofs become increasingly more sketchy and less mechanical. It is possible to extend the methods sketched below to obtain a *complete* system of number theory in the sense that any theorem in the usual axiom system of number theory can in theory be proved, but this aspect will not be considered here. For example, to achieve repeated inductions, one needs, besides LC, the general choice of a least ambiguous constant.

The truth-functional principles (including one for equality) are for the moment confined to the following:

TF1. If A and $A \supset B$, then B.

TF2. If A, C, $(A \wedge C) \supset B$, then B.

TF3. If $A \supset B$ and $\neg B$, then $\neg A$.

---

* See Refs.35 to 39.

** Compare Ref.37, p.221, and more specifically Ref.31.

TF4. If $A \supset (B \vee C)$ and $\neg C$, then $A \supset B$.

TF5. If $A \equiv B$ and A, then B (usually for definitions).

TF6. Substitute equals for equals.

TF7. If $\neg A$, delete $A \supset B$.

The basic instructions (oon the first level of complexity) are as follows:

LC. Write down conditions for least counter-examples.

AC. Create ambiguous constants and substitute them for free variables in earlier lines of the proof.

TFL. Apply TF1—TF7, as far as possible, to lines obtained so far in the proof which contain no free variables.

TFG. Apply TF1—TF7 on the basis of both the lines of the proof and SF, but with the restriction that at least one of the premises is in the proof and without free variable, and that the consequence contains no functions or predicates not occurring in the proof so far.

SC. Search for contradictions in the lines of the proof obtained thus far.

A number of mechanizable implicit conventions are assumed. For example, addition and multiplication are commutative and associative. $x \nmid y$ and $x \nless y$ for $\neg x/y$, $\neg x < y$. $x < y < z$ for $x < y \wedge y < z$. $a=b$ and $b=a$ are treated as the same.

For the initial SF, we take arbitrarily the following, with much room for improvement:

B1. $a \nless a$; B2. $(a < b \wedge b<c) \supset a < c$;

B3. $a \leqq a$; B4. $ab = ac \supset b = c$; B5. $x \leqq y \equiv (x=y \vee x < y)$; B6. $x = y \vee x < y \vee y < x$; B7. $a \leqq b \supset b \nless a$.

D1. $x \mid x$; D2. $(x \mid y \wedge y \mid z) \supset x \mid z$; D3. $x \mid y \equiv y = xu_{xy}$; D4. $xy=z \supset (x \mid z \wedge y \mid z)$; D5. $x \mid y \supset x \leqq y$.

P1. $(Px \wedge x \mid yz) \supset (x \mid y \vee x \mid z)$; P2. $y>1 \supset [\neg Py \equiv (1<x_y < y \wedge x_y \mid y)]$.

As a first round of an attempted proof, we proceed as follows. Start: (1) LC, (2) AC, (3) TFL, (4) SC, (5) TFG, (6) SC, (7) return to (2) and repeat. Exit either when a contradiction is found at SC or when AC yields no new substitutions. In the latter cases, more complex appeal to SF is called for.

ExN1. $x > 1 \supset (Ey)(Py \wedge y \mid x)$

By LC, we get:

$$m>1 \tag{1}$$

$$Pb \supset b \nmid m. \tag{2}$$

$$1 < a < m \supset (Py_a \wedge y_a \mid a). \tag{3}$$

Apply AC:

$$Pm \supset m \nmid m. \tag{4}$$

$$1<m<m \supset (Py_m \wedge y_m \mid m). \tag{5}$$

TFL and SC produce nothing. Apply

TFG. By TF7 and B1, (5) can be deleted.

$$\neg Pm, \text{ by (4), D1, and TF3.} \tag{6}$$

$$\neg Pm \equiv (1 < x_m < m \wedge x_m | m), \text{ by (1), P2, TF1.} \quad (7)$$

By (6), (7), TF5:

$$1 < x_m < m. \quad (8)$$

$$x_m | m. \quad (9)$$

SC produces no result. Return to AC:

$$Px_m \supset x_m \nmid m, \text{ by (2) with } x_m \text{ for b.} \quad (10)$$

$$1 < x_m < m \supset (Py_{x_m} \wedge y_{x_m} | x_m), \text{ by (3) with } x_m \text{ for a.} \quad (11)$$

Apply TFL:

$$Px_m, \text{ by (9), (10), TF3.} \quad (12)$$

$$Py_{x_m}, \text{ by (8), (11), TF1.} \quad (13)$$

$$y_{x_m} | x_m, \text{ by (8), (11), TF1.} \quad (14)$$

SC produces no result. Apply TFG.

$$\neg Px_m \equiv (1 < u_{x_m} < x_m \wedge u_{x_m} | x_m), \text{ by (8), P2, and TF1.} \quad (15)$$

$$1 < u_{x_m} < x_m, \text{ by (12), (15), TF5.} \quad (16)$$

$$u_{x_m} | x_m, \text{ by (12), (15), TF5.} \quad (17)$$

$$y_{x_m} | m, \text{ by (9), (14), D2, TF2.} \quad (18)$$

$$u_{x_m} | m, \text{ by (9), (17), D2, TF2.} \quad (19)$$

SC produces no result. Return to AC:

$$Py_{x_m} \supset y_{x_m} \nmid m. \quad (20)$$

$$Pu_{x_m} \supset u_{x_m} \nmid m. \quad (21)$$

$$1 < y_{x_m} < m \supset (Py_{y_{x_m}} \vee y_{y_{x_m}} | y_{x_m}). \quad (22)$$

$$1 < u_{x_m} < m \supset (Py_{u_{x_m}} \vee y_{u_{x_m}} | u_{x_m}). \quad (23)$$

Apply TFL:

$$y_{x_m} \nmid m, \text{ by (20), (13), TF1.} \quad (24)$$

$$\neg Py_{x_m}, \text{ by (20), (18), TF3.} \quad (25)$$

$$\neg Pu_{x_m}, \text{ by (21), (19), TF3.} \quad (26)$$

Apply SC and we find (24) contradicts (18), (25) contradicts (13). Hence, ExN1 is a theorem of number theory.

In order to prove the next theorem, we extend SF by adding:

P3. $y > 1 \supset [\neg Py \equiv (1 < x_y < y \wedge Px_y \wedge x_y | y)]$.

B8. $(a|b \wedge a|(b+c)) \supset a|c$.

F1. $a \leqq b \supset a|b!$; F2. $1 < a! + 1$; F3. $a < a! + 1$; F4. $\Box a! + 1 \leqq a$.

ExN2. $Px \supset (Ey)(Py \wedge x < y \leqq x! + 1)$.

By LC:

$$Pm. \quad (1)$$

$$Pa \supset (a \leqq m \vee m! + 1 < a). \quad (2)$$

The least part happens to be useless and, for brevity, we omit it.

Apply AC:

$$Pm \supset (m \leqq m \vee m! + 1 < m). \quad (3)$$

$$P(m! + 1) \supset (m! + 1 \leqq m \vee m! + 1 < m! + 1). \quad (4)$$

TFL and SC produce nothing except:

$$m \leqq m \vee m! + 1 < m, \text{ by (1), (3), TF1.} \quad (5)$$

Apply TFG:

$P(m!+1) \supset m!+1 \leqq m$, by (4), B1, TF4. (6)

$\neg P(m!+1)$, by (6), F4, TF3. (7)

$1 < m!+1$, by F2 (not justified by restricted TFG). (8)

$1 < x_{m!+1} < m!+1$, by (8), (7), P3, TF1, TF5. (9)

$Px_{m!+1}$. (10)

$x_{m!+1} \mid m!+1$. (11)

SC produces nothing. Return to AC:

$Px_{m!+1} \supset (x_{m!+1} \leqq m \vee m!+1 < x_{m!+1})$. (12)

Apply TFL:

$m!+1 \nless x_{m!+1}$, by (9) (not explicitly provided for). (13)

$Px_{m!+1} \supset x_{m!+1} \leqq m$, by (12), (13), TF4. (14)

$x_{m!+1} \leqq m$, by (14), (10), TF1. (15)

SC produces no result. Apply TFG.

$x_{m!+1} \mid m!$, by (15), F1, TF1. (16)

$x_{m!+1} \mid 1$, by (16), (11), B8, TF2. (17)

$x_{m!+1} \leqq 1$, by (17), D5, TF1. (18)

$1 \nless x_{m!+1}$ by (18), B7, TF1. (19)

But (19) contradicts (9).

The next two examples suggest the need of more complex strategies such as ordering all occurring ambiguous constants, breaking up products, considering the separate cases $a=b$, $a<b$, $b<a$; and so on.

ExN3. $Pw \supset wx^2 \neq y^2$.

The application of LC is ambiguous since there are more than one free variables. We take the following:

Pk. (1)

$km^2 = n^2$. (2)

$(a<m \wedge b<n) \supset ka^2 \neq b^2$. (3)

There are various ways of applying AC, mostly with trivial results. TFL is rather useless. Go to relaxed TFG with the guiding principle of breaking products.

$k|n^2$, by D4, (2), and TF1. (4)

$m^2|n^2$. (5)

$k|n$, by (1), (4), P1, TF2 (and convention of replacing $A \vee A$ by A). (6)

$n = x_{kn}k$, by (6), D3, TF5. (7)

$n^2 = k^2(x_{kn})^2$, using $a=b \supset a^2=b^2$, because n^2 occurs above. (8)

$km^2 = k^2(x_{kn})^2$, by (8), (2), TF6. (9)

$m^2 = k(x_{kn})^2$, by (9), B4, TF1. (10)

$m<n$, by (2), (1), and $(Pa \wedge ab^2=c^2) \supset b<c$. (11)

$x_{kn} < n$, by (7), (1), and $(Pa \wedge ab=c) \supset b<c$. (12)

Substitute x_{kn} for a, m for b in (3):

$(x_{kn} < m \wedge m<n) \supset k(x_{kn})^2 \neq m^2$. (13)

$k(x_{kn})^2 \neq m^2$, by (13), (11), (12), TF2. (14)

But (14) contradicts (10).

ExN4.

$$[x = \Pi_{i=1}^{u} y_i = \Pi_{i=1}^{v} z_i \wedge (i)_1^{u-1} (y_i \leqq y_{i+1} \wedge Py_i < Py_{i+1}) \wedge (i)_1^{v-1} (z_i \leqq z_{i+1} \wedge Pz_i \wedge Pz_{i+1})] \supset (u = v \wedge (i)_1^{u} y_i = z_i).$$

Take least counterexample $x = k$. If $y_1 = z_i$, then $\Pi_{i=2}^{u} y_i = \Pi_{i=2}^{v} z_i < k$. If $y_1 < z_1$, then $y_1 \Pi_{i=2}^{v} z_i < \Pi_{i=1}^{v} z_i$, and y_1 divides both sides. Hence, $y_1 \mid (z_1 - y_1) \Pi_{i=2}^{v} z_i$. Since Pz_1, $y_1(z_1 - y_1)$, and $y_1 z_2 \ldots z_n$. Hence $y_1 \Pi_{i=1}^{w} t_i = (z_1 - y_1) \Pi_{i=2}^{u} z_i < k$.

Examples from quantification theory

Instead of proving that a given formula is a theorem, we take its negation and derive a contradiction. We use similar notation for ambiguous constants as in the section covering examples from number theory. The principal example of this section is:

ExQ1. Derive a contradiction from the conjunction of:

$$m \neq n, \quad (1)$$

$$n \neq k, \quad (2)$$

$$k \neq m, \quad (3)$$

$$y = m \vee [Fym \equiv (Ez)(z \neq m \wedge z \neq y \wedge Fyz \wedge Fzy)], \quad (4)$$

$$y = n \vee [Fyn \equiv (Ez)(z \neq n \wedge z \neq y \wedge Fyz \wedge Fzy)], \quad (5)$$

$$y = k \vee [Fyk \equiv (y = m \vee y = n)]. \quad (6)$$

Two related but harder examples are:

ExQ2. Replace (2) and (3) in ExQ1 by (2′) $(n = k \vee k = m)$ and (3′) $y = j \vee (Fyj \equiv y = k)$.

ExQ3. Drop (2′) from ExQ2.

These harder examples will not be discussed here. For a more intuitive understanding of these examples, think of F as the membership relation $\in$.

There are three somewhat different methods of doing ExQ1. The first method would be to use the decision procedure for the special prefix. Thus bring the conjunction to the prenex form with prefix (Em) (En) (Ek) (y) (z) (Eu) (Ev). If one removes artificial restrictions from, and adds methods for dealing with equality to the program of Friedman[16], then ExQ1 should be manageable.

Second Method

Substitute m, n, k for y in (4), (5), (6) much as with AC before. Of the nine possible substitutions, three give trivial results (m for y in (4), etc.). We consider just 4 of the other 6, and simplify by (1), (2), (3), $k = k$, $n = n$:

$$Fkm \equiv (Ez)(z \neq m \wedge z \neq k \wedge Fkz \wedge Fzk), \text{ k for y in (4).} \quad (7)$$

$$Fkn \equiv (Ez)(z \neq n \wedge z \neq k \wedge Fkz \wedge Fzk), \text{ k for y in (5).} \quad (8)$$

$$Fmk, \text{ m for y in (6).} \quad (9)$$

$$Fnk, \text{ n for y in (6).} \quad (10)$$

We deal with (Ez) by trying to apply the equivalence of Gw with $(Ez)(z = w \vee Gz)$. From (6), we get:

$$(Fyc \wedge y \neq k \wedge y \neq m) \supset y = n;$$

$$(Fyc \wedge y \neq k \wedge y \neq n) \supset y = m.$$

Hence, we can augment the right side of (7) by $y = n$, and that of (8) by $y = m$. Thus, eliminating Ez in both cases, we get:

$$\mathrm{Fkm} \equiv (\mathrm{n} \neq \mathrm{m} \wedge \mathrm{n} \neq \mathrm{k} \wedge \mathrm{Fkn} \wedge \mathrm{Fnk}).$$
$$\mathrm{Fkn} \equiv \neg (\mathrm{m} \neq \mathrm{n} \wedge \mathrm{m} \neq \mathrm{k} \wedge \mathrm{Fkm} \wedge \mathrm{Fmk}).$$

By (1) and (2), these can be simplified.

$$\mathrm{Fkm} \equiv (\mathrm{Fkn} \wedge \mathrm{Fnk}). \qquad (11)$$
$$\mathrm{Fkn} \equiv (\neg \mathrm{Fkm} \vee \neg \mathrm{Fmk}). \qquad (12)$$
$$\mathrm{Fkn} \equiv \neg \mathrm{Fkm}, \text{ by (9) and (12).} \qquad (13)$$
$$\mathrm{Fkm} \equiv \mathrm{Fkn}, \text{ by (10) and (11).} \qquad (14)$$

By (13) and (14), Fkm≡Fkm, a contradiction.

Third Method

The quantifiers and ≡ in (4) and (5) are eliminated in a more mechanical manner to give:

$[\mathrm{y=m} \vee \mathrm{Fym} \vee (\mathrm{u=m} \vee \mathrm{u=y} \neg \mathrm{Fyu} \vee \neg \mathrm{Fuy})] \wedge [\mathrm{y}$
$= \mathrm{m} \vee \neg \mathrm{Fym} \vee (\mathrm{x}_y \neq \mathrm{m} \wedge \mathrm{x}_y \neq \mathrm{y} \wedge \mathrm{Fyx}_y \wedge \mathrm{F}x_y\mathrm{y})]$. (4′)

$[\mathrm{y=n} \vee \neg \mathrm{Fyn} \vee (\mathrm{v=n} \vee \mathrm{v=y} \vee \neg \mathrm{Fyv} \vee \neg \mathrm{Fvy})] \wedge [\mathrm{y}$
$= \mathrm{n} \vee \mathrm{Fyn} \vee (\mathrm{w}_y \neq \mathrm{n} \wedge \mathrm{w}_y \neq \mathrm{y} \wedge \mathrm{Fywy} \wedge \mathrm{Fw}_y\mathrm{y})]$. (5′)

The problem is to derive a contradiction from (1), (2), (3), (4′), (5′), and (6). We write every clause in the conjunctive normal form so that the result is a conjunction of many clauses:

(1);
(2);
(3);
(4.1) $\mathrm{y=m} \vee \neg \mathrm{Fym} \vee \mathrm{x}_y \neq \mathrm{m}$;
(4.2) $\mathrm{y=m} \vee \neg \mathrm{Fym} \vee \mathrm{x}_y \neq \mathrm{y}$;
(4.3) $\mathrm{y=m} \vee \neg \mathrm{Fym} \vee \mathrm{Fyx}_y$;
(4.4) $\mathrm{y=m} \vee \neg \mathrm{Fym} \vee \mathrm{Fx}_y\mathrm{y}$;
(4.5) $\mathrm{y=m} \vee \mathrm{Fym} \vee \mathrm{u=m} \vee u\mathrm{=y} \vee \neg \mathrm{Fuy} \vee \neg \mathrm{Fuy}$;
(5.1) $\mathrm{y=n} \vee \mathrm{Fyn} \vee \mathrm{w}_y \neq \mathrm{n}$;
(5.2) $\mathrm{y=n} \vee \mathrm{Fyn} \vee \mathrm{w}_y \neq \mathrm{y}$;
(5.3) $\mathrm{y=n} \vee \mathrm{Fyn} \vee \mathrm{Fyw}_y$;
(5.4) $\mathrm{y=n} \vee \mathrm{Fyn} \vee \mathrm{Fw}_y\mathrm{y}$;
(5.5) $\mathrm{y=n} \vee \neg \mathrm{Fyn} \vee \mathrm{v=n} \vee \mathrm{v=y} \vee \neg \mathrm{Fyv} \vee \neg \mathrm{Fvy}$;
(6.1) $\mathrm{y=k} \vee \mathrm{y} \neq \mathrm{m} \vee \mathrm{Fyk}$;
(6.2) $\mathrm{y=k} \vee \mathrm{y} \neq \mathrm{n} \vee \mathrm{Fyk}$;
(6.3) $\mathrm{y=k} \vee \mathrm{y=m} \vee \mathrm{y=n} \vee \neg \mathrm{Fyk}$.

Substitute m for y in (6.1), using m≠k and m=m:

$$\mathrm{Fmk}. \qquad (15)$$
$$\mathrm{Fnk}, \text{ n for y in (6.2), using } \mathrm{n} \neq \mathrm{k} \text{ and } \mathrm{n=n}. \qquad (16)$$
$$\mathrm{k=n} \vee \neg \mathrm{Fkn} \vee \mathrm{m=n} \vee \mathrm{m=k} \vee \neg \mathrm{Fkm} \vee \neg \mathrm{Fmk}, \text{ k for y and m for v in (5.5).} \qquad (17)$$
$$\neg \mathrm{Fkn} \vee \neg \mathrm{Fkm}, \text{ by (1), (2), (3), (15), (17).} \qquad (18)$$
$$\mathrm{Fkm} \vee \neg \mathrm{Fkn}, \text{ k for y and n for u in (4.5), by (1), (2), (3), (16)} \qquad (19)$$
$$\neg \mathrm{Fkn}, \text{ by (18), (19).} \qquad (20)$$

$$w_k \neq n,\ w_k \neq k,\ Fkw_k,\ Fw_kk,\ k \text{ for y in (5.1)—(5.4), by (2) and (20).} \quad (21)$$

$$w_k = m,\ w_k \text{ for y in (6.3), by (21).} \quad (22)$$

$$Fkm, \text{ by (21) and (22).} \quad (23)$$

$$x_k \neq m,\ x_k \neq k,\ Fkx_k,\ Fx_kk,\ k \text{ for y in (4.1)—(5.4), by (3) and (23).} \quad (24)$$

$$x_k = n,\ x_k \text{ for } y \text{ in (6.3), by (21).} \quad (25)$$

$$Fkn, \text{ by (24), (25).} \quad (26)$$

(20) and (26) contradict each other.

This proof is quite close in spirit to those obtained in Chinlund, et al.[4] Indeed, if we extend the method of Chinlund, et al.,[4] by adding a treatment of equality and eliminate wasted steps, the machine can probably come up with a proof more or less like the one given, in a reasonable time. Apart from the question of equality, there is, however, one major theoretical difference between the method of Chinlund, et al., and the above proof, *viz.*, the above proof does not use the prenex normal form. This improvement is suggested by Herbrand's thesis* and requires a somewhat more elaborate justification[10] than the method using the prenex form.

Work for this paper was supported in part by Bell Telephone Laboratories, Murray Hill, N.J.

References

1. S. Aanderaa, *A Deterministic Proof Procedure* (manuscript), Harvard, May 1964.
2. D. G., Bobrow, "Natural Language Input for a Computer Problem-Solving System," Ph.D. thesis, MIT, September, 1964.
3. W. S. Brown, *The ALPAK System*, MM-63-1214-3, Bell Laboratories, April 1963.
4. T. Chinlund, M. Davis, P. G. Hinman and D. McIlroy, *Theorem-Proving by Machine*, Bell Laboratories, Spring 1964.
5. G. E. Collins, *Computational Reductions in Tarski's Decision Method for Elementary Algebra*, IBM Corporation, Yorktown Heights, July 1962.
6. M. Davis, and H. Putnam, "A Computing Procedure for Quantification Theory," *Journal of the Association for Computing Machinery*, 7, 1960. pp.201—215.
7. M. Davis, G. Logemann and D. Loveland, "A Machine Program for Theorem-Proving," *Communications of the Association for Computing Machinery*, 5, 1962. pp.394—397.
8. M. Davis, "Eliminating the Irrelevant from Mechanical Proofs," *Proceedings of Symposium in Applied Mathematics*, American Mathematical Society, v. 15, 1963.
9. B. Dreben, P. Andrews and S. Aanderaa, "False Lemmas in Herbrand," *Bulletin of the American Mathematical Society*, v. 69, 1963. pp.699—706.
10. B. Dreben, and H. Wang, *A Refutation Procedure and Its Model-Theoretic Justification* (manuscript), Harvard University, November 1964.
11. B. Dunham, R. Fridshal and G. L. Sward, "A Nonheuristic Program for Proving Elementary Logical Theorems," *Proceedings of the First International Conference on Information Processing*, Paris, 1959; pub. Unesco, 1960. pp.282—285.

* Compare Refs. 9, 20, and 38.

12. B. Dunham, R. Fridshal and J. North, *Exploratory Mathematics by Machine, Recent Development in Information and Decision Processes*, Robert E. Machol and Paul Grey (eds.), Mac–Millan, N. Y., 1962. (*Proceedings of a Symposium at Purdue University*, April 1961.)
13. B. Dunham and J. H. North, "Theorem Testing by Computer," paper presented April 24—26, 1962, *Mathematical Theory of Automata*, Polytechnic Press, Brooklyn, 1963. pp.173—177.
14. J. Friedman, "A Semidecision Procedure for the Functional Calculus," *Journal of the Association for Computing Machinery*, v. 10, 1963. pp.1—24.
15. J. Friedman, "A Computer Program for a Solvable Case of the Decision Problem," *Journal of the Association for Computing Machinery*, v. 10, 1963. pp.348—356.
16. J. Friedman, *A New Decision Procedure in Logic and Its Computer Realization*, Ph.D. thesis, Harvard University, September 1964.
17. H. Gelernter, J.R. Hanson and D. W. Loveland, "Empirical Investigations of the Geometry Theorem Machine," *Proceedings of the Western Joint Computer Conference*, San Francisco, 1960. pp.143—149.
18. P. C. Gilmore, "A Proof Method for Quantification Theory—Its Justification and Realization," *IBM Journal of Research and Development*, v. 4, 1960. pp.28—35.
19. P. C. Gilmore, *An Examination of the Geometry Theorem Machine*, IBM Corporation, Yorktown Heights, April 1962.
20. J. Herbrand, "Recherches sur la Theorie de la Demonstration," *Travaux de la Société des Sciences et des Lettres de Varsovie*, Cl. III, Math. Phys., v. 33, 1930.
21. S. Kanger, "A Simplified Proof Method for Elementary Logic," *Computer Programming and Formal Systems*, P. Braffort and D. Hirschberg (eds.), North-Holland Publishing Co., Amsterdam, 1963. pp.87—94 (*Proceedings of Seminars at Blaricum, Holland in 1961*).
22. S. Kuroda, "An Investigation of the Logical Structure of Mathematics XIII, A Method of Programming Proofs in Mathematics for Electronic Computers," *Magoya Mathematical Journal*, v. 16, 1960. pp.195—203.
23. D. H. Lehmer, "Some High Speed Logic," *Proceedings Symposium in Applied Mathematics*, American Mathematical Society, v.15, 1963.
24. J. McCarthy, "Computer Programs for Checking Mathematical Proofs," *AMS Symposium on Recursive Function Theory*, New York, April 1961.
25. A. Newell, J. C. Shaw and H. A. Simon, Empirical Explorations of the Logic Theory Machine," *Proceedings of the Western Joint Computer Conference*, 1957.
26. A. Newell, and J. C. Shaw, *A Variety of Intelligent Learning in a General Problem Solver, in Self-Organizing Systems*, Marshall C. Yovits and Scott Cameron (eds.), New York, Pergamon Press, 1960. pp.153—189.
27. D. Prawitz, H. Prawitz and N. Voghera, "A Mechanical Proof Procedure and its Realization in an Electronic Computer," *Journal of the Association for Computing Machinery*, v. 7, 1960. pp.102—128.
28. D. Prawitz, "An Improved Proof Procedure," *Theoria* (a Swedish Journal of Philosophy and Psychology), v. 26, 1960. pp.102—139. (Widener, Phil 29—27(5W.))
29. A. Robinson, "On the Mechanization of the Theory of Equations," *Bulletin of the Research Council of Israel*, 9F, 1960. pp.47—70.
30. A. Robinson, "A Basis for the Mechanization of the Theory of Equations," *Computer Programming*

8. SOME DATA FOR ATP*

8.1 On axioms of conditional set existence[1)]

1. Outline of arguments

In what follows, the (restricted) predicate calculus with equality is assumed throughout. Let $(U!u)\, Hu$ be short for $(w)\,(u)\,((Hw \wedge Hu) \supset w=u)$, $(E!y)\, Hy$ be short for $(Ey)\, Hy \wedge (U!y)\, Hy$, $Func\ H$ be short for $(v)\,(U!u)\, Huv$. The convention is to think of Huv as $u=f(v)$, so that a one-many relation defines a function. The basic axioms of the extended ZERMELO set theory, commonly referred to as ZF. are given as:

A. Axiom of extensionality. $(z)\,(z \in x \equiv z \in y) \supset x=y$.

B. Axioms of conditional existence.

B1. The axiom of pairs. $(Ey)\,(x)\,(x \in y \equiv (x=a \wedge x=b))$. Briefly, $(Ey)\,(y=\{a, b\})$.

B2. The sum set (union) axiom. $(Ey)\,(u)\,(u \in y \equiv (Ev)\,(u \in v \wedge v \in x))$. Briefly, $(Ey)\,(y=\bigcup x)$.

B3. The power set axiom. $(Ey)\,(u)\,(u \in y \equiv u \subseteq x)$. Briefly, $(Ey)\,(y=Px)$.

B4. The axiom (schema) of replacement. $Func\ G \supset (Ey)\,(u \in y \equiv (Ev)\,(v \in x \wedge Guv))$.

Briefly, $FuncG \supset (Ey)\,(y=G''x)$.

C. Axiom of infinity (unconditional set existence).

The other axioms (regularity and choice) are generally regarded as more specialized.

The purpose of this note is to discuss the possibility of combining B1—B4 into an organic single schema. This is in part motivated by a wish to supply an analogue of the three axioms of type theory: extensionality, (the axiom of) comprehension, and infinity. The search is not completely successful because in each case the axiom of unit set is required as an auxiliary. It is not clear whether this is a natural need to answer to the notational distinction of different types.

The axiom of unit set is:

J. $(x)\,(Ey)\,(z)\,(z \in y \equiv z=x)$ or $(x)\,(Ey)\,(y=\{x\})$.

Two axiom schemata are considered:

K. $FuncH \supset (Ey)\,(y=\bigcup(H''(Px)))$, or briefly, $FuncH \supset (Ey)\,(y=\bigcup H''Px)$.

L. $FuncH \supset (Ey)\,(y=H''P\bigcup x)$.

More exactly, $y=\bigcup H''Px$ is:

$$(s)\,(s \in y \equiv (Eu)\,(s \in u \wedge (Ev)\,(v \subseteq x \wedge Huv))).$$

---

* First published in *Zeitschrift f. Mathematische Logik und Grundlagenforschung*, vol, 13, pp 183—188.

1) Work for this note was supported by NSF grant GP-228.

The main result of this note is:

Theorem. *If we assume the predicate calculus with equality and the axiom of extensionality* A, K (*or* L) *and J are equivalent to* B1—B4. *Consequently.* B1—B4 *can be replaced by* K (*or* L) *and* J *in the above system* ZF; *this remains true if we omit the axiom of infinity.*

In fact, the details of the derivations from K yield a slightly stronger result. Let A* and K* be:

A*. $(x)\,(E!y)\,(y=\{x\})$.

K*. $FuncH\supset(E!y)\,(y=\bigcup H''Px)$.

Corollary. *The axioms* A *and* B1—B4 *are equivalent to* A* *and* K*.

Intuitively the proofs are quite simple. The notation $\bigcup H''Px$ and $H''P\bigcup x$ makes it clear that K and L are derivable from B2—B4; J is a special case of B1. In the other direction, B1 is known (ZERMELO [3], p.31) to be derivable from B3 and B4. For each of B2, B3, B4, we choose a suitable H so that $\bigcup H''Px$ (or $H''P\bigcup x$) is the required set. Thus, for K and J, we have:

B2. $\bigcup x=\bigcup H_1''Px$, H_1uv being $v=\{u\}$.

B3. $Px=\bigcup H_2''Px$, H_2uv being $u=\{v\}$.

B4. $G''x=\bigcup H_G''Px$, $H_G uv$ being $(Ez)\,(Ew)\,(Gzw \wedge u=\{z\} \wedge v=\{w\})$.

For L and J, we have:

B2. $\bigcup x = (\{u\}=v)''P\bigcup x$.

B3. $Px=(u=v)''\,P\bigcup\{x\}$.

B4. $G''x= (Ew)\,(Guw \wedge v=\{w\})''P\bigcup\{x\}$.

The need of J in the case of L is very explicit, while the need of J in the case of K is more concealed. In fact, the writer erroneously thought that J is not needed and said so in [2]. The application of J is needed in replacing $(Ey)\,(y=\{x\} \wedge Fy)$ by $F\{x\}$, even when $\{x\}$ does not occur before $\in$. Thus, since $\{x\}=\{x\}$, if we can infer $(Ey)\,(y=\{x\} \wedge y=y)$, we obtain J right away; conversely, if $\{x\}$ does not occur before $\in$ in $F\{x\}$, we can use J to derive the desired equivalence. To get $F\{x\}$ from $(Ey)\,(y=\{x\} \wedge Fx)$ is elementary. On the other hand, $(y=\{x\} \wedge F\{x\}) \supset (Ey)\,(y=\{x\} \wedge Fy)$; therefore, $(Ey)\,(y=\{x\}) \supset (F\{x\} \supset (Ey)\,(y=\{x\} \wedge Fy))$; and the antecedent is dropped by J.

To obtain the corollary, we use the fact that in the detailed derivation of B2—B4 from K and L, only one special case of the axiom of extensionality is needed, viz., $(a=\{x\} \wedge b=\{x\}) \supset a=b$. We note first that Axiom A yields $(U!y)\,(y=\bigcup H''Px)$. On the other hand, we can parallel the derivation of B4 from K and J to derive $FuncG \supset (E!y)\,(y=G''x)$ from J* and K*. If we take $u=v$ as Guv, then clearly $FuncG$ and, therefore, $(E!y)\,(y=G''x)$, i.e., $(E!y)\,(u)(u\in y\equiv(Ev)\,(v\in x \wedge u=v))$, or $(E!y)\,(u)\,(u\in y\equiv u\in x)$. Hence, $(Ey)\,(u)\,(u\in y\equiv u\in x)$ and $(u\in y\equiv u\in z)\,((u\in y\equiv u\in x) \wedge (u\in z\equiv u\in x))$. But by the definition of $(U!y)\,Hy$, we have: $((u)\,(u\in y\equiv u\in x) \wedge (u)\,(u\in z\equiv u\in x))\supset y=z$. Therefore, simple rules about universal quantifiers give us Axiom A.

A result quite similar to the one in this note has been obtained previously by ONO in [1]. Even though his paper proposes to give a new approach to set theory, the main result would appear to be the following. He offers an axiom (schema) which is, in the notation here:

G. $Func\ H \supset (E!y)[y = H''\hat{z}(Ew)(z \subseteq w \wedge w \in x)]$.

He then proves that the axioms A, B, C are equivalent to G, J, C. It is possible to modify his proof to avoid the appeal to C in the derivation of B from G and J; but it is not entirely clear that he can avoid all cases of extensionality in developing his system with G, J, C.

In fact, ONO does not use the axiom J but rather employs the unit set operation $\{\ \}$ as a primitive symbol to replace $=$; this has the effect that the axiom J is absorbed into the predicate calculus through his axiom $x \in \{x\}$. His approach is also of interest insofar as he argues that $\{\ \}$ and J are a part of elementary logic rather than one of set theory proper. From his philosophical position, the intrusion of J would appear less puzzling.

Since we often make implicit uses of the axioms of set theory, it would seem desirable to supplement the above outline of the derivations of B1—B4 with pedantic details. Only the derivations from K and J will be considered.

2. Detailed derivations

In the development below, familiar devices and results, including the use of abstracts, will be assumed. To assist the checking for correctness, these definitions and theorems are listed below. It should be emphasized that none of these appeal to the axioms of set theory, but all are available in the predicate calculus with equality.

D1. For any formula Hx, $y \in \hat{x}Hx$ stands for Hy.

D2. If A and B are two abstracts or one variable and one abstract, $A = B$ stands for $(x)(x \in A \equiv x \in B)$.

D3. $\{x\}$ stands for $\hat{y}(y = x)$.

D4. $\{x, y\}$ stands for $\hat{z}(z = x \vee z = y)$.

D5. 0 stands for $\hat{x}(x \neq x)$.

D6. If A is an abstract, $A \in B$ stands for $(Ex)(x = A \wedge x \in B)$; if each of A and B is a variable or an abstract, $A \subseteq B$ stands for $(x)(x \in A \supset x \in B)$.

T1. *If a is an abstract and Hv is a formula, we can replace $(Ev)(v = a \wedge Hv)$ or $(v)(v = a \supset Hv)$ by Ha in any context; similarly, be the axiom of unit set* J, *if a is $\{b\}$, b is a variable, and a does not occur immediately before $\in$ in Ha, the same is true for a.*

T2. $u \in x \equiv \{u\} \subseteq x$, *i.e.*, $u \in x \equiv (v)(v = u \supset v \in x)$.

T3. $\{x\} = \{y\} \supset x = y$, *i.e.*, $(z)(z = x \equiv z = y) \supset x = y$.

T4. *If $Hx \supset G$ and x is not free in G*, $(Ey)Hy \supset G$.

T5. $(Ex)(Ey)Hxy \equiv (Ey)(Ex)Hxy$.

T6. $(v)(Gxy \equiv Hxy) \supset (F[Gxy] \equiv F[Hxy])$, *$v$ free in G and H but bound in F.*

T7. *If* $(Fx \wedge Gx) \supset Hx$, *then* $((x)Fx \wedge (x)Gx) \supset (x)Hx$.

T8. $((H \equiv G) \wedge (H \equiv F)) \supset (G \equiv F)$.

T9. $(x = y \wedge Hx) \supset Hy$.

T10. *If* $F \supset (G \supset H)$, *then* $G \supset (F \supset H)$ *and* $(F \wedge G) \supset H$.

T11. *If x is not free in G*, $(Ex)(Hx \wedge G) \equiv ((Ex)Hx \wedge G)$.

Derivation of B* from B1—B4.

Given x and H, $FuncH$.

By B3, there exists w,

$$(v)(v\in w\equiv v\subseteq x). \qquad (1)$$

Since $FuncH$, by B4, there exists z, $(u)(u\in z\equiv(Ev)(v\in w\wedge Huv))$.
Therefore, by (1) and T6,

$$(u)(u\in z\equiv(Ev)(v\subseteq x\wedge Huv)). \qquad (2)$$

By B2, $(Ey)(s)(s\in y\equiv(Eu)(s\in u\wedge u\in z))$.
Therefore, by (2) and T6, $(Ey)(s)(s\in y\equiv(Eu)(s\in u\wedge(Ev)(v\subseteq x\wedge Huv)))$. Hence, we get:

B*. $FuncH\supset(Ey)(s)(s\in y\equiv(Eu)(s\in u\wedge(Ev)(v\subseteq x\wedge Huv)))$.

Derivations of B1—B4 from B*.

B4. $FuncG\supset(Ey)(s)(s\in y\equiv(Ew)(w\in x\wedge Gsw))$.

Given x and G, $FuncG$.

Take $(Ez)(Ew)(Gzw\wedge u=\{z\}\wedge v=\{w\})$ as Huv in B*.
By T8, $((x\in v\equiv x=a)\vee(x\in v\equiv x=w))\supset(x=a\equiv x=w)$.
Therefore, by T7 (& D3, D2, D1) and T3,

$$(v=\{a\}\wedge v=\{w\})\supset a=w. \qquad (1)$$

Since $FuncG$,

$$Gsw\vee Gzw\supset s=z. \qquad (2)$$

By T9 (& D2), $(s=z\wedge t=\{s\}\wedge u=\{z\})\supset((x)(x\in t\equiv x=z)\wedge(x)(x\in u\equiv x=z))$.
By T7 and T8, $(s=z\wedge t=\{s\}\wedge u=\{z\})\supset(x)(x\in t\equiv x\in u)$.
By the axiom of extensionality,

$$(s=z\wedge t=\{s\}\wedge u=\{z\})\supset t=u. \qquad (3)$$

By T9,

$$(a=w\wedge Hsa\wedge Hzw)\supset(Hsw\wedge Hzw). \qquad (4)$$

Hence, by (1), (2), (3), (4),

$$(Hsa\wedge t=\{s\}\wedge v=\{a\})\supset((Hzw\wedge u=\{z\}\wedge v=\{w\})\supset t=u).$$

Therefore, by T4 and T10,

$$((Ez)(Ew)(Hzw\vee t=\{z\}\vee v=\{w\})\vee(Ez)(Ew)(Hzw\vee u=\{z\}\vee v=\{w\}))\supset t=u.$$

I.e., $FuncH$.

Hence, by B*, there exists y, $y=\bigcup H''Px$.

I.e., $s\in y\equiv(Eu)(s\in u\wedge(Ev)(v\subseteq x\wedge(Ez)(Ew)(Gzw\wedge u=\{z\}\wedge v=\{w\})))$.
By T11 and T5, $s\in y\equiv(Ew)(Ez)(Eu)(u=\{z\}\wedge s\in u\wedge(Ev)(v=\{w\}\wedge v\subseteq x\wedge Gzw))$.
By T1, $s\in y\equiv(Ew)(Ez)(s\in\{z\}\wedge\{w\}\subseteq x\wedge Gzw)$.
By T6 (& D1, D3), $s\in y\equiv(Ew)(Ez)(s=z\wedge\{w\}\subseteq x\wedge Gzw)$.
By T1, $s\in y\equiv(Ew)(\{w\}\subseteq x\wedge Gsw)$.
By T6 and T2, $s\in y\equiv(Ew)(w\in x\wedge Gsw)$, $s\in y\equiv s\in G''x$.
Hence, $G''x=\bigcup H''Px$.

Observe that (3) uses the axiom of extensionality A. In fact, the case A_1 needed here, viz. $(t=\{s\}\wedge u=\{s\})\supset t=u$, is the only consequence of axiom A that is required in the derivations of B1—B4. Thus given A_1, we can replace (3) by:
By T9, $(s=z\wedge t=\{s\}\wedge u=\{z\})\supset(t=\{s\}\wedge u=\{s\})$.
By A_1, $(s=z\wedge t=\{s\}\wedge u=\{z\})\supset t=u$.

B3. $(x)(Ey)(y=Px)$.

In B*, take $u=\{v\}$ as Huv. Clearly, $FuncH$ by axiom A (or the subcase A_1). By B*, there exists y, $y=\bigcup H''Px$.

I.e., $$\begin{aligned} s\in y &\equiv (Eu)(s\in u \wedge (Ev)(v\subseteq x \wedge u=\{v\})) \\ &\equiv (Ev)(s\in\{v\} \wedge v\subseteq x) \\ &\equiv (Ev)(s=v \wedge v\subseteq x) \\ &\equiv s\subseteq x \\ &\equiv s\in Px. \end{aligned}$$

B2. $(x)(Ey)(y=\bigcup x)$.

In B*, take $v=\{u\}$ as Huv. By T3, $FuncH$. Hence, by B*, there exists y, $y=\bigcup H''Px$, i.e.,

$$\begin{aligned} s\in y &\equiv (Eu)(s\in u \wedge (Ev)(v\subseteq x \wedge v=\{u\})) \\ &\equiv (Eu)(s\in u \wedge \{u\}\subseteq x) \\ &\equiv (Eu)(s\in u \wedge u\in x) \\ &\equiv s\in\bigcup x. \end{aligned}$$

B1. $(a)(b)(Ey)(y=\{a,b\})$.

Zermelo's argument ([3], p.31) is included here for completeness. Take $(u=v \wedge v\neq v)$ as Guv in B4. We get then the set 0 (D5). Take $PP0=\{0,\{0\}\}$ as x and $(u=a \wedge v=0)\vee(u=b \wedge v=\{0\})$ as Guv in B4. Then we get $\{a,b\}=G''PP0$.

References

[1] K. Ono, A set theory founded on unique generating principle. Nagoya Mathematical Journal **12**(1957), 151—159.

[2] Hao Wang. A universal axiom of conditional set existence. Notices of the American Mathematical Society, **10**(1963), 588.

[3] E.Zermelo, Über Grenzzahlen und Mengenbereiche. Fundamenta Mathematica **16**(1930), 29—47.

8.2 Natural hulls and set existence* 1)

1. Introduction

A set x is called a *partial hull*, briefly $PH(x)$, if x is a transitive set and closed with respect to the operation of forming power sets; it is a *natural hull*, briefly $NH(x)$, if it is a partial hull and also closed with respect to the operation of forming sum sets (unions). The natural hull ηa (or *partial hull* πa) of a set a is the intersection of all x such that $NH(x)$ (or $PH(x)$) and $a\in x$.

D1. $CP(x)$ for $(y)(y\in x\supset Py\in x)$.

D2. $CU(x)$ for $(y)(y\in x\supset\bigcup y\in x)$.

D3. $trans(x)$ for $(y)(z)((y\in z \wedge z\in x)\supset y\in x)$ or $\bigcup x\subseteq x$.

D4. $PH(x)$ for $trans(x) \wedge CP(x)$.

* First published in *Zeitschrift f. Mathematische Logik und Grundlagenforschung*, vol, 13, pp 172—182.

1) Work for this note was supported by NSF grant GP-228.

D5. $NH(x)$ for $PH(x) \wedge CU(x)$.
D6. $b\pi a$ for $(x)((PH(x) \wedge a\in x) \supset b\in x)$.
D7. $b\eta a$ for $(x)((NH(x) \wedge a\in x) \supset b\in x)$.

Intuitively it is clear that if w is the intersection of all x, such that $NH(x)$ and $a\in x$, then $NH(w)$ and $a\in w$. It requires, however, a formal proof that the intersection does exist in a given formal system. We shall be concerned exclusively with the familiar system ZF, the extended ZERMELO theory, and prove:

Theorem I. *In the system* ZF, *we can define a functional* η *such that for each set* a, *there is a unique set* ηa, *called the natural hull of* a, *for which we can prove in* ZF: (1) $b\in\eta a \equiv b\eta a$, *and, therefore,* $\eta a \subseteq x$ *if* $NH(x) \wedge a\in x$; (2) $NH(\eta a) \wedge a\in\eta a$. *Similarly, we can define the partial hull* πa *of every set* a *so that* $b\in\pi a \equiv b\pi a$, *and* $PH(\pi a) \wedge a\in\pi a$.

As a first application of the concepts of natural and partial hulls, we shall prove that if in the axiom of replacement, the replaced set x is substituted for by πx or ηx, the resulting schema is as strong as all axioms of set existence taken together. To express this fact more exactly, we state the new axioms explicitly and recall the basic axioms of ZF.

It is convenient to use some familiar abbreviations.

D8. $FuncH$ for $(u)(v)(w)((Huv \wedge Hwv) \supset u=w)$.
D9. $y=H''x$ for $(u)(u\in y \equiv (Ev)(Huv \wedge v\in x))$.
D10. $(E!y)\ Fy$ for $(Ey)\ Fy \wedge (y)(z)((Fy \wedge Fz) \supset y=z)$.

SE. Axiom of set existence.
If $FuncH$ (*i.e.,* H *is one-many*), *then*
$(Ey)(u)(u\in y \equiv (Ev)(Huv \wedge v\eta x))$, or, briefly, using Theorem I, $(Ey)(y=H''\eta x)$.

SE′. *If* $FuncH$, *then* $(Ey)(y=H''\pi x)$.

UE. Axiom of unique existence.
If $FuncH$, *then* $(E!y)\ (y=H''\eta x)$.

UE′ *If* $FuncH$, *then* $(E!y)(y=H''\pi x)$.

Roughly, SE says that given any functional H, and a set x, the result of replacing the sets v in the natural hull of x by their values $u=Hv$ form a set again. Similarly, UE says that these form one and only one set.

The basic axioms of ZF are:

A. Axiom of extensionality. $(z)(z\in x \equiv z\in y) \supset x=y$.
B. Sum set axiom. $(Ey)(\bigcup x \subseteq y)$ *or* $(Ey)(y=\bigcup x)$.
C. Power set axiom. $(Ey)(Px \subseteq y)$ *or* $(Ey)(y=Px)$.
D. Axiom of replacement. $FuncG \supset (Ey)(y=G''x)$.
E. Axiom of infinity. $(Ey)\ (0\in y \wedge (x)(x\in y \supset \{x\} \in y))$.

The axiom of pairs and the Aussonderungsaxiom are known to be derivable from C and D. The axioms of regularity and choice are regarded as less basic.

The exact result is:

Theorem II. *The axioms SE, SE′, UE, UE′ all are derivable in* ZF; *and within the (restricted) predicate calculus with equality, all the axioms* A-E *of* ZF *are derivable from* UE (*or* UE′), *all the existence axioms* B—E *of ZF are derivable from* SE (*or* SE′). *In other words,* SE *and* SE′ *each can replace all the existence axioms;* UE *and* UE′ *each is equivalent to all the axioms of* ZF *taken together.*

This theorem suggests as a corollary several alternative axiomatizations of the system ZF which can be summarized as follows.

AN. Axiom of natural hulls. $(Ey)(y=\eta x)$ or $(Ey)(z)(z\in y\equiv z\eta x)$.

AN′. Axiom of partial hulls. $(Ey)(y=\pi x)$.

D*. Axiom of unique replacement. $FuncG\supset(E!y)(y=G''x)$.

Theorem III. *Given the predicate calculus with equality, the system* ZF *can be axiomatized by any of the following combinations:* (1) UE; (2) UE′; (3) SE *and extensionality;* (4) SE′ *and extensionality;* (5) *unique replacement and* AN; (6) *unique replacement and* AN′; (7) AN, *extensionality, and replacement;* (8) AN′, *extensionality, and replacement.*

2. Natural Hulls

In the proof of Theorem I, we are free to use familiar facts about the system ZF such as the possibility of developing natural numbers and using inductive definitions over natural numbers. Once Theorem I is proved, it will be obvious that SE, SE′, UE, UE′ can also be derived in ZF.

In order to be explicit and to prepare for the derivations of axioms of ZF from the new axioms, we recall that $\in$ and $=$ are the only primitive predicates of ZF, other symbols being introduced by definitions, often contextually. We give some examples needed in this note.

D11. $x\subseteq y$ for $(z)(z\in x\supset z\in y)$.

D12. $y\in Px$ for $y\subseteq x$.

D13. $y\in\bigcup x$ for $(Ez)(y\in z\wedge z\in x)$.

D14. $v=Px$ for $(y)(y\in v\equiv y\in Px)$.

D15. $v=\bigcup x$ for $(y)(y\in v\equiv y\in\bigcup x)$.

D16. $\bigcup x\subseteq x$ for $(y)(y\in\bigcup x\supset y\in x)$.

D17. $Px\in y$ for $(Eu)(u=Px\wedge u\in y)$.

D18. $\bigcup x\in y$ for $(Eu)(u=\bigcup x\wedge u\in y)$.

D19. $y\in\{x\}$ for $y=x$.

D20. $A=\{x\}$ for $(y)(y\in A\equiv y\in\{x\})$.

D21. $\{x\}\in z$ for $(Ey)(y=\{x\}\wedge y\in z)$.

A slight complication in defining ηa and πa by induction over natural numbers is the fact that while $CP(x)$, $CU(x)$ go from one set to one set, $trans(x)$ goes from one set to all of its members. To make the conditions uniform, we introduce two natural new operations.

D22. $\bigcup^*x$ for $\{\bigcup y\mid y\in x\}$, i.e., $u\in\bigcup^*x$ for $(Ey)(y\in x\wedge u=\bigcup y)$.

D23. P^*x for $\{Py\mid y\in x\}$, i.e., $u\in P^*x$ for $(Ey)(y\in x\wedge u=Py)$.

With the help of these notions, we can correlate with each given set a, a unique corresponding function f_a (or g_a) from natural numbers to arbitrary sets, and define πa (or ηa) in terms of it.

Definition I. $f_a(1)=\{a\}$, $f_a(2n)=\bigcup f_a(n)$, $f_a(2n+1)=P^*f_a(n)$.
$g_a(1)=\{a\}$, $g_a(3k-2)=\bigcup^*g_a(k)$, $g_a(3k-1)=P^*g_a(k)$, $g_a(3k)=\bigcup g_a(k)$.

Definition II. $b\in\pi a$ for $(Ei)(b\in f_a(i))$ or $b\in\{y\mid(Ei)(y\in f_a(i))\}$.
$b\in\eta a$ for $(Ei)(b\in g_a(i))$ or $b\in\{y\mid(Ei)(y\in g_a(i))\}$.

12 Ztschr. f. math. Logik

Theorem 1. $a=b\supset\pi a=\pi b$; $a=b\supset\eta a=\eta b$.

The first thing is to prove that $\eta a(\pi a)$ is indeed a natural (partial) hull containing a as a member.

Theorem 2. $NH(\eta a)\wedge a\in\eta a$; $PH(\pi a)\wedge a\in\pi a$.

Proof. Since $a\in\{a\}=g_a(1)$, $a\in\eta a$.

If $u\in\eta a$, then there is some i, $u\in g_a(i)$. Then $\bigcup u\in\bigcup^* g_a(i)=g_a(3i-2)$, and $Pu\in P^* g_a(i)=g_a(3i-1)$. Hence, if $u\in\eta a$, then $\bigcup u\in\eta a$ and $Pu\in\eta a$.

If $u\in v\in g_a(i)$, then $u\in\bigcup g_a(i)=g_a(3i)$. Hence, $trans(\eta a)$.

Therefore, $NH(\eta a)\wedge a\in\eta a$.

Similarly, we can prove: $PH(\pi a)\wedge a\in\pi a$.

Theorem 3. $(NH(x)\wedge a\in x)\supset\eta a\subseteq x$; $(PH(x)\wedge a\in x)\supset\pi a\subseteq x$.

Proof. If $NH(x)\wedge a\in x$ and $u\in\eta a$, we wish to show that $u\in x$. If $u\in\eta a$, then there is some i, $u\in g_a(i)$. We prove by induction on i that if $u\in g_a(i)$, then $u\in x$.

If $i=1$, then $u\in g_a(i)=\{a\}$, and, therefore, $u=a$. Since $a\in x$, $u\in x$ when $i=1$.

If $i=3k-2$, then $u\in\bigcup^* g_a(k)$, *i.e.*, $(Ey)(y\in g_a(k)\wedge u=\bigcup y)$. By induction hypothesis, if $y\in g_a(k)$, then $y\in x$. But $y\in x\supset\bigcup y\in x$. Hence, $u\in x$.

If $i=3k-1$, then $u\in P^* g_a(k)$ and there is y, $y\in g_a(k)$ and $u=Py$. Since $y\in x$ by induction hypothesis and $CP(x)$, we get again: $u\in x$.

If $i=3k$, then $u\in\bigcup g_a(k)$ and there is y, $u\in y\in g_a(k)$. By induction hypothesis, $y\in x$. Since $trans(x)$, $u\in x$.

This completes the proof of the first half of Theorem 3. The proof of the second half is entirely similar.

Theorem 4. $b\in\eta a\equiv b\eta a$; $b\in\pi a\equiv b\pi a$.

Proof. By Theorem 3, D7, and D6.

This completes the proof of Theorem I.

Moreover, this also shows that SE, SE′, UE, UE′ all are derivable in ZF. Thus, given a set x, we can first get ηx and πx in ZF, and then apply the axiom of replacement to get SE and SE′. In order to get UE and UE′ then, we simply apply the axiom of extensionality. Hence, in order to complete the proof of Theorem II, we only have to derive the axioms of ZF from the new axioms. Before doing this, we give some examples of natural hulls and partial hulls.

Example 1. Let F be the set of all finite sets built up from the empty set 0, i.e., $0\in F$, $x\cup\{y\}\in F$ if $x, y\in F$, and nothing else belongs to F. Then $\eta a=\pi a=F$ if $a\in F$. If x is πa or ηa, clearly $x\subseteq F$ since $NH(F)\wedge a\in F$. On the other hand, $0\in x$ since $a\in x$ and $trans(x)$. Hence, $P0$, $PP0$, etc. all belong to x since $CP(x)$. Therefore, $F=\bigcup_{i=0}^{\infty}P^i 0\subseteq x$.

Example 2. If $a=F$, $\eta a=\pi a=\bigcup_{i=0}^{\infty}P^i F$.

Example 3. Let l^0x, l^1x, l^2x, etc. stand for x, $\{x\}$, $\{\{x\}\}$, etc., and $Inf=\bigcup_{i=0}^{\infty}l^i 0$, $a=\bigcup_{i=0}^{\infty}l^{2i+1}0$, $b=\bigcup_{i=0}^{\infty}l^{2i}0$. Then $\pi a=F\cup\bigcup_{i=1}^{\infty}P^i a$. Since $0\in\bigcup a$, $F\subseteq\pi a$ by $CP(\pi a)$; since $a\in\pi a$, Pa, PPa, etc. all belong to πa and, by $\bigcup\pi a\subseteq\pi a$, $\bigcup_{i=1}^{\infty}P^i a\subseteq\pi a$. Conversely, since $PH(F)$, we have only to consider the infinite sets. If $y\in P^i a$, then

$Py \in P^{i+1}a$; if $y \in z \in Pa$, then $y \in F$; if $y \in z \in P^{i+1}a$. then $y \in P^i a$. Hence, $\pi a = F \cup \bigcup_{i=1}^{\infty} P^i a$. Since $b = \bigcup a$, $b \in \eta a$ although $b \notin \pi a$. Moreover, if c is $\{0, l0\}$, then $\{a, c\} \notin \pi a$ although $a \in \pi a$ and $c \in \pi a$. In fact, $\eta a = F \cup \bigcup_{i=1}^{\infty} P^i\, Inf$, and $\{a, c\} \notin \eta a$.

We note incidentally that the formal definition of ηa and πa is compatible with the axiom of regularity being false, because infinite descending chains $\cdots \in x_n \in x_{n-1} \in \cdots \in x_2 \in x_1$ are not excluded by the construction. In fact, since the axiom of regularity is known to be independent of the axioms A—E of ZF, and UE, SE, UE′, SE′ are derivable from A—E, it follows that the axiom of regularity is also independent of the new axioms.

Using the notions $\bigcup^*$ and P^*, we can also express natural hulls in an alternative form:

Theorem 5. $\bigcup^* x \subseteq x \equiv CU(x)$; $P^* x \subseteq x \equiv CP(x)$; $trans(x) \equiv \bigcup x \subseteq x$; $NH(x) \equiv (\bigcup x \subseteq x \wedge \bigcup^* x \subseteq x \wedge P^* x \subseteq x)$; $PH(x) \equiv (\bigcup x \subseteq x \wedge P^* x \subseteq x)$.

3. Derivation of the usual axioms

In fact, we get four new axiomatizations of ZF according as we use SE, SE′, UE, or UE′. In every case, we assume the predicate calculus with equality. Then the four systems are simply: (1) A, SE; (2) A, SE′; (3) UE; (4) UE′.

We note first that within the predicate calculus, UE′ follows from UE, and SE′ follows from SE.

Lemma 1. $b\pi a \supset b\eta a$; $(b\pi a \wedge b\eta a) \equiv b\pi a$.

Proof. $NH(x) \supset PH(x)$. Hence:

$$((PH(x) \wedge a \in x) \supset b \in x) \supset ((NH(x) \wedge a \in x) \supset b \in x).$$

Lemma 1 is proved by D6 and D7.

If we take $Huv \wedge v\pi x$ in place of Huv in SE (or UE), we obtain immediately, by Lemma 1, SE′ (or UE′).

Hence, it is sufficient to derive the usual axioms from SE′ and UE′. To avoid tedious explanations, we give just the derivation of A—E from UE′.

In other words, we use a system containing only the predicate calculus with equality and the single axiom (schema):

UE′. *If* $FuncH$, *then* $(E!y)(u)(u \in y \equiv (Ev)(Huv \wedge v\pi x))$.

The axiom can be expressed in the primitive notation of ZF by D8, D6, D4, D3, D1, D17, D14, D12, D11.

We are now to derive the axioms A—E from UE′. Once the derivation is done, the proof of Theorem II will be complete.

Lemma 2. $b \in a \supset b\pi a$.

Proof. $(b \in a \wedge \bigcup x \subseteq x \wedge a \in x) \supset b \in x$. But $PH(x) \supset \bigcup x \subseteq x$. Hence, Lemma 2 is proved by D6.

Axiom D (Replacement). $FuncG \supset (Ey)(y = G''x)$.

This is derivable from SE′, and, hence, also from UE′.

Let Huv be $Guv \wedge v \in x$. Then $FuncH$. By SE′,

$$(Ey)(u)(u \in y \equiv (Ev)(Guv \wedge v \in x \wedge v\eta x)).$$

By Lemma 2,

$$(v\in x \wedge v\eta x)\equiv v\in x.$$

Hence,

$$(Ey)(u)(u\in y\equiv (Ev)(Guv \wedge v\in x)).$$

Therefore, $FuncG\supset (Ey)(y=G''x)$.

Lemma 3 (Aussonderungsaxiom). $(Ey)(u)(u\in y\equiv (u\in x \wedge Fu))$.

Proof. Take $u=v \wedge Fv$ *as* Guv in Axiom D, we get Lemma 3.

So far we have not made use of the axiom of extensionality A. We derive now Axiom A from UE′.

Axiom A (Extensionality). $(z)(z\in x\equiv z\in y)\supset x=y$.

Proof. If we use UE′ instead of SE′, we can prove Axiom D and Lemma 3 in the stronger form with $E!$ replacing E. In particular, if $u=u$ is taken as Fu in the strengthened Lemma 3, we get:

$$(E!y)(u)(u\in y\equiv u\in x).$$

Hence,

$$((u)(u\in y\equiv u\in x) \wedge (u)(u\in w\equiv u\in x))\supset y=w \tag{1}$$

and

$$(Ey)(u)(u\in y\equiv u\in x). \tag{2}$$

Therefore, by simple rules of logic,

$$(u)(u\in y\equiv u\in w)\supset(u)(u\in w\equiv u\in x).$$

Hence,

$$((u)(u\in y\equiv u\in x) \wedge (u)(u\in y\equiv u\in w))\supset((u)(u\in y\equiv u\in x) \wedge (u)(u\in w\equiv u\in x))$$

and, by (1),

$$\supset y=w.$$

Hence, by (2), $(u)(u\in y\equiv u\in w)\supset y=w$, the axiom of extensionality.

To be entirely exact, we should either develop a fairly extensive theory to deal directly with set terms such as Px, $\bigcup x$ introduced by contextual definitions, or expand the contexts into primitive notations. Since, however, such details are rather well known and tend to make the proofs less easy to follow, we shall be fairly informal.

Lemma 4. $b\subseteq a\supset b\pi a$.

Proof. $(b\subseteq a \wedge c=Pa)\supset b\in c$,

$$(\bigcup x\subseteq x \wedge b\in c \wedge c\in x)\supset b\in x.$$

Hence,

$$(c=Pa \wedge c\in x)\supset((b\subseteq a \wedge \bigcup x\subseteq x)\supset b\in x),$$
$$Pa\in x\supset((b\subseteq a \wedge \bigcup x\subseteq x)\supset b\in x),$$
$$(CP(x) \wedge a\in x)\supset Pa\in x.$$

Therefore,

$$b\subseteq a\supset((PH(x) \wedge a\in x)\supset b\in x).$$

Lemma 4 follows by D6.

Lemma 5. $(Ey)(u)(u\in y\equiv u\pi x \wedge Fu))$.

Proof. Since SE′ follows from UE′, we can use SE′. Take $u=v \wedge Fv$ as Huv in SE′, we get Lemma 5.

Axiom C (Power set axiom). $(Ey)(y=Px)$.

Proof. Take $u\subseteq x$ as Fu in Lemma 5 and apply Lemma 4 to delete $u\pi x$.

Lemma 6. $b\in\bigcup a\supset b\pi a$.

Proof. $(c\in a \wedge a\in x \wedge \bigcup x\subseteq x)\supset c\in x$,

$(b\in c \wedge c\in x \wedge \bigcup x\subseteq x)\supset b\in x$.

Therefore,

$$b\in\bigcup a\supset((\bigcup x\subseteq x\vee\ a\in x)\supset b\in x).$$

Lemma 6 follows by D6.

Axiom B (Sum set axiom). $(Ey)\,(y=\bigcup x)$.

Proof. Take $u\in\bigcup x$ as Fu in Lemma 5 and apply Lemma 6 to delete $u\pi x$.

We have thus far derived A, B, C, D from UE′. It remains to prove Axiom E which calls for a more elaborate argument. Roughly speaking, the set F of all finite sets built up from the empty set is a subset of every nonempty partial hull. Hence, $F\subseteq\{b\mid(a)b\pi a\}=Inf$ and, in particular, $0\in Inf$, $(x)(x\in Inf\supset\{x\}\in Inf)$. The following derivations make this rough sketch a little more formal.

Lemma 7. $(Ey)(y=\{b\mid(a)b\pi a\}$ or $(Ey)(u)(u\in y\equiv(v)u\pi v)$.

Proof. Take $(v)\ u\pi v$ as Fu in Lemma 5 and delete $u\pi x$ by the relation: $(v)u\pi v\supset u\pi x$.

Lemma 8. $(Ey)(u)(u\in y\equiv u\neq u)$. *Briefly*, $(Ey)(y=0)$.

Proof. Take $u\neq u$ as Fu in Lemma 5 and delete $u\pi x$ by the relation: $u\neq u\supset u\pi x$.

Lemma 9. $(v)0\pi v$.

Proof. $0\subseteq y$. By Lemma 4, $(y)0\subseteq y\supset(y)0\eta y$.

Lemma 10. $(v)(x\pi v)\supset(v)(Px\pi v)$.

Proof. If $PH(w)$ and $x\in w$, then $Px\in w$. Hence:

$$((PH(w)\vee\ v\in w)\supset(x\in w))\supset((PH(w)\ \wedge v\in w)\supset(Px\in w)).$$

Lemma 10 follows by D6.

Lemma 11. $(Ey)(y=\{x\})$.

Proof. Take $u=x$ as Fu in Lemma 5 and delete $u\pi x$ by the relation: $u=x\supset u\pi x$.

Lemma 12. $(v)x\pi v\supset(x)(\{x\}\pi v)$.

Proof. Since $\{x\}\subseteq Px$, by Lemmas 4 and 10, we get Lemma 12.

Axiom E (Axiom of infinity). $(Ey)\,(0\in y\wedge(x)\,(x\in y\supset\{x\}\in y))$.

Proof. Take $y=\{u|(a)u\pi a\}$. By Lemmas 9 and 12, we get E.

8.3 A theorem on definitions of the Zermelo-Neumann ordinals * 1)

1. Introduction and summary

According to the definition of ordinals by ZERMELO and VON NEUMANN, every ordinal is the set of all smaller ordinals. Thus, 0 must be the empty set since there is no smaller ordinal than 0, x' or $x+1$ must be $x\cup\{x\}$, and a limit ordinal must be the union

---

* First published in *Zeitschrift f. Mathematische Logik und Grundlagenforschung*, vol, 13, pp 241—250. © VEB Deutscher Verlag, 1967. Reproduced by permission.

1) Work for this note was supported in part by NSF grant GP-228.

16 Ztschr. f. math. Logik

Σw of an infinite set w of ordinals having Σw as the least upper bound. Intuitively it is clear that the ordering relation for these ordinals is just the membership relation $\in$, or, alternatively, proper inclusion $\subset$.

Given this intuitive conception, it remains a nontrivial task to define formally the class of all ordinals in an axiom system of set theory, e.g., the extended ZERMELO theory, commonly referred to as ZF.

There are in the literature at least four equivalent elegant definitions which can be stated briefly with the help of standard abbreviations.

D1. $a \subseteq b$ for $(x)(x \in a \supset x \in b)$.

D2. $a \subset b$ for $a \subseteq b \wedge a \neq b$.

D3. x' for $x \cup \{x\}$.

D4. 0 for $\{y | y \neq y\}$.

D5. $\Sigma a \subseteq a$ or *trans*(a) (*a is transitive*): $(x)(y)((x \in y \wedge y \in a) \supset x \in a)$.

D6. *wfd*(a) (*a is well-founded*): $(c)(0 \subset c \subseteq a \supset (Eb)(b \in c \wedge b \cap c = 0))$.

D7. *conn*$_\in(a)$ (*a is connected by* $\in$): $(x)(y)((x \in a \wedge y \in a \wedge x \neq y) \supset (x \in y \vee y \in x))$.

The four definitions are:

DZ. ZERMELO 1915 ([2], p.6 and p.10; [3], p.88; [1], p.19).

$Od(a)$ iff $(a = 0 \vee 0 \in a) \wedge (b)(b \in a \supset (b' \in a \vee b') = a)) \wedge (c)(c \subseteq a \supset (\Sigma c \in a \vee \Sigma c = a))$.

Briefly, $Od(a)$ iff $(b)(b \in a \supset b' \in a') \vee (c)(c \subseteq a \supset \Sigma c \in a')$.

DG. GÖDEL 1937 ([2], pp.9—10; [1], p.20).

$Od(a)$ iff $\Sigma a \subseteq a \wedge wfd(a) \wedge (b)(b \in a \supset \Sigma b \subseteq b)$.

Briefly, $Od(a)$ iff $(b)(b \in a' \supset \Sigma b \subseteq b) \wedge wfd(a)$.

DR. R.M. ROBINSON 1937 ([1], p.20; [3], p.80).

$Od(a)$ iff $\Sigma a \subseteq a \wedge conn_\in(a) \wedge wfd(a)$.

DB. BERNAYS 1941 ([2], p.10; [1], p.20).

$Od(a)$ iff $\Sigma a \subseteq a \wedge (b)((b \subset a \wedge \Sigma b \subseteq b) \supset b \in a)$.

Briefly, $Od(a)$ iff $(b)((\Sigma b \subseteq b \subseteq a) \equiv b \in a')$.

Formally, DB appears to be the shortest definition. If we assume the Fundierungsaxiom, then wfd(a) can be deleted from DG and DR, and the shortest definition is (b) $(b \in a' \supset \Sigma b \subseteq b)$. The proof of equivalence of these definitions is summed up in BACHMANN'S book ([1], pp.19—21). For the present purpose, it is convenient to set down this result more exactly:

Lemma. *With extensionality, Aussonderung, and self-adjunction (i.e., the successor axiom* (x) (Ey) $(y = x \cup \{x\})$), *the definitions* DZ, DG, DR, DB *of Od are all equivalent.*

As a result, if we wish to show that a predicate $On(a)$ defines the class of all ordinals, it is sufficient to derive $On(a) \equiv Od(a)$, with Od construed according to any of the definitions. In fact, we shall prove a theorem by using DG to get $On(a) \supset Od(a)$ and DR to get $Od(a) \supset On(a)$. In order to state the theorem, we shall speak of transfinite induction for a definition On, when we mean the following particular schema.

STI. Strengthened transfinite induction for On:

$$[On(x) \wedge (v)(Fv \supset Fv') \wedge (w)(w \subseteq F \supset F(\Sigma w))] \supset Fx.$$

The main theorem of this note is:

Theorem I. *Let $On(a)$ be a formula of ZF and* S *be a subsystem of ZF containing extensionality, Aussonderung, and self-adjunction. If* STI *for On can be proved in* S, *then* $\vdash_s On(x) \supset Od(x)$; *if, furthermore,* $\vdash_s Od(x) \supset On(x)$, *then* $\vdash_s On(x) \equiv Od(x)$, *and $On(a)$ is an adequate definition of ordinals as conceived by* ZERMELO *and* VON NEUMANN.

The interest of this theorem lies in the fact that the definitions of *Od* listed above do not reveal the intuitive picture of the class of ordinals, and that this theorem enables us to give "genetic" definitions of the class, patterned after the FREGE–DEDEKIND device of ancestrals for finite ordinals, which seem intuitively more appealing. A perhaps more important advantage of the genetic approach is that it depends less on the particular successor function we choose.

The natural definition in this approach would read: a is an ordinal if and only if

$$(y)([0 \in y \wedge (z)(z \in y \supset z' \in y) \wedge (w)(w \subseteq y \supset \Sigma w \in y)] \supset a \in y).$$

But this is not acceptable because we could find no set y in the set theory ZF which would satisfy the antecedent of this definition. It is, therefore, necessary to find ways of avoiding the need of excessively large collections. $0 \in y$ can be absorbed in $(w)(w \subseteq y \supset \Sigma w \in y)$ since, for all y, $0 \subseteq y$, and $\Sigma 0 = 0$.

Two definitions, suggested by previous methods for avoiding infinite sets in defining finite ordinals, are:

DI. $On(x)$ for $(u)\{[(v)((v \in u \wedge v' \in x') \supset v' \in u) \wedge (w)((w \subseteq u \wedge \Sigma w \in x') \supset \Sigma w \in u)] \supset x \in u\}$.

*DI**. $On(x)$ for $(u)\{[x \in u \wedge (v)(v' \in u \supset v \in u)] \supset (Ew)(\Sigma w \in u \wedge w \subseteq x \wedge u \cap w = 0)\}$.

Since it is easy to prove STI for *On* according to either DI or DI* and it is fairly clear that every ordinal does satisfy both definitions, a more or less immediate corollary of Theorem I is:

Theorem II. *The theory of transfinite ordinals can be developed on the basis of* DI *or DI*, using extensionality, Aussonderung, and self-adjunction; when Aussonderung is strengthened to replacement, we can also obtain definitions by transfinite recursion.* Finally, we specify DI, DI* and Theorem II to finite ordinals.

2. Proof of Theorem I

Let S be the system having as axioms extensionality, Aussonderung, and self-adjunction; and $On(x)$ be an arbitrary predicate for which we assume STI. We shall prove that if $On(x)$, then x is an ordinal by DG. Essential use will be made of the fact that the successor function x' is $x \cup \{x\}$. The procedure is to prove first in S the induction hypotheses for $\Sigma x \subseteq x$, $(b)(b \in x \supset \Sigma b \subseteq b)$, and $wfd(x)$; then we get $On(x) \supset Od(x)$ by STI. We shall use the convenient notation $w \subseteq F$ to stand for $(z)(z \in w \supset Fz)$.

Lemma 1. $\Sigma v \subseteq c \supset \Sigma v' \subseteq v'$.

Proof. Since $v' = v \cup \{v\}$, we have:

$$(y \in z \wedge z \in v') \supset ((y \in z \wedge z \in v) \vee (y \in z \wedge z = v))$$

Therefore, $(trans(v) \wedge (y \in z \wedge z \in v')) \supset y \in v \supset y \in v'$.

Lemma 2. $w \subseteq trans \supset trans(\Sigma w)$, *or* $(z)(z \in w \supset \Sigma z \subseteq z) \supset \Sigma(\Sigma w) \subseteq \Sigma w$.

Proof. $u \in y \in \Sigma w \supset (Ez)(u \in y \in z \in w)$.

Since $w \subseteq trans$, we have:

$$(u \in y \in z \wedge z \in w) \supset (u \in z \wedge z \in w) \supset u \in \Sigma w.$$

In proving the other induction hypothesis, it is sometimes necessary to append the condition of transitivity.

Lemma 3. $(\Sigma v \subseteq v \wedge (b)(b \in v \supset \Sigma b \subseteq b)) \supset (\Sigma v' \subseteq v' \wedge (b)(b \in v' \supset \Sigma b \subseteq b))$.

Proof. By L.1, we only have to derive $(b)(b \in v' \supset \Sigma b \subseteq b)$ from $\Sigma v \subseteq v$, $(b)(b \in v \supset \Sigma b \subseteq b)$. By D3, $b \in v' \supset (b \in v \vee b = v)$. But

$$(\Sigma v \subseteq v \wedge b = v) \supset \Sigma b \subseteq b, \ ((b)(b \in v \supset \Sigma b \subseteq b) \wedge b \in v) \supset \Sigma b \subseteq b.$$

Hence,

$$b \in v' \supset \Sigma b \subseteq b.$$

Lemma 4. $(z)[z \in w \supset (b)(b \in z \supset \Sigma b \subseteq b)] \supset (b)(b \in \Sigma w \supset \Sigma b \subseteq b)$.

Proof. $b \in \Sigma w \supset (Ez)(b \in z \in w)$. By the antecedent, $(z \in w \vee b \in z) \supset \Sigma b \subseteq b$. Hence, $b \in \Sigma w \supset \Sigma b \subseteq b$. The proof of L.4 is complete.

Lemma 5. $(\Sigma v \subseteq v \wedge wfd(v)) \supset (\Sigma v' \subseteq v' \wedge wfd(v'))$.

Proof. By L.1, we have to derive only $wfd(v')$ from $\Sigma v \subseteq v$ and $wfd(v)$. By D3, $v' = v \cup \{v\}$. Therefore:

$$0 \subset c \subseteq v' \supset (0 \subset c \subseteq v \vee 0 \subseteq c - \{v\} \subseteq v) \supset (0 \subset c \subseteq v \vee c = \{v\} \vee 0 \subset c - \{v\} \subseteq v). \quad (1)$$

Since $wfd(v)$,

$$0 \subset c \subseteq v \supset (Ey)(y \in c \wedge y \cap c = 0) \quad (2)$$

$$c = \{v\} \supset [(v \in c \wedge v \cap \{v\} = 0) \vee (v \cap \{v\} = c = \{v\})]. \quad (3)$$

Clearly,

$$c = \{v\} \wedge v \in c \wedge v \cap \{v\} = 0 \supset (Ey)(y \in c \wedge y \cap c = 0) \quad (4)$$

$$v \cap \{v\} = c = \{v\} \supset v \in v \supset v' = v \text{ (by D3)}.$$

By hypothesis, $wfd(v)$. Therefore, $wfd(v')$. Hence, since $c = \{v\} \supset 0 \subset c \subseteq v'$, we have:

$$(c = \{v\} \wedge v \cap \{v\} = c = v) \supset (Ey)(y \in c \wedge y \cap c = 0). \quad (5)$$

By (3), (4), (5), we get:

$$c = \{v\} \supset (Ey)(y \in c \wedge y \cap c = 0). \quad (6)$$

Finally, we consider the last alternative of (1).

$$y \cap (c - \{v\}) = 0 \supset (y \cap c = 0 \vee (v \in y \wedge v \in c)). \quad (7)$$

$$(y \in c - \{v\} \wedge c - \{v\} \subseteq v) \supset y \in v. \quad (8)$$

Since $\Sigma v \subseteq v$, $(y \in v \wedge v \in y) \supset v \in v$.

By D3,

$$(y \in v \vee v \in y) \supset v = v' \quad (9)$$

$$(y \in c - \{v\} \wedge y \cap c = 0) \supset (y \in c \wedge y \cap c = 0) \supset (Ey)(y \in c \wedge y \cap c = 0). \quad (10)$$

By (7), (8), (9), (10),

$$(y \in c - \{v\} \wedge y \cap (c - \{v\}) = 0) \supset ((Ey)(y \in c \wedge y \cap c = 0) \vee (c - \{v\} \subseteq v \supset v = v')).$$

But by $wfd(v)$,

$$0 \subset c - \{v\} \subseteq v \supset (Ey)(y \in c - \{v\} \wedge y \cap (c - \{v\}) = 0).$$

Hence,

$$0 \subset c - \{v\} \subseteq v \supset (Ey)(y \in c \vee y \cap c = 0) \vee v = v'. \quad (11)$$

If $v = v'$, then, by hypothesis $wfd(v)$, we have $wfd(v')$. Hence:

$$(v = v' \wedge 0 \subset c - \{v\} \subseteq v) \supset (wfd(v') \wedge 0 \subset c \subseteq v') \supset (Ey)(y \in c \wedge y \cap c = 0). \quad (12)$$

By (11) and (12):

$$0 \subset c - \{v\} \subseteq v \supset (Ey)(y \in c \wedge y \cap c = 0). \tag{13}$$

By (1), (2), (6), (13), we have, using $wfd(v)$ and $\Sigma v \subseteq v$: $wfd(v')$, i.e.,

$$0 \subset c \subseteq v' \supset (Ey)(y \in c \wedge y \cap c = 0).$$

This completes the proof of L.5.

Lemma 6. $(w \subseteq trans \wedge w \subseteq wfd) \supset (trans(\Sigma w) \wedge wfd(\Sigma w))$.

Proof. By L.2, we only have to derive $wfd(\Sigma w)$ from $w \subseteq trans$ and $w \subseteq wfd$. In other words, given $0 \subset c \subseteq \Sigma w$, we wish to prove: $(Ey)(y \in c \wedge y \cap c = 0)$.

Suppose there where c, $0 \subset c \subseteq \Sigma w$ and:

$$(y)(y \in c \supset y \cap c \neq 0). \tag{1}$$

Since $c \neq 0$, let $y_1 \in c$. Then, since $c \subseteq \Sigma w$, there is some z, $y_1 \in z \in w$. Hence, $y_1 \in c \vee y_1 \in z$, and $c \cap z$ is a nonempty subset of c. By (1), $(y)(y \in z \cap c \supset y \cap c \neq 0)$; i.e.,

$$y \in z \cap c \supset (Et)(t \in y \wedge t \in c). \tag{2}$$

By hypothesis, $w \subseteq trans$. *Since* $z \in w$, $\Sigma z \subseteq z$ and:

$$t \in y \in z \supset t \in z. \tag{3}$$

By (2) and (3),

$$y \in z \cap c \supset (Et)(t \in y \wedge t \in z \cap c) \supset y \cap (z \cap c) \neq 0.$$

Hence, under the assumption (1), there is a set z:

$$z \in w$$

and

$$0 \subset z \cap c \subseteq z \wedge \neg (Ey)(y \in z \cap c \wedge y \cap (z \cap c) = 0).$$

In other words, under the assumption (1), we get a counterexample to the premiss $w \subseteq wfd$. Hence, $wfd(\Sigma w)$ is a consequence of $w \subseteq trans$ and $w \subseteq wfd$, and the proof of L.6 is complete.

It should be emphasized that thus far our considerations depend only on the particular successor function $x' = x \cup \{x\}$ and do not depend on how the whole class of ordinals is formally defined.

If now we make use of the other hypothesis of Theorem I, i.e., that STI is available for *On*, then we immediately complete the proof of Theorem I.

STI. $[On(x) \wedge (v)(Fv \supset Fv') \wedge (w)(w \subseteq F \supset F(\Sigma w))] \supset Fx$.

Let Fx be $\Sigma x \subseteq x \wedge (b)(b \in x \supset \Sigma b \subseteq b) \wedge wfd(x)$, i.e., $Od(x)$ according to DG. By L.3, L.5, we have clearly: $Fv \supset Fv'$. By L.4, L.6, we have: $(w)(w \subseteq F \supset F(\Sigma w))$. Hence, by STI, we get: $On(x) \supset Fx$. In other words, we have:

Theorem 1. $On(x) \supset Od(x)$.

This completes the proof of Theorem I.

3. Proof of Theorem II

We shall first prove STI for *On* as defined by DI, using extensionality, Aussonderung, and self-adjunction; hence, by Theorem I, the class *On* by DI is included in the class *Od*. Next, we shall, using the same axioms, prove that the class *On* in DI* is included in the class *On* in DI. Finally, we shall show, using the same axioms, that the class *Od* is included in the class *On* by DI*. Clearly this will prove the three classes coextensional and establish Theorem II, since the development of transfinite recursion (compare [3]) can be repeated with the help of the axiom of replacement.

To avoid notational confusion, we restate DI and DI* with indices on On.

DI. $On_1(x)$ for $(u)\{[(v)((v'\in x' \wedge v\in u) \supset v'\in u) \wedge (w)((\Sigma w\in x' \wedge w\subseteq u)\supset \Sigma w\in u)] \supset x\in u\}$.

DI*. $On_2(x)$ for $(y)[(x\in y \wedge (v)(v'\in y\supset v\in y)) \supset (Ew)(\Sigma w\in y \wedge w\subseteq x \wedge y\cap w=0)]$.

Theorem 2. $[On_1(x) \wedge (v)(Fv\supset Fv') \wedge (w)(w\subseteq F\supset F(\Sigma w))]\supset Fx$.

Proof. By Aussonderung and self-adjunction, there is, for each set x, a set $u = \{t\,|\,(Ft \wedge t\in x')\}$.

Hence, by DI, if we assume $On_1(x)$, we have:

If (a) $(v)((Fv \wedge v\in x' \wedge v'\in x')\supset(Fv' \wedge v'\in x'))$ and

(b) $(w)[((z)(z\in w\supset(Fz\vee z\in x')) \wedge \Sigma w\in x')\supset(F(\Sigma w)\vee \Sigma w\in x')]$,

then $x\in x'$ and Fx.

But, if $(v)(Fv\supset Fv')$, then (a); and if $(z)(z\in w\supset (Fz\vee z\in x'))$ then $w\subseteq F$, and therefore, if $(w)[w\subseteq F\supset F(\Sigma w)]$, then (b).

Hence, the proof of Th.2 is complete.

Remark. If we use a different successor function, say πx as x', in DI, we can similarly derive Th.2 from DI, with self-adjunction replaced by a successor axiom for the different successor function. The condition $F0$ follows from the last hypothesis of Th.2 and is, therefore, left out for convenience. It is perhaps of interest that in contrast with the derivation of L. 1—L.6, which depends heavily on the particular successor function used, the derivation of transfinite induction is "invariant" relative to the successor functions.

Hence, by Theorem I, we have:

Theorem 3. $On_1(x)\supset Od(x)$.

Next we prove:

Theorem 4. $On_2(x)\supset On_1(x)$.

Proof. Put $y=\{t|(t\in x' \wedge t\subseteq x' \wedge t\notin u)$ in DI*. Such a set y exists for each given set x by self-adjunction and Aussonderung. Hence, if $On_2(x)$, then:

If $x\in x' \wedge x\subseteq x' \wedge x\notin u$, and

(1) $(v)((v'\in x' \wedge v'\subseteq x' \wedge v'\notin u)\supset (v\in x' \wedge v\subseteq x' \wedge v\notin u))$, then

(2) $(Ew)[\Sigma w\in x' \wedge \Sigma w\subseteq x' \wedge \Sigma w\notin u \wedge w\subseteq x \wedge (z)((z\in x' \wedge z\subseteq x' \wedge z\notin u)\supset z\notin w)]$.

Clearly $x\in x'$ and $x\subseteq x'$ are true and can be dropped. By contraposing, we get:

(A) $x\in u$ if (1) and

(2*) $(w)\{[\Sigma w\in x' \wedge \Sigma w\subseteq x' \wedge w\subseteq x \wedge (z)((z\in x' \wedge z\subseteq x' \wedge z\in w)\supset z\in u)]\supset \Sigma w\in u\}$.

Now (1) is equivalent to the conjunction of:

(1a) $(v)((v'\in x' \wedge v'\subseteq x' \wedge v'\notin u) \supset (v\in x' \wedge v\subseteq x'))$, and

(1b) $(v)((v'\in x' \wedge v'\subseteq x' \wedge v\in u)\supset v'\in u)$.

Since $v\in v'$ and $v\subseteq v'$, $v'\subseteq x'\supset(v\in x' \wedge v\subseteq x')$ and (1a) is true. Therefore, (1) is equivalent to (1b) and we have:

(3) $(v)((v'\in x' \wedge v\in u)\supset v'\in u)\supset(1)$.

In (2*), $z\subseteq x'$ can be dropped because $z\in w$ and $\Sigma w\subseteq x'$ yield it as a consequence. Hence, (2*) can be written briefly as:

(2*) $(w)[(\Sigma w\in x' \wedge \Sigma w\subseteq x' \wedge w\subseteq x \wedge x'\cap w\subseteq u)\supset \Sigma w\in u]$.

But $x\subseteq x'$ and, therefore, $(w\subseteq x \wedge x'\cap w\subseteq u)\supset w\subseteq u$. Hence, we get:

$$(\Sigma w\in x' \wedge \Sigma w\subseteq x' \wedge w\subseteq x \wedge x'\cap w\subseteq u)\supset(\Sigma w\in x' \wedge w\subseteq u).$$

Therefore:

(4) $(w)((\Sigma w\in x' \wedge w\subseteq u)\supset \Sigma w\in u)\supset(2^*)$.

By (A), (3), and (4), we have:

$$(u)\{[(v)((v'\in x' \wedge v\in u)\supset v'\in u) \wedge (w)((\Sigma w\in x' \wedge w\subseteq u)\supset \Sigma w\in u)]\supset x\in u\},$$

i.e., $On_1(x)$. Hence, $On_2(x)\supset On_1(x)$.

Finally, we prove:

Theorem 5. $Od(x)\supset On_2(x)$.

Here we can draw on standard developments of ordinals and use the Lemma about the equivalence of various definitions of *Od*. In fact, since it is intuitively clear that ordinals do possess the property On_2, it is reasonable to accept Th.5 as provable even before formal details are supplied.

For the sake of completeness, we give a derivation of Th.5 by using the development of *Od* by BERNAYS ([3], pp.80—89) together with the lemma just mentioned. Thus we are entitled to use the following theorems on *Od*:

T1. $Od(a)\supset[\Sigma a=a \vee (Ey)(a=y'\wedge\Sigma a=y)]$ ([3], p.88, 2.14 and 2.12).

T2. $((w)[(Od(w) \wedge w\subseteq F)\supset Fw] \wedge Od(x))\supset Fx$ ([3], p.86, 1.17).

T3. $Od(x)\supset(\Sigma x\subseteq x \wedge \Sigma x'\subseteq x')$ ([3], p.80, 1.4, and p.87, 2.2).

Before proving Th.5, we use a similar but simpler argument to prove STI for *Od*:

T4. $Od(x) \wedge (v)(Fv\supset Fv') \wedge (w)(w\subseteq F\supset F(\Sigma w))\supset Fx$.

Proof. It is sufficient to derive the hypothesis of T2 from that of T5.

$$[(w\subseteq F\supset F(\Sigma w)) \wedge \Sigma w=w]\supset(w\subseteq F\supset Fw). \quad (1)$$

$$[(w=y' \wedge \Sigma w=y) \wedge (Fy\supset Fy') \wedge (w\subseteq F\supset F(\Sigma w))]\supset(w\subseteq F\supset Fw). \quad (2)$$

By T1,

$$Od(w)\supset[\Sigma w=w \vee (Ey)(w=y' \wedge \Sigma w=y)]. \quad (3)$$

Combining (1), (2), (3), we get:

$$[Od(x) \wedge (v)(Fv\supset Fv') \wedge (w)(w\subseteq F\supset F(\Sigma w))]\supset[Od(x) \wedge (w)((Od(w) \wedge w\subseteq F)\supset Fw)].$$

Hence, T4 is proved.

Remark. In view of the derivation of T4 from T1 and T2, it follows from Theorem I that instead of STI for *On*, we can require that T1 and T2 be provable for *On* (i.e., replace *Od* by *On* in them). Hence, it is sufficient to find a condition *cond*(*x*) such that we can derive T2 for *On* when *On*(*x*) is taken to be $cond(x) \wedge [\Sigma x=x \vee (Ey)(y'=x \wedge \Sigma x =y)]$.

If we drop the clause $w\subseteq x$ from the consequent of DI*, we get a more elegant definition which W.V.QUINE and the writer will elsewhere prove to be adequate, with the help of the axiom of power set. For this simpler definition, a counterpart of Th.5 is easily forthcoming.

T5. $[Od(x) \wedge x\in u \wedge (v)(v'\in u\supset v\in u)]\supset(Ew)(\Sigma w\in u \wedge u\cap w=0)$.

Proof. Put $t \notin u$ for Ft in T4 and contrapose.

It does not seem possible to apply T4 to shorten the proof of Th.5. Rather T2 is used directly, in conjunction with T1.

We can restate Th.5 more explicitly:

Theorem 5. $[Od(x) \vee x \in u \vee (v)(v' \in u \supset v \in u)] \supset (Ew)(\Sigma w \in u \vee w \subseteq x \vee u \cap w = 0)$.

Proof. Put $t \in x' \supset t \notin u$ for Ft in T2 and simplify. We get:

$$(x \in u \vee Od(x)) \supset (Ew)(Od(w) \vee (w \cap x') \cap u = 0 \vee w \in x' \vee w \in u). \tag{1}$$

It is sufficient to derive the consequent (c) of Th.5 from $x \in u$, $Od(x)$, (a) $(v)(v' \in u \supset v \in u)$, with the help of the consequent of (1): briefly, (b) $(Ew)(Od(w) \vee Hwxu)$.

By T3, $(Od(x) \vee w \in x') \supset w \subseteq x$, because $w \in x' \supset (w = x \vee w \in x)$. Hence, $w \cap x' = w$ and we have:

$$(Od(x) \vee w \in x' \vee (w \cap x') \cap u = 0) \supset (w \subseteq x \vee w \cap u = 0). \tag{2}$$

Moreover, $$(\Sigma w = w \vee w \in u) \supset \Sigma w \in u. \tag{3}$$

$$(w = y' \vee \Sigma w = y \vee (y' \in u \supset y \in u) \vee w \in u) \supset \Sigma w \in u. \tag{4}$$

By (2) and (3), $(Od(x) \vee \Sigma w = w \vee Hwxu) \supset (c)$.

By (2) and (4), $((a) \wedge (Ey)(w = y' \wedge \Sigma w = y) \vee Hwxu) \supset (c)$.

Hence, $(Od(x) \wedge (a) \wedge (Ew)((\Sigma w = w \wedge (Ey)(w = y' \wedge \Sigma w = y)) \wedge Hwxu)) \supset (c)$. (5)

By T1, $(b) \supset (Ew)((\Sigma w = w \vee (Ey)(w = y' \wedge \Sigma w = y)) \wedge Hwxu)$. (6)

By (1), (5), (6), $(x \in u \wedge Od(x) \wedge (a)) \supset (c)$.

Hence, $Od(x) \supset On_2(x)$, and the proof of Th.5 is complete.

When we combine Th.3, Th.4, Th.5, we get: $Od = On_1 = On_2$. Since all the proofs are done in the predicate calculus with equality with the help only of extensionality, Aussonderung, and self-adjunction, the proof of Theorem II is complete.

4. Finite Ordinals

If we specialize DI and DI* to finite ordinals, then the only nonsuccessor case is 0, and we get:

DN. $Nn_1(x)$ for $(u)\{[(0 \in x' \supset 0 \in u) \wedge (v)((v' \in x' \wedge v \in u) \supset v' \in u)] \supset x \in u\}$.

DN*. $Nn_2(x)$ for $(u)\{[x \in u \wedge (v)(v' \in u \supset v \in u)] \supset 0 \in u\}$.

DN is new, while DN* is a definition proposed by W.V.Quine and developed by K.R.Brown. Brown's direct development of DN* contains many interesting features. Previously, Quine had already given a thorough treatment of a predecessor of DN* with $\{x\}$ instead of $x \cup \{x\}$ as x' (see [4]). It is, however, of interest to observe here that we can derive results about Nn_1 and Nn_2 rather quickly from a comparison with Bernays' definition of finite ordinals ([3], p.89).

DB. $Nu(x)$ for $Od(x) \wedge (y)(y \in x' \supset (y = 0 \vee (Ez)(Od(z) \wedge y = z')))$.

In fact, we can prove:

Theorem 6. *$Nn_1(x) \equiv Nn_2(x) \equiv Nu(x)$; or briefly, $Nn_1 = Nn_2 = Nu$.*

Hence, since Bernays has developed arithmetic (Peano axioms and recursive definitions) on the basis of Nu with the help of extensionality, Aussonderung, and adjunction, i.e., $(x)(y)(Ez)(z = x \cup \{y\})$, we can derive the following:

Theorem III. *Arithmetic can be developed on the basis of* DN *or* DN*, *using* *extensionality, Aussonderung, and adjunction; self-adjunction is sufficient if recursive definitions are not required.*

The part on DN* was proved first by BROWN using an elegant direct argument (not going through infinite ordinals); he had previously already a similar result where $v' \in u$ is replaced by $(v' \in u \wedge v' \in x)$.

By the way, $0 \in x' \supset$ can be dropped in DN, if we use adjunction to get induction.

The key lemma for proving Th.6 is:

Lemma 7. $Nn_1 \subseteq On_1$; $Nn_2 \subseteq On_2$.

Proof. Immediate from DI and DN (or DI* and DN*) since $0 \in x' \supset 0 \in u$ if $(w)[(w \subseteq u \wedge \Sigma w \in x') \supset \Sigma w \in u]$, and $(Ew)(\Sigma w \in u \wedge w \subseteq x \wedge u \cap w = 0)$ if $0 \in u$ (seeing that $\Sigma 0 = 0$, $0 \subseteq x$, *and* $u \cap 0 = 0$).

Hence, if $Nn_1(x)$ or $Nn_2(x)$, x has all properties proved to hold for ordinals. For example:

Lemma 8. $Nn_2(x) \supset \Sigma x' \subseteq x'$.

Mathematical induction is immediate for Nn_1 and Nn_2.

Lemma 9. $[F0 \wedge (v)(Fv \supset Fv') \wedge Nn_1(x)] \supset Fx$.

Proof. Take $\{v|(Fv \wedge v \in x')\}$ as u in DN and simplify.

Lemma 10. $[F0 \wedge (v)(Fv \supset Fv') \wedge Nn_2(x)] \supset Fx$.

Proof. Taking $\{v|(v \in x' \vee -Fv)\}$ as u in DN* and assuming $Nn_2(x)$, we get:

$$[x \in x' \wedge -Fx \wedge (v)((v' \in x' \wedge -Fv') \supset (v \in x' \wedge -Fv))] \supset (0 \in x' \wedge -F0).$$

Dropping $x \in x'$ as true, breaking up $p \supset (q \wedge r)$ into $(p \supset q) \wedge (p \supset r)$, and contraposing, we get:

$$[F0 \wedge (v)((v' \in x' \wedge -Fv') \supset v \in x') \wedge (v)((v' \in x' \wedge Fv) \supset Fv')] \supset Fx.$$

By L.8, $(Nn_2(x) \vee v' \in x') \supset v \in x'$, since $v \in v'$. Hence, we can drop the second clause as true. But the third clause follows from $(v)(Fv \supset Fv')$. Hence, L.10 is proved.

The following is now immediate:

Lemma 11. $Nn_1(x) \supset Nu(x)$; $Nn_2(x) \supset Nu(x)$.

Proof. By L.7 and the previously established fact that $Od = On_1 = On_2$, we have: $Nn_1(x) \supset Od(x)$, $Nn_2(x) \supset Od(x)$. Let, now, Fx be

$$(y)(y \in x' \supset [y = 0 \vee (Ez)(Od(z) \wedge y = z')])$$

in L.9 or L.10. Clearly $F0$, since $y \in 0' \supset y = 0$. Assume Fv, i.e.,

$$(y)(y \in v' \supset [y = 0 \vee (Ez)(Od(z) \wedge y = z')])$$

Since $y \in v'' \supset (y \in v' \vee y = v')$, Fv' follows from Fv. This is clear when $y \in v'$. When $y = v'$, we use the fact $v \in v'$; by Fv, $v = 0$ or $(Ez)(Od(z) \wedge y = z')$. In either case $Od(v)$ by 2.1 and 2.2 of [3] (p.87).

Finally, since induction is available for Nu ([3], p.90, 3.4), we can derive:

Lemma 12. $Nu(x) \supset Nn_1(x)$; $Nu(x) \supset Nn_2(x)$.

Proof. Putting $t \in x' \supset t \in u$ for Ft in the induction principle for Nu, we get:

$$[Nu(x) \wedge (0 \in x' \supset 0 \in u) \wedge (v)(((v \in x' \supset v \in u) \wedge v' \in x') \supset v' \in u) \wedge x \in x'] \supset x \in u.$$

The third clause can be broken up into $(v)(v' \in x' \subset (v' \in u \vee v \in x'))$ and $(v)((v \in u \wedge v' \in x') \supset v' \in u)$. The first is true by the fact that $Nu(x) \supset \Sigma x' \subseteq x'$ (from [3]). Hence, $Nu(x) \supset Nn_1(x)$.

Putting $t \notin u$ for Ft and contraposing, we immediately get; $Nu(x) \supset Nn_2(x)$.

This completes the proof of Th. 6 and therewith the proof of Theorem III.

References

[1] H. BACHMANN, Transfinite Zahlen, Berlin 1955.
[2] P. BERNAYS, A system of axiomatic set theory, Part II, Journal of Symbolic Logic **6** (1941), 1—17.
[3] P. BERNAYS and A. FRAENKEL, Axiomatic Set Theory, Amsterdam, 1958.
[4] W. V. QUINE, Set Theory and its Logic, Cambridge, Mass. 1963.

9. PROVING THEOREMS BY PATTERN* RECOGNITION, II

Theoretical questions concerning the possibilities of proving theorems by machines are considered here from the viewpoint that emphasizes the underlying logic. A proof procedure for the predicate calculus is given that contains a few minor peculiar features. A fairly extensive discussion of the decision problem is given, including a partial solution of the (x) (Ey) (z) satisfiability case, an alternative procedure for the (x) (y) (Ez) case, and a rather detailed treatment of Skolem's case. In connection with the (x) (Ey) (z) case, an amusing combinatorial problem is suggested in Section 4.1. Some simple mathematical examples are considered in Section VI.

9.1 A survey of the decision problem

1.1 *The Decision Problem and the Reduction Problem*

With regard to any formula of the predicate calculus, we are interested in knowing whether it is a theorem (the problem of provability), or equivalently, whether its negation has any model at all (the problem of satisfiability). Originally this decision problem was directed to the search for one finite procedure which is applicable to all formulae of the predicate calculus. Since it is known that there can be no such omnipotent procedure, the main problem is to devise procedures effective for classes of formulae which satisfy suitable conditions.

The complementary problem of reduction is to give effective procedures which reduce broader classes to narrower ones while preserving provability or satisfiability. In this way, a decision procedure for a smaller class can be made to apply to a larger one. Thus far, most work on the reduction problem has been directed to the special case of finding procedures which reduce all fromulae of the predicate calculus to members of some special class(e.g, those in the Skolem normal form). Each such class is called a reduction class relative to satisfiability or provability according to whether satisfiability or provability is preserved by the transformations (Ref. 2, p. 32). It follows automatically that the corresponding decision problem for each reduction class is unsolvable.

The reduction classes and the procedures employed to obtain them are, being

* First published in **_Bell System Technical Journal_**, vol, 40, pp, 1—41.

concerned with undecidable cases, only of indirect use for the problem of discovering positive results on the decision problem. More directly relevant are reduction procedures which are applicable when the reduced class is not a reduction class and may in particular be a decidable class. Some very preliminary results on this more general aspect of the reduction problem will be described in Section V.

For both the decision problem and the reduction problem, there is, beyond the "yes or no" as to satisfiability, a further question of determining all models and devising transformation procedures which preserve all models. Such questions have been studied to a certain extent (Ref. 3, p. 23), but will be disregarded in what follows.

It is customary to characterize reduction classes and decidable classes in terms of formulae in the prenex normal form, i.e., with all quantifiers at the beginning. Sometimes, with regard to satisfiability (oor provability), conjunctions (or disjunctions) of formulae in the prenex normal form are considered. We shall call this the extended prenex form.

In Section V, a procedure will be given for reducing any formula to a finite set of generally simpler formulae in the extended prenex form such that the original formula is provable if and only if all formulae in the reduced set are. In this and the next few sections, we shall only be concerned with formulae in the extended prenex form. Furthermore, we shall give in Section V a proof–decision procedure for the quantifier–free logic, obtained from the propositional calculus by adding equality, function symbols and individual constants. Any theorem in it is called a quantifier–free tautology, as an extension of the notion of a propositional tautology. We shall make use of the fact that we can always decide whether a given formula is a quantifier–free tautology.

1.2 *A Brief Formulation of the Predicate Calculus*

1.2.1 *Primitive Symbols*

1.2.1.1 Variables x, y, z, etc. (an infinite set).
1.2.1.2 Individual constants (a finite or infinite set).
1.2.1.3 Propositional (Boolean) operations: $\sim$, $\vee$, &, $\supset$, $\equiv$.
1.2.1.4 Predicate letters (a finite or infinite set).
1.2.1.5 Function letters (a finite or infinite set).
1.2.1.6 Equality: = (a special predicate symbol).
1.2.1.7 Quantification symbols: ( ), (*E*).
1.2.1.8 Parentheses.

1.2.2 *Inductive Definition of Terms and Formulae*

1.2.2.1 A variable or an individual constant is a term.
1.2.2.2 A function symbol followed by a suitable number of terms is a term.
1.2.2.3 A predicate followed by a suitable number of terms is a formula (and an

atomic formula); in particular, if α, β are terms $=(\alpha, \beta)$ or $\alpha = \beta$ is a formula (and an atomic formula).

or $\alpha = \beta$ is a formula (and an atomic formula).

1.2.2.4 If φ, ψ, are formulae and α is a variable, then $(\alpha)\varphi$, $(E\alpha)\varphi$, $\sim \varphi$, $\varphi \vee \psi$, $\varphi \,\&\, \psi$, $\varphi \supset \psi$, $\varphi \equiv \psi$ are formulae.

1.2.3 *Inductive Definition of Theorems*

1.2.3.1 A quantifier–free tautology is a theorem.

1.2.3.2 If a disjunction D of n alternatives is a theorem, $\varphi\alpha$ is one of the alternatives and β is a variable, then:

(a) If α is a term, then the result of replacing $\varphi\alpha$ by $(E\beta)\varphi\beta$ in D is a theorem;

(b) if α is a variable free in $\varphi\alpha$ but not free in the other alternatives and β is α or does not occur in $\varphi\alpha$, then the result of replacing $\varphi\alpha$ by $(\beta)\ \varphi\beta$ in D is a theorem.

1.2.3.3 If $\varphi \vee \ldots \vee \varphi$ is theorem, so is also φ.

The above formulation is complete only with respect to formulae in the extended prenex form.

1.3 *The Fundamental Theorem of Logic*

The main purpose of the next few sections is to study the decision problem on the theoretical foundation of the fundamental theorem of logic, an approach initiated by Skolem[4] and Herbrand,[5] and recently revived by Church,[6,7] and by Klaua[8] and Dreben.[9,10]

Suppose $Mxyz$ is a quantifier–free matrix:

1.3.1 $$(x)(Ey)(z)Mxyz,$$

1.3.2 $$(Ex)(y)(Ez) \sim Mxyz.$$

Let now D_n be $M_1 \vee \ldots \vee M_n$ and M_i be $M1ii'$, i' being an abbreviation for $i+1$. The fundamental theorem, when applied to 1.3.1, states:

1.3.3 The following three conditions are equivalent:

(a) 1.3.1 is a theorem of the predicate calculus; (b) for some n, D_n is a quantifier–free tautology; (c) 1.3.2 is not satisfiable.

If D_n is a quantifier–free tautology, then, by 1.2.3.1, both it and the result of substituting distinct variables for distinct numbers in it are theorems. For example, suppose the result is:

1.3.4 $$Maab \vee Mabc \vee Macd.$$

We have: by 1.2.3.2(b),

$$Maab \vee Mabc \vee (z)Macz;$$

by 1.2.3.2(a),

$$Maab \vee Mabc \vee (Ey)(z)Mayz.$$

Similarly,

$$Maab \vee (Ey)(z)Mayz \vee (Ey)(z)Mayz,$$
$$(Ey)(z)Mayz \vee (Ey)(z)Mayz \vee (Ey)(z)Mayz,$$

by 1.2.3.3,

$$(Ey)(z)Mayz;$$

by 1,2.3.2(b),

$$(x)(Ey)(z)Mxyz.$$

Hence, condition (b) implies conditions (a) and (c) in 1.3.3.

On the other hand, if no D_n is a quantifier–free tautology, then there is, for each D_n, some interpretation of the function and predicate symbols on the set $\{1, ..., n'\}$ which satisfies $\sim D_n$. By a well–known argument, there is then an interpretation on the domain of all positive integers which satisfies $\sim D_1$, $\sim D_2$, etc. simultaneously. This, however, means that under the interpretation each finite segment of the infinite conjunction

1.3.5 $\sim M112 \ \& \sim M123 \ \& \sim M134 \ \& \ ...$

is true. But then there is an integer x, viz. 1, such that for every integer y, there is an integer z, viz. y', such that $\sim Mxyz$. In other words, 1.3.2, the negation of 1.3.1, is true under the interpretation. Hence, the negation of condition (b) implies the negations of conditions (a) and (c).

If we take 1.3.5 as a model of 1.3.2, it seems natural to regard y as an independent variable z as a dependent variable and x as an initial variable (the limiting case of a dependent variable, a function of zero arguments). The general principle of constructing M_n from 1.3.1 may be summarized by saying that each initial variable gets a constant number, the independent variables taking on all possible positive integers as values and the dependnet variables always taking on numbers not used before.

In the general case, we must consider a disjunction (for provability) or conjunction (for satisfiability) of formulae with arbitrary strings of quantifiers. Then we can again construct the related quantifier–free formulae in the same way, with the numbers in each clause proceeding independently.

Thus, if we wish to study the satisfiability problem, we consider any formula of the form:

1.3.6 $$\varphi_1 \ \& \ ... \ \& \ \varphi_n \qquad (n \geqq 1),$$

where each φ_1 is of the form, with $d_1 \geqq 0$, $e_c \geqq 0$, $c \geqq 1$, e_1, d_2, e_2, ..., $d_c \geqq 1$:

1.3.7 $$(Ey_1^1) \ldots (Eyd_1^1)(x_1^1) \ldots (x_e{}^1) \ldots (Ey_1^c) \ldots$$

$$(Ey_{d_c}^c)(x_1^c) \ldots (x_{e_c}^c)\, My_1^1 \ldots x_{e_c}^c.$$

One familiar way of obtaining M_1, M_2, etc. for the formula 1.3.7 begins by replacing the dependent variables (those with the letter y) each with a function (sometimes called a "Skolem function") of all the preceding independent variables (those with the letter x), and then dropping all the quantifiers. Let the result be M^*. In particular, the initial (dependent) variables are replaced by distinct constants which may be viewed as trivial function. Suppose $e_1 + ... + e_c = p$, $d_1 + ... + d_c = q$ in 1.3.7.

The Skolem functions are any functions $g_1, ..., g_q$ which, taken together, satisfy the following conditions:

1.3.8 (a) For each g_i, $g_i(u_1, ..., u_m) \neq u_j$, $j = 1, ..., m_2$ $i = 1, ..., $ q.

(b) For each g_i, $g_i(u_1, ..., u_m) = g_i(v_1, ..., v_m)$ only when $u_1 = v_1, ..., u_m = v_m$.

(c) For any g_i, g_j, $i \neq j$, $g_i(u_1, ..., u_m) \neq g_j(v_1, ..., v_n)$, for all $u_1, ..., u_m, v_1, ..., v_n$.

Then we can take the smallest domain which contains the constants for the initial (dependent) variables (or an arbitrary constant when there is no such initial variable) and is closed with respect to the Skolem functions. Once such an (enumerable) domain is available, we can some-how enumerate all the p-tuples of members of the domain. Then, for each i, M_i is simply the result obtained from M^* when the independent variables are replaced respectively by members of the ith p-tuple.

The satisfiability problem of 1.3.7 is then reduced to that of the infinite conjunction:

1.3.9 $$M_1 \ \& \ M_2 \ \& ...$$

Similarly, the satisfiability problem of 1.3.6 can be handled by reducing each φ_i separately and then taking the conjunction of the n infinite conjunctions of the form 1.3.9.

It is customary to use the positive integers as the domain, fix some enumeration of the p-tuples, and specify the Skolem functions in a natural manner. One familiar enumeration of the p-tuples is the following:

1.3.10 $(a_1,...,a_p)$ precedes $(b_1, ..., b_p)$. if either

(a) they are permutations of each other but $(a_1,...,a_p)$ precedes $(b_1, ..., b_p)$ in the lexicographic order; or

(b) $\max(a_1,...,a_p) = \max(b_1, ..., b_p)$, $\Sigma a_i = \Sigma b_i$, but $(a_1,...,a_p)$, rearranged according to nondecreasing magnitude, precedes $(b_1, ..., b_p)$, similarly rearranged, in the lexicographic order; or

(c) $\max(a_1,...,a_p) = \max(b_1, ..., b_p)$, but $\Sigma a_i < \Sigma b_i$; or

(d) $\max(a_1,...,a_p) < \max(b_1, ..., b_p)$.

The Skolem functions are usually chosen by going through the infinite conjunction 1.3.9 from left to right and using each time the smallest unused integer for the next functional expression not yet evaluated. Thus, e. g., $y_1^1, ..., y_{d_1}^1$ in 1.3.7 get the constant values 1, ..., d_1, and M_1 is:

$$M1 \ ... \ d_1^1 1 \ ... \ 1 d_1' \ ... \ (d_1 + d_2) \ ... \ (q - d_c + 1) \ ... \ \mathrm{q}1 \ ... \ 1.$$

Fach time a functional expression gets a value, the value is substituted in all later occurrences of the same expression.

In this way we arrive at a form of the fundamental theorem of logic as a generalization of 1.3.3.

It is natural to observe that the infinite conjunction 1.3.9 can be divided into sections (Ref. 4, p. 138):

1.3.11 The first section is the set of those M_i's in which the p-tuples replacing the independent variables are made p of integers in the set $\{1, ..., d_1\}$, or the set $\{1\}$ if $d_1 = 0$; the $(n + 1)$th section is the set of those M_i's not belonging to the nth section in which the p-tuples are made up of integers which occur in the union of the first n sections.

The notion has been used by Skolem in explaining some decision procedures (see Section II below).

1.4 *Special Cases of the Decision Problem*

The principal known decidable classes are, with regard to satisfiability the following:

I. *The monadic case*. The class of all formulae which contain only monadic predicate letters and no function symbols.

II. *The EA satisfiability case (the AE provability case)*. The class of all formulae in the prenex form with prefixes of the form $(Ey_1) \ldots (Ey_m)(x_1) \ldots (x_n)$, $m, n \geq 0$, and no function symbols [or the form $(y_1) \ldots (y_m)(Ex_1) \ldots (Ex_n)$ for provability].

III. *The conjunctive satisfiability case*. Every formula in the prenex form with a matrix which is a conjunction of atomic formulae and their negations. (Equivalently, the disjunctive provability case.)

IV. *The Skolem case*. Every formula in the prenex form with no function symbols such that it has a prefix ending with $(Ey_1) \ldots (Ey_n)$, $n > 0$, and every atomic formula occurring in the matrix contains either one of the variables $y_1, \ldots, y_n$, or all the independent variables. [For provability, $(y_1) \ldots (y_n)$ at the end.]

V. *The EA_2E satisfiability case (the AE_2A provability case)*. Every formula containing no function symbols in the prenex form with a prefix $(Ey_1^1) \ldots (Ey_m^1)(x_1)(x_2)(Ey_1^2) \ldots (Ey_n^2)$.

VI. *The Ackermann case*. For satisfiability, every formula which contains no function symbols, no equality sign, only a single dyadic predicate (G say), and has the form $(x)(Ey)Gxy \; \& \; (x_1) \ldots (x_m) \, Mx_1 \ldots x_m$, $m \leqq 4$, M quantifier-free.

In addition to these, two other cases may be mentioned:

VII. *The $A_1E_1A_1$ satisfiability case*. Every formula with the prefix $(x_1)(Ey)(x_2)$ and with no function symbols.

VIII. *The Surányi normal form case*. For satisfiability, every formula which has no equality sign, no function symbols, only dyadic predicate symbols, and has the form $(x_1)(x_2)(x_3) \, Mx_1x_2x_3 \; \& \; (x_1)(x_2)(Ey_3)$, $Nx_1x_2y_3$, M, N quantifier-free.

It may be noted that in all the cases, with the single exception of III, no function symbols are permitted. Indeed, very little is known about the decision problem of formulae containing function symbols (compare Ref. 3, pp. 98—107). Unless otherwise stated, we shall always assume that no function symbols occur.

In what follows, cases I and VI will not be considered. So far as the monadic case without equality (a subcase of I) is concerned, it is possible to obtain a decision procedure from one for case II. Some of the problems suggested by the Ackermann case are also encountered by the $A_1E_1A_1$ case, while other implications of this case seem to call for a closer examination of certain arithmetic perdicates.

Formulae under case VIII form a reduction class in the sense that there is an effective procedure by which every formula, possibly containing $=$ and function symbols, can be reduced to one in the class with satisfiability preserved (Ref. 2, p. 60). It follows that there exists no decision procedure for this case. It is, however, desirable to find some "semidecision procedure" for the class which is a decision procedure for some subclass of it that is not specified explicitly in advance. It is thought that such semidecision procedures are a useful way of extending the range of formulae decidable by a predetermined finite set of procedures. A brief discussion is included in Section IV to point to the sort of thing which can be done along this line. It should be of interest to design semidecision procedures for case VIII, as well as for other reduction classes.

The case VII is perhaps the best known unsettled case; it has been mentioned in vrious connections (see, e.g., Ref. 11, p. 576 and Ref. 12, p. 420). In Section IV a procedure will be given which may be a decision procedur for the whole case but has only been shown to terminate for certain special cases. A proof of finiteness of the procedure is wanting. It is thought that, incomplete as the solution is, it is quite suggestive for further works on the decision problem. Some rather amusing combinatorial problems are also related to the considerations on this case.

An alternative decision procedure for the much-studied case V will be given in Section III in the equivalent form $\mathbf{A}_2\mathbf{E}$ (for satisfiability).

The Skolem case will be examined in considerable detail in Section II, using ideas proposed by Skolem[4](p.138) and Church[6] (p.264). Remarks relevant to machine realizations of the procedure will also be included.

The Skolem case includes the following special cases:

IVa. The A_1E satisfiability case. Because every atomic formula has to include some variable and there is only one independent variable.

IVb. For satisfiability, every formula whose prefix ends with $(Ey_1)\ \ldots\ (Ey_n)$, and in which every atomic formula contains at least one of the variables $y_1,\ldots,y_n$.

IVc. For satisfiability, every formula whose prefix is

$$(Ey_1{}^1)\ldots(Ey_m{}^1)(x_1)\ldots(x_n)(Ey_1{}^2)\ldots(Ey_k{}^2)$$

and in which every atomic formula contains either all of $x_1,\ldots,x_n$ or at least one of $y_1{}^2,\ldots,y_k{}^2$.

IVd. For satisfiability, every formula in the Skolem normal form, i.e., with prefix$(x_1)\ldots(x_m)(Ey_1)\ldots(Ey_n)$, such that every atomic formula contains at least m distinct variables.

For the extensive literature on the decision problem, the reader is referred to the bibliographies in Refs. 2 and 3. The writer has not been able to study carefully much of the relevant literature, and is not certain that the procedures deseribed in Sections II and III may not turn out to be inferior to existing ones. Recently, the writer noticed that ideas along the line of the solution of the E_1A provability case given in Section 3 of Part I[1] are contained in Skolem's writings (e.g., Ref. 4, p. 135).

Of the two remaining cases, II and III, some brief comments will suffice.

1.5 *Two Simple Cases*

The *EA* satisfiability case II has agreeable decision procedures not dependent on the fundamental theorem of logic (see Ref. 13, p. 13). It is also easy to devise a decision procedure on the basis of the fundamental theorem. Consider

1.5.1 $$(Ey_1)...(Ey_m)(x_1)...(x_n)My_1...x_n.$$

This is in fact equivalent to:

1.5.2 $$M_1 \& ... \& M_k,\ k = m^n, \text{ or } 1 \text{ when } m = 0.$$

In fact, this is a limiting case of the fundamental theorem because no Skolem functions are needed, so that the m constants for the initial variables are all we need for fabricating a model. In other words, either the negation of 1.5.2 is a quantifier–free tautology, and the negation of 1.5.1 is a theorem; or 1.5.2 has a model, and 1.5.1 has a model too. The presence of the equal sign is permitted, but the presence of function symbols in 1.5.1 would invalidate the procedure.

The conjunctive satisfiability case III was originally solved by Herbrand (Ref. 5, pp. 44—45). Suppose the matrix is:

1.5.3 $$A_1 \& ... \& A_m \& \sim B_1 \& ... \& \sim B_n,$$

or, in different notation:

1.5.4 $$A_1,...A_m \nrightarrow B_1,...,B_n.$$

Assume first that neither equality nor function symbols occur. If no predicate letter occurs both on the left side and on the right side, then we can simply choose to make all predicates occurring on the left side true of all numbers and those on the right false for all numbers, and then the infinite conjunction corresponding to the given formula is true under the interpretation.

Whenever there is one clause on the left and one on the right which contain the same predicate letter, e.g., A_i is *Gabc* and B_j is *Guvw*, we compare them and ask whether it is possible to assign the same integers to their arguments in some M_s and M_t respectively. If the answer is yes, the original formula can have no model, because the infinite conjunction must be always false. If the answer is no for every such pair, then the original formula has a model.

To compare A_i and B_j, we examine the three pairs of corresponding variables. If both variables in some pair are distinct dependent variables, then the two clauses A_i and B_j can never get the same numbers. When this is the case for none of the pairs, we can decide the question by asking whether there are positive intgegers s, t such that $a(s) = u(t)$, $b(s) = v(t)$ and $c(s) = w(t)$, where, for each variable α in the original formula, $\alpha(n)$ is a function giving the number which replaces α in M_n. It is possible to give a scheme to generate such function for each given formula. When there are solutions for some pair of clauses, the original formula is not satisfiable.

If the formula 1.5.4 contains function symbols but not $=$, then the comparison of A_i and B_j has to take functions into considerations sometimes. We may have to ask whether $f(a(s)) = g(u(t))$, instead of $a(s) = u(t)$, has a solution. In such cases, there is a solution only when f and g are the same function, because otherwise we can always give different values to $f(a(s))$ and $g(u(t))$ to avoid the incompatibility of M_s and M_t.

When the equals sign also occurs, we have to list all the equations among $A_1,...,A_m$, if there is any, and complete the list by using transitivity. If there are none, we need only to proceed as before, except that we can also reject satisfiability on the ground of, e. g., having an equation $u = v$ among $B_1, ..., B_n$, and $u(p) = v(p)$ has a solution in p. In the general case, we must compare A_i and B_j, which have the same predicate letter, in a more complicated manner. One way to do this is to give an effective survey of all the equalities obtainable in $M_1, ..., M_t$, for every t. And then the question of comparing *Gabc* and *Guvw* is reduced to the following: whether there are $p, q. t$ such that, with the help of the equalities obtainable from $M_1, ..., M_t$, we have $a(p) = u(q)$, $b(p) = v(q)$, $c(p) = w(q)$. Since these considerations are only subsidiary for the main purpose of the paper, details for this and other steps sketched above will not be supplied.

9.2 The skolem case

2.1 *Outline of a General Method*

The subcase IVb, where every atomic formula contains at least one of the last string of dependent variables, is particularly simple. Thus, in every M_k, each such variable always gets replaced by some new number so that no atomic formula in M_k can have occurred in any of $M_1,...,M_{k-1}$. *Hence, a formula of such a form is satisfiable if and only if* $\sim M_1$ *is not a quantifier-free tautology.*

In the general Skolem case, we make use of the definition of sections given above in 1.3.11. Let $(a_1^k, ..., a_p^k)$ be the p-tuple which replaces the dependent variables in M *to get* M_k.

Given any member M_i of the nth section, the only related instances in the nth section are those M_k for which$(a_1^k, ..., a_p^k)$ is a permutation of $(a_1^i, ..., a_p^i)$, and the only related instances in the (n + 1) th section are those M_j for which $(a_1^j, ..., a_p^j)$ include only numbers occurring in M_i and at least one number not in the set $\{a_1^j,..., a_p^j\}$.

Hence, it is possible to get a decision procedure by determining whether there exists any set of possibilities which includes models for the instances of the first section, as well as models for all related instances M_k any M_j for every model for M_i in the set.

When the formula is in the Skolem normal form or the form of IVc, somewhat more is true:

2.1.1 If M_j belongs to the $(n + 1)$th section, then it can have common atomic formulae with only at most one M_j in the nth section.

This is so because each atomic formula in M_j either contains a new number not occurring in any member of the nth section, or otherwise contains all of $\{a_1^j, ..., a_p^j\}$ with at least one number (say a_i^j) which appeared for the first time in one specific member (say M_i) of the nth section. In the first case the atomic formula in M_j does not occur in any member of the nth section. In the second case, M_j can contain no common atomic formula with any member of the nth section except possibly M_i, since a_i^j does not occur in any of the other members of the nth section.

Detailed considerations will be confined to the treatment of a simple special case.

2.2 *An Explicit Procedure for a Special Case*

We consider a very simple special case in which the matrix contains no equals sign (and of course no function symbols), and a single dyadic predicate G:

2.2.1 $$(x)(y)(Ez)\,Mxyz.$$

As an illustration, we use the negation of Example (2) of Part I:[1]

2.2.2 $(x)(y)(Ez)[(Gxy \,\&\, Gyx \,\&\, {\sim} Gxz \,\&\, {\sim} Gzy \,\&\, {\sim} Gzz) \quad \vee \quad (Gxz \,\&\, Gzy \,\&\, Gzz \,\&\, {\sim} Gxy \,\&\, {\sim} Gyx)]$.

In an alternative notation, the matrix is:

2.2.3 $$Cxy,\ Gyx \rightarrow Gxz,\ Gzy,\ Gzz;$$
$$Gxz,\ Gzy,\ Gzz \rightarrow Gxy,\ Gyx.$$

We construct a truth table of all the possibilities which can satisfy the above matrix:

2.2.4

Gxy	*Gyx*	*Gxz*	*Gzx*	*Gyz*	*Gzy*	*Gzz*
t	t	f			f	f
f	f	t			t	t

The blanks may take either t or f as values. Hence, there are eight rows in all.

For the prefix $(x)(y)(Ez)$, the numbers to substitute for (x,y,z) in M_1, M_2, M_3, M_4, etc., are (1,1,2), (1,2,3), (2,1,4), (2,2,5), etc. In order to decide whether a formula of the form 2.2.1 has a model, we ask whether it is possible to make *M*112, *M123*, *M214*, etc., simultaneously true, or, in other words, whether we can find for each M_i one row from the above table according to which M_i is true, such that these infinitely many rows are all compatible in the sense that the same atomic formula always gets the same truth value (t or f).

Among the number triples we can distinguish two classes, those in which x and y get the same numbers, such as (1,1,2), and those in which they get different numbers, such as (2,1,4). The conditions under which a model is possible are roughly: (i) to satisfy *Maab*, a row in the truth table has to behave in a way that x and y are interchangeable; (ii) for each row satisfying *Mabc*, there must be a related row satisfying *Mbac;* (iii) for the two types of row, two corresponding patterns of continuation must be possible, e. g.,

These conditions can be formalized more exactly and applied, in particular, to show that 2.2.2 has a model, and therefore its negation is not a theorem. For this purpose, we assume a formula of the form 2.2.1 for which a truth table *T* like 2.2.4 is constructed. When, for example, *Gxy* in a row *R* of *T* gets the same value as *Gzz* in a row *S* of *T*, we shall use the brief notation $R_{xy} = S_{zz}$.

2.2.5 A row *S* in the table *T* is a uniform row if $S_{xy} = S_{yx}$, $S_{xz} = S_{yz}$, $S_{zx} = S_{zy}$.

Clearly, for a row to satisfy *M*112, it is necessary that it be uniform. If there is no uniform row, then there is no model for the original formula.

2.2.6 A row *S* in the table *T* is an heir of a row *R* in *T* if *S* is a uniform row and $R_{zz} = S_{xy}$.

2.2.7 A row in *T* is trivial if it has no heir.

Since a row having no heir cannot be continued, we may. cross out all trivial rows and be concerned only with nontrivial rows. This is not theoretically necessary because further requirements would cross out trivial rows anyhow, but it makes for efficiency.

2.2.8 A row *R* in the table *T* is an ordinary row if there is a row *S* such that $R_{xy} = S_{yx}$, $R_{yx} = S_{xy}$, $R_{xz} = S_{yz}$, $R_{zx} = S_{zy}$, $R_{yz} = S_{xz}$, $R_{zy} = S_{zx}$. *R* and *S* are said to be mates of each other.

This is the condition under which *R* and *S* can satisfy (*M*123, *M*214) or (*M*214, *M*123) respectively.

In the table 2.2.4 for the formula 2.2.2, it is easily verified that only the two following rows are uniform rows or rodinary rows:

	Gxy	*Gyx*	*Gxz*	*Gzx*	*Gyz*	*Gzy*	*Gzz*
α	t	t	f	f	f	f	f
β	f	f	t	t	t	t	t

In fact, α and β are the only uniform rows, as well as the only ordinary rows. Each of α and β is only a mate of itself.

2.2.9 A uniform row *R* is permanent if (i) it has an heir which is permanent, and (ii) there is a permanent ordinary row *S* such that $R_{yz} = S_{xy}$, $R_{zy} = S_{yx}$. *S* is said to be a subordinate of *R*.

2.2.10 An ordinary row *R* is permanent if (i) it has an heir which is a permanent (uniform) row, (ii) it has a mate that is a permanent ordinary row, and (iii) there are two permanent ordinary rows *P* and *S* such that $R_{xz} = P_{xy}$, $R_{zx} = P_{yx}$, $R_{yz} = S_{xy}$, $R_{zy} = S_{yx}$. *P* and *S* are asid to be a pair of subordinates of *R*.

The two definitions 2.2.9 and 2.2.10 embody a simultaneous recursion. Condition (ii) in 2.2.9 is necessary, if, e.g., *R* is to satisfy *M*112 and *S* is to satisfy *M*123. Condition (iii) in 2.2.10 is necessary if, e.g., *R* is to satisfy *M*123, *P* is to satisfy *M*136 and *S* is to satisfy *M238*.

2.2.11 The formula 2.2.1 has a model if and only if its truth table *T* contains a permanent uniform row.

This assertion will be justified in 2.3. We observe first that both α and β are permanent uniform rows for the example 2.2.2. In fact, we have various models for the formula, which are determined, in outline, by the following patterns of continuation:

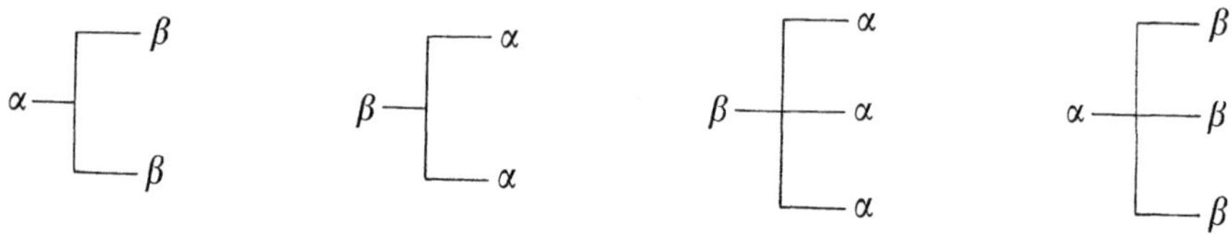

More exactly, choose. e.g., α as a model of *M*112. As a continuation of this, β satisfies *M*123 and *M*225; since β is its own mate in the sense of 2.2.8, β also satisfies *M*214. Similarly, since α is its own mate, as a continuation of β satisfying *M*123, α satisfies *M*136, *M*317, *M*238, *M*329, and *M*33(10). In this particular case, the model β of *M*214 can be continued in the same way. Moreover, the model β of *M*225 can be continued by the row α, and, e.g., the model α of *M*136 can be continued by the row β, and so on.

In the general case, a symmetry argument is needed to show that if a medel of, e.g., *M*123 can be continued, then a model of *M*214 can also be continued. For example, if (*R*, *S*) satisfy (*M*123, *M*124) respectively, and (*A*, *B*, *C*, *D*) satisfy respectively the continuation (*M*136, *M*317, *M*238, *M*329) of *M*123, then it is easy to see that (*B*, *A*, *D*, *C*) satisfy the corresponding extension of *M*214. This means that condition (ii) of 2.2.10 can be weakened to require a mate that is an ordinary row with a permanent heir.

The decision procedure implicit in the above definitions may be described explicitly thus:

2.2.12 The decision procedure:

1. Construct a truth table *T*.
2. Find all uniform rows.
3. Cross out all trivial rows.

Let U_0 be the set of remaining uniform rows, V_0 be the set of remaining ordinary rows. Each time, assume U_n and V_n are given and continue the following four steps:

4. Eliminate every uniform row from U_n which has no subordinate row in V_n, thus obtaining U_{n+1} from U_n and V_n.

5. Eliminate from V_n every ordinary row which has no mate or no pair of subordinate rows in V_n , thus obtaining V_{n+1} from V_n.

6. Eliminate every uniform row from U_{n+1} which has no heir in U_{n+1}, thus obtaining U_{n+2} from U_{n+1}.

7. Eliminate every ordinary row from V_{n+1} which has no heir in U_{n+2}, thus obtaining V_{n+2} from V_{n+1} and U_{n+2}.

8. The steps 4 through 7 are repeated until one of two things happens: either at some stage we obtain an empty U_i and an empty V_i, then we stop and conclude that the original formula 2.2.1 has no model; or else after a whole round of the steps 4 and 7, we find U_{n+2} and V_{n+2} remain the same as U_n and V_n, then we stop and conclude that the original formula 2.2.1 has model.

In practice, it is more efficient to perform, if possible, each of the steps 4 through 8 repeatedly, before going to the next step.

The procedure is clearly finite, since U_0 and V_0 are finite, and each round of steps 4 through 8 must reduce the size of U_n or V_n if the procedure has not come to a stop yet. Moreover, the final sets U_i and V_i must be both empty or both nonempty.

2.3 *Justification of the Procedure*

As a Skolem case, the formula 2.2.1 must not contain *Gxx* and *Gyy*. It is, however, not obvious that we are justified in not including two columns *Gxx* and *Gyy in the truth tables such as 2.2.4. For a model constructed on the basis of such reduced tables, it is not evident that, for some positive integer a, Gaa* might not be compelled to take on the value t at one place, and the value f at another. However, we can prove the following: 2.3.1 In every model obtained on the basis of a truth table not including columns for *Gxx* and *Gyy*, for every number *a*, *Gaa* is never compelled to take on two different values.

Take, for example, *G*22. If *Gzz* occurs in the original formula, *G*22 is compelled to take a fixed value in a model with a row *R* for *M*112. In the same model, if *S* is the row for *M*225, then $R_{zz} = S_{xy} = S_{yx}$. Hence, it is harmless that S_{xx} and S_{yy} are compelled to take the same value as both R_{zz} and S_{xy}(oor S_{yx}). In all other cases, the values for *G*22 can always be given the value of R_{zz} because there is no other place where *G*22 is independently compelled to take a certain turth value.

For the same reason, if neither an atomic formula nor any one obtainable from it by permuting the variables occurs, we may leave out the columns for them. For example, if *Gzz* does not occur, we can leave it out. If neither *Gxy* nor *Gyx* occurs, we can leave both of them out.

On the other hand, if e.g., *Gxy* and *Gzy* occur but *Gyx* does not, we still must include a column for *Gyx*. Otherwise, since we do not record the value of *Gyx*, it may happen that *R* satisfies *M*112, with R_{zy} = t, and *S* satisfies *M*214 with S_{xy} = f. Then no row *P* can satisfy *M*123, because P_{yx} is compelled to take both the value t and the value f, and this is not recorded without a column for *Gyx*.

To prove 2.2.11, we remark first that there are three types of instances illustrated by *M*112, *M*123, *M*214. For the first kind, an M_i of the form *Maab*, the only $M_j, j > i$, which have common atomic formulae with M_i *are Mbbc, Mabd, Mbae,* because these are the only ways in which both the independent variables *x* and *y* can be replaced by numbers occurring in M_i, and having only one of the two arguments from M_i yields no common atomic formula. Similarly, if $M_i$is *Mabc*, $a < b$, there are only five $M_j, j > i$ which have common atomic formula with M_i. By the symmetry argument preceding 2.2.12, the mate *Mbae* is also taken care of.

Hence, if there is any permanent uniform row, we canfind a model for all instances M_1, M_2, etc., such that each has some common atomic formula with an earlier one, or, in other words, all those occurring on an infinite tree beginning at M_1. This does not exhaust all the instances. For example, M_{14} and M_{15} [i.e., *M*34(15) and *M*43(16)] are not included. Since, however, they contain no common atomic formulae with the instances already interpreted, we can take two permanent ordinary rows which are mates and get a model for another sequence of instances. In this way, it is seen that, if there is a permanent uniform row in the table *T*, then one can so interpret the predicate *G* in the domain of the positive integers that the whole sequence M_1, M_2, etc., are simultaneously satisfied.

The converse is quite obvious. If there is no permanent uniform row, then no interpretation of *M*112 can be continued indefinitely, and there is an *i*. such that M_1 & ... & M_i is true under no interpretation.

2.4 *Questions of Efficiency*

When doing an example by hand, there are shortcuts we find natural to use. These may be viewed as more refined methods which can be mechanized by additional efforts. We give some informal illustration of the type of quick method we tend to use.

Consider the negation of Example (3) given in Part I.[1]

2.4.1 $(x)(y)(Ez)\{[Gxy \;\&\; (\sim Gyz \vee \sim Gzz)] \vee [(Gxy \;\&\; Hxy) \;\&\; (\sim Hxz \vee \sim Hzz)]\}$.

In the alternative notation, the matrix of the above formula is:

2.4.2 $Gxy \nrightarrow Gyz$; $Gxy \nrightarrow Gzz$; $Gxy, Hxy \nrightarrow Hxz$; $Gxy, Hxy \nrightarrow Hzz$.

The truth table for this is:

2.4.3	*Gxy*	*Hxy*	*Gyx*	*Hyx*	*Hxz*	*Hzx*	*Gyz*	*Gzy*	*Gzz*	*Hzz*
α	t						f			
β	t								f	
γ	t	t			f					
δ	t	t								f

Although the formual contains two predicates instead of just one, it is easy to see that the procedure described above can be extended to cover the case in a very straight–forward manner.

Since there are many blanks in the table, it is essential for efficiency that we do not expand the table by filling in the blanks (there would be 2^{24} rows), until we are compelled to do so. In other words, we try to carry out the decision procedure by treating each row containing blanks as a single row and make expansion only when we are not able to eliminate them as single rows.

We observe that for every row, in particular, every uniform row, *Gxy* gets the value t. It follows that row β, or more exactly, all the 2^7 rows obtainable from β are trivial by 2.2.7, since an heir of β must have *Gxy* take the value of *Gzz* in β, which is f. Hence we may delete row β altogether.

In order that row α, or any specification *R* of α, be permantnt (uniform or ordinary), it is necessary, by 2.2.9 and 2.2.10, that there is a subordinate row *S*, such that *Gxy* gets the same value in *S* as *Gyz* in *R*, or $R_{Gyz} = S_{Gxy}$. But this is impossible because R_{Gyz} is f in every row obtainable from α, but S_{Gxy} is *t* in every row. Hence, we can delecte row α altogether, and be concerned only with the rows γ and δ.

Since *Hxy* gets t in all the remaining rows and *Hzz* gets the value f in δ, every row obtainable from δ has no heir, and the whole row δ can be deleted.

However, no permanent ordinary row can be obtained from γ alone bacause, by 2.2.10, for any such row R there must be a subordinate row P such that $R_{Hxz} = P_{Hxy}$, but in row γ, Hxz is always f and Hxy is always t. Hence, there can also be no permanent uniform row, and, by 2.2.11, the formula 2.4.1 has no model. Therefore, Example (3) in Part I,[1] the negation of 2.4.1, is a theorem.

Another method of deciding 2.4.1 is the following. We begin with M_1, which is a disjunction of conjunctions, and choose M_i, M_j, etc., which contain common atomic formula with M_1, in the hope that M_1 & M_i & M_j & ... as multiplied out into a disjunction of conjunctions will include in each conjunction some atomic formula and its negation. The process may have to be continued.

As we observed before, only M_2, M_3, M_4 can have common atomic formulae with M_1. Of these three, on account of the special structure of 2.4.1, M_3 has no common part with M_1. Hence, we need to consider, to begin with, only M_1, M_2, M_4:

	(i)	(ii)	(iii)	(iv)
$M112$	$G11 \nrightarrow G12$;	$G11 \nrightarrow G22$;	$G11, H11 \nrightarrow H12$;	$G11, H11 \nrightarrow H22$
$M123$	$G12 \nrightarrow G23$;	$G12 \nrightarrow G33$;	$G12, H12 \nrightarrow H13$;	$G12, H12 \nrightarrow H33$
$M225$	$G22 \nrightarrow G25$;	$G22 \nrightarrow G55$;	$G22, H22 \nrightarrow H25$;	$G22, H22 \nrightarrow H55$

By the row for $M123$, (i) of $M112$ can be deleted because (i) contains $\sim G12$ (i.e., after $\nrightarrow$), while each clause in the row for $M123$ contains $G12$. It can be seen then that every row in column(i) can be deleted in the same way, Similarly, (ii) of the row for $M112$ can be deleted because it contains $\sim G22$, while each clause in the row for $M225$ contains $G22$; therefore, the whole column (ii) can be deleted eventually, and we need only consider the columns (iii) and (iv). But then (iii) of the row for $M112$ can also be deleted because it contains $\sim H12$, and all the remaining columns of the row for $M123$ contain $H12$. Finally, we have only column (iv) left. Now, however $\sim H22$ occurs in the row of $M112$ and $H22$ occurs in the row for $M225$. Hence, the conjunction of the three rows of column (iv) is a contradiction, and 2.4.1 has no model.

2.5 *The Inclusion of Equality*

The decision procedure in 2.2 can be extended to deal with cases where the equal sign occurs in the given formula:

2.5.1 $\qquad (x)(y)(Ez)\,Mxyz, \qquad$ with $=$ occurring.

Additional considerations are needed to take care of the special properties of $=$. First we bring $Mxyz$ into a disjunction of conjunctions of atomic formulae and their negations, in the usual manner. Then we modify the resulting matrix to take care of the properties of $=$. (a) Each conjunction that contains an inequality of the form $v \neq v$, v being x or y or z, is deleted. (b) In each conjunction, a clause of the form $v = v$ is deleted. (c) Within each conjunction, if $u = v$ is a clause with distinct variables u and v, we add also, as new clauses (if not occurring already), $v = u$ and the result of replacing any number of occurrences of u by v(oor v by u) in each clause of the conjunction; this is repeated for every equality until no new clause is generated. (d) Repeat the steps (a) and (b) on the result obtained by step (c); in addition, any conjunction which contains both an atomic formula and its negation is deleted.

We now construct the truth table on the basis of the new matrix (in a disjunctive normal form). Uniform rows, ordinary rows and permanence can be defined in a similar manner as before, except that a uniform row has to satisfy the additional condition that $x = y$ and $y = x$ both get the truth value t (not only that they just get a same value). In this way, we can obtain a decision procedure for all formulae of the form 2.5.1.

It is believed that the same type of consideration can be used to extend all the cases considered in this paper to include also the equal sign. In the next two sectings, equality will be left out and attention will be confined to formulae not containing the equals sign (nor, of course, function symbols).

9.3 The A_2E satisfiability case

We give an alternative treatment of this case which, it is conjectured, is in general more efficient than the method of Schütte[11] as reformulated by Klaua.[8] The method will be explained with the special case when only one dependent variable and only one dyadic predicate G occur:

3.1 $$(x)(y)(Ez)Mxyz.$$

The main difference between this case and the case solved in 2.2 above is that Gxx and Gyy are permitted to occur in $Mxyz$. As a result, for example, $M123$ may contain common atomic formula with any $Mabc$ in which a or b is one of 1, 2, 3.

As an example, we choose arbitrarily the following:

3.2 $$(x)(y)(Ez)[\sim Gxx \mathbin{\&} (Gxy \supset \sim Gyx) \mathbin{\&} Gxz \mathbin{\&} (Gzy \supset Gxy)].$$

The matrix may be rewritten as:

3.3 $$Gxz \nrightarrow Gxx,\ Gxy, Gzy;\ Gxz, Gxy \nrightarrow Gxx, Gyx;$$
$$Gxz \nrightarrow Gxx, Gyz, Gzy.$$

The truth table is:

3.4

Gxx	*Gxy*	*Gyx*	*Gyy*	*Gxz*	*Gyz*	*Gzx*	*Gzy*	*Gzz*
f	f			t			f	
f		f		t			f	
f	t	f		t				

The problem is, as before, to decide whether there is a model that satisfies M_1, M_2, etc., simultaneously. The conditions are rather similar to those in 2.2 except that for any two rows R and S which, say, satisfy $Mabc$ and $Mdef$ in a model, there must be two rows which satisfy $Mcfg$ and $Mfch$ in the model. There is also a related requirement for a row satisfying M_1, because the number 1 is never used to replace a dependent variable. The various conditions may be stated:

3.5 A row R is uniform if $R_{xx} = R_{xy} = R_{yx} = R_{yy}$, $R_{xz} = R_{yz}$, $R_{zx} = R_{zy}$.

3.6 A row S is an heir of a row R if S is uniform and $R_{zz} = S_{xx}$.

3.7 Two rows R and S form a parallel pair if $R_{xx} = S_{yy}$, $R_{xy} = S_{yx}$, $R_{yx} = S_{xy}$, $R_{yy} = S_{xx}$, $R_{xz} = S_{yz}$, $R_{yz} = S_{xz}$, $R_{zx} = S_{zy}$, $R_{zy} = S_{zx}$.

Two rows of a parallel pair are said to be mates of each other.

If R and S are to satisfy *Mabc* and *Mbae*, it is necessary that they form a parallel pair. In general, for a row satisfying *Mabc*, there must also be two parallel pairs of related rows satisfying *Macd*, *Mcae*, *Mbcf*, *Mcbg*, When $a = b$, the two parallel pairs become one. This, plus the requirement that every row in a model must have an heir may be summarized in the following condition.

3.8 A row R is normal if the following conditions are all satisfied:

3.8.1 It has a normal row as a mate;

3.8.2 It has an heir which is a normal row;

3.8.3 There are two normal rows P and S such that $R_{xx} = P_{xx}$, $R_{xz} = P_{zy}$, $R_{zx} = P_{yx}$, $R_{zz} = P_{yy}$, and $R_{yy} = S_{xx}$, $R_{yz} = S_{xy}$, $R_{zy} = S_{yx}$, $R_{zz} = S_{yy}$. Such rows P and S are said to be subordinates of R.

A uniform row is its own mate, although a self–mated row is not always a uniform row. For a uniform row, 3.8.1 is a redundant condition, and P and S coincide in 3.8.3. The definition 3.8 of normality is clearly recursive.

In the table 3.4, we observe that, because *Gxx* always takes the value f, *Gzz* can only take the value f in order that the row has an heir. Moreover, since *Gxx* always gets the value f and *Gxz* always gets the value t, in order that a row has a mate, *Gyy* must always take the value f and *Gyz* always t. Hence, we need consider only the following eight rows which result from filling the remaining gaps:

3.9

	Gxx	*Gxy*	*Gyx*	*Gyy*	*Gxz*	*Gyz*	*Gzx*	*Gzy*	*Gzz*
α	f	f	f	f	t	t	f	f	f
β_1	f	f	t	f	t	t	f	f	f
β_2	f	t	f	f	t	t	f	f	f
a	f	f	t	f	t	t	t	f	f
b	f	t	f	f	t	t	f	t	f
c	f	f	f	f	t	t	t	f	f
d	f	t	f	f	t	t	t	f	f
e	f	t	f	f	t	t	t	t	f

Row e has no mate, beacuse of the columns 5 to 8. Rows c and d have no mate, because b, the only row satisfying the condition on *Gzx* and *Gzy*, does not satisfy the condition on *Gxy* and *Gyx*. Neither row a nor row b has subordinates as required by 3.8.3. Hence, we have only the remaining rows α, β_1, β_2 to, consider.

α is the only uniform row, (β_1, β_2) form a parallel pair, and β_2 is both P and S in 3.8.3 for all the three rows α, β_1, β_2. Hence, we have, for example:

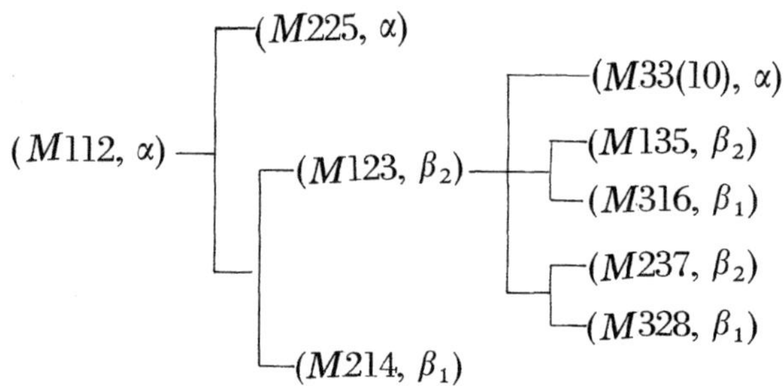

In particular, $(M124, \beta_1)$ can be continued in the same way as $(M123, \beta_2)$. Indeed, continuation in every branch can be made similarly. In other words, α, β_1, β_2 are all normal by 3.8. This, however, does not yet secure a model for the formula 3.2. There are, for example, those instances in which (1, 5), (5, 1), (3, 4), (4, 3), etc., replace (x, y) of $Mxyz$; they also have common atomic formulae with the instances shown in the above graph.

3.10 A formula 3.1 has a model if and only if (a) it has a nonempty table of normal rows, (b) this table has a nonempty subtable T such that:

3.10.1 For every pair (R, S) in T', there is a parallel pair (P, Q) in T' such that $P_{xx} = R_{zz}$, $Q_{xx} = S_{zz}$.

3.10.2 There is a uniform row R in T' such that for every row S in T', there is a parallel pair (P, Q) in T', for which $P_{xx} = R_{xx}$, $Q_{xx} = S_{zz}$.

There are the additional requirements mentioned after 3.4. In the example under consideration, the table consisting of all the three normal rows α, β_1, β_2 satisfies the requirements on T'. Hence, 3.2 does have models. One model for the predicate G is the relation $<$ among positive integers. That is, however, not the only model, because the model of G does not have to be transitive. For example, $G15$ and $G51$ can be (t, f) or (f. t) or (f, f).

It can be verified that the conditions in 3.10 are indeed necessary and sufficient.

9.4 The $A_1E_1A_1$ Satisfiability case

4.1 *A Generalized Game of Dominoes*

The study of the decision problem of the present case has suggested a related abstract mathematical problem which can easily be stated in everyday language. The problem appears to be of interest even to those who are not concerned with questions in mathematical logic.

Assume we are given a finite set of square plates of the same size with edges colored, each in a different manner. Suppose further there are infinitely many copies of each plate (plate type). We are not permitted to rotate or reflect a plate. The question is to find an effective procedure by which we can decide, for each given finite set of plates, whether we can cover up the whole plane (or, equivalently, an infinite quadrant thereof) with copies of the plates subject to the restriction that adjoining edges must have the same color.

For example, suppose a set consists of the three plates:

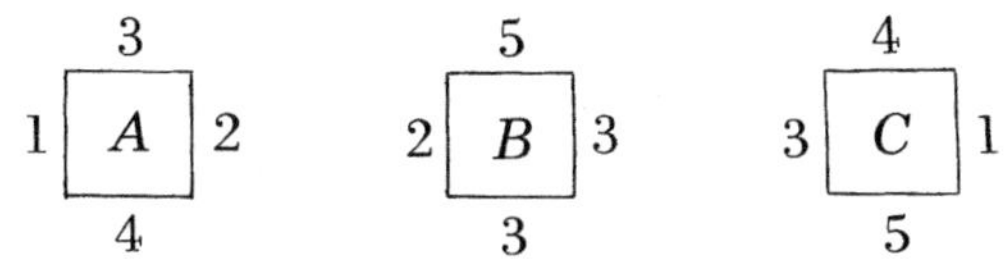

Then we can easily find an infinite solution by the following argument. The following configuration satisfies the constraint on the edges:

$$\begin{matrix} A & B & C \\ C & A & B \\ B & C & A \end{matrix}$$

Now the colors on the periphery of the above block are seen to be the following:

$$\begin{matrix} & 3 & 5 & 4 & \\ 1 & & & & 1 \\ 3 & & & & 3 \\ 2 & & & & 2 \\ & 3 & 5 & 4 & \end{matrix}$$

In other words, the bottom edge repeats the top edge, and the right edge repeats the left edge. Hence, if we repeat the 3×3 block in every direction, we obtain a solution of the given set of three plates. In general, we define a "cyclic rectangle".

4.1.1. Given any finite set of plates, a cyclic rectangle of the plates is a rectangle consisting of copies of some or all plates of the set such that: (a) adjoining edges always have the same color; (b) the bottom edge of the rectangle repeats the top edge; (c) the right edge repeats the left edge.

Clearly, a snfficient condition for a set of plates to have a solution is that there exists a cyclic rectangle of the plates.

What appears to be a reasonable conjecture, which has resisted proof or disproof so fam is:

4.1.2 *The fundamental conjecture:* A finite set of plates is solvable (has at least one solution) if and only if there exists a cyclic rectangle of the plates; or, in other words, a finite set of plates is solvable if and only if it has at least one periodic solution.

It is easy to prove the following:

4.1.3 If 4.1.2 is true, we can decide effectively whether any given finite set of plates is solvable.

Thus, we proceed to build all possible rectangles from copies of the plates of different sizes, using smaller ones first. If 4.1.2 is true, the process will always terminate in one of two ways: either at some stage we arrive at a cyclic rectangle and, therefore, the original set is solvable; or else we arrive at a size such that there is no rectangle of that size in which adjoining edges always have the same colors. The latter alternative is in fact a necessary and sufficient condition under which the original set is not solvable. However, if 4.1.2 is not true, it would be possible that a set has a solution, but we can never see this fact by the latter criterion at any finite stage: there would always be the possibility that for the next size there exist no rectangles with same–colored adjoining edges.

There is a naturally uneasy feeling about the effectiveness of such a procedure. The argument is essentially the familiar one that if a set and its complement are both recursively enumerable, then the set is recursive. It shows that the procedure always terminates (provided 4.1.2 is true) but gives no indication in advance as to how long it might take in each case.

If 4.1.2 is proved, it seems likely that it would be proved in a stronger form by exhibiting some simple recursive function f with the following property. For any set of plates with m distinct colors and n distinct plates, if the set is solvable, there is a cyclic square of the size $k \times k$, where $k = f(m, n)$. If that happens, or even if we have not exhibited such a function f but 4.1.2 can be proved by fairly elementary arguments, we would have some estimate in advance of how long the procedure takes in each case.

As it is, we can make the testing procedure quite systematic even though we do not know whether 4.1.2 is true. The procedure would be a decision procedure and presumably quite an efficient one, if 4.1.2 is true. If 4.1.2 should turn out to be false, then the procedure would only be a semidecision procedure. In fact, it is possible to show that the procedure does work in several classes of cases, e.g., when a set has unique solution apart from translations, or whenever either horizontally or vertically no color can be followed by different colors. But we shall not delay over such partial results.

If 4.1.2 should be false, then there would be two possibilities: either the set of all solvable finite sets of plates is not recursive, or it is recursive but requires a more complex decision procedure.

The problem can clearly be generalized to higher dimensions: for example, to cubes with colored surfaces instead of squares with colored edges.

We return now to the $A_1E_1A_1$ satisfiability case.

4.2 *Preliminary Definitions and an Example*

The general form of the case is:

4.2.1 $$(x)(Ey)(z)\, Mxyz,$$

where M is a quantifier–free matrix containing neither function symbols nor the equality sign. From the fundamental theorem, it follows that 4.2.1 is satisfiable (solvable) if and only if each finite subset of the infinite set of matrices $Miij$ ($i, j = 1, 2, \ldots$) is solvable (not contradictory). Since the second number is always the successor of the first, we shall write Mij for $Mii'j$.

We illustrate the general case by considering the special case where $Mxyz$ contains only a single dyadic predicate G. The negation of Example (4) given in the introduction of Part I[1] will be the concrete example:

4.2.2 $$(x)(Ey)(z)\,[\sim Gxx \;\&\; Gxy \;\&\; (Gyz \supset Gxz)].$$

In the alternative notation, the matrix is

4.2.3 $$Gxy,\ Gxz + \rightarrow Gxx;\ Gxy + \rightarrow Gyz,\ Gxx.$$

The truth table is:

4.2.4

Gxx	Gxy	Gyx	Gyy	Gxz	Gzx	Gyz	Gzy	Gzz
f	t			t		t		
f	t			t		f		
f	t			f		f		

Since there are five blank columns, there are altogether 3×2^5 or 96 rows. The problem now is to decide whether we can choose one row for each matrix $Mij(ij = 1, 2, \ldots)$ such that , taken, together, all the matrices come out true. This really involves both the problem of finding the pieces and the problem of putting them together. Thus, if j is distinct from i and i', any row can satisfy Mij alone, if we substitute i, i', j for x, y, z in the truth table; but a row can satisfy Mij when j is i or i' only in case certain related columns get the same truth values. This is the problem of finding the pieces. When there are such pieces, there is the harder problem of putting them together. For example, if there are rows satisfying $M11$ and $M12$ separately, there may yet be no pair of rows which satisfy $M11$ and $M12$ simultaneously because the common atomic formulae in both matrices must get identical values.

Since the putting–together part is quite complex, it seems natural to combine small pieces into blocks first. For this purpose, we consider row pairs and row quadruples (i.e., pairs of pairs).

D4.1 Two rows P, Q in the truth table T form a basic row pair (P, Q) if, for some i, they can simultaneously satisfy Mii' and Mii respectively. More explicitly, the conditions are:

i. $P_{yy} = P_{yz} = P_{zy} = P_{zz}$, $P_{xy} = P_{xz}$, $P_{yx} = P_{zx}$;

ii. $Q_{xx} = Q_{xz} = Q_{zx} = Q_{zz}$, $Q_{xy} = Q_{zy}$, $Q_{yx} = Q_{yz}$;

iii. $P_{xx} = Q_{xx}$, $P_{xy} = Q_{xy}$, $P_{yx} = Q_{yx}$, $P_{yy} = Q_{yy}$.

In the table 4.2.4, it is easy to verify that there are only two basic row pairs (α, β) and (γ, δ):

4.2.5

	Gxx	Gxy	Gyx	Gyy	Gxz	Gzx	Gyz	Gzy	Gzz
α	f	t	f	f	t	f	f	f	f
β	f	t	f	f	f	f	f	t	f
γ	f	t	f	t	t	f	t	t	t
δ	f	t	f	t	f	f	f	t	f

Obviously basic row pairs are necessary for building a model of 4.2.1. In fact, given any formula 4.2.1, if its truth table T contains no basic row paris, then it has no model and, indeed, the conjunction of $M11$ and $M12$ is a contradiction.

We shall consider pairs of row pairs, called row quadruples, which are useful in chaining row pairs together.

D4.2. Given any two row quadruples $(A, B; C, D)$ and $(P, Q; R, S)$, if $C = P$, $D = Q$, then the former is a predecessor of the latter and the latter is a successor of the former.

D4.3 Four rows P, Q, R, S form a basic row quadruple $(P, Q; R, S)$ if, for some i, they satisfy simultaneously Mii', Mii, $M'ii''$, $Mi'i'$, respectively, or, more explicitly, if:

i. (P, Q) and (R, S) are basic row pairs;

ii. $P_{yy} = R_{xx}$;

iii. $(P, Q; R, S)$ has a successor which is a basic row quadruple.

In the table 4.2.4, there is only one basic row quadruple, viz. $(\alpha, \beta; \alpha, \beta)$. The quadruple $(\alpha, \beta; \gamma, \delta)$ satisfies i and ii, but not iii. It is easy to see that, given any formula 4.2.1, if its truth table T contains no basic row quadruples, then it has no solution and, indeed, the conjunction of $M12$, $M11$, $M23$, $M22$, $M34$, $M33$ is a contradiction.

Clearly, if a row R satisfies Mij in a model, then there must be one row S which satisfies Mji, one basic row quadruple $(A, B; C, D)$ which satisfies Mii', Mii, $Mi'i''$, $Mi'i'$, and one basic quadruple which satisfies Mjj', Mjj, $Mj'j''$, $Mj'j'$. In particular, when j is i, we get the basic row pairs which occur in some basic quadruple.

D4.4 Two rows R, S form an ordinary row pair (R, S) if

i. $R_{xx} = S_{zz}$, $R_{xz} = S_{zy}$, $R_{zx} = S_{yz}$, $R_{zz} = S_{yy}$;

ii. There is a basic quadruple $(A, B; C, D)$ such that $A_{xx} = R_{xx}$, $A_{zy} = R_{xy}$, $A_{yx} = R_{yx}$, $A_{yy} = R_{yy}$;

iii. There is a basic quadruple $(P, Q; K, L)$ such that $P_{xx} = S_{xx}$, $P_{zy} = S_{xy}$, $P_{yx} = S_{yx}$, $P_{yy} = S_{yy}$.

In the table 4.2.4, since the only basic quadruple is $(\alpha, \beta; \alpha, \beta)$, it is relatively simple to find all the rows which do occur in ordinary row pairs. Since every row which is to satisfy some Mij in any solution must occur as one row in some ordinary row pair, we tabulate all such rows together and, from now on, confine our attention to them. It happens in this example that all these rows have in common five columns:

Gxx	*Gxy*	*Gyx*	*Gyy*	*Gzz*
f	t	f	f	f

Therefore, we only have to list the remaining columns:

4.2.6

	Gxz	*Gzx*	*Gyz*	*Gzy*	*ordinary pairs*
α	t	f	f	f	(α, β)
β	f	f	f	t	(β, α)
δ_1	t	t	t	t	(δ_1, δ_1)

δ_2	f	f	f	f	(δ_2, δ_2)
δ_3	t	f	f	t	(δ_3, δ_3)
δ_4	t	f	t	f	(δ_4, δ_5)
δ_5	f	t	f	t	(δ_5, δ_4)
δ_6	t	f	t	t	(δ_6, δ_7)
δ_7	t	t	f	t	(δ_7, δ_6)

In fact, if only the four columns have to be considered, there are 12 rows in the original table 4.2.4, and the two rows (R, S) in each ordinary row pair satisfy the condition: $R_{xz} = S_{xy}$, $R_{zx} = S_{yz}$. Hence, it is easy to get the above table. Briefly, the relevant information for the example is the nine ordinary pairs given above and the basic quadruple $(\alpha, \beta; \alpha, \beta)$.

Thus far we have been concerned only with rather elementary properties of the rows in the truth table. The more involved part is to design a scheme of extending recursively the construction of models. In order to explain how this is done, we introduce a chart.

4.2.7 Chart for $(x)(Ey)(z)\ Mxyz$:

Basic Pairs (Mii', Mii) *Cyclic Pairs* $(Mii'', Mi'i)$ *Common Row Pairs* $(Mi(i''+k), M(i'+k)i)$

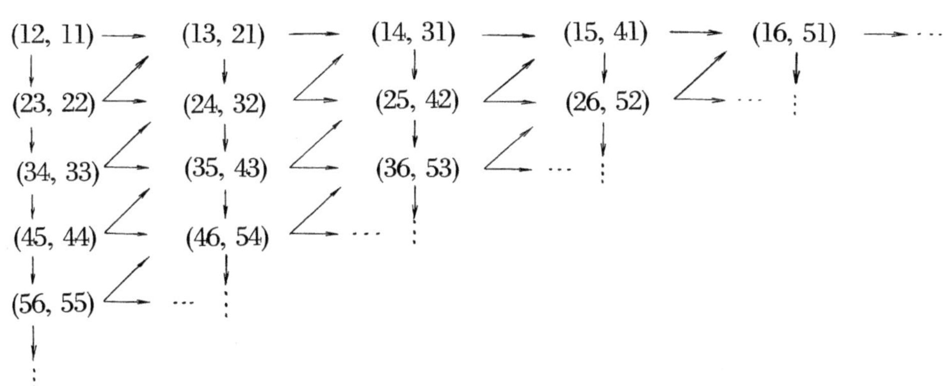

In the chart, the ordinary row pairs satisfying (Mij', Mji) are divided into three classes: basic when $i = j$, cyclic when $i' = j$, common otherwise. The general plan of the procedure is as follows. The existence of basic row quadruples assures that we can find a model for all the matrices $M12$, $M11$, $M23$, $M22$, etc. in the first column. Similarly, we can define cyclic quadruples to give an effective condition for the existence of a model for all matrices appearing in the second column of the chart, and so on. But in order that these models can be combined to give a model for all the matrices and therewith for a given formula 4.2.1, each column must be related to the column on its left in a suitable manner. This situation with two infinite dimensions seems to be the chief cause of the complexity of the $A_1E_1A_1$ case.

In the chart of 4.2.7, each row pair (R, S) that is not basic is subordinate to a quadruple $(A, B; C, D)$ made up of the two row pairs (A, B), (C, D) on its left with arrows leading to it. The quadruple is said to be superior to the pair (R, S).

D4.5 An ordinary row pair (R, S) is a subordinate of a quadruple $(A, B; C, D)$ if
i. $R_{xx} = A_{xx}$, $R_{xy} = A_{xy}$, $R_{yx} = A_{yx}$, $R_{yy} = A_{yy}$, $R_{yz} = C_{xz}$, $R_{xy} = C_{zx}$, $R_{zz} = C_{zz}$;
ii. $S_{xx} = D_{xx}$, $S_{xy} = D_{xy}$, $S_{yx} = D_{yx}$, $S_{xz} = A_{zx}$, $S_{zx} = A_{xz}$.
Aquadruple $(R, S; P. Q)$ is subordinate to a row sextuple $(A, B; C, D; K, L)$ if (R, S) is subordinate to $(A, B; C, D)$, and (P, Q) to $(C, D; K, L)$.

D4.6 Two rows R, S form a cyclic row pair $(R. S)$ if
i. $(R. S)$ is an ordinary row pair;
ii. $R_{xy} = S_{zx}$, $R_{yx} = S_{xz}$, $R_{yy} = S_{xx}$, $R_{yz} = S_{xy}$, $R_{zy} = S_{yx}$.
Obviously, given 4.1, if its table contains no two rows forming a cyclic pair, then the conjunction, briefly C_6, of $M12$, $M11$, $M23$, $M22$, $M13$, $M21$ is a contradiction.

In the table 4.2.6, there are, among the nine ordinary row pairs, only one that is cyclic, (δ_4, δ_5). Since there are only one basic quadruple, each has only one superior. This is of course not always the case, it is only due to special features of the example 4.2.2.

In order to find out whether there is any succession of cyclic pairs which will satisfy all rows of the column for cyclic pairs in the chart, we study cyclic quadruples.

D.4.7 Four rows P, Q, R, S form a cyclic quadruple (P, Q, R, S) if
i. $(P. Q)$ and (R, S) are cyclic row pairs;
ii. $Q_{xx} = R_{xx}$, $Q_{xy} = R_{xy}$, $Q_{yx} = R_{yx}$, $Q_{yy} = R_{yy}$;
iii. There is a basic sextuple $(A, B; C, D; K, L)$; which is respectively superior to $(P, Q; R, S)$;
iv. $(P, Q; R, S)$ has a successor which is also a cyclic quadruple.
Obviously, given a formula 4.2.1, if its table contains no rows that form a cyclic quadruple, then the conjunction of C_6, $M34$, $M33$, $M24$, $M32$ is a contradiction.

The existence of a cyclic quadruple certainly assures that we can satisfy all the rows of the second column of the chart simultaneously. It assures a bit more: the two pairs (P, Q), (R, S) of a cyclic quadruple are always compatible with any three pairs (A, B), (C, D), (K, L) which form two basic quadruples, respectively superior to them. This is, however, insufficient to secure that all the rows in the first two columns of the chart can be simultaneously satisfied, because it is possible that no cyclic quadruple beginning with (R, S) is subordinate to any quadruple begining with (K, L). In other words, the blocks might not fit together.

As it happens, this problem does not arise with the example 4.2.2. Since there is only one cyclic pair (δ_4, δ_6), there can be at most one cyclic quadruple, viz, $(\delta_4, \delta_5; \delta_4, \delta_6)$. It can be verified by D4.7 that this is indeed a cyclic row quadruple. Since there is only one basic quadruple $(\alpha, \beta; \alpha, \beta)$, we see immediately that by using (α, β) for (Mii', Mii) $(i = 1, 2, ...)$ and (δ_4, δ_5) for $(Mii'', Mi'i)$ $(i = 1, 2, ...)$, all these matrices (oof the first two columns of the chart) are simultaneously satisfied. Moreover, this is the only possible model for the two initial infinite columns of matrices.

We shall first define common row quadruples, settle 4.2.2, and then come back to the more general question.

D.4.8 Two ordinary row pairs (R, S), (P, Q) form a common quadruple $(R, S; P, Q)$ of order k[i.e., in the $(2+k)$ th column of the chart] if

i. When $k=1$, there is a cyclic row sextuple which is superior to $(R, S; P, Q)$; or when $k=n+1$, for some positive integer n, there is a common row sextuple of order n which is superior to $(R, S; P, Q)$.

ii. $(R, S; P, Q)$ has a successor which is a common quadruple of order k.

By this definition, we can successively find the common row quadruples of orders 1, 2, etc. In the actual procedure, we examine each time to determine whether we have already enough information to decide the original formula. Only when this is not the case do we find the common quadruples of the next order.

In the case of 4.2.2, since $(\delta_4, \delta_5; \delta_4, \delta_5)$ is the only cyclic quadruple, it is easy to verify, by 4.2.6 and D4.5 that $(\delta_4, \delta_5; \delta_4, \delta_5)$ is the only common quadruple of order 1. Thus, by D4.5, if (R, S) is subordinate to the cyclic quadruple $(\delta_4, \delta_5; \delta_4, \delta_5)$, $R_{yz} = (\delta_4)_{xz} = t$, $R_{zy} = (\delta_4)_{zx} = f$, and $S_{xz} = (\delta_4)_{zx} = f$, $S_{zx} = (\delta_4)_{xz} = t$. By 4.2.6, (R, S) must be (δ_4, δ_5).

From this, it follows that, for every n, there is exactly one common quadruple of order n, viz. $(\delta_4, \delta_5; \delta_4, \delta_5)$. This is an immediate consequence of D4.8 and the above transition from the cyclic column to the first common column in the chart. Hence, we have obtained a model for 4.2.2. It is easy to verify that the model for G is just the usual ordering relation $<$ among positive integers.

This completes the solution of the example 4.2.2, which, however, is not a sufficient illustration of the general case. We have to discuss a procedure by considering more complex situations.

4.3 *The Procedure*

One possible procedure is to add one infinite column at a time. Thus, it is possible to represent all possible solutions of each column by a graph, and to represent the solutions satisfying all the initial n' columns by a finite set of graphs if it is possible so to represent all solutions satisfying the initial n columns. Since the common columns enjoy a measure of uniformity, simultaneous solutions for all the columns would be assured if suitable repetitions occur. An exact explanation of such a procedure would be quite lengthy. In any case, a successful choice of patterns of repetition has not been found to assure that for every solvable table, such repetition always occurs.

Instead of elaborating the above procedure, we transform the problem to something similar to the abstract question of 4.1. Thus, given any formula of the form 4.2.1, we can, as in 4.2, construct its truth table and find all the common row pairs in the table. Among the common row pairs, some are also cyclic row pairs and some are also basic row pairs.

If now we take the common row pairs a, b, c, d, etc., as elementary units which are to fill up the infinite quadrant as shown it the chart given under 4.2.7, then the following scheme appears to be feasible. Suppose the points in the infinite quadrant are to be filled by a_{ij}, $i,j = 1, \delta, \cdots$, then we may consider instead all the 2×2 matrices:

$$\begin{pmatrix} a_{ij} & a_{ij'} \\ a_{i'j} & a_{i,j'} \end{pmatrix}, \qquad \text{for all } i, \ j = 1, 2, \cdots$$

In other words, given the common row pairs, we can form all possible 2×2 matrices of them which satisfy the relations of subordination. These 2×2 matrices are then the basic pieces from which we are to obtain an infinite solution subject to the conditions: (a) consecutive rows or colunms from two matrices are the same; (b) only basic and cyclic row pairs are permitted in the first two columns.

It can be verified that the problem of finding a model for the original formula is equivalent to that of finding a way to fill up the infinite quadrant by such derived 2×2 blocks of row pairs.

The abstract problem is: given any finite set of 2×2 matrices of the form

$$\begin{pmatrix} a & c \\ b & d \end{pmatrix},$$

to decide whether it is possible to fill up the infinite quadrant with copies of these pieces. This is not quite the same as the problem of colored plates described in 4.1, because here what is done amounts to coloring the corners, or imposing connections between neighboring sides within a same square.

Any set of such 2×2 matrices can also be construed as a set of colored plates. Conversely, given any set of colored plates, we can also find in a systematic manner a corresponding set of such matrices such that the solvability problems for them are equivalent. For example, we may replace a colored plate by a block of nine 2×2 matrices so that the restriction on neighboring sides no longer operates.

It is possible to use a procedure similar to the one described roughly in 4.1. Some change is needed to take care of the additional conditions on the first two columns. Thus, a sufficient condition is to get a cyclic rectangle $m \times n$ on which we can attach a frill of two columns on the left to obtain a rectangle $m \times (2+n)$ such that: (a) the tops of the first two columns are the same as the bottoms; (b) the additional requirements of being basic or cyclic are satisfied by the frills.

4.4 *Further Problems*

The discussions so far seem to have barely scratched the surface of a group of rather difficult problems, among which the basic one is probably that of measuring the complexity of formulae in the predicate calculus.

One may measure the complexity of a formula in many different ways. The "simplest" model of a formula may be taken as a semantic measure. The quantifier prefix or graph of a formula may be taken as a syntactic measure. In addition, for formulae with a same prefix, we may also classify the possible matrices by the truth tables. Our knowledge on using these criteria to give detailed classifications seems very limited. One example of the ignorance is the following open problem (Ref. 2, p. 177): whether there is any class of formulae which is neither decidable, nor a reduction class. It appears reasonable to conjecture that there must be such *classes, although the first examples which one will get are likely to be artificial ones.*

Some of the reduction classes are, formally speaking, surprisingly simple. For example, from the Surányi normal form given above as case VIII, it follows that, for satisfiability, one reduction class is:

4.4.1 Formulae with prefix $(x)(y)(Ez)(w)$ $Mxyzw$, where M contains neither function symbols, nor = nor predicate letters which are not dyadic.

Since each matrix M is effectively determined by a truth table on the atomic formulae in M, the class may be viewed as a union of a simple sequence of finite classes C_1, C_2, etc., where C_n is the subclass of formulae each containing exactly n predicates (or, equivalently, the first n predicates in some enumeration). There is a sense in which the decision problem for each finite set of formulae is solvable, and yet usually we as a matter of fact only solve the problem as a corollary to a solution for some infinite class.

To obtain a semidecision procedure for the class VIII or 4.4.1, we need more complicated arrangements of triples or quadruples of positive integers than the case $A_1E_1A_1$. Take, for example, the class in case VIII. We have to consider not only the triples (a, b, c) with $b = a'$, but all the triples for the first half of the formula, and among them those for the A_2E_1 case are used simultaneously for the second half of the formula.

An example is:

4.4.2 $\quad (x)(y)(z)(\sim Gxy \vee \sim Gyz \vee Gxz) \,\&\, (x)(y)(Eu)(\sim Gxx \,\&\, \mathrm{Gyu}).$

If we use the Skolem function g of the A_2E_1 case, we can rewrite the above as

4.4.3 $\quad (\sim Gxy \vee \sim Gyz \vee Gxz) \,\&\, (\sim Gxx \,\&\, Gygxy).$

In general, we are concerned with deeiding the satifiablity of formulae of the form

4.4.4 $\quad Mxyz \,\&\, Nxygxy.$

As (x, y, z) runs through all triples of positive integers, we get an infinite sequence from 4.4.4, and a semidecision procedure is to decide, for certain cases, whether such an infinite sequence can be simultaneously satisfied.

For example, we may throw together all permutations of a given triple, and confine ourselves to the triples (a, b, c) with $a \leqq b \leqq c$, assigning each of them a lattice point:

$$f(x, y, z) = (x - 1, z - x, z - y),$$
$$f^{-1}(x, y, z) = (x + 1, x + y, x + y + z).$$

The correlation uses all lattice points (x, y, z) of nonnegative integers. For instance, (1, 3, 5) gets the point (0, 2, 2).

We might try to create different types of cubes each with eight vertices from (i, j, k) to (i', j', k') and piece them together. But it is not easy to see how to find a procedure analogous to that described in 4.1 which would at the same time take into consideration the second half of the formula.

9.5 A proof procedure for the predicate calculus

5.1 *The Quantifier Free Logic F*

Given the definition of formulae in 1.2, we can define sequents, antecedents, consequents, as in Ref. 13, p. 5 The sequents in F are those containing no quantifiers and the rules for F are exactly the same as those for P. (Ref. 13, p. 8), except for containing not only variables but also functional expressions as terms.

Example1.$1 \neq x',\ \ x = x + 1 \to 1 \neq x + 1$

By the rules *P*2a and *P*2b (Ref. 13, p. 5), this is a theorem if the following is:

$$1 = x + 1,\ \ x' = x + 1 \to 1 = x'.$$

This is a theorem be *P*7 and *P*8 (Ref. 13, p. 8).

Example 2.$x + y' = (x + y)',\ \ y \neq x + y,\ \ y' = v' \supset y = v,\ \ v = x + y \to y' \neq x + y$

By *P*2a, *P*2b, and *P*5b, this is a theorem if the following two sequents are:

i. $x + y' = (x + y)',\ \ y = v,\ \ v = x + y,\ \ y' = x + y \to y = x + y;$

ii. $x + y' = (x + y)',\ \ v = x + y,\ \ y' = x + y' \to y' = v',\ \ y = x + y.$

i. is a theorem by *P*7 and *P*8 since we can replace y and $x + y$ by v.
ii. is also a theorem because we can replace v' by $(x + y)'$ and then y' by $x + y'$ in the first clause of the consequent and the result is a theorem by *P*1.
These rules in fact yield a decision procedure for all quantifier–free sequents. In order to see this, we use a more efficient method to speed up applications of P7 and P8.
Given an atomic sequent which contains equality but is not yet a theorem by *P*1 or *P*7. List every pair (a, b) if $a = b$ occurs in the antecedent. Extend repeatedly the set of pairs by symmetry and transitivity. Join each pair by the equals sign and add all of them to the antecedent. Now compare each clause in the antecedent with each clause in the consequent to see whether there is a pair of clauses which can be obtained from each other by substituting equals for equals; moreover, examine each equality in the consequent to see whether it can trun into $\alpha = \alpha$ by substituting equals for equals. If either case occurs, the sequent is a theorem. If neither is the case, then we can find an interpretation of the functions and predicates so that the antecedents are all true but the consequents are all false.

5.2 *The Rules for Quantifiers.*

In the present formulation of the predicate calculus, one emphasis is on separating out rever*sible rules of proof which serve to supply decision procedures as well, because they have the property that not only the premises imply the conclusion but also conversely.*

The rules governing quantifiers were given in Part I.[1*]

The justification of the reduction to subproblems (Part I, T2.1) is obvious because all truth–functional rules are reversible and $(x)(Gx \;\&\; Hx)$ is a theorem if and only if $(x)Gx$ and $(x)Hx$ both are.

Usually T2.2 (Part I) is true, but restrictions are necessary, as the following example would show:

$$(x)(Ey)[(z)Gyz \;\&\; Hxy].$$

Although x does not occur in the scope of (z), there is no way to bring (z) out of the scope of (x) because the variable y ties up the two clauses in the formula. There are several possible alternatives: one may make exact the restrictions needed, or record the scope of each quantifier in the usual manner, or use the easy simplification that when a quantifier governs a formula with two halves joined by a logical connective but the variable of the quantifier occurs only in one of the two halves, the scope is just that half.

The test of connectedness of variables and functors (Part I, T2.3) is meant as a device to simplify the interconnections between quantifiers. In particular, the test gives a method for ascertaining that certain apparently complex sequents fall under the *AE* provability case. In order, however, actually to bring such a set of sequents into the *AE* form, we need in general transformations similar to those used in reducing a sequent to the miniscope form. Since the process can be tedious, one may prefer an alternative method of not carrying out the transformation but merely determining a bound k such that either the original sequent is a theorem or has a counter–model with

---

*"S4. When the input problem contains quantifiers, the following preliminary simplifications are made: (i) All free variables are replaced by numbers, distinct numbers for distinct variables. (ii) Vacuous quantifiers, i.e., quantifiers whose variables do not occur in their scopes, are deleted. (iii) Different quantifiers are to get distinct variables; for example, if (x) occurs twice, one of its occurrences is replaced by (z), z being a new variable. This last step of modification is specially useful when occurrences of a same quantifier are eliminated more than once at different stages.

"S5. After the above preliminary simplifications, each problem is reduced to as many subproblems as possible in the following manner: (i) Eliminate in the usual manner every truth–functional connective which is not governed by any quantifiers. (ii) Drop every initial positive quantifier (i.e., universal in the consequent or existential in the antecedent that is not in the scope of any other quantifier) and treat its variable as free, i.e., replace all its occurrences by those of a new number. (i) and (ii) are repeated for as long as possible. As a final result of this step, each problem is reduced to a finite set of subproblems such that the problem is a theorem if and only if all the subproblems are .

"T2.1 The original problem is a theorem if and only if all its subproblems (in the above sense) are.

"T2.2 We can separate out Q and its scope from those quantifiers whose variables do not occur in the scope of Q.

"T2.3 If two symbols, each a functor or a variable, are not connected in the final matrix, we can always so transform the original sequent as to separate the two quantifiers which give way to them."

no more than k objects. If this alternative is chosen, a method for calculating the bound k has to be devised.

In any case, when we have a finite set of atomic sequents and a set of governing relations among the variables and functors, we should further simplify the matrix, i.e., the set of atomic sequents by the familiar methods of dropping repetitions and immediate consequences.

If there are two subsets of the set of atomic sequents which contain neither common variables nor common functors, then they can be separated.

Moreover, each atomic formula that contains neither variables nor functors can be eliminated by the familiar method of replacing $F(p)$ by $F(t)$ & $F(f)$. In other words, it can simply be dropped on the ground of the following consideration. E.g., take

$$Guv,\quad G11 \rightarrow Gvk.$$

This is equivalent to the conjunction of:

$$Guv,t \rightarrow Gvk;$$
$$Guv,f \rightarrow Gvk.$$

But the second sequent is always true and can be dropped; the t in the first sequent can be dropped, so that we have

$$Guv \rightarrow Gvk.$$

After all the above steps, we arrive at a finite set of finite sets of atomic sequents which, taken together, are equivalent to the original problem. We may consider each finite set of atomic sequents separately and proceed according to the governing relations between their variables and functors.

We can view the set as a formula in the prenex form with a matrix in a conjunctive normal form. Or, if we prefer, we may replace $\rightarrow$ by $\nleftrightarrow$ and construe the variables as universal quantifiers, the functors as existential quantifiers. Then we get a negation of the formula in prenex form with a matrix in the disjunctive normal form.

In either case, the remaining problem is to be handled by considerations such as those explained in Sections II through IV.

There is an easily mechanizable procedure by which we can, in theory, not only prove all provable formulae, but also refute all formulae which have finite countermodels. All we have to do is test, besides the sequence M_1, M_2, M_3, etc., whether a formula is satisfiable in a domain with one object, or two objects, or etc. For example, given

$$(x)(y)(Ez)Mxyz, \tag{1}$$

if some of $M112$, $M123$, $\cdots$ is contradictory, then the negation of (1) is a theorem; if relative to some finite domain, (1) can be satisfied, then the negation of (1) is not a theorem. For example, (1) is satisfiable in a domain with one object if and only if $M111$ is satisfiable; with two objects, if and only if

$$(x)(y)(Mxy1 \vee Mxy2)$$

or

$$(x)[(Mx11 \vee Mx12) \ \& \ (Mx21 \vee Mx22)]$$

or

$$[(M111 \vee M112) \ \& \ (M121 \vee M122)] \ \& \ [(M211 \vee M212) \ \& \ (M221 \vee M222)]$$

is satisfiable.

9.6 Remarks on mathematical disciplines

Besides the contrast between proving and calculating, there is contrast between symbol manipulation and number manipulation. There are problems such as proving trigonometric identities, factorization, differentiation and integration, which all appear to be mechanizable. In numerical calculations, it appears likely that the process of choosing one or another method of ealculation can also be mechanized in many cases.

There is the problem of applying the methods considered so far to deal with concrete examples.

One example referred to in Part I[1] (p. 231) is Hintikka's derivation of a contradiction from his own formal system. [14]Here, intuitive understanding is required to select from the set of all axioms suitable members which are sufficient to produce contradictions. Experience, however, shows that, even after a reasonable selection is made, to actually give an exact derivation of a contradiction remains quite a dreary affair. In such a case, the sort of procedure discussed in this paper can be useful.

In fact, Hintikka uses five axioms to derive a contradiction. Write briefly:

$$Hayz \text{ for } z \neq a \ \& \ z \neq y \ \& \ z \in y \ \& \ y \in z.$$

The conjunction of the axioms is:

$$(Ex)(Ey)(x \neq y) \ \& \quad (Ea)(Eb)(Ec)(Ed)(y)\{[y \neq a \supset (y \in a \equiv (Ez)Hayz)] \ \& \\ [y \neq b \supset (y \in b \equiv \sim (Ez)Hbyz)] \ \& \qquad (2) \\ [y \neq c \supset (y \in c \equiv (y = a \vee y = b))] \ \& \ [y \neq d \supset (y \in d \equiv y = c)]\}.$$

The assertion is that (2) leads to a contradiction. In other words, (2) has no model, and its negation is a theorem of the predicate calculus. To decide whether this assertion is true, we only have to test (2) by essentially the method of Section III because (2) can be transformed into a formula with EA_2E prefix. Such a method yields also a proof or a refutation of the assertion that (2) gives a contradiction.

In a different direction, we may consider some simple examples in the arithmetic of positive integers.

First, we consider the example, $x' \neq x$. We wish, in other words, to prove, with the help of induction, that this is a consequence of the axioms:

$$x' \neq 1,$$
$$x' \neq y' \rightarrow x \neq y.$$

As a general principle, we try to use induction. Since there is only one variable, we reduce the problem to:

$$(x)x' \neq 1, (x)(y)(x' = y' \supset x = y) \rightarrow 1' \neq 1, \tag{3}$$
$$(x)x' \neq 1, (x)(y)(x' = y' \supset x = y), \quad x' \neq x \rightarrow x'' \neq x'. \tag{4}$$

These can be dealt with by the program described in Part I, except that, to avoid confusion, we use now a, b, c, etc., instead of numerals to replace the positive variables. We have:

$$1' = 1, \quad u = v \rightarrow x' = 1,$$
$$1' = 1 \rightarrow x' = 1, \quad u' = v',$$
$$u = v, \quad a'' = a' \rightarrow x' = 1, \quad a' = a,$$
$$a'' = a' \rightarrow x' = 1, \quad a' = a, \quad u' = v'.$$

These sequents are all true by substitution: 1 for x in the first two; a' for u and a for v in the last two.

As a somewhat more complex example, we take the commutaitvity of addition. In order to prove $x + y = y + x$, we may use induction either on x or on y. We arbitrarily take the earliest variable:

$$1 + y = y + 1, \tag{5}$$
$$x + y = y + x \rightarrow x' + y = y + x'. \tag{6}$$

To prove $1 + y = y + 1$, we make induction on y:

$$1 + 1 = 1 + 1,$$
$$1 + a = a + 1 \rightarrow 1 + a' = a' + 1.$$

The first is a theorem by the property of equality. To prove the second, we use another general principle, viz., when a defined symbol occurs, we make use of the definition. In this particular case, we make use of the recursive definition of addition, and try to prove

$$u + 1 = u', \quad u + v' = (u + v)', \quad 1 + a = a + 1 \rightarrow 1 + a' = a' + 1.$$

In order to derive the consequent from the antecedent, we start from $1 + a'$ and $a' + 1$, use the equalities in the antecedent to transform them, and attempt to find a chain to join them. Thus, we may try to make all possible applications of the three equalities in the antecedent:

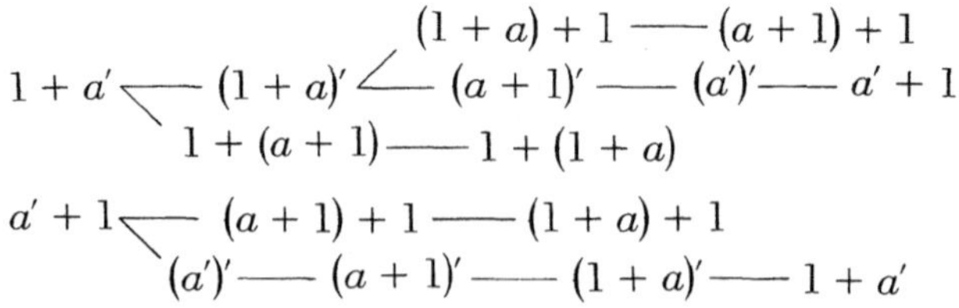

In general, we may begin two trees simultaneously from both sides of the equality, do not write down any term which has already occurred in the same tree, and stop when a common term appears on both trees. When we get to the more complicated situations, we have to investigate two additional things. First, it would take too long to search through trees, so that it is desirable to organize available informations in forms which are more quickly accessible. Second, we may exhaust two trees and still fail to get a common term. Then we need to prove some lemma which would join up the two trees.

For example, the above graphs give us a proof of (5). To prove the other induction hypothesis, viz. (6), we may try to do the same with:

$$u + 1 = u'u + v' = (u + v)', \quad a + b = b + a \rightarrow a' + b = b + a',$$

$$a' + b \text{ --- } (a + 1) + b$$

$$b + a' \text{ --- } (b + a)' \text{ --- } (a + b)' \text{ --- } a + b' \text{ --- } a + (b + 1)$$

$$b + a' \text{ --- } b + (a + 1) \qquad (b + a)' \text{ --- } (b + a) + 1 \text{ --- } (a + b) + 1$$

In this way, we have exhausted the applicable cases of the equalities in the antecedent. Since we have proved the first induction hypothesis (5), we can add it to the antecedent. Then we get some further extensions:

$$(a + 1) + b \text{ --- } (1 + a) + b,$$
$$b + (a + 1) \text{ --- } b + (1 + a),$$
$$a + (b + 1) \text{ --- } a + (1 + b).$$

At this stage, we would ask whether any other given theorem can be used to join up the two trees for $a' + b$ and $b + a'$, or, if not, what a reasonable lemma would be. If the associative law has been proved, we may observe that the missing link is supplied by:

$$(a + 1) + b = a + (1 + b). \tag{7}$$

Otherwise we should try to make a "reasonable" selection of some suitable lemma and prove it. If, for example, we have chosen (7), we would try to establish it by induction on a or on b.

It is possible that the quantifier–free theory of positive integers, including arbitrary simple recursive definitions, can be handled mechanically with relative ease, and yield fairly interesting results. The restriction to quantifier–free methods means that we are concerned only with quantifier–free theorems to be proved without using quantifiers in, e.g., applying the principle of mathematical induction. It is clear from works in the literature that this restricted domain of number theory is rather rich in content. It goes beyond logic in an essential way because of the availability of (quantifier–free) mathematical induction.

With regard to the general questions of using machines to assist mathematical research, there is a fundamental contrast between problem and method. While it seems natural to choose first the objective (e.g., number theory or geometry) and then look for methods, it is likely that a more effective approach is to let the methods lead the way. For example, since the known interesting decidable classes of formulae of the predicate calculus either do not contain function symbols or do not contain quantifiers, we are led to the simple examples above: quantifier–free number theory or function–free set theory.

References

1. Wang, H., Proving Theorems by Pattern Recognition—I, Comm. Assoc. Comp. Mach., 3, 1960, p. 220.
2. Surányi, J., *Reduktionstheorie des Entscheidungsproblems*, Budapest, 1959.
3. Ackermann, W., *Solvable Cases of the Decision Problem*, North–Holland, Amsterdam, 1954.
4. Skolem, T., Über die mathematische Logik, Norsk Matematisk Tidsskrift, 10, 1928, p. 125.
5. Herbrand, J., Sur le probléme fondemental de la logique mathématique, Sprawozdania z posiedzen Towarzystwa Naukowege Warszawskiego, Wydz. III, 24, 1931, p. 12.
6. Church, A., *Introduction to Mathematical Logic*, vol. I, Princeton Univ. Press, Princeton, N. J., 1956.
7. Church, A., Special Cases of the Decision Problem, Revue philosophique de Louvain, 49, 1951, p. 203; 50, 1952, p. 270.
8. Klaua, D., Systematische Behandlung der lösbaren Fälle des Entscheidungsproblems für den Prädikatenkalkül der ersten Stufe, Zeitschrift für mathematische Logik und Grundlagen der Mathematik, 1, 1955, p. 264.
9. Dreben, B., On the Completeness of Quantification Theory, Proc. Nat. Acad. Sci. U. S. A., 38, 1952, p. 1047.
10. Dreben, B., Systematic Treatment of the Decision Problem, Summer Institute of Symbolic Logic, Cornell Univ., 1957, p. 363.
11. Schütte, K., Untersuchungen zum Entscheidungsproblem der mathematischen Logik, Mathematische Annalen, 109, 1934, p. 572.
12. Ackermann, W., Beiträge zum Entscheidungsproblem der mathematischen Logik, Mathematische Annalen, 112, 1936, p. 419.
13. Wang, H., Toward Mechanical Mathematics, IBM J. Res. Dev., 4, 1960, p. 2.
14. Hintikka, K. J. J., Vicious Circle Principle and the Paradoxes, J. Symb. Log., 22, 1957, p. 245.

PART THREE

DECIDABILITY AND COMPLEXITY

10. GAMES, LOGIC AND COMPUTERS*

A close kinship among them is demonstrated by a game of solitaire played with colored "dominoes." Whether or not the game can be won is analogous to whether or not a problem can be solved by computer.

Today much of the work once done by human muscles and brains is being delegated to machines, and people in all walks of life are asking: What human abilities are irreplaceable? What can machines not do? It may surprise the reader to learn that, whereas the first question has no definite answer, the second has a straightforward mathematical solution.

Even before the first modern computing machine was built the late British logician Alan Turing asked the question: What can computers not do? In his attempt to create a theory of what can be computed and what cannot, Turing devised a slow and simple imaginary computer that he proved to be theoretically capable of performing all the operations of *any* computer. He used his machine to demonstrate the close kinship of computer theory and logic, branches of mathematics that are both concerned with mathematical proof and with notations that can present our thoughts in exact form. This article will undertake to illustrate some fundamental concepts in the area of overlap between computer theory and logic by means of games.

The human mind can grasp only relatively small numbers and quantities. The discipline of mathematics, on the other hand, is primarily concerned with infinity. Finite mathematical operations and infinite mathematical entities present a significant and fascinating contrast. The smooth transition from intuitively comprehensible indicidual cases to unrestricted general situations is a remarkable achievement of the human intellect. Abstract considerations concerning games can introduce us to this phenomenon quite naturally.

Finding the sequence of moves most likely to lead to victory in a game such as ticktacktoe presents a logical problem precisely analogous to finding the series of steps that will yield a solution to any mathematical problem of a given class. In certain games there is no optimum strategy that will guarantee victory; in certain classes of problems there is no algorithm–no general method of supplying a series of steps leading to a solution. Since a computer program is simply an algorithm designed for execution by machine, this means that there are classes of problems that computers cannot solve. Before considering the difficulties of constructing algorithms for solving problems (also of devising programs for arriving at solutions and of working out optimum strategies for winning games), let us examine why it is that their construction not only

* First published in *Scientific American*, vol. 213, no. 5 (November), pp 98—106.

is useful but also represents an ultimate goal of mathematics.

Obviously it would take infinite time and energy to memorize the multiplication table if, instead of just including the products of all single-digit numbers taken two at a time, the table included the products of all multidigit numbers taken in this way. Man has made this infinite multiplication table unnecessary by memorizing, along with the multiplication table for single-digit numbers, a list of steps involving "carrying" and the addition of partial products that will yield the product of any two multidigit numbers.

We know that the operations of elementary arithmetic involve formal rules. and most of us recall that certain other operations, such as the extraction of a square root, can be done according to a fixed list of sequential steps. As we get to problems of greater complexity it becomes less clear that they can be solved by algorithm. Consider the following problem: "Given the two positive integers 6 and 9, find their largest common divisor." The reader will immediately see the answer: 3. If the two numbers were 68 and 153, readers who are inclined to try various possibilities might still find the answer (17). If it could be shown, however, that the general problem "Given two positive in tegers a and b, find their greatest common divisor" can be solved by algorithm, then anyone or any machine capable of performing the specified operations could solve it for any a and b. Such an algorithm exists; it was devised by Euclid [*see illustrations at top of page 100*].

DOMINO PROBLEM involves assembling three colored tiles called domino types to form a block that can be infinitely extended with colors matching on all adjacent edges (*opposite page*). It is assumed that the player has an infinite quantity of each domino type and that no domino can be rotated in two dimensions. The problem is solved by finding a rectangular block in which the color sequence on the top edge is the same as that on the bottom edge and the sequence on the left edge is the same as that on the right. Such a unit(*heavy outline on opposite page*) can be repeated in all directions to fill an infinite plane.

The usefulness of algorithms in equally apparent in the realm of games, where they provide instructions for the most advantageous moves. The mathematician, of course, is less interested in winning a game than in understanding the abstract structure of that class of games. By considering the existence or nonexistence of a winning strategy, he gains insight into the abstract structure of the game and those of the same kind.

Take, for example, the game known as nim. Any number of objects, say six matches, are arranged in three piles. Two players, A and B, draw in turn, each taking any number of matches from any one pile. Whoever takes the last match is the winner. Since the finite quantity of matches will in the end be exhausted, it is obvious that the game allows no draw. It is significant that only one of the two players has a winning

```
      :R
GCD   :R      THIS RØUTINE CØMPUTES THE GREATEST CØMMØN
      :R      DIVISØR ØF TWØ INTEGERS A AND B.
      :R
      EXTERNAL FUNCTIØN (A,B)
      NØRMAL MØDE IS INTEGER
      ENTRY TØ GCD.
LØØP  REMAIN = B - A*(B/A)
      WHENEVER REMAIN.E.0,FUNCTIØN RETURN .ABS.(A)
      B = A
      A = REMAIN
      TRANSFER TØ LØØP
      END ØF FUNCTIØN

      TYPE INPUT                   TYPE INPUT
      A=8,B=12*                    A=12345678,B=87654321*
      THE GCD ØF A AND B IS        THE GCD ØF A AND B IS
            4                            9
```

ALGORITHM IN COMPUTER LANGUAGE provides a series of steps by which the largest common divisor of any two numbers can be found. At bottom are computed solutions for two pairs of numbers. Procedure is in the language called MAD (for Miehigan algorithm decoder). Letter *O* is printed with a line through it to distinguish it from zero.

1. Consider two positive integers, *a* and *b*. Proceed to next instruction.
2. Compare the two numbers under consideration (determine if they are equal, and if not, which is larger). Proceed to next instruction.
3. If the numbers are equal, each is the answer; stop. If not, proceed to next instruction.
4. Subtract the smaller number from the larger one and replacê the two numbers under consideration by the subtrahend and the remainder. Proceed to instruction *2*.

SAME ALGORITHM is stated in ordinary language. The process of division is rendered as repeated subtraction. The series of steps is known as the Euclidean algorithm.

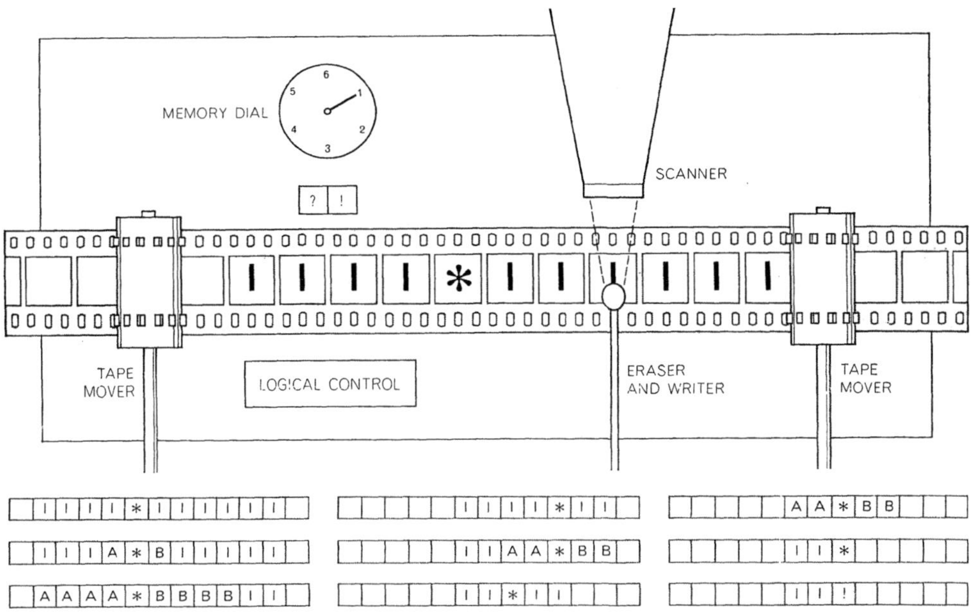

TURING MACHINE designed to perform steps of the Euclidean algorithm displays the numbers 4 and 6 on its tape in this schematic illustration. (Each digit is represented by a strokc; asterisk signals separation of numbers.) The logical control of the machine consists of instructions determined by a mark on the square of tape being scanned and the position of the memory dial. Steps of the Euclidean algorithm lead to changes on the tape illustrated is sequence at bottom. The machine first determines which number is larger by a "comparison loop" in which it replaces strokes to right and left of asterisk with symbols (*"A" at left and "B" at right*). When one set of strokes is exhausted, the machine begins a "subtraction loop," erasing the symbols of the smaller number and converting the symbols of the larger number back into strokes. These are separated from the two strokes representing the remainder of the subtraction by an asterisk. The process of comparison and subtraction is repeated with 4 and 2 on tape, and then with 2 and 2; finally, 2 and 0 appear on tape. As comparison loop begins, blank tape on one side of asterisk evokes a halt signal (!) from logical control, and the machine stops with the answer (2) on its tape. Machine is idealized because its tape is potentially infinite.

strategy, depending on the size of the initial three piles and who moves first. In the particular game defined by piles of one, two and three matches the winning strategy belongs to B, the player who moves second. This can be proved by means of a schematic tree with the node of each set of branches representing a situation at a stage of the game, and the branches from each node the possible moves a player can make in that situation [*see top illustration at right*].

Suppose we are to play this game with three piles containing 10 million, 234 and 2,729 matches. It is theoretically feasible to tabulate all the possible sequences of moves with these three piles and then to tell by inspection whether A or B has a winning strategy. No one, however, would be willing or able to undertake such a tabulation. The mathematician would undertake a systematic search for shortcuts to make the operations easier and to achieve economy of thought. A search of this kind has in fact been made for the game of nim, and a simple recipe has been worked out for determining which player has the winning strategy. The recipe states that A can always win if, when the numbers of objects in the three piles are expressed in binary notation, they add up to a figure that contains an odd number. Such a recipe can represent a dazzling but unimportant stunt or a way of achicving significant mathematical insight, depending on the nature of the game and the directness of its relation to major mathematical and logical problems.

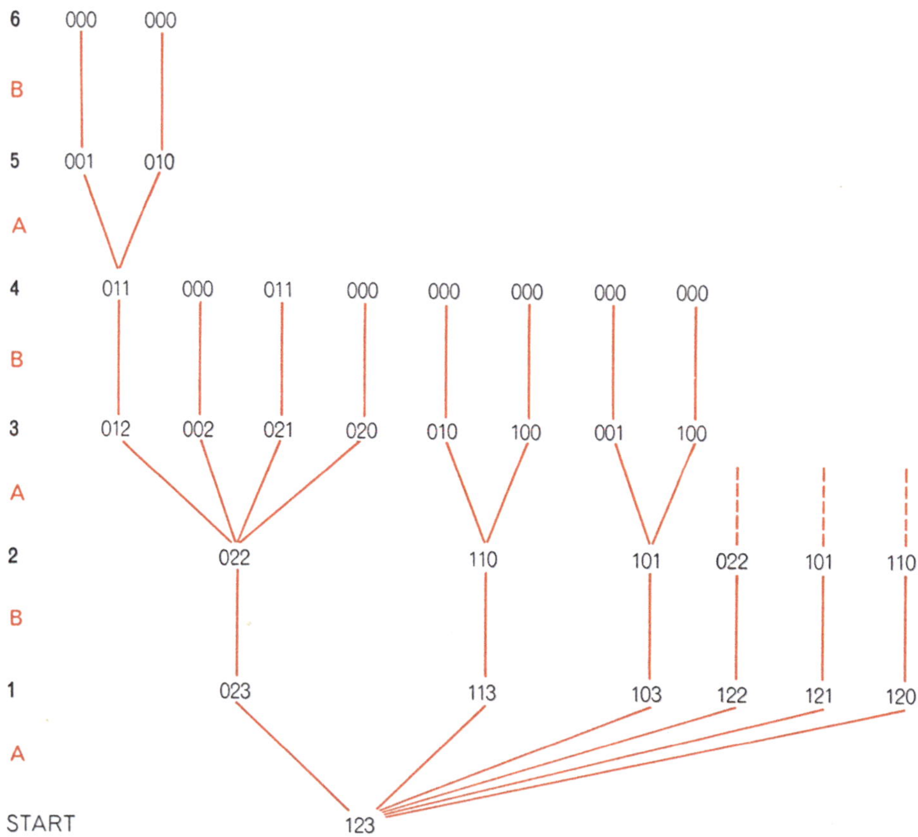

TREE FOR GAME OF NIM indicates that the player who moves second, B, has a winning strategy. At beginning of game (*bottom*) there are piles of one, two and three matches (*digits at each node of tree give number of matches remaining in piles*). Players remove one or more matches from a single pile until someone wins by removing the last match. Branches show all the possible moves for A and the unanswerable response of B to each of them.

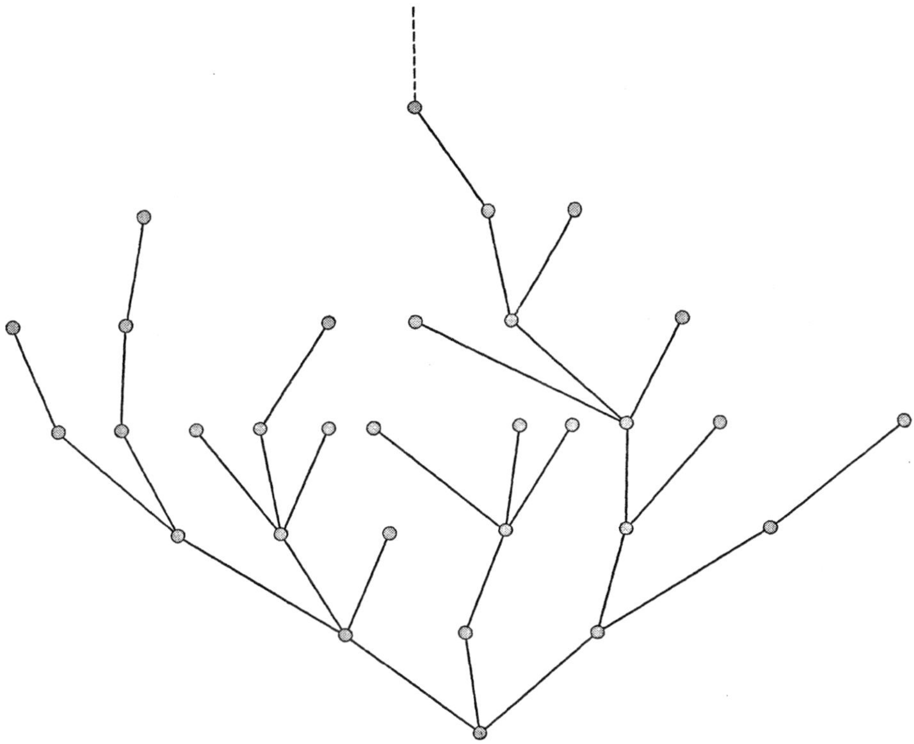

INFINITY LEMMA is suggested by a tree that continues at top. The lemma is a proposition to the effect that if there are infinitely many connected branches in a tree, and only finitely many branches from each of its nodes, then there must be one node at every level from which branches extend indefinitely upward; these, taken together, form an infinite path through the tree. The lemma can be paraphrased: "If the human species never disappears, there exists today someone who will at any future time have a living descendant."

A game such as nim is said to be "unfair," because one player always has a winning strategy. A game such as ticktacktoe is said to be "futile," because each player has a nonlosing strategy that eliminates the possibility of a winner. These characterizations can be restated as a theorem: Every game is either futile or unfair if there is a fixed, finite upper boundary to the length of each path on its tree and only finitely many branches come directly from each node. The theorem holds because if there are no endless games and only finitely many legal moves at each stage, then the total number of possible sequences of moves is finite. If we represent all the permissible plays by a tree, we see that if neither player has a winning strategy, then each player has a nonlosing strategy.

The theorem does not apply directly to the game of chess, because there are no precise rules to prevent endless matches. Let us assume, however, that we could

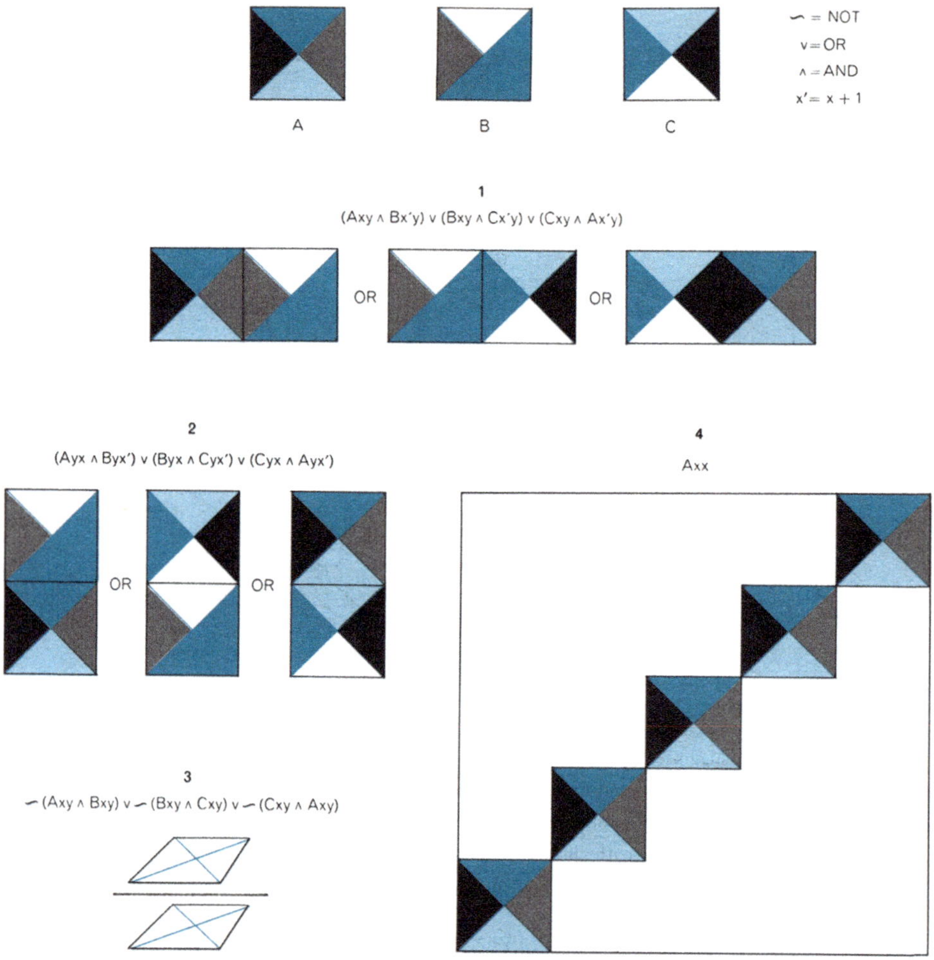

RULES FOR DOMINO PROBLEMS are set forth in the formal shorthand used by students of mathematical logic (*glossary is at top right*). At top center is a set of dominoes: *A*, *B* and *C*. The first expression states that colors must match on left and right edges, second that colors must match on top and bottom edges. The third rule is that dominoes must not be placed one atop another. The fourth expression, a constraint typical of those used to complicate games in approximating difficult problems of computation, states that only *A* can lie on the main diagonal of the plane. The positions on the plane are described by Cartesian coordinates. In designation such as "*Ayx*" domino's position on horizontal axis is given by the first variable, *y*, and vertical position by the second.

Cover a section of the Cartesian plane with black and white tiles so that no block (of the size outlined at right or larger) has edges at left and right and top and bottom that match. Is there a method of filling an infinite plane in this way?

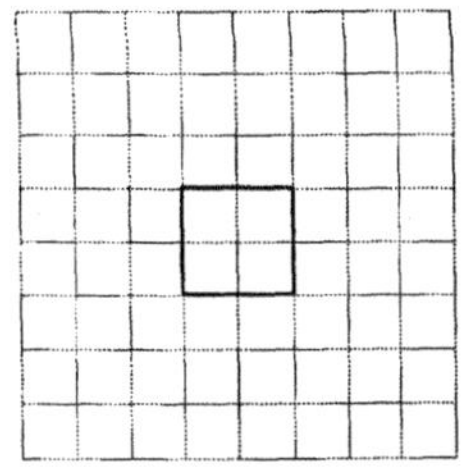

FOUR PROBLEMS are presented by the author to the resolute reader. Only this problem and the problem at bottom have known solutions. Solution to this problem is on next page.

We can make a string of 0's and 1's yield "progeny" by these rules:
1. If the string has fewer than three symbols, stop.
2. If the string begins with 0, delete the first three symbols and append 00 to the end.
3. If the string begins with 1, delete the first three symbols and append 1101 to the end.

Is there an algorithm to determine whether one of two given strings is a progeny of the other?

011010001001	101110110011
01000100100	1101100111101
0010010000	11001111011101
001000000	011110111011101
00000000	11011101110100
0000000	111011101001101
000000	0111010011011101
00000	101001101110100
0000	0011011101001101
000	10110100110100
00	1101001101001101
(stop)	(progeny continue)

SECOND PROBLEM is to find an algorithm that shows whether two strings of 0's and I's are related. The problem is complicated by the fact that certain strings may give rise to infinite progeny. An alternative solution would be to prove that no such algorithm can exist.

Is there an algorithm to decide if a polynomial equation with integral coefficients has roots that are integers?

Equations of this type include

$$x^2 - 4x + 3 = 0$$

and

$$a^2 + b^2 - c^2 = 0 .$$

The first equation has only one unknown, x. It thus has the form

$$a_n x^n + a_{n-1} x^{n-1} + \ldots + a_1 x + a_0 = 0$$

and for such equations the desired algorithm is known:
1. Find all the divisors of a_0.
2. Substitute each for x and calculate the resulting values for the left side of the equation.
3. If any yields the value 0, it is a root. If none do, the equation has no roots that are integers.

The problem is to devise such an algorithm for equations, such as the second, that contain more than one unknown.

THIRD PROBLEM has been known as "Hilbert's 10th problem" since the German mathematician David Hilbert listed it in 1900 as an outstanding problem confronting mathematics.

introduce rules to exclude endless chess matches, without imposing a limit on the absolute number of moves allowed in a complete match. We would then be able to apply to chess a proposition known as the infinity lemma, which states that if there are infinitely many connected branches in the tree of a game and only finitely many branches from each node, then there is an infinite path.

Given the infinity lemma and the assumption that new rules have excluded endless matches, it follows that the tree representing the game of chess has only finitely many branches. Otherwise, in view of the fact that there are only finitely many branches from each node, there would be an infinte path (an endless game). Hence there are only finitely many possible sequences of moves and, as we showed earlier, chess is either unfair or futile.

The proof of the infinity lemma is fairly straightforward. Take the node at the very bottom of the tree. Since we are assuming infinitely many branches but only finitely many branches directly from any one node, at least one of the nodes on the next level must be the bottom of a subtree with infinitely many branches. Let us call this node X. Our hypothesis states, however, that there are only finitely many branches directly from X. Therefore one of the nodes on the next level above X must be the root of an infinite subtree. By repeating this argument we see that on every level there is at least one node that is the root of an infinte subtree and that these roots together determine an infinite path through the tree. An anthropomorphic way of applying the infinity lemma would be to state that if the human species never disappears, there exists today someone who will at any future time have a living descendant.

One chooses to examine a game, of course, according to the importance of the mathematical questions it raises. In 1960, while I was studying certain problems in logic at the Bell Telephone Laboratories, I devised a new game of solitaire played with "dominoes" that are actually colored tiles. More recently my colleagues and I at the Harvard Computation Laboratory have found some surprising and significant applications of this game. Several problems that arise in the domino game are exact analogues of problems that Turing machines are designed to solve. The conditions under which a domino game is played can be made to correspond to the computations of Turing machines, so that working with dominoes grants us another view–sometimes a particularly revealing one of certain mathematical problems.

In a domino game we are given a finite set of square tiles (the dominoes); the tiles are all the same size but each edge of a tile has a stipulated color and the colors are combined in several specified ways. We assume that we have infinitely many copies of each type of domino, and that we are not permitted to rotate a domino in two dimensions. The object of the game is to cover an infinite plane with dominoes in such a way that adjoining edges have the same color. If the plane can be covered with a given set of dominoes, the set is said to be solvable.

Consider the set of three dominoes in the illustration on page 98. The set is solvable because it can be assembled into a nine–domino block that satisfies the rule with respect to edges and can be repeated in every direction. Given a solution for the whole plane, we can obviously chop off three quadrants, or quarters, to get a solution over one quadrant. The converse is less obvious, but it can be established with the help of the infinity lemma. Since there exists a solution over an infinite quadrant, there exist

TEN–DOMINO PROBLEM calls for arranging these tiles in a block in which the color scheme is the same for top and bottom and left and right edges. Solution is on page 106.

a

a b

a b b a

a b b a b a a b

a b b a b a a b b a a b a b b a ...

SOLUTION TO FIRST PROBLEM on preceding page is illustrated. To repeat its construction, let *a* represent a black tile, *b* a white one (*left*). Write *a* and replace it with *ab*. Replace *b* with *ba* and continue to replace *a* and *b* in this way. Transcribe this sequence onto top row of plane and copy each symbol along diagonal from top right toward bottom left.

partial solutions of *its* quadrants–solutions of any area *n* by *n*. We can make an infinite tree out of such partial solutions and show by the infinity lemma that there is an infinite path in the tree that yields a solution over the whole plane. Thus if it is possible to fill a quadrant of the infinite plane, it is possible to fill the whole plane.

We can use dominoes to simulate various Turing machines and to create an equivalent of Turing's important "halting problem." This is more easily done if we specify what domino goes at the origin of the plane–what domino we put down first. With greater effort we can accomplish the same thing either by specifying that certain dominoes occur on the main diagonal or by omitting any restriction other than those mentioned earlier. Let us consider this equivalence between the halting problem and the domino game in greater detail.

Turing devised his simple computer to emulate a human calculator. A man solving a mathematical problem is likely to use pencil and paper for writing and erasing numbers; he may also have a collection of mathematical facts in the form of a book of tables and, contained in his mind or in the book, a set of instructions for performing the proper steps in the proper sequence. The imaginary Turing machine also has a marking device and an eraser for writing numbers according to instructions from a logical control unit that follows a prepared table of commands. The numbers are written in the form of single strokes on square cells of an infinitely long tape serving as the memory unit. (Since no actual machine can have an infinite memory, the Turing machine is idealized.)

One square of the tape at a time is considered by a scanner that relays the symbol on the square to the control [*see bottom illustration on page 100*]. The control then consults its internal instructions by means of a dial that points to a location designated "Current instruction." Depending on the symbol at hand, the instruction specifies one of four commands: (1) Print a mark on the square, erasing it if necessary, (2) Move the tape one square to the right, (3) Move the tape one square to the left, (4) Halt! Then the instruction indicates the next instruction location. A Turing machine can be endowed with the requisite number of instruction locations and commands, as well as a tape with infinitely many squares, to solve any specified mathematical problem (if that problem is in a class solvable by algorithm).

Turing devised the halting problem to exemplify a problem for which no program could yield all the correct solutions. He surpassed the power of any possible single machine by formulating a question about all Turing machines. He was able to show that although each machine, depending on its tape, would either halt or continue operating indefinitely, there is no general algorithm to determine this behavior, no recipe equivalent to the one for determining the invariable winner of a given game of nim. Now, it is possible to find for each Turing machine a set of dominoes such that the machine will eventually halt if and only if the set of dominoes does not have a solution. It is then a direct consequence that the domino problem is unsolvable. If we could solve the domino problem, we could solve the halting problem; we cannot solve the halting problem and so we cannot solve the domino problem. In other words, there is no general method for deciding if any given set of dominoes has a solution.

The domino problem is an example of an infinite decision problem of the kind that frequently turns up in logic, in computer theory and in mathematics in general. It is an infinite problem in the following sense. Any solution to the domino problem must be a single method that provides the correct yes or no answer to an infinite number of questions in the form: "Does a certain set of dominoes cover the plane?" Whereas any specified set of domino types is finite, there is of course an infinite set of such sets, and therefore an infinite number of questions.

We have thus reduced problems about sets of dominoes to problems about machines, and we have established results about dominoes by appealing to known results about machines. The next step is to reduce the question of interpreting a formula in logic to the problem of solving a set of dominoes. Since the condition that a set of dominoes has a solution can be expressed by a simple formula in logic, this reduction yields an answer to a longoutstanding decision problem in logic.

If we wish to express the condition that a set of three domino types has a solution in the first quadrant of our infinite plane, we think of the familiar Cartesian

SOLUTION TO 10–DOMINO PROBLEM is a rectangular block of 36 dominoes, two of which are separated by heavy black line through center of illustration. The solution is not unique, that is, other configurations of the 10 dominoes are possible and equally acceptable.

coordinates for the positions of dominoes in the quadrant and represent each domino by a predicate: *Axy*, for example, indicates that domino *A* occurs at position (x, y). If we use x' for $x+1$, the required condition can be given by a number of clauses that require very few quantifiers (constructions with "for all" and "there is"). We are generous only with such finite operations of formal logic as "not." "and" and "or" [*see illustration on page 102*]. We can conclude that for any given set of dominoes we can find a corresponding "*AEA* formula" at sentenec beginning "For all x there is a y such that for all $z. \cdots$." followed by a logical combination of predicates without quantifiers such that the set has a solution if and only if the formula is not self contradictory. In other words. we can translate a domino question into a logical formula by specifying certain constraints and then determine if the donino set is solvable by seeing if the formula is or is not self contradictory. Therefefe since the general domino problem is unsolvable then is no general method for deciding if an arbitrary *AEA* formula is self contradictory.

The result is useful because the complevity of formulas in logic is to a Large extent measured by the number and order of quantifiers, and the formulas of logic are often put into different classes according to the structure of quantifiers. It is surprising that as simple a class as that of *AEA* fornulas (with three quantifiers only) is undecidable. In fact. with this result the deeision problems for all quantifier classes are answered. Given any string of quantifiers we can now tell if the class of formulas determined by it is decidable.

The decision problem of logic is signifieant because all mathematieal then ries can be fornulated in the framework of elementary logic. The question of whether or not a formula (*F*) can be derived from a set of axioms (*A*) is reduced to deciding if the logical fornla "*A* but not *F*" is not sell-contradictory In this sense all mathematics is redueible to logic. Indced one measure of the complexity of a mathematical problem is given by the structure of its cornsponding fornmla in logic. It is therefore an important cnterprise to cletermine the complexity of various classes of logical formnlas.

We can justifiably say that all mathematies can be reduced. by means of Turing machines to a game of solitine with dominoes. In most instances the reduction does not make a mathermatical problem any casier to hamdle. Nevertheless, proving certain problems to be unsolvable by computer can be facilitated by reducing them to domino problems.

Appendix: Notes on a class of tiling problems*

Abstract. *The class of problems considered here was at first called the "domino problems" and has found extensive applications with regard to the decision problem of the predicate calculus. This paper includes, apart from a brief survey of work related to this class of problems, a number of isolated results which have been obtained over the years. These results mostly have little direct connection with mathematical logic but may be, for that very reason, of some interest to a wider circle of mathematicians.*

The class of tiling problems deals with the following general situation. Suppose that we are given a finite set of unit squares with colored edges, placed with their edges horizontal and vertical. We are interested in tiling the plane with copies of these tiles obtained by translatives only. The tiles are to be placed with their vertices at lattice points, and abutting edges must have the same color. The first question, the unrestricted tiling problem, is whether there is a general method of deciding which finite sets of colored squares are solvable (i.e., can be used to tile the plane in this way). The second question (closely related to the first, see below) is whether every solvable set has a periodic solution (i.e. yields a square of some size which repeats to cover the plane). If we think of the first quadrant instead of the whole plane, it is more convenient to speak also of the origin-constrained (i.e. the tile at the origin is restricted to a given subset) and the diagonal-constrained (i.e. the tiles along the main diagonal are restricted) tiling problems. These apparently frivolous problems have led to various interesting investigations. And it is my purpose here to give a number of fragmentary results mostly obtained in discussions with colleagues and students some time ago. In particular, several of the basic ideas are due to Edward F. Moore. I shall begin with a brief historical survey of some of the results in the literature.

Around the beginning of 1960, while continuing my work on the mechanization of mathematical arguments, I was diverted into a study of the theoretical problem of deciding the class of sentences with the simple quantifier prefix AEA in elementary logic. After a period of effort, I succeeded in transforming the decision problem into the easily understandable combinatorial or geometrical tiling problem (called the "domino problem" by a colleague). This greatly facilitated not only my communication with my colleagues at the industrial laboratory who were mostly ignorant of mathematical logic, but also the ability to focus attention on the mathematical core of the original decision problem. At this time, I also discovered that the origin-constrained tiling problem is unsolvable, because operations of any Turing machine can be simulated by a particular tiling problem. The formulation of the tiling problems was written up in May 1960 and

* First published in *Fundamenta mathematicae*, vol, 82, pp. 295—305. Polska Akad. Nauk. Warszawa, 1975. Reproduced by permission of the author.

published in January 1961 ([17]), and the result on the origin-constrained problem was written up in August 1961 for circulation (see [18]).

In the autumn of 1961, I lectured on these things. And in collaboration with Kahr and Moore, I was able to show that the diagonal-constrained tiling problem was also unsolvable for much the same reason as the origin-constrained problem. We were able to infer that the decision problem for the AEA case is unsolvable. In fact, the AEA sentences with only dyadic predicates form a reduction class. These results were published in [11]. Shortly afterwards, Kahr further refined the result to eliminate all but one dyadic predicate, using only monadic predicates otherwise. A summary of this last result is included in [10] and [19], and the full proof is given in Kahr's MIT Ph. D. dissertation (June 1962). Afterwards Berger, another student of mine, demonstrated that the (original) unrestricted tiling problem is also unsolvable (in his Harvard dissertation, June 1964; a briefer version appeared as [1]).

Over the years, there have appeared a number of papers related to these tiling problems and decision problems. The following items have come to my attention. There are three pairs of natural subclasses of the AEA sentences with dyadic predicates only. Two of these three pairs have been shown to be decidable (see [2] and also [19]). The remaining pair is, surprisingly, shown to be undecidable by S. Aanderaa (in his Harvard dissertation, August 1966). Elaborations, extensions, and simplifications of [10], [11], and [19] are contained in the papers by Genenz, Hermes, and Maslov ([3], [4], [8], [9], and [12]). On the question of nonperiodicity, Berger's published proof of the unsolvability of the tiling problem contains a complex solvable set of tiles with no periodic solution. In his dissertation, he includes a simpler set with 104 tiles. In April 1966, H. Lauchli sent me a nonperiodic set with 40 tiles which, as far as I know, has not been published. In [15], Robinson has gone into the solvability and the periodicity problems carefully and obtained more economical solutions. In a somewhat different direction, Hanf has shown in [6] that, under the origin constraints, there is a finite solvable set of tiles which has no recursive solution. This was extended by Myers (in [13]) to the unrestricted case. The tiling obtained by Hanf can be described by a 1–trial predicate (a concept of Putnam [14]). Carl Jockusch has found a solvable finite set of tiles which has no m–trial tiling for any m. Hanf's work was aimed at proving Conjecture I of [7], but these results have failed to settle his conjecture.

Let N be the class of unsolvable tile sets, F be the class of tile sets with periodic solutions, and J be the class of solvable sets without periodic solutions. It is proved in [5] that the classes N, F, J are pairwise recursively inseparable.

I proceed to list a number of fragmentary results, often omitting complex constructions.

With the origin constraint, it is possible to force special solutions with amusing properties. One example is to distinguish prime from composite numbers, first done by Edward F. Moore and then simplified by M. Fieldhouse.

1. There is a set of 30 tiles, including three tiles A, P and C, such that if A is required to appear at the origin, the set has a unique solution in which P and C occur respectively at the prime and composite positions in the first row.

A variant formulation of the tiling problem is to color the corners rather than the edges so that each tile can naturally be represented by a 2 by 2 matrix giving the colors of the four corners. It is easy to show that this formulation is equivalent to the other one in an obvious sense.

It can be verified that permitting all rotations or all reflections would make all sets of tiles solvable. If, however, we use regular hexagons instead of squares and allow reflections and rotations, we have:

2. Given a set of square plates, we can find effectively a set of regular hexagons such that there is a one-to-one correspondence between the solutions of the two sets. Conversely, given a set of regular hexagons, we can also find effectively a corresponding set of squares.

Given a set of squares $A_1, \cdots, A_n$, we introduce a set of $2n + 1$ hexagons with $6 + n$ new colors as follows. A "cementing" hexagon:

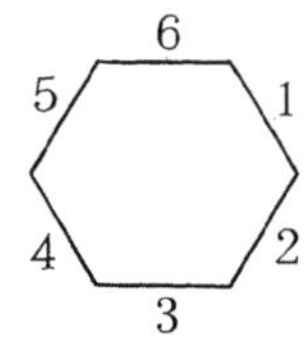

Suppose A_i is:

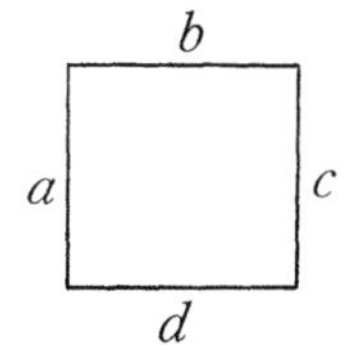

Introduce two hexagons with a new color a_i:

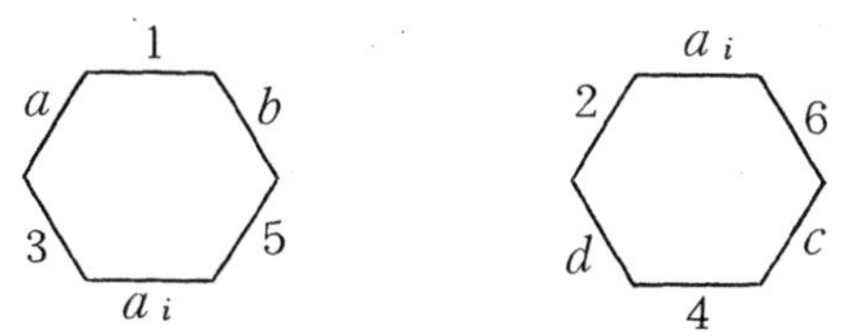

It can be seen that on account of the cementing piece, each pair of hexagons can only be used as a unit thus:

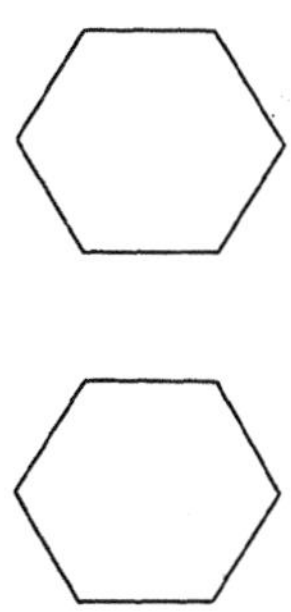

Conversely, given a set of regular hexagons $H_1, \cdots, H_m$, we give one block of three squares for each of the 12 positions for each H_i, using two new colors for each position. For example, if H_i is

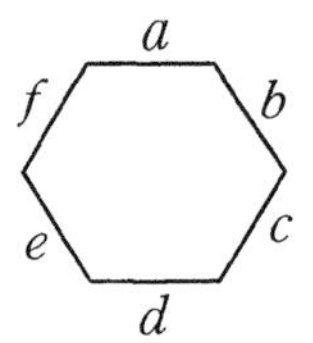

we use for the particular position the obvious combination of the three squares:

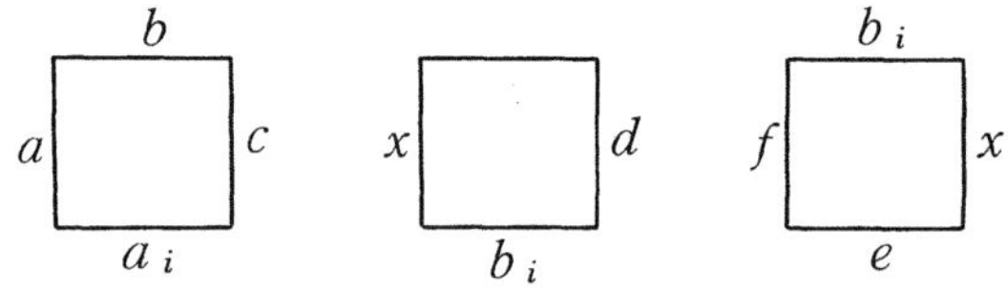

The letter x indicates a new color used for every triple.

Hence, we get a set of $36m$ squares with $24m + 1$ new colors.

We omit the detailed proofs of the two halves of Proposition 2.

It is convenient to speak of a "torus" when we have a square or rectangle of some size such that the top agrees with the bottom (in colors) and the left edge agrees with the right edge. As we have noted before, Berger and others have given solvable sets which yield no tori because they have no periodic solutions. An easier question had been considered and answered earlier, viz. to find a set which has a solution in which no torus occurs.

This uses an interesting construction due to Thue (see [16]), according to which the union R of the infinite sequence obtained from a by successively replacing a by ab, b by ba contains no part UyV with $Uy = yV$. Such a two–way infinite Thue sequence R can be applied to design a pattern of a's and b's on the plane in which there is no "torus". We put the sequence R horizontally on the plane and copy each symbol across the diagonal bisecting the first and the third quadrants. Then there can be no rectangle block which forms a torus. For, suppose there were such a block, say

$$\begin{matrix} x_4 x_3 x_2 x_1 \\ x_3 x_2 x_1 y_1 \\ x_2 x_1 y_1 y_2 \end{matrix}$$

Then $x_4x_3x_2x_1 = x_2x_1y_1y_2$ and, in particular, $x_4x_3x_2 = x_2x_1y_1$. Therefore, $x_4x_3x_2x_1$ y_1 would be a part Ux_2V such that $Ux_2 = x_2V$.

In order to find a set of tiles with a solution in which no torus occurs, we represent a and b by several tiles, taking into consideration the two neighbours in the same row. Since we do not permit aaa or bbb, we need only six tiles:

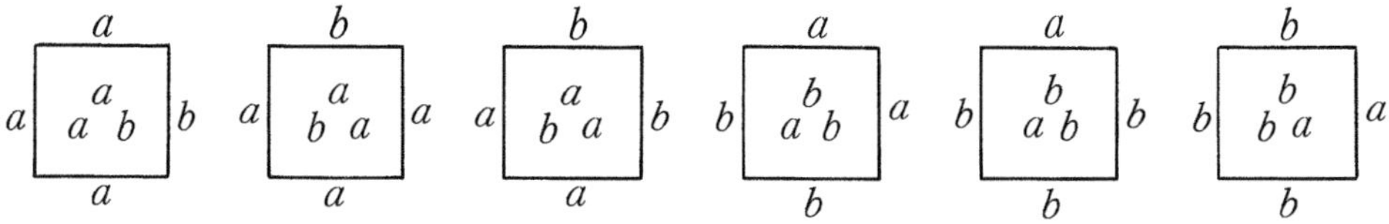

It can be verified that the pattern of a's and b's described above can be simulated to any size. Hence, applying 5.1 to be proved below, we obtain:

3. There exists a set of six tiles which has a solution in which no torus occurs.

Incidentally, Thue also found that, for any alphabet with three or more letters, there are sequences in which no part is of the form DD. A sequence of this kind with three letters has been found which is simpler to describe than Thue's:

4. The sequence T obtained from the sequence R described above by substituting ca for aa, cb for bb contains no part DD.

We recall that R is the union of:

$R_1 = a$, $R_2 = ab$, $R_3 = abba$, $R_4 = abbabaab$, $\cdots$

Consider now the even terms of R beginning with R_4: R_4, R_6, etc. We observe that, because neither aaa nor bbb occurs in R, the results of substituting ca for aa and cb for bb are composed of occurrences of only the following 4-letter sequences:

$$A = acba, \qquad B = bcab, \qquad C_1 = acbc, \qquad C_2 = bcac.$$

We distinguish different occurrences of C and use c_1 in a context bca, c_2 in a context acb. Define T_1, T_2, T_3, etc. by:

$$T_1(a, b, c_1, c_2) = ac_2babc_1ab,$$

$$T_{n+1}(a, b, c_1, c_2) = T_n(A, B, C_1, C_2).$$

Then T is the union of the T_n's and T_n is obtained from R_{2n+2} by substituting ca for aa, cb for bb.

If DD occurs in T, then there is some T_n such that DD occurs in T_n and T_n contains more letters following DD. Let M be obtained from D by substituting A, B, C_1, C_2 for a, b, c_1, c_2. Then MM occurs in T_{n+1}, and M begins with a or b. Suppose $M = xN(x = a$ or $b)$. We then have $xNxN$ in T_{n+1}. The letter following $xNxN$ must also be x because x follows N once in T_{n+1} in $xNxN$. For example, if x is a, then M must begin with A or C_1, but C_1 can only follow B, and A can only follow C_1 or B. Hence, M must end with C_1 or B. But C_1 can only be followed by A, and B can only be followed by A or C_1. Therefore, the part in T_{n+1} after $xNxN$ must begin with a (i.e., as head of A or C_1). Hence, we have $xNxNx$ in T_{n+1} and, therefore, $xKxKx$ in R_{2n+2}, contradicting the fact that R contains no part UxV with $Ux = xV$.

There are a number of more or less direct consequences of the infinity lemma concerned with solutions of sets of tiles.

5.1. For a given tile set P, if, for every n, there is a solution of size n by n, then P has a solution.

We consider partial solutions of size $2n - 1$ for $n = 1, 2, \cdots$ and make a tree such that a block K of size $2n - 1$ by $2n - 1$ leads directly only to blocks of size $2n + 1$ by $2n + 1$ with K in the center. The hypothesis and the fact that any part of a partial solution is also a partial solution ensures an infinite tree. Hence, the infinity lemma yields an infinite path in the tree which represents a solution of P.

5.2. A tile set is solvable over the whole plane if and only if it is solvable over a quadrant.

If it is solvable over the plane, we can of course get a solution over a quadrant by deleting the other quadrants from a given solution. Conversely, if it is solvable over a quadrant, then it has a solution of size n by n for every n. Hence, by 5.1, it has a solution over the whole plane.

5.3. Given a tile set P for which it is possible to form two adjacent infinite rows such that for every m, any 1 by m block occurring in the top row also occurs in the bottom row. Then the set P is solvable.

Let A be the given top row, B be the given bottom row. For each m, and a 1 by m block C_m in A, there is a D_m in A such that D_m can be correctly put on top of C_m, because C_m also occurs in B. Repeating this process with D_m, and so on, we can obtain a partial solution of size m by m. Hence, by 5.1, the set P has a solution.

A solvable tile set is said to be minimal solvable if in every solution every tile of the set occurs.

5.4. Given a minimal solvable set, there exists an integer n such that every tile of the set occurs in every n by n block in every solution.

Assume 5.4 false and we have for every n some solution S and some tile T_i such that there is an n by n block not containing T_i. If the number of tiles is k, make k trees $K_1, \cdots, K_k$ as in 5.1 so that K_i includes all the $2n - 1$ by $2n - 1$ $(n = 1, 2, \cdots)$ blocks which does not include T_i in some solution. By our assumption, at least one of the trees, say K_i, must be infinite. This determines a solution in which T_i does not occur at all, contrary to the hypothesis that the given set is minimal solvable.

5.5. If the (unrestricted) tiling problem is unsolvable, then there is a solvable set with no periodic solution.

This was observed in [17]. If every solvable set had periodic solutions, we would have the following situation. Either a set is solvable, there would then be some n such that there is a torus of size n by n, Or a set is unsolvable, then, by the infinity lemma, there would be some n such that there is no solution of size n by n, Hence, we would be able to test successively for each n, whether there is a solution or a torus of size n by n. This process must terminate at some finite stage, and we would have a decision method for the tiling problem.

Of course, as mentioned before, we now know that the tiling problem is unsolvable and we possess also relatively simple examples of solvable sets with no periodic solutions.

5.6. Given a solvable tile set and an integer n, there exists a solution in which every occurring finite block of size no bigger than n by n occurs infinitely often.

Consider any given solution. Since there are only a finite number of tiles, there must be at least one which occurs infinitely often. Let M_1 be the set of all tiles which occur infinitely often. Let M_2 be the set of all 2 by 2 blocks of tiles each of which occurs infinitely often in the solution and, in addition, consists only of members of M_1. M_2 is again not empty, since the set of 2 by 2 blocks which occur finitely often can only take up a finite area in the plane. In general, given M_n, the set M_{n+1} of $n + 1$ by $n + 1$ blocks each of which occurs infinitely often and contains only members of $M_1, \cdots, M_n$ is not empty. Hence, by the infinity lemma, there is a solution with a member of M_1 in the center, a member of M_2 as the central 2 by 2 block, etc. This gives a solution in which every finite block occurs infinitely often in the original solution. To make sure that we get the desired solution for a given n, we repeat the process of eliminating blocks which occur only finitely often. Since there is only a finite number of distinct blocks of size no bigger than n by n, this process must come to an end.

5.7. Every solvable set has a solution S such that every finite block occurring in S occurs infinitely often in S.

Given a solution T and the set K of all finite blocks occurring in T, consider the set L of all subsets of K such that a subset A of K belongs to L if there is a solution covered by A, i.e., in which all the occurring finite blocks (or, equivalently, just squares) belong to A. The set L is not empty because K belongs to it and it has minimal members. Take any minimal member B and any solution S covered by B. If there is any block in B which appears only finitely often in S, we can eliminate it by 5.6, and B would not be minimal. Hence, every block in B must appear infinitely often in every solution S covered by B.

5.8. If a solvable set P has no periodic solutions, then it has as many distinct solutions as there are real numbers.

Since the set P is solvable, it has, by 5.7, a solution S in which every occurring finite block occurs infinitely often. Hence, if an n by n block occurs in S, it must have two nonoverlapping occurrences. Begin with two occurrences in S, it must have two nonoverlapping occurrences. Begin with two occurrences in S of a single tile T. There must be some n_1, such that the n_1 by n_1 blocks with T at the center at the two places are different. Otherwise, the two infinite columns C and D containing the two occurreces of T (or rows if they are in the same column) must be the same at corresponding positions, in which case we join the two occurrences of T by a staircase and consider all analogous staircases between the two columns. Since there are infinitely many staircases, at least two must be identical. But then we can take the region R bounded by C, D and two identical staircases and repeat it up and down to get an infinite strip S bounded by the modified columns C' and D'. Since C and D are identical at corresponding places, the two vertical parts V_1 and V_2 bounding R are identical. Therefore, we can also repeat the infinite strip S and cover the plane. Since each column consists of repetitions of V_1 (or, what is the same thing, V_2), there must be two infinite columns which are identical (i.e., without any staircase shift). Since there are infinitely many segments of rows bounded by the two columns, two of them must be identical. Therefore, we would have a torus, contradicting the hypothesis of nonperiodicity.

Hence, beginning with T and its two occurrences, we can expand to two different n_1 by n_1 blocks. Each n_1 by n_1 block has two nonoverlapping occurrences which, for similar reasons, can be extended to two distinct bigger blocks. Hence, we have, by repeating the process, a full binary tree with as many infinite paths as there are real numbers. But each infinite path determines a solution.

Given an infinite set K of solutions of a tile set P, a solution S of P is said to be a *limit solution* of P and K if every finite block in S agrees with infinitely many solutions in K over that block.

5.9. Every infinite set of solutions of a tile set has a limit solution; in other words, if there are infinitely many infinite paths, then there is a path on which every node appears in infinitely many infinite paths.

Suppose K is an infinite set of solutions of the tile set P. Form a tree as follows: The nodes on the nth level of the tree consist of all $2n - 1$ by $2n - 1$ partial solutions of P which coincide with infinitely many members of K when these members are restricted to the $2n - 1$ by $2n - 1$ block centered at the origin. A node on the nth level is connected to a node on the $(n + 1)$–st level if and only if the smaller partial solution comprises the center of the larger one. To see that the resulting tree is infinite, we need only verify that there is at least one node on every level. But this is true because there are infinitely many members of K, but (for fixed n) only finitely many blocks of size $2n - 1$ by $2n - 1$. Finally it is clear that only finitely many branches spring from each node. Hence, by the infinity lemma, the tree has an infinite path. This path describes a solution, which is a limit solution to the set K; i.e. every finite block of the solution coincides with the corresponding block in infinitely many members of K.

We mention incidentally an application of the infinity lemma in a different context.

5.10. If a Turing machine halts for every initial state and every initial tape (which may contain infinitely many marked squares), then there is a number N such that the machine always halts before N steps.

Consider at each moment t the pair (q_t^i, S_t^i), where q_t^i is the state and S_t^i is the symbol under scan at t. At the initial time $t = 1$, we have only a finite number of (q_1^i, S_1^i). At each moment, from each (q_t^i, S_t^i), we have only a finite number of (q_{t+1}^i, S_{t+1}^i). Since the machine always stops, the tree contains no infinite path. But, by the infinity lemma, if there are altogether infinitely many finite paths, there is some infinite path. Hence there can be only a finite number of finite paths, and hence a finite bound N to the height of all paths. The theorem is proved.

References

[1] R. Berger, *The undecidability of the domino problem*, Mem. Amer. Math. Soc. 66 (1966), pp.72.

[2] B. Dreben, A. S. Kahr, and H. Wang, *Classification of* AEA *formulas by letter atoms*, Bull. Amer. Math. Soc. 68 (1962), pp. 528——532.

[3] J. Genenz, *Reduktionstheorie nach der Methode von Kahr–Moore–Wang*, Münster 1964, pp. 88.

[4] J. Genenz, *Untersuchungen zum Entscheidungsproblem*, Münster 1965, pp. 44.

[5] Gurjevits and Korjakov, *Remarks on a paper of Berger on a problem of dominoes* (in Russian), Sib. Math. Journal 13 (2) (1972), pp. 459—463.

[6] W. Hanf, *Nonrecursive tilings of the plane I*, submitted to J. Symb. Logic.

[7] —— *Model-theoretic methods in the study of elementary logic*, Theory of Models 1965, pp. 132—145.

[8] H. Hermes, *Entscheidungsproblem und Dominospiele*, Selecta Mathematica II (1970), pp. 114—140.

[9] —— *A simplified proof for the unsolvability of the* AEA *case*, Logic Colloquium 1969, (1971), pp. 307—309.

[10] A. S. Kahr, *Improved reductions of the Entscheidungsproblem to subclasses of* AEA *formulas*, Mathematical Theory of Automata 1963, pp. 57—70.

[11] —— E. F. Moore, and H. Wang, *Entscheidungsproblem reduced to the* AEA *case*, Natl. Acad. Sci. US 48 (1962), pp. 365—377.

[12] S. Ju. Maslov, *The inverse method for logical calculi*, Trudy Mat. Inst. Steklov. 98 (1968), pp. 26—87 (see § 12.3).

[13] D. Myers, *Nonrecursive tilings of the plane II*, to appear.

[14] H. Putnam, *Trial and error predicates*, J. Symb. Logic 30 (1965), pp. 49—57.

[15] R. M. Robinson, *Undecidability and nonperiodicity for tilings of the plane*, Invent. Math. 12 (1971), pp. 177—209.

[16] A. Thue, *Über die gegenseitige Lage gleicher Teile gewisser Zeichenreihen*, Kristiania, 1913, pp. 67.

[17] H. Wang *Proving theorems by pattern recognition II*, Bell Systems Technical J. 40 (1961), pp. 1—41.

[18] —— *An unsolvable problem on dominoes*, Harvard Computation Laboratory report no. BL-30 (II-15), duplicated August 1961, bound Jan., 1962.

[19] —— *Dominoes and the* AEA *case of the decision problem*, Mathematical Theory of Automata 1963, pp. 23—56.

11. DOMINOES AND THE AEA CASE OF THE DECISION PROBLEM*

It has recently been established that the AEA case is unsolvable and forms a reduction class. Several people have looked into possible directions along which the result can be strengthened, using in part earlier methods developed by Buchi for the F & AEA case. There are three different aspects. First, unsolvable AEA subcases such as restrictions on the number of dyadic predicates, on the form of the quantifier free component, on the complexity of the models (e.g., finite, essentially periodic, etc.). Second, solvable AEA subcases. Third, the detailed structure of the reduction of the general case to the AEA case. A survey of these questions is presented.

I. INTRODUCTION AND SUMMARY

Since all mathematical theories can be formulated within the framework of the predicate calculus (quantification theory, elementary logic), Hilbert spoke of "the" decision problem when he was referring to the problem of finding a general algorithm to decide, for each given formula of the predicate calculus, whether it is satisfiable in some non-empty domain (or, has a model). He called this the main problem of mathematical logic. It is familiar today that this problem in its general form is unsolvable in a technical sense which is widely accepted as implying unsolvability according to the intuitive meaning. An interesting problem is to investigate the limits of decidable subdomains and the underlying reasons of the phenomenon of undecidability.

Recently, the general problem has been reduced to the formally simple case of formulas of the form AxEuAyMxuy, where M is quantifier-free and contains neither the equality sign nor function symbols.[8] It is, therefore, of special interest to study the AEA case in greater detail. Moreover, the simplicity of the AEA prefix makes it possible to confine our attention to the essential problems without distraction by extraneous complexities.

The first published attempt to settle the AEA case appeared in a paper by the author (reference 14, pp. 23—32), in which steps toward a positive solution were reported, and a domino problem (the unrestricted domino problem) was formulated in connection with a procedure for deciding some AEA formulas. During the spring of 1960, when this part was written, it was also shown that the origin-constrained domino problem is unsolvable, although the result was not included in this part.

Immediately after the appearance of reference 14, Buchi studied the domino

* First published in *Mathematical Theory of Automata*, pp. 23—55. Polytechnic Press, 1963. Reproduced by permission of the author.

problems, independently obtained a different proof of the unsolvability of the origin–constrained problem, and combined it with a fundamentally new application of Lowenheim's theorem to derive the unsolvability of the E $\wedge$AEA case.[3,4] This last result improves the method used by Turing[13] and Bernays.[2] Although Buchi failed to obtain the decisive result on the AEA case which has superseded previous results on the reduction problem, his work greatly clarified the relation between Turing machines and the unsolvability aspect of the decision problem, and suggested for the first time that the AEA case might turn out to be unsolvable. Finally it was shown by Kahr, Moore and Wang[8] that the diagonal–constrained domino problem is unsolvable and thence the unsolvability of the AEA case was derived.

Section II of this part contains a review of the formulation of the domino problems and a proof of the unsolvability of the originconstrained domino problem along the line originally taken. This should be useful for an understanding of reference 8. Section III digresses into a closer examination of the relation between AEA formulas and the unrestricted domino problem. Section IV deals with the E $\wedge$AEA case. Section V gives, as an alternative to the treatment in reference 8, a different proof of the unsolvability of the AEA case, with an additional result that in reducing an arbitrary formula to an AEA formula, finite models are preserved. It then follows easily that there is no effective method for deciding whether an AEA formula has finite models. In Section VI, we give a rather inadequate sketch of Kahr's further reduction to the AEA case with formulas containing, beyond monadic predicates, only a single dyadic predicate, a result announced in reference 7. Finally, we supply in Section VII an alternative treatment of results on solvable AEA subcases proved in a paper by Dreben, Kahr and Wang.[6] While Sections II, IV, V, and VI are closely interrelated and all deal with unsolvability results, Section III is a digression, and Section VII deals with positive solutions and can be read directly since it does not presuppose the other sections of this part.

II. THE GAME WITH DOMINOES

Assume given a finite set P = $\{D_1, \cdots, D_M\}$ of quadruples (a, b, c, d) of positive integers. We wish to study assignments A of these quadruples to all the lattice points of the first infinite quadrant of the Cartesian plane, or mappings A of the set N^2 of ordered pairs of non–negative integers into the set P, such that:

2.0. $\quad a(Axy') = c(Axy),$
$\quad b(Ax'y) = d(Axy),$

where x′ is short for x⌢1 and a(D), b(D), c(D), d(D) are respectively the first, second, third, fourth members of the quadruple D. The first question is whether there is a general procedure by which, given any finite set P, we can decide whether there exists an assignment satisfying the condition 2.0. More graphically, we can describe this and related questions as follows.

We assume there are infinitely many square plates (the domino types) of the same size (say, all of the unit area) with edges colored, one color on each edge but different edges may have the same color. The type of a domino is determined by the colors on its edges and we are not permitted to rotate or reflect any domino. There are infinitely many pieces of every type. The game is simply to take a finite set of types and try to cover up the whole first quadrant of the infinite plane with dominoes of these types so that all corners fall on the lattice points and any two adjoining edges have the same color.

2.1. A (finite) set of domino types is said to be solvable if and only if there is some way of covering the whole first quadrant by dominoes of these types.

It is natural to use the ordinary Cartesian coordinates and identify each unit square with the point at its lower left–hand corner. Then we can speak of the origin (0, 0), the main diagonal $x = y$, etc.

The following general questions on these games have been considered:

2.2. The (unrestricted) domino problem. To find an algorithm to decide, for any given (finite) set of domino types, whether it is solvable.

2.3. The origin–constrained domino problem. To decide, for any given set P of domino type and a member C thereof, whether P has a solution with the origin occupied by a domino of type C.

2.4. The diagonal– (row–, column–) constrained domino problem. To decide, for any given set P of domino types and a subset Q thereof, whether P has a solution with the main diagonal (the first row, the first column) occupied by dominoes of types in Q.

These three problems will be considered in different parts of this paper.

In order to prove the unsolvability of the origin–constrained domino problem, we shall reduce to it the following familiar unsolvable halting problem of Turing machines.

(HB) To decide, given any Turing machine, whether it eventually halts if the initial tape is blank.

To keep our ideas fixed, we shall assume one special formulation of Turing machines, although it will be clear that similar considerations are applicable to other formulations. We use a one–way infinite tape, take q_1 as the initial state, the leftmost square of the tape as the initially scanned square, two tape symbols S_0 (blank) and S_1 (marked), the basic acts R (shift the reading head right one square), L (shift the reading head left one square), S_1 (print S_1), S_0 (print S_0). Each machine has a finite number of states $q_1, \cdots, q_n$, and, at each moment, the present state and the content of the scanned square together determine the acts (one print act and one shift act) to be taken, as well as the state at the next moment.

An example which will be used for illustration is:

Machine X

$q_1S_0S_1Rq_2$	$q_1S_1S_1Rq_1$
$q_2S_0S_0Rq_3$	$q_2S_1S_1Lq_3$
$q_3S_0S_1Lq_4$	$q_3S_1S_0Lq_4$
$q_4S_0S_0Lq_1$	

We shall give a general method by which, given any Turing machine X, we can find a corresponding domino set P_X containing a distinguished type D such that X halts on an initially blank tape if and only if P has no solution with a domino of type D at the origin. Essentially, we choose P_x so that in a solution of P, for every y, the y–th row contains the whole situation of X (tape, state, and scanned square) at time y. As a result X eventually halts if and only if P_x has no solution.

Thus, for the example X above, we choose P_X as follows.

2.5. P_X consists of the following domino types:

2.5.1. Two domino types for each tape symbol: $[S_0]$, $[LS_0]$, $[S_1]$, $[LS_1]$.

2.5.2. One domino type for each permissible kind of scanned square (state and symbol): $[q_i S_j]$, $i = 1, 2, 3, 4$; $j = 0, 1$; $(i, j) \neq (4, 1)$.

2.5.3. One domino type for the next scanned square (symbol and next state) after a left shift: $[Lq_iS_j]$, $i = 1, 3, 4$; $j = 0, 1$.

2.5.4. One type for the next scanned square after a right shift: $[Rq_iS_j]$, $i = 1, 2, 3$; $j = 0, 1$.

2.5.5. Four domino types for the initial row and column: [D] for the origin, [B] for the beginning of the tape, [↑] for initial row, [→] for the initial column.

If we use the x–coordinate to represent tape positions and the y–coordinate to represent time, a simulation of the particular machine X should be given by a partial solution of P_X as in Fig. 1. Machine X halts at $y = 8$, because the reading head ends up scanning S_1 in state q_4, and it is understood that the machine halts if the reading head scans S_j in state q_i but there is no entry in the machine table beginning with q_iS_j, or there is a left shift while scanning the beginning of the tape.

There remains the problem of specifying the four numbers or colors of each domino type in P_x to exclude undesired solutions.

Each solution is an assignment or mapping A of a domino type to a member of N^2. It is natural to write $Axy = D_i$ briefly as D_ixy. We can now state the conditions needed for the simulation.

2.6. The conditions on P_X:

2.6.1. The origin constraint. (Ex) [D]xx.

In other words, the type [D] must occur somewhere. This position is treated as the origin (0, 0). By choosing the colors on D suitably, we can make it impossible for any domino to occur to its left or below it.

2.6.2. The initial row and the initial column are the boundary:

2.6.2.1. $[D]xy \supset [B]x'y$.

2.6.2.2. $([B] \vee [\uparrow])xy \supset [\uparrow]x'y$.

2.6.2.3. $([D] \vee [\rightarrow])yx \supset [\rightarrow]yx'$.

2.6.3. The next row above the initial row simulates the initial configur ation:

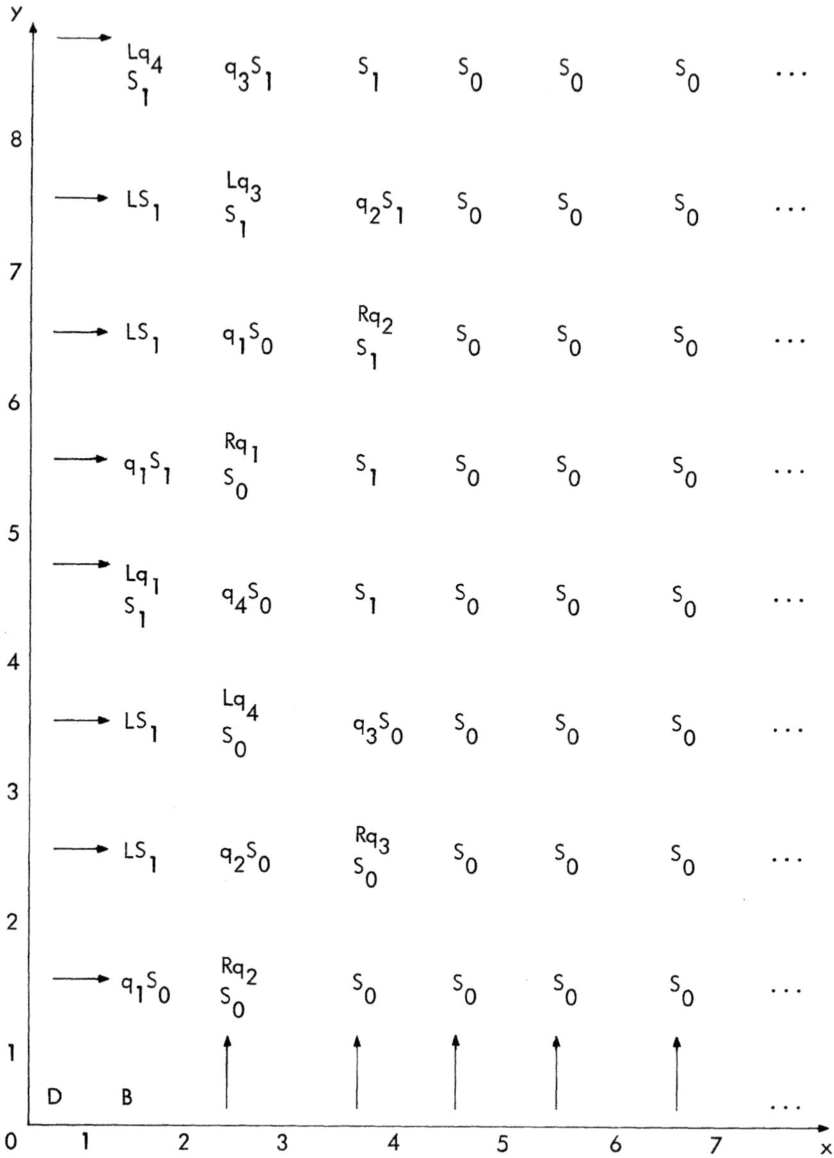
y
Lq_4 S_1 q_3S_1 S_1 S_0 S_0 S_0 ...
8
LS_1 Lq_3 S_1 q_2S_1 S_0 S_0 S_0 ...
7
LS_1 q_1S_0 Rq_2 S_1 S_0 S_0 S_0 ...
6
q_1S_1 Rq_1 S_0 S_1 S_0 S_0 S_0 ...
5
Lq_1 S_1 q_4S_0 S_1 S_0 S_0 S_0 ...
4
LS_1 Lq_4 S_0 q_3S_0 S_0 S_0 S_0 ...
3
LS_1 q_2S_0 Rq_3 S_0 S_0 S_0 S_0 ...
2
q_1S_0 Rq_2 S_0 S_0 S_0 S_0 S_0 ...
1
D B ...
0 1 2 3 4 5 6 7 x

Figure 1

2.6.3.1. $[B]yx \supset [q_1S_0]yx'$.

2.6.3.2. $[\uparrow]yx \supset ([Rq_2S_0] \vee [S_0])yx'$.

2.6.4. The left or right neighbor of the scanned square at time y is in part determined by a left or right shift and embodies information for the scanned square at time y′. It is convenient to write briefly $[Lq_i]$ for $[Lq_iS_0] \vee [Lq_iS_1]$, $[Rq_i]$ for $[Rq_iS_0] \vee [Rq_iS_1]$.

2.6.4.1. $[q_iS_j]x'y \supset [Lq_k]xy$; (i, j, k) = (2, 1, 3), (3, 0, 4), (3, 1, 4), (4, 0, 1).

2.6.4.2. $[q_iS_j]xy \supset [Rq_k]x'y$; (i, j, k) = (1, 0, 2), (1, 1, 1), (2, 0, 3).

2.6.5. The state and scanned square at time y′ are determined by $[Lq_i]$ or $[Rq_i]$ at time y.

2.6.5.1. $[Lq_iS_j]yx \supset [q_iS_j]yx'$; i = 1, 3, 4, j = 0, 1.

2.6.5.2. $[Rq_iS_j]yx \supset [q_iS_j]yx'$; i = 1, 2, 3, j = 0, 1.

2.6.6. The tape symbol at time y′ and position x is determined by the tape symbol at (x, y).

2.6.6.1. $[S_i]yx \supset ([S_i] \vee [Rq_1S_i] \vee [Rq_2S_i] \vee [Rq_3S_i])yx'$; i = 0, 1. $[LS_i]yx \supset ([LS_i] \vee [Lq_1S_i] \vee [Lq_3S_i] \vee [Lq_4S_i])yx'$; i = 0, 1.

2.6.6.2. $[q_2S_0]yx \supset [Lq_4S_0]yx'$,
$[q_3S_1]yx \supset [S_0]yx'$,
$[q_4S_0]yx \supset ([Rq_1S_0] \vee [Rq_2S_0])yx'$,
$[q_1S_0]yx \supset ([LS_1] \vee [Lq_3S_1])yx'$,
$[q_1S_1]yx \supset [LS_1]yx'$,
$[q_2S_1]yx \supset [S_1]yx'$,
$[q_3S_0]yx \supset [S_1]yx'$,

2.6.7. In each row, we have to distinguish $[S_i]$ and $[LS_i]$.

2.6.7.1. $([Rq_1] \vee [Rq_2] \vee [Rq_3] \vee [q_2S_1] \vee [q_3S_0] \vee [q_3S_1] \vee [q_4S_0] \vee [S_0] \vee [S_1])xy \supset ([S_0] \vee [S_1])x'y$.

2.6.7.2. $([Lq_1] \vee ([Lq_3] \vee [Lq_4] \vee [q_1S_0] \vee [q_1S_1] \vee [q_2S_0], \vee [LS_0] \vee [LS_1]\,x'y \supset ([LS_0] \vee [LS_1] \vee [\rightarrow])xy$.

2.6.8. The halting conditions.

2.6.8.1. $\neg [q_4S_1]xy$.

We can either exclude the type $[q_4S_1]$ or choose its colors so that no domino can occur above it.

2.6.8.2. $[\rightarrow]xy \supset \neg([Lq_1] \vee [Lq_3] \vee [Lq_4])xy$.

This could be deleted if we had included the condition that no two types can be assigned to the same place.

These conditions determine the colors on the domino types in the following manner.

2.7. Colors on the domino types in P_X are given in Fig. 2.

By 2.6.5, the top of Lq_iS_j or Rq_iS_j is the same as the bottom of q_iS_j. By 2.6.3.1, the top of B is the same as the bottom of q_1S_0.

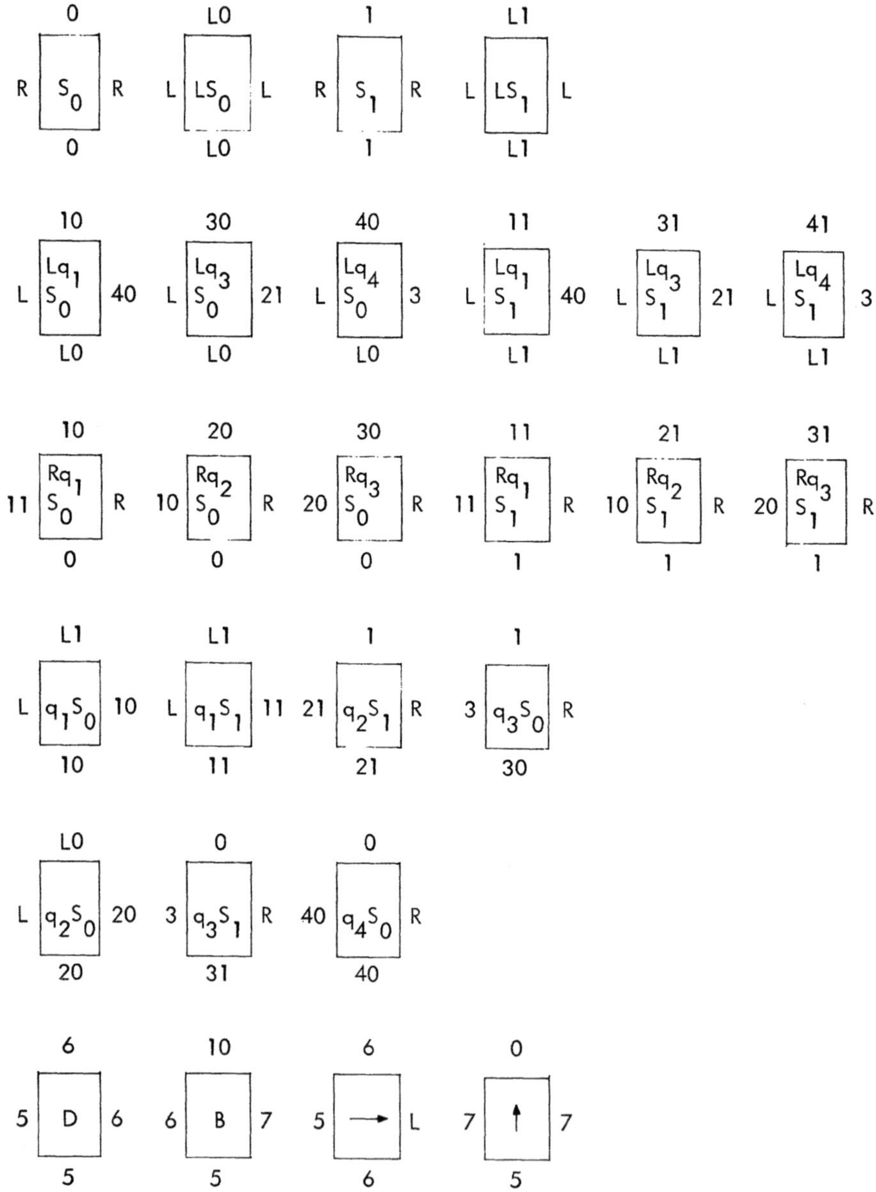

Figure 2

By 2.6.6.1, the bottom of the first three rows and the top of the first row are filled. By 2.6.3.2, the top of [↑] is determined. By 2.6.6.3, the top of the third and the fourth rows is determined.

By 2.6.7.1 and 2.6.7.2, the left of the first two rows, the right of the first and the third rows, the L's and R's in the third and the fourth rows, and the right of [→] are determined.

By 2.6.4, the remaining edges in the middle four rows are determined.

Finally, by 2.6.2, the edges of the last row are determined.

The conditions 2.6 do not exclude the possibility that several domino types occur at the same place, although the condition 2.6.1 does yield the requirement that at least one domino type occurs at each place. To assure uniqueness, we may either add an explicit condition to such an effect or use $\equiv$ instead of $\supset$. If we replace $\supset$ by $\equiv$, we have to put together occurrences of the same basic formula, such as $[q_1S_0]yx'$ in 2.6.3.1 and 2.6.5.1:

$$([Lq_1S_0] \vee [Rq_1S_0] \vee [B])yx \equiv [q_1S_0]yx'.$$

Such a treatment is perhaps less easy to follow.

Since the example is sufficient to illustrate the general situation, we have proved:

Theorem I: The origin–constrained domino problem is unsolvable; in fact, the halting problem (HB) is reducible to it.

III. FORMULAS FOR THE UNRESTRICTED PROBLEM

Given a set $P = \{D_1, \cdots, D_M\}$, we may divide the set into four sets, P_1, P_2, P_3, P_4, of subsets such that two dominoes belong to the same subset in P_1 if and only if their first members (bottom edge) are the same (oof the same color), and so on. Then we can match up subsets of P_1 with those of P_3, subsets of P_2 with those of P_4 so that, for example, a subset $\{D_e, \cdots, D_f\}$ in P_4 is matched up with a subset $\{D_g, \cdots, D_h\}$ in P_2 if the fourth member of D_e is the same as the second member of D_g.

3.1. A mapping of N^2 into P gives a solution if and only if:

3.1.1. Every pair (x, y) in N^2 gets a unique quadruple from P; or, writing D_ixy as short for $Axy = D_i$:

3.1.1.1. $D_1xy \vee \cdots \vee D_Mxy$.

3.1.1.2. $[D_1xy \supset (\neg D_2 \wedge \cdots \wedge \neg D_M)xy] \wedge \cdots \wedge [D_Mxy \supset (\neg D_1 \wedge \cdots \wedge \neg D_{M-1})xy]$. Briefly, 3.1.1 (that is, the conjunction of 3.1.1.1 and 3.1.1.2) is also written as: $\nabla(D_1xy, \cdots, D_Mxy)$.

3.1.2. $b(Ax'y) = d(Axy)$. Or, using truth functions only: $[(D_e \vee \cdots \vee D_f)xy \equiv (D_g \vee \cdots \vee D_h)x'y] \wedge \cdots \wedge [(D_p \vee \cdots \vee D_q)xy \equiv (D_s \vee \cdots \vee D_t)x'y]$.

3.1.3. $a(Ayx') = c(Ayx)$, with a similar truth functional expression of the form $V(D_1yx, \cdots, D_Myx; D_1yx', \cdots, D_Myx')$.

More exactly, we have assumed that in the set P, for every domino D_i, there are D_j, D_k, such that $b(D_j) = d(D_i)$, $a(D_k) = c(D_i)$, because we can effectively eliminate domino types D_i for which the above condition does not hold. The fact that we are concerned with the first quadrant rather than the whole plane presents a slight complication insofar as a type D_i for which there is no D_j, $a(D_i) = c(D_j)$, or no D_k, $b(D_i) = d(D_k)$ may occur in the first row or the first column but not elsewhere. It is, however, clear that all members of each $P_k (k = 1, 2, 3, 4)$ are mutually exclusive and jointly exhaust P. Moreover, it is true in every case that in 3.1.2 and 3.1.3, for each D_i, D_ixy or $D_ix'y$ or D_iyx or D_iyx' appears in at most one disjunct because, e.g., if $c(D_i) = a(D_j) = a(D_k)$ and $c(D_p) = a(D_j)$, we must also have $c(D_p) = a(D_k)$.

In condition 3.1, we can, by the help of 3.1.1.2, replace $\equiv$ by $\supset$ in 3.1.2 and 3.1.3. Another possibility is to replace $\equiv$ by $\vee$, and $\vee$ by $\vee$ in 3.1.2 and 3.1.3. Then we can delete 3.1.1.1 as an independent condition.

From 2.1, it follows immediately that to every set P of domino types, there is a corresponding formula of the form:

$$U(G_1xy, \cdots, G_Kxy; G_1x'y, \cdots, G_Kx'y)$$
$$\wedge V(G_1yx, \cdots, G_Kyx; G_1yx', \cdots, G_Kyx'), \qquad (1)$$

or briefly,

$$U(xy, x'y) \wedge V(yx, yx'), \qquad (1^*)$$

where U and V are truth–functional combinations of the components such that P is solvable if and only if (1) has a model. That is, there is an interpretation in the domain N of non–negative integers of $G_1, \cdots, G_k$ which makes (1) true. This fact was first pointed out by Buchi.

Note, incidentally, that we can generally use fewer dyadic predicates than dominoes. Given M domino types, let $K = \mu n(2^n \geqslant M)$. We can then use K dyadic predicates to represent the M domino types, since, for any x and y, each of $G_1xy, \cdots, G_Kxy$ can be true or false and we can identify each D arbitrarily with one of these 2^k distributions of truth values; for example, we can replace D_ixy by $G_exy \wedge \cdots \wedge G_fxy \wedge \neg G_gxy \wedge \cdots \wedge \neg G_hxy$. Using any such representation, we can restate 3.1, for example, by:

3.2 The set $\{D_1, \cdots, D_M\}$ is solvable if and only if we can assign values to $G_1, \cdots, G_K$ over all (x, y) in N^2, such that:

3.2.1. $[(D_exy \vee \cdots \vee D_fxy) \wedge (D_gx'y \vee \cdots \vee D_hx'y)] \vee \cdots \vee [(D_pxy \vee \cdots \vee D_qxy) \wedge (D_sx'y \vee \cdots \vee D_tx'y)]$

3.2.2. Similarly for (y, x) and (y, x').

The uniqueness of 3.1.1.2 is now dispensable since the truth distributions of G_1xy, $\cdots$, G_Kxy are automatically mutually exclusive. Moreover, as just noted, the existence of condition 3.1.1.1 also follows from 3.2.1 (or 3.2.2) since, no matter which disjunct of 3.2.1 (or 3.2.2) is true, some D_i must be true of (x, y).

We shall now prove that, conversely, given a formula F of the form (1), we can find a corresponding domino set P_F such that P_F is solvable if and only if F has a model in N.

We assume F contains K dyadic predicates $G_1, \cdots, G_K$ and U, V are in the fully developed disjunctive normal form so that U (or V) is a disjunction of conjunctions each of which is of the form

$$\pm(G_1xy \wedge \cdots \wedge \pm G_kxy \wedge \pm G_1x'y \wedge \cdots \wedge \pm G_kx'y, \qquad \text{or} \tag{2}$$

$$\pm G_1yx \wedge \cdots \wedge \pm G_kyx \wedge \pm G_1yx' \wedge \cdots \wedge \pm G_kyx'. \tag{2*}$$

If pq is ambiguously xy, yx, x′y, or yx′, we separate out the sign pattern of each occurring K–termed conjunction $\pm G_1pq \wedge \cdots \wedge \pm G_Kpq$ by writing it as C_ipq, one number i for each pattern. Each of the patterns C_i is taken as a color and we define the set P_F of dominoes as the set of all quadruples $D_k = (C_i, C_i, C_j, C_k)$ such that $[(C_iyx \wedge C_jyx') \supset V(yx, yx')]$ and $[(C_ixy \wedge C_kx'y) \supset U(xy, x'y)]$ are truth–functional tautologies.

In this way, each sign pattern C_i gives rise to $m \times n$ domino types if it can be followed on the top by m sign patterns, and on the right by n sign patterns. Had we taken each C_i as a single domino type. we would not be able to exclude in general the situation that, e.g., C_1 can be followed by C_2 and C_3 on the right, but C_4 can only be followed by C_2 on the right, violating the requirement on colors.

If F has a model, then the conjunction of all instances of $U(xy, x'y) \wedge V(yx, yx')$ for all (x, y) in N^2 is true for some selection of a conjunction (2) and a conjunction (2*), for each pair (x, y). This then yields a solution of P_F.

Conversely, if P_F has a solution, we obtain from the solution two conjunctions (2) and (2*) for each pair (x, y). All these selections yield together a model for the given formula F. Hence, we get Theorem II.

Theorem II: Given a domino set P we can find a formula F_P of the form (1) such that P has a solution if and only if F_P has a model. Conversely, given a formula F of the form (1), we can find a domino set P_F such that F has a model if and only if P_F has a solution. Hence, the unrestricted dominoproblem is undecidable if and only if the decision problem of the class of all formulas of the form (1) is unsolvable.

IV. UNSOLVABILITY OF THE E $\wedge$ AEA CASE.

The relation between the domino problems and certain simple classes of formulas of the predicate calculus is brought out by Buchi's application of the familiar Lowenheim theorem (see Skolem[10]) as a lemma which for our purpose may be stated thus:

Lemma 1: A formula $EzKz \wedge AxEuAyMxuy$, in which K and M are quantifier–

free, is satisfiable if and only if Ko ∧ AxAyMxx′y is satisfiable in the domain of non–negative integers; similarly if we delete EzKz and Ko.

The shortest proof of this lemma uses the axiom of choice to give a function f(x) which gives the corresponding u for each x in any given model. Since EzKz, let a be an object such that Ka; or, when EzKz is absent, take any object a of the model. Using the function f, we get a domain {a, f(a), f(f(a)), ⋯} closed with respect to f. This may be identified with {0, 1, 2, ⋯}. It is important that in thus using the familiar domain of non–negative integers, we do not assume that in the model the conditions $x' \neq 0$ and "$x' = y'$ implies $x = y$" are satisfied. Hence, finite models are not excluded.

Using this lemma, Buchi proved:[3,4]

Theorem III: The class of E ∧ AEA formulas with monadic predicates and three dyadic predicates is unsolvable.

Clearly, once we have an unlimited supply of dyadic predicates, we may use some of them as monadic predicates by allowing, for example, G to occur only in contexts Gvv, for some variable v. Hence, this is stronger than the result, which can be derived directly from Lemma 1 and Theorem I, that permits an unbounded number of dyadic predicates.

We reproduce here Buchi's proof in a slightly different form in order to make some of the details more explicit.

For a machine with K states, three dyadic predicates and K + 3 monadic predicates are used which have the following intuitive meaning:

Zz: z is the distinguished element 0.
Sxy: The square x is marked at time y.
Kxy: The square x is under scan at time y.
Jxy: The square x′ is under scan at time y.
Q_iy: The state at time y is $q_i (i = 1, \cdots, K)$.
Ly: A left move is made at time y.
Ry: A right move is made at time y.

The procedure for writing out an E ∧ AEA formula from each Turing machine consists of a specification of the transitions and acts of a given machine, plus a general form applicable to all machines with the specified parts as parameters.

4.1. The specification for the machine (X) above (preceding 2.5).

4.1.1. The transition formula $\mathcal{Q}$yxx′ from time x to time x′:

$$\begin{aligned} Kyx \supset (\{Syx \supset [(Q_1x \supset Q_1x') \wedge (Q_2x \supset Q_3x') \\ \wedge (Q_3x \supset Q_4x')]\} \\ \wedge \{\neg Syx \supset [(Q_1x \supset Q_2x') \wedge (Q_2x \supset Q_3x') \\ \wedge (Q_3x \supset Q_4x') \wedge (Q_4x \supset Q_1x')]\}). \end{aligned}$$

4.1.2. The moves of the different states are summarized:

$\mathcal{L}yx$ for $[\neg Syx \wedge (Q_2x \wedge Q_3x)] \wedge [\Box Syx \wedge (Q_3x \wedge Q_4x)]$
$\mathcal{R}yx$ for $[K\neg Syx \wedge (Q_1x \wedge Q_2x)] \wedge (Syx \wedge Q_1x)$.
S_0yx for $[\neg Syx \wedge (Q_2x \wedge Q_4x)] \wedge (Syx \wedge Q_3x)$.
S_1yx for $[\neg Syx \wedge (Q_1x \wedge Q_3x)] \wedge [Syx \wedge (Q_1x \wedge Q_2x)]$.

4.1.3. The halting condition is given by:
$\mathcal{K}yx$ for $Kyx \wedge [(Syx \wedge Q_4x) \vee (Zx \wedge \mathcal{L}yx)]$.

Relative to any specification of $\mathcal{Q}yxx'$, $\mathcal{L}yx$, $\mathcal{R}yx$, S_0yx, S_1yx, $\mathcal{K}$ yx, the corresponding formula of the given machine is given as a conjunction of conditions.

4.2. The formula for a given machine.
4.2.1. Specification of Z: $EzZz \wedge \neg Zx'$.
4.2.2. Exactly one state at each time (see 3.1.1): $\nabla(Q_1x, \ldots, Q_Kx)$.
4.2.3. Condition on Jxy: $Kx'y = Jxy$.
4.2.4. Initial configuration: $Zx \supset [(Q_1x \wedge Kxx \wedge \neg Jyx) \wedge \neg Syx]$.
4.2.5. Transition of states: $\mathcal{Q}yxx'$.
4.2.6. No model if halting condition emerges: $\neg\, \mathcal{K}yx$.
4.2.7. The scanned square at x′:

$$[(Kyx \wedge \mathcal{L}yx) \supset Lx] \wedge [(Kyx \wedge \mathcal{R}yx) \supset Rx] \wedge \neg(Lx \wedge Rx).$$
$$Lx \supset (Jyx \equiv Kyx').$$
$$Rx \supset (Kyx \equiv Jyx').$$

4.2.8. The contents of tape squares at x′:

$$Syx' \equiv [(Kyx \wedge S_1yx) \vee (\neg Kyx \wedge Syx)].$$

The above formula is of the form $EzKz \wedge AxAyMxx'y$, and has the property that it has a model if and only if the given machine does not halt eventually when beginning with a blank tape. Hence, by Lemma 1, Theorem III is proved.

It may be noted that the description of the machine by the above formula does not depend on the fact that the integers are all distinct. In fact, whether or not the formula has any finite models depends on the behaviors of K, S, J, $Q_1, \ldots, Q_K$. If an axiom of infinity were included as part of the condition, there would, as with Turing and Bernays, never be any finite models.

It is clear that the same argument can also be used to derive the unsolvability of E∨ AEA from Theorem I, only without a bound on the number of dyadic predicates. While Buchi's proof is more elegant in this case, the unsolvability of AEA^8 was obtained originally by extending Theorem I, which also has certain advantages when we try to find a bound on the number of dyadic predicates in an unsolvable class of AEA formulas.

At this stage it appears obvious that the AEA case is unsolvable if and only if the diagonal–constrained domino problem is unsolvable, since we can, in expressing the diagonal problem, replace 2.6.1 (viz., (Ez) [D] zz) by a condition of the form $([D_a] \vee \cdots \vee [D_b])_{xx}$ and avoid the extra quantifier (Ez).

V. UNSOLVABILITY OF THE AEA CASE

In order to extend the proof of Theorem I ot the unsolvability of the row–constrained or diagonal–constrained domino problem, we are faced with the difficulty that we do not have a fixed origin. This obstacle is removed in reference 8 by the use of relative origins (oor barriers) so that, corresponding to each moment of operation of the simulated Turing machine X, infinitely many copies of the effective part of the simulation of X in the origin–constrained case are present in the solution. In this way, although we do not get one fixed origin, we are assured of an infinite sequence of equally spaced relative origins. There is the additional complication that we need, in general, an unbounded number of tape squares in simulating a Turing machine. Since, however, at each time t, we need only be concerned with, say no more than t′ squares of the tape, we get enough room by doubling the distance between two barriers as we come to the next row or diagonal. This is achieved by using a notion of parity such that a barrier is preserved in the next row or diagonal only when it has the active parity, and within each row or diagonal, the parity changes at each barrier. The proof is carried out in reference 8 for the diagonal–constrained problem. A similar but slightly simpler argument would establish the unsolvability of the row or column–constrained domino problem. We can also give a general method by which, given P and Q, we can find P* and Q* such that P has a column or row–constrained solution relative to Q, if and only if P* has a diagonal–constrained solution relative to Q*.

Theorem IV: The diagonal–(row–, column–) constrained domino problem is unsolvable. In fact, we can find, for each Turing machine X, a set of domino types P_X such that X eventually halts on an initial blank tape if and only if P_X has no constrained solution.

From Lemma 1 and Theorem IV, the unsolvability of the AEA case follows immediately. In fact, somewhat stronger results are proved[8] by classifying AEA formulas Mxx′y with dyadic predicates only according to the forms of the occurring atomic formulas and the structure of M. Thus, if $G_1, \ldots, G_K$ are the dyadic predicates, there are nine possible forms G_ixx, G_ixx', $G_ix'x$, $G_ix'x'$, G_iyy, G_ixy, G_iyx, $G_ix'y$, G_iyx' for each G_i. We have, therefore, altogether 2^9—1 possible classes determined by including or excluding formulas of each of these forms. We can furthermore specify the possible contexts in which formulas of given forms occur. The following theorem is proved in reference 8:

Theorem V: Any AEA class including all formulas which contain three of the four forms (xy, yx, x′y, yx′) is undecidable: the class of all AEA formulas of the form Wxx $\wedge$ U(xy, x′y) $\wedge$ V(yx, yx′), that of the form U(xy, x′y) $\wedge$ V(xy, yx), that of the form U(yx, yx′) $\wedge$ V(xy, yx), are all undecidable, where W, U, V are, as in Eq. (1) of Section III, truth–functional expressions. Moreover, all these classes are reduction classes.

The unsolvability of the AEA case completely settles the question of decidable and undecidable subclasses of the predicate calculus insofar as these are determined solely by the prefix forms of their component formulas in the prenex normal form. This follows directly from common knowledge in the literature.[8] In fact, this is ture even when we permit extended prenex forms (see, e.g., reference 14, p. 2 and p.5, and reference 3), that is, formulas which are conjunctions of formulas in the prenex normal form.

Theorem VI: An extended prefix form class is a reduction type (and undecidable) if and only if either the prefix of at least one conjunct contains AEA or AAAE as a (order-preserving but not necessarily consecutive) substring, or there are two conjuncts of which the prefixes contain AAA and AE respectively. Moreover, it is decidable if and only if it contains no axioms of infinity, i.e., formulas which have only infinite models.

When commenting on an earlier draft of reference 8, Richard Buchi and Dana Scott independently suggested that an alternative proof of the unsolvability of the AEA case be used which does not employ the dominoes in such a conspicuous manner. Such a proof would be an extension of the proof of Theorem III rather than one of Theorem I.

It turns out that either way of proving the theorem can be modified to have additional useful properties in connection with the question of preserving finite models in the reduction procedure and finding a bound on the number of dyadic predicates. Since the modifications are similar in both cases, we present here a proof along the alternative approach which includes the additional features.

Instead of reducing (HB) as before, we shall use:

(HI) For a suitably chosen machine (usually a universal Turing machine), to decide, given any initial tape containing a consecutive finite string of marked squares, whether it eventually halts.

We shall reduce (HI) to the diagonal-constrained domino problem and make use of monadic predicates to take care of the inputs. Moreover, we shall assume that the Turing machine employed may have a distinguished state q_h, entry into which puts the machine into a suitable form of repetitive behavior. Later on we shall describe a general method by which, given an arbitrary formula F in the predicate calculus, we get a corresponding Turing machine T_F such that, among other things, if F has a finite model, then a state q_h occurs in T_F. The feasibility of this depends on the fact that the formulas having finite models form a recursively enumerable set.

We shall aim at a proof that is easy to follow, and make no attempt to economize the number of dyadic predicates or to avoid redundant conditions. For definiteness, we use the particular machine given in Section II and assume that initially there is a string of k consecutive marked squares at the beginning of the tape. We shall arbitrarily take q_5 as the distinguished state q_h so that in the simulation special conditions are introduced for q_5, which are to yield certain desired properties in the corresponding AEA formula.

5.1. The machine X; same as before, except as a state q_5 is permitted.

Instead of simulating the configuration at time t by the row $y = t$, we use the t-th diagonal $x = y + t$. The initial configuration is simulated on the main diagonal.

5.2. We use 2k + 4 monadic predicates to set up a basic frame:

$$M_i x \equiv M_{i+1} x' (i = 0, \ldots, 2k+2), \quad M_{2k+3} x \equiv M_0 x'.$$
$$\nabla(M_0 x, \ldots, M_{2k+3}),$$
(∇ means exactly one; see 3.1 above).

We set up a period of length 2k + 4 on the main diagonal and this is doubled as we go from one diagonal to the next so that the period at time t is of length $2^{t+1}(k + 2)$.

5.3. We use a number of predicates with the following intended meaning:

Bxy: the point (x, y) is a barrier (relative origin), which is followed by the beginning of the tape at time x–y.

Pxy: the point (x, y) belongs to the active part at time x–y.

Kxy: if (u, v) is the nearest point on the same diagonal as (x, y) such that Buv and v < y, then the (y–v)–th square of the tape is under scan at time x–y.

Sxy: if (u, v) is as before, then the (y–v)–th square of the tape is marked at x–y.

Lxy: the move at x–y is shift left.

Rxy: the move at x–y is shift right.

S_0xy: the third symbol in the quintuple used at x–y is S_0.

S_1xy: the third symbol in the quintuple used at x–y is S_1.

Hxy: the machine stops at x–y.

Q_ixy: the machine is at state q_i at x–y(i = 1, 2, 3, 4, 5).

Cxy: the machine begins operation at x–y.

5.4. To make the exposition easy, we use also Bx′y′, Px′y′, ..., Kx′y′ and even Kx″y′. These can be eliminated by additional predicates so that y′ and x″ do not occur. For example, we can replace Kx′y′ by $K^*x'y$, and Kx″y′ by $K^+x'y$, and add two conditions: $K^*yx \equiv Kyx'$, and $K^+xy \equiv K^*x'y$.

5.5. The set of conditions for a given machine.

5.5.1. Depending on the given machine, we give the transition of states and different moves:

$(Pxy \wedge Kxy \wedge Sxy) \supset [(Q_1xy \supset Q_1x'y) \wedge (Q_2xy \supset Q_3x'y)$
$\quad \wedge (Q_3xy \supset Q_4x'y)]$.

$(Pxy \wedge Kxy \wedge \neg Sxy) \supset [(Q_1xy \supset Q_2x'y) \wedge (Q_2xy \supset Q_3x'y)$
$\quad \wedge (Q_3xy \supset Q_4x'y) \wedge (Q_4xy \supset Q_1x'y)]$.

$(Pxy \wedge Kxy \wedge Sxy \wedge Q_4xy) \supset Hx'y$.

$(Px'y' \wedge Kx'y' \wedge Lx'y' \wedge Bxy) \supset Hx'y$.

$\{Pxy \wedge Kxy \wedge [((Q_2xy \vee Q_3xy) \vee Sxy) \vee ((Q_3xy \vee Q_4xy)$
$\quad \wedge \neg Sxy)]\} \supset Lxy$.

$\{Pxy \wedge Kxy \wedge [(\neg Sxy \wedge Q_2xy) \vee Q_1xy]\} \supset Rxy$.

$\{Pxy \wedge Kxy \wedge [(\neg Sxy \wedge (Q_2xy \vee Q_4xy)) \vee (Sxy \wedge Q_3xy)]\} \supset S_0xy$

$\{Pxy \wedge Kxy \wedge [(\neg Sxy \wedge (Q_1xy \vee Q_3xy)) \vee (Sxy$
$\quad \wedge (Q_1xy \vee Q_2xy))]\} \supset S_1xy$.

5.5.2. Some auxiliary conditions:

$Q_i xy \equiv Q_i x'y'$, $(i = 1, \ldots, 5)$, $\nabla(Q_1 xy, \ldots, Q_5 xy)$.
$Lxy \equiv Lx'y'$, $Rxy \equiv Rx'y'$, $S_0 xy \equiv S_0 x'y'$,
$S_1 xy \equiv S_1 x'y'$,
$Hxy \equiv Hx'y'$, $\neg(Lxy \wedge Rxy)$, $\neg(S_0 xy \wedge S_1 xy)$.

5.5.3. Determination of square under scan and contents of tape squares at next moment.

$(Pxy \wedge Lxy) \supset (Kx'y' \equiv Kx'y)$.
$(Pxy \wedge Rxy) \supset (Kxy \equiv Kx''y')$.
$(\neg Pxy \wedge Bxy) \supset (\neg Kx'y \vee Cx'y)$.
$S_0 xy \supset \neg Sx'y$; $S_1 xy \supset Sx'y$; $(Pxy \wedge \neg Kxy) \supset (Sxy \equiv Sx'y)$.

5.5.4. Conditions on parity:

$Bx'y' \supset [(Pxy \equiv \neg Px'y') \vee Cx'y]$.
$\neg Cx'y \supset [Bx'y \equiv (Bxy \wedge Pxy)]$.
$(\neg Pxy \wedge \neg Cx'y) \supset (\neg Sx'y \wedge \neg Kx'y)$.

5.5.5. Conditions on the diagonal:

Cxx.
$Cxy \supset [(M_0 y \vee \ldots \vee M_{k+1} y) \equiv Pxy]$.
$Cxy \supset [(M_0 y \vee M_{k+2} y) \equiv Bxy]$.
$Cxy \supset Q_1 xy$.
$Cxy \supset \{(Kxy \equiv M_1 y) \wedge [Sxy \equiv (M_1 y \vee \ldots \vee M_k y)]\}$.

5.5.6. Conditions on stopping and the distinguished state q_5:

$\neg Hxy$
$Cxy \equiv Cx'y'$; $Q_5 xy \equiv Cx'y$.
This is the end of the conditions.

By the remark in 5.4, the conjunction of the above conditions is a for mula F of the form Mxx′y with monadic predicates and a finite number of dyadic predicates determined by the number of states of the simulated Turing machine. The conjunct $\neg Hxy$ ensures that the formula has no model if and only if the machine stops. By appealing to auxiliary considerations in reference 8, we can applying Lemma 1, obtain an alternative proof of Theorem V and therewith, one of Theorem VI. In fact, using considerations similar to those in Section III, it is also possible to get an alternative proof of Theorem VI in this way. If there is a distinguished state q_h, then the formula F gets a model which repeats the initial periodic diagonal (the main diagonal) on all diagonals ks for some s, with the help of the predicate C. In the other satisfiable case, we get only infinite models because of the doubling.

A familiar fact in logic is:

Lemma 2: There is an effective partial procedure by which, given a formula of the predicate calculus, we can test whether it has no model, a finite model, or only infinite models. The procedure terminates in the first two cases but does not terminate in the last case.

Hence, given a formula, we can construct a Turing machine which stops if the formula has no model, gets into a distinguished state q_h if the formula has a finite model, and behaves otherwise is the formula has only infinite models. The reason why the second case yields finite models may be seen from Fig. 3. Thus, since the parallelograms bounded by solid lines are identical, and the remaining triangles along the y–axis may be viewed as parts of identical parallelograms, the squares are also identical and we can geta finite model.

5.6. Definition: Consider classes of formulas of the predicate calculus. For any class X, let N(X), I(X), F(X) be the subclasses of X which contain all formulas in X which have respectively no model, only infinite models, finite models. If R is a reduction procedure which reduces a given class Y to Y* and every subclass Z of Y to Z*, then R is said to be a conservative reduction procedure for Y, if (F(Y))* = F(Y*).

Since R is a reduction procedure, (N(Y))* = N(Y*), and $(I(Y) \cup F(Y))^* = I(Y^*) \cup F(Y^*)$. Hence, any effective mapping R from Y is a conservative reduction for Y if and only if (N(Y))* = N(Y*), (I(Y))* = I(Y*), (F(Y))* = F(Y*). If one wishes, the notion of conservative reduction can be further refined by distinguishing within I(X) those formulas which have recursive models from those which have only nonrecursive models. We shall not include this further distinction.

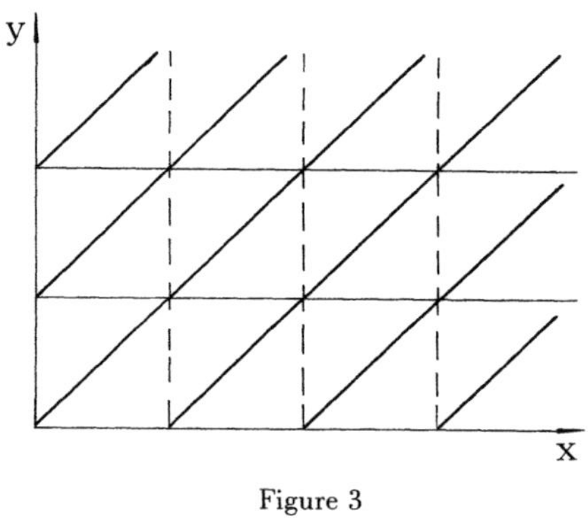

Figure 3

Theorem VII: If K is the class of all formulas of the predicate calculus and R is a conservative reduction procedure for K, then no two of the three classes N(K*), I(K*), F(K*) are recursively separable.

This depends on the known result of Trachtenbrot[12] that N(K) and F(K) are recursively inseparable. Thus, since R is conservative, N(K*) and F(K*) are recursively inseparable, for if a recursive P separates N(K*) and F(K*), then $Q = \hat{A}(R(A) \in P)$ is recursive and separates N(K) and F(K). On the other hand, if N(K*) and I(K*) were separable by a recursive P, then we would be able to test each member of P to decide whether it has a finite model or has no model. By Lemma 3, the process always terminates, since P does not contain any member of I(K*). Similarly, if F(K*) and I(K*) were recursively separable, we could do the same.

Using this theorem and Lemma 2, we get from the proof in the preceding section:

Theorem VIII: If Z is the class of AEA formulas, then no two of the three classes N(Z), I(Z), F(Z) are recursively separable.

Thus, for each formula A of the predicate calculus, we find a Turing machine T such that T halts on a blank tape if and only if A has no model; T gets into a distinguished state q_h if and only if A has finite models; T never halts or gets inot q_h if and only if A has only infinite models. If we use a universal Turing machine T, we can correlate each formula A with a suitable initial input and use the same machine for all formulas. In this way, we obtain an unsolvable class of AEA formulas which contain only monadic predicates and a fixed finite number of dyadic predicates. The question of reducing the number of dyadic predicates will be discussed at length in the next section.

VI. REDUCTION TO Δ_1, THE AEA CLASS WITH ONE DYADIC PREDICATE

Recently Kahr[7] reduced the predicate calculus to a set Δ_1 of formulas with the prefix AEA such that each formula of the set contains only monadic predicates and a single dyadic predicate. This is similar to Suranyi's result[11] for the AAA $\wedge$ AAE prefix (which may be written as AAEA or AAAE).

Kahr reduces Δ_1 to the reduction class obtained in reference 8 of asymmetric diagonal–constrained domino problems (i.e., problems all of whose solutions A are such that, for all distinct x and y, Axy $\neq$ Ayx). A procedure is given such that for each such problem C = $\langle P, Q\rangle$, we can find a corresponding formula F_c such that F_c is satisfiable if and only if C has some (asymmetric) solution. If C contains K dominoes, then F_c contains a single dyadic predicate D and 6K monadic predicates $M_0, \ldots, M_{6_k - 1}$.

Let $P = \{D_i | 0 \leqslant i \leqslant K-1\}$, $Q = \{D_{d_i} | 1 \leqslant i \leqslant t\}$, $\oplus$ be addition modulo 6K, and

(i, j): Axy stand for $(M_i x \vee M_j y) \supset Axy$.

B(i, j) stand for $\bigwedge_{m=0}^{K-1} [(i \oplus m, j \oplus m): (Dyx \equiv Dxy)]$.

k(i, j) stand for $\bigwedge_{m=0}^{K-1} \bigwedge_{n=0}^{K-1} [(i \oplus m \oplus n, j \oplus m): (Dyx \equiv Dx'y)]$.

Define further Hab, for $0 \leqslant a \leqslant K-1$, $0 \leqslant b \leqslant 2K-1$. If $0 \leqslant b \leqslant K-1$ and the right edge of domino D_a and the left edge of D_b are of the same color, then Hab $\equiv$ Dx′y $\wedge$ Dyx; if $K \leqslant b \leqslant 2K-1$, and the bottom of D_a and the top of D_{b-k} are the same, then Hab $\equiv$ Dx′y $\wedge$ Dyx; otherwise Hab $\equiv$ Dxy $\wedge$ $\neg$Dxy. This gives a scheme for specifying the part of F_C which depends on characteristics of each given particular domino set.

The monadic predicates are employed to set up grid boxes of size $36K^2$ on the plane.

6.1 $\nabla(M_0x, ..., M_{6K-1}x) \wedge (M_0x \equiv M_1x') \wedge ... \wedge (M_{6K-1}x \equiv M_0x')$.

Within each grid box, we can think of the point (x, y) such that M_ix and M_jy as ⟨i, j⟩. In this way each point in the first infinite quadrant gets both a global representation (x, y) and a local representation ⟨i, j⟩.

Each domino D_i is simulated by several grid boxes depending on the choice of the neighboring dominoes. But, in each case, the box is identified with D_i by the fact that within the box, at ⟨2K, 0⟩, ..., ⟨2K, K − 1⟩ the unique dyadic predicate is true at ⟨2K, i⟩ only. The K–tuple $(a_0, ..., a_{k-1})$ in which a_i is true and all others are false, is called the name of D_i. In order to do this, a device is introduced by which exactly one of D⟨2K. 0⟩, ..., D⟨2K, K − 1⟩ is true in each box:

6.2. $$[(0, 2K): \neg Dxy] \wedge [(K, 2K): Dxy] \wedge \bigwedge_{i=0}^{K-1} \{(i. 2K): [((Dxy \equiv Dx'y) \wedge \neg Dyx) \wedge (\neg Dxy \wedge \; Dx'y \vee \; Dyx)]\}.$$

Thus, it is first required that in each box B, ¬D⟨0, 2K⟩ but D⟨K, 2K⟩. Hence, there must be some i_0, $0 \leqslant i_0 \leqslant K$ such that ¬D⟨i, 2K⟩ but D⟨i', 2K⟩. The last clause requires that in the grid box which is the mirror image $\hat{B}$ of B (cf. reference 8), D ⟨2K, 0⟩, ..., D 2K, K − 1⟩ are all false except at ⟨2K, i_0⟩, when D does apply. In particular, it excludes the possibility ¬D⟨i, 2K⟩ but D⟨i', 2K⟩, $0 \leqslant i \leqslant K$, in the original box B.

Within each box $\hat{B}$, it is possible, with the help of points in $\hat{B}$ above the diagonal of B, to convey (see 6.5 below) the assignments to D at the stripe ⟨2K, 0⟩, ..., ⟨2K, K − 1⟩ to the two stripes ⟨3K, K⟩, ..., ⟨4K − 1, K⟩ and ⟨3K, 2K⟩, ..., ⟨4K − 1, 2K⟩. Taking this for granted, we force to be realized along the stripes R_h = (⟨4K, K⟩, ..., ⟨4K, 2K − 1⟩) and R_v = (⟨4K, 2K⟩, ..., ⟨4K, 3K − 1⟩), only names of D_j and D_k, respectively, such that D_j can occur to the right of the domino D_i represented by the given box, D_k can occur above D_i:

6.3. $$\bigwedge_{j=K}^{3K-1} \{[(3K, j): \neg Dxy] \wedge \bigwedge_{i=3K}^{4K-1} \{(i, j): [Dxy \equiv Dx'y) \vee H(i - 3K, j - K)]\}\}.$$

By the first clause of 6.3, ¬D⟨3K, K⟩, ..., ¬D(3K, 3K − 1). By the second clause, for each j, $K \leqslant j \leqslant 3K - 1$, once D⟨i, j⟩ is true for some i in the region $3K \leqslant i \leqslant 4K - 1$, it is true for all greater i in the region. Moreover, we can get D⟨i, j⟩ after ¬D ⟨i, j⟩ only when D_{j-K} can occur to the right or D_{i-2} can occur above, by the specification of H. That we cannot use several different permissible values of D_{j-K} or D_{j-2} does not follow from the specification of H, but is rather a consequence of additional connections in 6.5 below which tie up the stripes R_h and R_v with the names of the neighboring dominoes.

The diagonal constraint is given by:

6.4. $$\bigwedge_{j=1}^{t} [(r_i, r_i): (Dxy \supset \neg Dyx)].$$

Finally, there are various conditions which transmit names of dominoes without change. Formally, these are the most complex of the conditions and are not easy to grasp without some graphical illustration:

6.5. $$5K(5K, 0) \wedge N(0, 5K) \wedge 4K(2K, 4K) \wedge (3K-1)(0, 3K) \wedge B(0, 3K) \wedge$$
$$B(0, 4K) \wedge B(0, 5K) \wedge B(K, 5K) \wedge B(4K, 2K) \wedge$$
$$\bigwedge_{i=4N}^{6N-1} \bigwedge_{j=N}^{3N-1} [(i, j): (Dxy \equiv Dx'y)].$$

The conjunction of 6.1 to 6.5 gives the required formula F_C in which D occurs only in the contexts Dxy, Dyx, Dx′y.

From an exact proof along the above line, Kahr establishes a theorem which is an exact analog of the extended main theorem of reference 8 with dyadic predicates replaced by monadic predicates plus a single dyadic predicate. This further reduction does not, however, preserve the form of the whole formulas in terms of W, U, V as given in Theorem V.

The above outline is only an inadequate summary of Kahr's oral and written explanations with drawings to assist. For a more adequate treatment of the above result, please consult Kahr's paper in this volume.

It is possible to modify the above proof and extend Theorem VIII to Δ_1 (see reference 9).

The reduction to Δ_1 does not yield automatically Suranyi's result with the prefix AAAE and a single dyadic predicate, because in reducing (x) (Eu) (y) Mxuy to the Skolem form, we go through (x) [(u) (y) (Gxu ⊃ Mxuy) ∧ (Eu)Gxu] and use a new dyadic predicate G.

To get Suranyi's result, we have to modify the above proof in such a way that, e.g., we can always choose F_C with the property that if Mxuu holds then u = x′. In that case, we may replace Gxu by Mxuu and obtain a reduction class with a single dyadic predicate with the prefix AAA ∧ AE.

VII. SOLVABLE AEA SUBCASES

In view of the fact that formulas of the form Mxx′y form a reduction class, it is natural to study the decision problem of subclasses of this undecidable class. We shall confine our attention to formulas containing only monadic and dyadic predicates, partly because in our undecidability results, we use only these predicates, partly because the following lemma can be established on the basis of a remark by Buchi.

Lemma 3: Given and AEA formula A, there is a corresponding AEA formula B which contains only dyadic predicates such that A is satisfiable or satisfiable in finite domains if and only if B is.

Essentially, since x and y are the only independent variables in an AEA formula, we can replace combinations of x, x′, y with a polyadic predicate by suitably related dyadic predicates.

Suppose, for example, A contains a triadic predicate H. Since each place can be filled by x or x′ or y, there are 27 possible atomic formulas containing H. We can now replace in A, Hxx′b, Hxbx′, Hx′xb, Hx′bx, Hbxx′, Hbx′x, Haab, Haba, Hbaa, respectively, by G_1xb, G_2xb, G_3xb, G_4xb, G_5xb, G_6xb, G_7ab, G_8ab, G_9ab, and add conditions: $G_1xx \equiv G_6xx \equiv G_8xx'$, $G_1xx' \equiv G_2xx' \equiv G_9x'x$, $G_2xx \equiv G_5xx \equiv G_7xx'$, $G_3xx \equiv G_4xx \equiv G_9xx'$, $G_3xx' \equiv G_5xx' \equiv G_8x'x$, $G_4xx' \equiv G_7x'x \equiv G_6xx'$, $G_7xx \equiv G_8xx \equiv G_9xx$.

Since all possible ordered triples which can be obtained by using x, x′, y are included, each occurrence of H must fall into one of the classes shown. When a predicate contains more than three places, we can make similar replacements. The cardinality of the models is not affected by these replacements.

Although the lemma is used as a justification of our choosing to confine our attention to dyadic (and monadic) predicates, it does not mean that we might not get more useful classifications of AEA formulas by permitting other predicates.

We shall primarily use here a classification according to the forms of occurring atomic formulas, and give an alternative treatment of the results given in reference 6. For example, Theorem II suggests a more detailed classification which we do not consider here.

Since a dyadic predicate may be followed by xx, xx′, x′x, x′x′, yy, xy, x′y, yx, or yx′, any atomic formula occurring in a given formula Mxx′y is of one of these nine forms. In terms of these forms we shall specify subclasses of the class U of all formulas of the form Mxx′y containing only dyadic predicates.

Consider the four forms xy, yx, x′y, yx′. First take any three of them. From reference 8 or Theorem V, we know that any subclass of U which includes all formulas whose atomic formulas are in just these three forms is a reduction class and hence is undecidable. Now take any two of the four forms. Combining them with the other five forms yields a subcalss of U. In this way we obtain six subclasses of U which divide into three pairs:

$$J = \{xy, x'y\}, \qquad J^* = \{yx, yx'\},$$
$$L = \{xy, yx\}, \qquad L^* = \{x'y, yx'\},$$
$$Q = \{xy, yx'\}, \qquad Q^* = \{yx, x'y\}.$$

The results to date may be summarized thus[6]:

Theorem IX: With the exception of subsets of Q and those of Q*, a class determined only by forms of occurring atomic formulas is decidable if and only if it contains at most two of the four forms xy, yx, x′y, yx′; it contains an axiom of infinity if and only if it contains three forms including either xy and x′y, or yx and yx′.

The conjecture is that Theorem IX is true exen when the clause on the exception is removed.

To illustrate our method of approach, we consider first two simple sucbclasses of J and J*: J_1 containing only atomic formulas of the two forms $G_i$xy and G_ix′y, for arbitrary i; J_2 containing only those of the two forms $G_i$yx and $G_i$yx′. Since the arguments for J_1 and J_2 are analogous we discuss only J_1.

7.1. The classes J_1 and J_2 contain no axioms of infinity. In fact, a formula with K predicates either has no model or is satisfiable with 2^{2K} or fewer objects.

Let Mxx′y, or briefly Mxy, be a member of J_1, containing K dyadic predicates. Then there are at most 2K atomic formulas, $d \leqslant 2^{2K}$ truth–value assignments (called matrix vectors) to them which make Mxy true. If we think of (x, y) as the Cartesian coordinates of a lattice point in the first quadrant, then each matrix vector generally affects two points, viz. (x, y) and (x′, y), and we shall say that each matrix vector contains two point vectors Bxy, Bx′y. The formula Mxy has a model if and only if we can choose a simultaneous assignment A of matrix vectors to all points (m, n) such that each point gets the same point vector in all the relevant matrix vectors. The class J_1 is especially simple and each point (m′, n) is affected only by the two vectors assigned to Mmn and Mm′n.

Take the set of all d matrix vectors (the sign patterns of disjuncts in the developed disjunctive normal form) of a given formula. Each matrix vector $(a_1, ..., a_K; b_1, ..., b_K)$ is such that a_i is the truth value (t or f) for $G_i$xy, and b_i is that for G_ix′y.

Cross out each matrix vector $R = (a_1, \cdots, a_K; b_1, \cdots, b_K)$ if there isno matrix vector in the set which begins with $b_1, \cdots, b_k$. We cannot assign R to any Mmn to get a model, since we would not be able to find any assignment for Mm′n that gives the same point vector Bm′n as R. We continue this process of reducing the set of matrix vectors until we have either eliminated all matrix vectors, or else each remaining matrix vector can be continued by a matrix vector that remains. In the first case, there is no model. In the second case, there is a model, and indeed there is a finite model.

Define the rank of a matrix vector R in the resulting set as the smallest number of steps needed, beginning from R, to continue up to a first repetition of a matrix vector along the path. Then call the smallest rank of all matrix vectors the rank of the formula. Since there are at nost 2^{2K} matrix vectors for a formula with K predicates, its rank is at most 2^{2K}.

We can assign K–tuples to every row in the quadrant in a uniform manner since different rows of the quadrant do not interact. Take a matrix vector whose rank r is the rank of the formula, so that we have r matrix vectors $(P_1; P_2), \cdots, (P_r; P_1)$, each $P_i$being a point vector. Take the path $(P_1, \cdots, P_r, P_1)$. Then we can assign P_i, to (x, y) whenever $x = kr + i, k = 0, 1, 2, \cdots, 0 \leqslant i \leqslant r$. From this, it is easy to define a model with r objects for the original formula. In fact, the relations $G_1, \cdots, G_k$ are defined over $\{1, \cdots, r\}$ as follows: $G_j$bc is t or f according as the jth term of P_b, is t or f, independently of c. This completes the proof of 7.1.

The argument, incidentally, illustrates an obvious lemma about finite models:

Lemma 4: An AEA formula with K dyadic predicates has a finite model if and only if there is an assignment of matrix vectors to all instances of Mxy which assigns a unique point vector to every lattice point such that the whole assignment is periodic; i.e., there exist k and j, such that (x_1, y_1) and (x_2, y_2) get the same assingment if $|x_1-x_2|$ is a multiple of k and $|y_1-y_2|$ is a multiple of j.

Given a model over m objects $\{1, ..., m\}$, we can clearly assign K–tuples to the square region $0 \leqslant x, y < m$, and then assign the value of (c, d) to$(am + c, bm + d)$, a, b

$= 0, 1, 2, ..., 0 \leqslant c, d < m$.

Conversely, given the periodic assignment, we can take the least common multiple of k and j, say m, and define a model with m objects, identifying $(am + c, bm + d)$ with (c, d), $0 \leqslant c, d < m$.

To decide the whole class J or J*, we need more complex considerations because the diagonal forms xx, x′x, xx′, x′x′, yy make the interconnection of assigned matrix vectors more involved.

Given a formula Mxy in J with K dyadic predicates, there are at most 7K atomic formulas, $d < 2^{7K}$ matrix vectors, and $e = 2^{d} - 1$ non-empty subsets of the set of matrix vectors. We shall show that Mxy has a model if and only if we can find a consistent set of assignments $\{Axy; xy\}$ over the region $0 \leqslant x, y \leqslant e$. This will yield a decision procedure because we can check by finite means whether such a (finite) set exists.

If there is no such set, then there is clearly no model for Mxy. Assume, therefore, there is such a set $S = \{A00; 00, \cdots, Aee; ee\}$ and let $T = \{A00, ..., Aee\}$. Consider now, for each fixed value c of x, the subset T_C of T consisting of all Acy, $0 \leqslant y \leqslant e$. By the pigeon-hole principle, there must be two values, $x = a$, $x = b$, $0 \leqslant a < b \leqslant e$, such that $T_a = T_b$. It follows that Aaa = Abb, since, in each T_c, every Acy contains the same point vectors Bcc, Bc′c, Bcc′, Bcc′, which make up Acc. For each Aay in T_a, there is some Abv in T_b, Aay = Abv; for each Aby in T_b, there is some Aau in T_a, Aby = Aau. Let now:

$$\begin{array}{lll}
\eta(y) = \mu_v(v \leqslant e \vee \ Aby = Aav), & & 0 \leqslant y \leqslant e \\
\chi(y) = \mu_u(u \leqslant e \vee \ Aay = Abu), & & 0 \leqslant y \leqslant e \\
\eta(Aby) = Aa\eta(y), & & 0 \leqslant y \leqslant e \\
\chi(Aay) = Ab\chi(y), & & 0 \leqslant y \leqslant e \\
S(y, Axy) = Ax'y, & \text{for } a \leqslant x \leqslant b, & 0 \leqslant y \leqslant e \\
P(y, Ax'y) = Axy, & \text{for } a \leqslant x \leqslant b, & 0 \leqslant y \leqslant e \\
& b - a = p &
\end{array}$$

From $T_a = T_b$, it follows that for every y, $0 \leqslant y \leqslant e$, we have an ordered set of p + 1 matrix vectors Aay, S(y, Aay) = Aa′y, ..., $S^{p}(y, Aay) = S(SP^{-1}(Aay)) = Aby$ $= \eta(Aby) = Aa\eta(y)$. Similarly, for P, Aby, P(y, Ay), ..., $P^{p}(y, Aby) = Aay = Ab\chi(y)$.

We now define an assignment A* of matrix vectors to all the instances of Mxy as follows:

(i) Diagonal region:

$$A^*(kp + i, kp + j) = A(a + i, a + j), \qquad k \geqslant 0, 0 \leqslant i, j \leqslant p.$$

Since Aaa = Abb, the repetition when $i = j = p$ is immaterial. It is clear here that $A^*(kp + p, kp + j) = A(b, a + j) = \eta(A(b, a + j))$, and $A^*(kp, kp + j) = A(a, a + j)$ $= \chi(A(a, a + j))$.

(ii) For $x < y$, $np < x - y \leqslant n'p$, $n \geqslant 1$, we define A^* by induction on n. When $n = 1$, we see from (i) that $A^*(kp + p, kp + j) = \eta(A(b, a + j)) = Aa\eta(a + j)$. Since $a + j$ and $\eta(a + j)$ are between 0 and e, $S(\eta(a + j), Aa\eta(a + j))$, ..., $S^p(\eta(a + j), Aa\eta(a + j))$ are defined and $S_p(\eta(a + j), Aa\eta(a + j)) = Ab\eta(a + j) = Aa\eta^2(a + j)$. We define now $A^*(y + p + i, y + j) = S_i(\eta(a + j), Aa\eta(a + j)) = A(a + i, \eta(a + j))$ for $y = kp$. Clearly, we can reiterate the process and define A^* for higher values of n.

(iii) For $y < x$, $np < y - x \leqslant n'p$, $n \geqslant 1$, we can similarly define A^* by induction on n.

To prove that we indeed have a model, we remark that any A^*xy affects the points xx, x′x, xx′, x′x′, yy, xy, x′y, and that consistency is assured on account of the consistency in the original stripe, $a \leqslant x \leqslant b$, $0 \leqslant y \leqslant e$. Thus, if (x, y) is in the diagonal region, consistency is assured locally in each $(p + 1) \times (p + 1)$ square. Below the diagonal, $x > y$, consistency is assured by repeated applications of S and η, because when Axy occurs in T_b, $\eta(Axy)$ is always defined, and $(\eta(y), \eta(y))$ gets the same point vector as (y, y). Similarly, with P and χ above the diagonal.

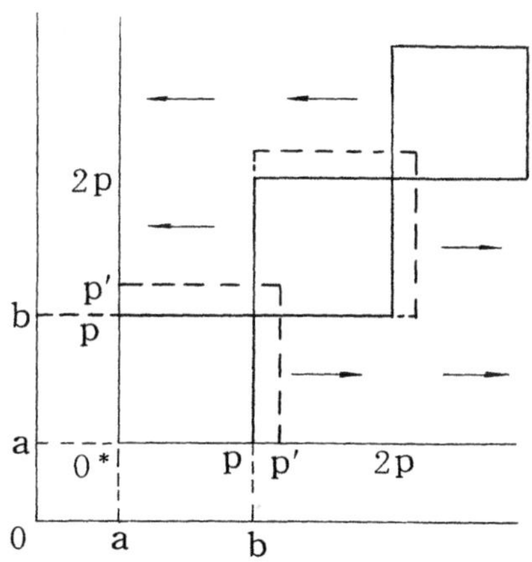

Figure 4

The procedure may become a little clearer if the proof is read in parallel with Fig. 4. Hence we have proved that J is decidable, and we have:

7.2. The classes J and J* are decidable.

These two decidable classes are of special interest because they contain axioms of infinity, while previously known decidable infinite classes do not contain axioms of infinity. We give some examples of axioms of infinity. in J and J*.

7.3. Any three forms including xy and x′y, or yx and yx′ give axioms of infinity.

Let a stand ambiguously for x, x′, or y, and be stand ambiguously for xx′ or x′x. The following are axioms of infinity.

7.3.1. $\neg Gaa \wedge (Gxy \supset Gx'y) \wedge Haa \wedge (Hxy \equiv Gx'y)$ (aa, xy, x′y).
7.3.2. $Gbc \wedge (Gcy \supset Gby) \wedge \neg Hbc \wedge (Hby \equiv Gcy)$ (bc, xy, x′y).
7.3.3. $\neg Gaa \wedge (Gyx \supset Gyx') \wedge Haa \wedge (Hyx \equiv Gyx')$ (aa, yx, yx′).
7.3.4. $Gbc \wedge (Gyb \supset Gyc) \wedge \neg Hbc \wedge (Hyc \equiv Gyb)$ (bc, yx, yx′).

To verify that these are axioms of infinity, we use the following simple fact.

7.4. If A is an axiom of infinity, B has some model, and B implies A, then B is also an axiom of infinity.

Thus, if B implies A, then any (finite) model of B is also a (finite) model of A.
Since the considerations in all cases of 7.3 are similar, we illustrate the line of proof by taking 7.3.1 with $a = x$. This formula B is satisfied if we take $x > y$ as Gxy, $x' > y$ as Hxy:

$$x \not> x \wedge (x > y \supset x' > y) \wedge x' > x \wedge (x' > y \equiv x' > y).$$

On the other hand, from Hxx and $Hxy \equiv Gx'y$, we can deduce Gx′x. Hence B implies the axiom of infinity A:

$$\neg Gxx \wedge Gx'x \wedge (Gxy \supset Gx'y).$$

That A is indeed an axiom of infinity is familiar. Thus G is satisfied by $>$. If t is an object, consider all the objects t, t′, t″, etc., which are all distinct. Given any of them, say s, since $\neg Gss$ and Gs′s, $s \neq s'$. Since $Gxy \supset Gx'y$, Gs″s, Gs‴s, G‴s′, etc. But $\neg Gss$, $\neg Gs's'$, etc., hence, $s'' \neq s$, $s''' \neq s$, $s''' \neq s'$, etc.
Incidentally, in the unsolvable subclasses of U, we have quite simple axioms of infinity. For example:

7.5. $(Gxy \supset \neg Gyx) \wedge (Gxy \wedge Gyx')$; $(Gx'y \supset \neg Gyx') \wedge (Gxy \wedge Gyx')$;
$(Gxy \supset \neg Gyx) \wedge (Gyx \wedge Gx'y)$; $(Gx'y \supset \neg Gyx') \wedge (Gyx \wedge Gx'y)$.

We turn now to the classes L and L* and prove:

7.6. The classes L and L* are decidable and contain no axioms of infinity.

We consider the class L and assume given a formula F in L with K dyadic predicates. There are $d \leqslant 2^{2K}$ matrix vectors. Among these, in order that one can be assigned to Mmm for some m, we must require that for every i, G_ixx, G_ixy, G_iyx, G_iyy get the same truth value. Any such matrix vector is called a diagonal matrix vector. If there are e such vectors, then clearly $e \leqslant 2^{4K}$. Hence, if there is any consistent assignment A of matrix vectors to all instances of Mxy over $0 \leqslant x, y \leqslant e$, then, by the pigeon–hole principle, there must be a and b, $0 \leqslant a \leqslant b \leqslant e$, such that Maa and Mbb are assigned the same diagonal matrix vector. It follows that Mbb′ and Mb′b can be assigned, respectively, the same matrix vector as Maa′ and Ma′a. We shall construct a finite model on the basis of the assignment A to the instances of Mxy for $a \leqslant x, y \leqslant b$.

Let $b - a = p$. We define A* as follows.

$$A^*(mp + i, np + j) = A(a + i, a + j), \qquad m, n \geqslant 0, 0 \leqslant i, j \leqslant p.$$

By Lemma 4, if A* determines a model of the given formula, then it gives a finite model with p objects. We prove that A* indeed gives a model, i.e., the assignments of matrix vectors give a unique point vector to every lattice point.

We consider first the assignments to Mxy, when $x = y$, $x' = y$, or $x = y'$. For a fixed number k, A*(Mkk), A*(Mkk′), A*(Mk′k) affect the seven lattice points (k, k), (k, k′), (k′, k), (k′, k′), (k″, k′), (k′, k″),(k″, k″). Hence, the assignments of A* to Mkk, Mkk′, Mk′k, when $k = 0, \cdots, p$, affect the points on the three central diagonals up to and including pp′, p′p, and p″p″. By hypothesis, A is consistent over Mxy, $a \leqslant x, y \leqslant b$, and the point vectors Bbb, Bbb′, Bb′b, Bb′b′ are the same as Aaa, Aaa′, Aa′a, Aa′a′. Hence, A* is consistent over all Mxy, when (x, y) is on one of the three central diagonals.

If $x \neq y$, $x' \neq y$, $y' \neq x$, then the point vectors assigned to (x, y) occur only in the matrix vectors assigned to Mxy and Myx. Conversely, A*(Mxy) and A*(Myx) are only constrained by each other and by A*(Mxx) and A*(Myy). It is easy to verify that A* does indeed satisfy these constraints. In particular, observe that if A*(Mxx) = A*(Myy), it is always permissible to choose A*(Mxy) and A*(Myx) such that A*(Mxy) = A*(Mxx) = A*(Myx).

The class L* is somewhat different but can be treated similarly. The difference is that, in general, A*(Mx′y) and A*(Myx′) are constrained by each other and by A*(Mxx), A*(Mx′x), A*(Myy), A*(My′y′).

We have so far not yet solved the classes Q and Q*. We have the following partial results.

7.7. We can delete xx′ and x′x′ from Q, x′x and x′x′ from Q* and the resulting classes can be decided or, further, contain no axioms of infinity if and only if the same is true of Q and Q*.

Thus, for example, for Q, we can replace Gxx′ by Hxx, Gx′x′ by Hx′x, and add a clause $Hxy \equiv Gyx'$.

7.8. The class Q_1 containing only the forms xy, yx′ (but not any of the other five forms of Q) and the class Q_2 containing only the forms yx and x′y are decidable, and contain no axioms of infinity.

Finally, we mention the possibilities of solving subclasses of Kahr's undecidable class Δ_1. This is challenging insofar as the class in question is, formally speaking, very simple. It is well known that if we delete the single dyadic predicate, the resulting class is decidable and contains no axioms of infinity. Moreover, if we delete the monadic predicates, then the remaining class can contain only three atomic formulas, say, Dxy, Dyx, Dx'y. Hence, we can have at most $2^3 = 8$ matrix vectors and $2^8 = 256$ sets of matrix vectors. This means we have essentially only 256 formulas in the class. The class does contain axioms of infinity, and we can actually exhibit a decision procedure for the class. This, by the way, is not entirely trivial, since, for example, the statement that Fermat's conjecture can be derived from the Peano axioms could be written as a single formula in the Predicate calculus. This suggests a new criterion of classification which, when applied to Δ_1, yields subclasses determined by the matrix vectors of formulas containing monadic predicates as well as D. Thus, each subclass is determined independently of the occurring monadic predicates by the partial matrix vector of a given formula relating to Dxy, Dyx, and Dx'y. C.W. Henson has recently decided some infinite subclasses of Δ_1, including axioms of infinity. He is investigating how we can get broader decidable subclasses if we impose on the monadic predicates restrictions which correspond to the rather weak form in which they are used in the reduction to Δ_1.

We do not have any idea of whether there is some natural undecidable set of formulas of the predicate calculus with a decision problem that is not of the maximum recursively enumerable degree.

References

1. W. Ackermann, *Solvable Cases of tbe Decision Problem* (Amsterdam: North–Holland Publishing Co., 1954).
2. P. Bernays, "Remarques sur le Probleme de la Decision en Logique Elementaire," *Edition de Centre Nat. de la Recb. Scient.* (Paris), vol. 13, pp. 39—44 (1958).
3. J. R. Buchi, "Turing Machines and the Entscheidungsproblem." *Matbematiscbe Ann.*, vol. 148, pp. 201—213 (1962).
4. J. R. Buchi, Abstract of Ref. 3, *Natices Amer. Matb. Soc.*, vol. 8, p. 354 (1961).
5. B. Dreben, "Solvable Suranyi Subclasses," *Ann. Harvard Computation Lab.*, 1962.
6. B. Dreben, A. S. Kahr, and Hao Wang, "Classification of AEA Formulas by Letter Atoms," *Bull. Amer. Matb. Soc.*, vol. 68, pp. 528—532 (1962).
7. A. S. Kahr, "A Reduction to a Class of AEA formulas Containing One Dyadic Predicate," *Notices Amer. Matb. Soc.*, vol. 9, p. 129, 1962.
8. A. S. Kahr, E. F. Moore, and Hao Wang, "Entscheidungsproblem Reduced to the AEA Case," *Proc. Nat. Acad. Sci., U. S. A.*, vol. 48, pp. 365—377, 1962.
9. A. S. Kahr and Hao Wang, "A Remark on the Reduction Problem with an Application to the AEA Formulas," *Notices Amer. Matb. Soc.*, vol. 9, p. 130, 1962.
10. Th. Skolem, "Sur la Portee du Theoreme de Lowenheim–Skolem," *Les Entretiens de Zuricb*, ed. F. Gonseth (Zurich: UNESCO, 1941), pp. 25—47.
11. J. Suranyi, *Reduktionstbeorie des Entscbeidungsproblem* (Budapest: Verlag der Ungarischen Akademie, 1959).
12. B. A. Trachtenbrot, "O Rekursionoj Otdelimosti," *Doklady Akademie Nauk SSSR*, vol. 88, pp. 953—956 (1953).

13. A. M. Turing, "On Computable Numbers, with an Application to the Entscheidungs-problem," *Proc. London Math. Soc.*, vol. 42, pp. 230—265 (1936); vol. 43, pp. 544—546 (1937).

14. Hao Wang, "Proving Theorems by Pattern Recognition, II," *Bell Syst. Tech. J.*, vol. 40, pp. 1—41 (1961).

12. TOWARDS FEASIBLE SOLUTIONS OF THE TAUTOLOGY PROBLEM*

This part studies the problem of testing Boolean validity in polynomial time. A number of hitherto unnoticed elegant properties of Boolean expressions are established in Section 12.3 to yield generally more efficient methods for many expressions. Combinatorial metatheorems concerning the tautology problem are proved in Section 12.4 Special partial methods are developed in Sections 12.6 and 12.7 which yield efficient solutions to two sets of hard examples which defeat existing methods in the literature. Finally, two general approaches in terms of size calculations and composite expressions with symbolic abbreviations which are under investigation are briefly indicated in Section 12.7

12.1 Computational complexity and Boolean validity

For many purposes, the vague notion of a "feasible" (or "quick") method of computation has come to be identified with the sharper concept of computability in polynomial time. A given infinite set of problems is computable in polynomial time if there is a plynomial f such thar for each problem (whose expression is) of length n, the problem is answered in $\leqslant f(n)$ steps (oor equivalently, there is a fixed N, such that the problem is answered in $\leqslant n^N$ steps). If there is no polynomial bound, then, for every general method, there are infinitely many n, such that the decision of some problem of length n requires more than polynomial time (e.g. $\geqslant 2^{\sqrt{n}}$). In that case, it is widely accepted that the set of problems has no feasible general solution.

When we consider the familiar "tautology problem" of quickly deciding whether a Boolean or truth–functional expression is valid (or tautologous), we now have a sharper formulation, viz. can this be done universally in polynomial time or does every general method require exponential time for an infinite number of special cases? While in some sense the area of truth–functional expressions has been intensively studied, it is our opinion that, with regard to the complexity of decision procedures, the area remains wide open and not enough results have been obtained thus far even to warrant a conjecture as to the answer to this general question. In what follows, we shall, apart from briefly surveying familiar techniques, list a number of new results which illustrate the type of closer examination that can be made.

A rather surprising development is the result of Cook [1] and Karp [8] according to which many familiar combinatorial problems are demonstrably of the same degree of complexity as the tautology problem, i.e. they are all P–reducible (polynomial–

* First published in *Annals of Mathematical Logic*, vol. 10. pp 117—154.

reducible) to one another (in particular, if any one set of problems has a feasible general solution, then all do). This result naturally broadens the range of people to whom the tautology problem is of interest.

We digress to clarify a point which is slightly confusing to those who are not familiar with certain technical terminology. This relates to the use of nondeterministic Turing machines which play a role in Cook's proofs. While an algorithm gives a string of steps, a nondeterministic machine gives a tree structure and behaves like a proof procedure rather than a decision provedure. A set of problems belongs to P if there is a general algorithm which solves every problem in polynomial time; it belongs to NP if there is a nondeterministic "method" with a polynomial f such that for each problem of length n, there is a path in the solution tree (a shortest proof) which is no longer than $f(n)$. Cook's theorem says that every set of problems belonging to NP is P–reducible to the tautology problem. This implies that if any of the P–equivalent combinatorial problems (centered around the tautology problem) is in P, then P = NP. In fact, it is currently the common practice to refer to each of the equivalent problems by the general consequence: Is P = NP?

It is obvious that P is closed under complementation, i.e., if A is in P then its complement is also in P, since a decision procedure gives a yes or no answer to each individual question. It is by no means obvious that NP is also closed under complementation. In fact, whether this is so has been shown (in [2]) to be equivalent to the existence of a quick proof system for tautologies. This relates to a distinction which is somewhat complex. The satisfiable Boolean expressions are directly seen to be in NP, because, given a Boolean expression H, we can eliminate its k (say) variables one by one and form a binary tree. And H is satisfiable if and only if there is a branch (of length k) which gets the value true. But it is an open question whether the tautologies also belong to NP, i.e., whether they possess even a quick proof system. Present knowledge does not exclude the possiblity that the former is in NP (but not in P) and the latter is not even in NP. But if the former is in P, so is the latter. Moreover, if the tautology problem is not in NP (and therefore not in P), then the satisfiablity problem is not in P and hence P ≠ NP. For the purpose of this part which works towards estabishing P = NP, we shall justifiably disregard the subtle distinction.

Our purpose is to work toward finding a quick decision procedure for Boolean validity. We shall in this part confine our attention to several partial methods and not attempt to organize the different methods into an organic whole, but rather leave open the detailed instructions on the exact order in which they are to be (repeatedly) applied. In particular, we shall offer quick solutions to a revelatory infinite set of examples from Tseitin [10], which demonstrably defeats familiar general methods.

The main results in this part include the substitution theorem (12.3.2), the separation theorem (12.3.3), and the decomposition theorem (12.3.6); the general combinatorial lemma 12.4.6 and lemma 12.4.8 on size calculations; as well as the several partial decision procedures in Sections 12. 6 and 12.7..

Two more comprehensive and more sophisticated methods are briefly indicated in Section 12.7, but we are obliged to postpone systematic explorations and expositions of these methods until a future occasion.

12.2 A brief overview with some general observations

The general problem is to decide whether an arbitrary truth–functional expression is valid (or satisfiable of contradictory). The familiar connectives are: not, and or, only if, exclusive or, if and only if (iff). Of these, "p only if q" is the same as "not p or q"; "p (exclusive) or q" is the same as "p iff $\bar{q}$. Hence, we may confine our attention to not, and, or,iff. It is sommon to eliminate iff also, but this is not always desirable because iff has many elegant properties and its elimination gives way to quite complex expressions in terms of not, and, or. Given the choice of connectives, there is also the question of the extent to which normal forms are to be used in our investigation.

The two most familar normal forms are the conjunctive and the disjunctive, in terms of not, and, or. For example, an expression is in conjunctive normal form if it is a conjunction of disjunctions of literals (i.e. variables and their negations). A first observation is that the validity of an expression in conjunctive normal form can always be tested quickely because it is valid only if every disjunction in it is valid (i.e., there is some variable p such that both p and $\bar{p}$ appear in it). Symmetrically, satisfiablity of an expression in disjunctive normal form can be tested quickly. We recall that an expression A is valid if and only if its negation $\bar{A}$ is not satisfiable (i.e. contradictory). Rather the open problem for expressions in conjunctive (resp. disjunctive) form is to test quickly whether it is satisfiable or contradictory (resp. valid). This familar observation illustrates the fact that we cannot assume without additional argument that expressions are given in any normal form, because, for example, the question of testing validity quickly of arbitrary expressions is reduced to the question whether converting an arbitrary expression into the conjunctive normal form can be done in polynomial time.

It is pointed out in [10] that we can, by suitably introducing new variables, quickly turn an expression A into an expression B in disjunctive form such that A is valid if and only if B is. Therefore, if we are studying validity, we can confine our attention to expressions in disjunctive form. Similarly, we can assume expressions turned into conjunctive normal form when studying satisfiability. In fact, we shall sometimes make such an assumption, although we use also other forms and even when we begin with only expressions of a given normal form, we shall not always adhere to the form in the process of transforming the expressions.

Suppose we are to decide generally whether a given expression is valid. (Similarly, if we were to decide satisfiability.) For this purpose, we are justified in applying any rule which preserves validity. This means that we can transform A into B provided the general rule assures us that either A and B are both valid or neither of them is valid. It is not necessary that A and B be equivalent in the sense that they contain the same variables and become true under exactly the same assignment of truth values to the variables. Clearly this fact permits greater freedom in designing rules to simplify an expression.

Given an expression in disjunctive normal form, the question of its validity amounts to asking whether the clauses jointly cover the whole truth table. An expression is in *canonical* disjunctive normal form if all the variables in the expression occur exactly once in each clause (a conjunction of literals). For such expressions, there is no difficulty in deciding quickly (i.e. in polynomial time) whether it is valid. Each clause represents a row in the truth table and we only have to check whether all the rows are there. A disjunctive normal form that is not canonical is more economical in that each clause may cover many rows. For example, if $p_1, \cdots, p_n$ are all the variables, the clause $p_1\bar{p}_2$ covers 2^{n-2} rows in the truth table (viz., all the rows in which p_1 gets 1 and p_2 gets 0, no matter what values are assigned to $p_3, \cdots, p_n$). Generally, an expression in disjunctive normal form is much shorter than its canonical form and may require exponentiation in the expansion. Hence, the obvious suggestion of expanding an expression into the canonical form does not help. In fact, the problem is to work directly with more economical representations and determine, without exponential explosion, whether they cover the whole truth table.

We review briefly some of the familiar rules and methods for testing Boolean validity (or, symmetrically, satisfiability or contradictoriness).

12.2.1 *The truth table method*

Given any expression A in disjunctive form containing the variables $p_1, \cdots, p_n$, we list all the 2^n possible ways of assigning 0 or 1 to each variable as rows in a table, and determine whether each row is covered by some clause in A. A is valid if and only if all rows are covered. For example, $p_1\bar{p}_2$ cover the 2^{n-2} rows in which p_1 gets 1 and p_2 gets 0. It is easily seen that when this method is applied mechanically, the test generally requires exponential time. For instance, if A is $p_1 \vee \cdots \vee p_n$, this method would still require the listing of the 2^n rows.

12.2.2 *Some familiar rules of simplification*

12.2.2.1 Subsumption. If A is an expression in disjunctive form, and a clause B subsumes a clause C (i.e. every literal in C occurs in B), we can drop B. Call the result A^*. If A^* is valid, then of course A is. On the other hand, every assignment making A true also makes A^* true, because if it makes B true it also makes C true and otherwise it makes a clause true which occurs in both A and A^*.

12.2.2.2 *Factorization* Any two clasues pB and $\bar{p}B$ can be simplified to B.

12.2.2.3 *Single literal clauses.* When a disjunctive expression A is $p \vee B(p)$, reduce it to $B(0)$; when it is $\bar{p} \vee B(p)$, reduce it to $B(1)$. Thus, when p gets 1, $p \vee B(p)$ is true and can be dropped; when p gets 0, we have $0 \vee B(0)$ which is the same as $B(0)$.

12.2.2.4 *Variables in partial state.* Given an expression in the disjunctive form, if a variable (p say) occurs only positively or only negatively, we can drop all clauses containing it (i.e. p or $\bar{p}$) without affecting validity. Consider, for example, a formula $A \vee B$ such that neither p nor $\bar{p}$ occurs in A, and each clause in B contains p (but none contains $\bar{p}$). We see that $A \vee B$ is valid if and only if A is. Of course, if A is valid, then $A \vee B$ is. Suppose now A is not valid. There is then an assignment to all variables in A

(not including p since by hypothesis neither p nor $\bar{p}$ occurs in A) such that A comes out false (has the value 0). Use the same assignment and in addition assign 0 to p. Then all the clauses in B get 0, and, therefore, $A \vee B$ gets 0. The variable p is said to be *in partial state* and the general rule could be described simply by saying that it can be eliminated without branching (just replace its occurrences by 0).

This rule and the following method of variable elimination were first observed independently in [5] and [3]. The first paper also defines variables in partial state for expressions not in normal form and containing iff.

12.2.3 *The method of variable elimination*

Given an expression $A(p)$ containing a variable p and its negation $\bar{p}$, two new expressions are formed. In one, p is replaced by 1; in the other, 0. The expression $A(p)$ is valid if and only if the conjuction of $A(1)$ and $A(0)$ is valid. In terms of the truth table, in elimination splits the table in two. If the expression A is in disjunctive form, it is generally of the form $Bp \vee C\bar{p} \vee R$. And the result of the split is $(B \vee R)(C \vee R)$, or, alternatively, $BC \vee R$. When we eliminate the variables one by one, we finally reach a simple evaluation of an expression with a single variable.

There are different ways of applying this method. First, the order in which the different variables are eliminated makes a difference to the speed. Second, for each variable eliminated, the familiar practice is to multiply out the conjunction BC and return $BC \vee R$ to an expression in the disjunctive form; in more sophisticated applications, we may choose to retain the composition clause BC. Third, there is a choice whether to simplify or consolidate the result at each stage by rules such as those listed under 12.2.2 above. It should be noted that the immediate advantage of such simplifications is often deceptive in that other reductions are thereby blocked.

Basically this method involves a successive test with the truth table. When we can examine an expression only a part at a time (for example, if it derives from an Herbrand expansion), a technique is developed in [7] which permits successive elimination of variables without duplicating earlier work done. Variables in partial state are treated in a special way.

12.2.4 *The method of consensus or resolution*

This concept of consensus was first introduced in the context of minimizing representations of a given truth function. The consensus of two clauses contradicting each other at exactly one variable is the conjunction of the two clauses with the variable and its negation deleted. For example, prs is the consensus of pqr and $p\bar{q}s$. The possiblity of using this operation to test validity is mentioned in [6]. In the most elementary form, the consensus method is pplied to an expression A in disjunctive form in the following manner. Form all possible consensi of the clauses in A; add the new clauses and continue untill either (1) we have obtained two clauses p and $\bar{p}$, for some variable p, or (2) we have not reached (1) but no more new consensi can be formed. In case (1), A is valid; in case (2), A is not valid.

The dual of forming the consensus of two clauses is the familiar "cut rule": form

$B \vee \bar{p}$ and $p \vee C$, infer $B \vee C$. When we begin with an expression A in conjunctive normal form, we can test whether A is contradictory by making all possible applications of the cut rule to all clauses at each stage until either we reach two conclusions p and $\bar{p}$ for some variable p or, failing that, can no longer apply the cut rule to get any new clauses. This is commonly referred to as the resolution method and attributed to Robinson [9], which introduces the term "resolution principle" to refer to a related way of testing Herbrand expansions in the predicate calculus. $B \vee C$ is said to be the resolvent of $p \vee B$ and $\bar{p} \vee C$.

The method of variable elimination is related to the method of consensus (or resolution) in a fairly direct manner. Consider an expression A in disjunctive form containing $p_1, \cdots, p_n$. Consider first those clauses containing p_1 or $\bar{p}_1$. We can put A in the form

$$Bp_1 \vee C\bar{p}_1 \vee R.$$

On the one hand, if we eliminate p_1, we get, as noted before, the result $BC \vee R$. Let D be the disjunctive normal form of BC. On the other hand, if we take all the consensi with regard to p_1 and $\bar{p}_1$, the result is also D. After adding all the consensi relative to p_1, the whole expanded expression is $Bp_1 \vee C\bar{p}_1 \vee D \vee R$. Later on, we shall prove the "substitution theorem" according to which Bp_1 and $C\bar{p}_1$ can be deleted after D is added. In that case, the parallelism between the methods of consensus and variable elimination is fairly direct.

12.2.5 *The clause elimination theorem* [6]

In forming the consensus of PA and $\bar{p}B$, *if* A and B are not consistent (e.g. if they are $\bar{q}C$ and qD), the consensus is useless. Therefore, it is natural to restrict the process to cases where two clauses E and F contradict each other in exactly one variable (p say); in that case we say that p in E complements $\bar{p}$ in F. Using this concept, one can state the clause elimination theorem which specifies a fundamental property of valid expressions in disjunctive normal form.

12. 2. 5. 1 **Theorem.** *Given an expression in disjunctive normal form, any clause which contains a literal nor complemented (by its oppostie in any other clause) can be dropped, without loss of validity.*

When no more clauses can be dropped by this theorem, the result is called a *closed residue* in which every literal in every clause is complemented. When all clauses are dropped by the theorem, we say there is no closed residue and the original expression is seen to be nonvalid. On the other hand, there are also closed residues which are not valid. An example is the disjunction of $p\bar{q}$, $\bar{p}q$, $p\bar{r}$, $\bar{p}r$, $q\bar{r}$, $\bar{q}r$.

We emphasize that the process of finding the closed residue of any expression in the disjunctive form can be done quickly by using "registers" which list all the relations of complementation. Consider, for example, the disjunction of the first five clauses above. Only a small amount of work is necessary to make the following register.

(1)	(2)	(3)	(4)	(5)
$p(4);\bar{q}(5)$	$\bar{p}(3);q$	$p(2);\bar{r}$	$\bar{p}(1);r(5)$	$q(1);\bar{r}(4)$

Since q in (2) and r in (3) are not complemented, we can delete (2) and (3). The remaining three clauses form a closed residue. In general, the deletion of clauses would entail the deletion of all references of these deleted clauses and there would be repeated deletions. But it is clear the procedure is not slow.

We note that Theorem 12.2.5.1 generalizes the previous remark about variables in partial state in a pleasing way and is clearly useful in simplifying expressions in disjunctive form. After such clauses are eliminated, we can try to separate the resulting expression into parts if the relation of complementation is weak or nonexistent between parts of the expression. Several separation theorems along such a line can be proved. Whether such separations are possible or not, we can also begin to add consensi with the additional method that generally the original clauses can be dropped after sufficiently many consensi are added. An exact formulation of the possiblities will be given by the substitution theorem to be proved in the next section.

12.3 Some basic properties of Boolean validity

Over the years, we have developed and on occasion lectured on several theorems related to the clause elimination theorem and the concept of closed residue. These theorems often help the speedy decision of validity, reveal certain nice properties of Boolean expressions, and possess fairly interesting proofs. We use this opportunity to publish these proofs for the first time.

12. 3. 1 **Theorem.** *Every expression in canonical disjunctive form is valid if and only if it is a closed residue.*

Proof. Of course, if the expression is valid, it is a closed residue since every full clause is in it. If a canonical expression is a closed residue, it must be valid (and contain all possibilities) by the following argument. Being a closed residue, the expression contains at least one clause. We wish to show that it contains all the canonical clauses. Suppose first that the absolutely positive clause $p_1 \cdots p_n$ is missing. Since every clause contains the same n variables, every clause Ap_i with an unnegated variable p_i must interact with a clause $A\bar{p}_i$. Hence, every clause with one negated variable must be missing. But then every clause with two negated variables must be missing. And so on. Therefore, all clauses must be missing. Suppose now some arbitrary clause is missing. In that case there must be one clause with one less negated variable missing, one with two less negated variables missing, and so on. Hence, the absolutely positive clause must be missing. But then all clause must be missing. Therefore, the closed canonical expression must contain all the clauses.

As we noted before, every valid expression in disjunctive from must contain a (nonempty) closed residue. We do not yet get a generally efficient method because a nonvalid expression may also contain a closed residue. But it is always equivalent to a disjunctive expression not containing a closed residue and this can often be seen fairly directly. For example, while $p\bar{q} \vee q\bar{r} \vee p\bar{r}$ is a closed residue, the equivalent expression with $p\bar{q}$ replaced by $pq\bar{r} \vee p\bar{q}r$ contains no closed residue. We are currently studying general ways of finding such equivalent expressions quickly.

A more substantial result is the substitution theorem mentioned before. The theorem is useful in rendering possible the elimination of variables with only local expansion or no expansion at all. It combines well with the clause elimination theorem in that they can often be applied repeatedly one after the other, because a closed residue often becomes open after applications of the substitution theorem.

12. 3. 2 **Substitution theorem.** *Let A be an expression in the disjunctive normal form. If a literal a in a clause, say Ba, of A is complemented by its opposite a' only in the clauses $C_1a', \cdots, C_ka'$, then we can replace Ba by $BC_1 \vee \cdots \vee BC_k$, without affecting validity.*

Proof. Observe that if a literal a and its opposite a' occur infrequently in A, the variable a(or a') can be eliminated with little or no expansion by this theorem because once all occurrences of a are eliminated, the clauses containing a' all drop out by the clause elimination theorem.

Suppose A satisfies the hypotheses of the theorem. Let A^* be the result obtained from A by substituting $BC_1 \vee \cdots \vee BC_k$ for Ba. Our goal is to prove that A is valid if and only if A^* is.

It is relatively easy to prove that if A^* is valid, then A is valid. Obviously if A^* is valid, then $A \vee BC_1 \vee \cdots \vee BC_k$ is valid, since A^* is nothing but the last expression with the clause Ba dropped. We need only show that A is valid if the last expression is. This follows from the known fact that adding the consensi $BC_1, \cdots, BC_k$ to A does not affect validity. The proof of this fact is as follows. Each $BC_i(i = 1, \cdots, k)$ implies $Ba \vee C_ia'$ because if BC_i gets the value 1 then $Ba \vee C_ia'$ becomes $a \vee a'$ which always has the value 1. Therefore, it can never happen that BC_i gets the value 1 but $Ba \vee C_ia'$ does not get the value 1. Hence, if $A \vee BC_1 \vee \cdots BC_k$ is valid, then A is also valid. Therefore, A is valid if A^* is.

We proceed to show that A^* is valid if A is. Suppose the clause Ba is $b_1 \cdots b_ma$, so that B is $b_1 \cdots b_m$. Our strategy is to break up both A and A^* into 2^m expressions with the subscripts 1 to 2^m, by considering all the 2^m possible truth values (0 or 1) which $b_1 \cdots, b_m$ can take. In particular, we take A_1 and A_1^* to be the results obtained when $b_1, \cdots, b_m$ all get the value 1. Clearly A is valid if and only if the conjunction of $A_1, \cdots, A_{2m}$ is; similarly with A^* and $A_1^*, \cdots, A_{2m}^*$. For $2 \leqslant i \leqslant 2^m$, at least one of $b_1, \cdots, b_m$ gets the value 0, so that B gets the value 0. Hence, the clause Ba in A and the clauses $BC_1, \cdots, BC_k$ in A^* all drop out. As a result, A_i and A_i^* are the same for all i, $2 \leqslant i \leqslant 2^m$. Hence, all we have to establish is that A_1^* is valid if A_1 is. Since $b_1, \cdots, b_m$ all get the value 1, the clause B in A_1 becomes a. Similarly, the clauses $BC_1, \cdots, BC_k$ in A_1^* become $C_1, \cdots, C_k$. Hence, A_1^* differs from A_1 only by containing $C_1 \vee \cdots \vee C_k$ in place of the clause a.

Suppose A_1 is $a \vee D(a)$. Then A_1^* is $C_1 \vee \cdots \vee C_k \vee D(a)$. If A_1 is valid, then $D(0)$ is valid, because we can transform A_1 into the conjunction of $1 \vee D(1)$ and $0 \vee D(0)$ which is simplified to $D(0)$. The remaining task is to prove that A_1^* is valid if $D(0)$ is. For this it is sufficient to show that all clauses of $D(0)$ are contained in A_1^*. Of course the only clauses which would make a difference are those containing a or a'. All clauses in A_1 which contain a get dropped in $D(0)$, because a gets the value 0. The clauses $C_1a', \cdots, C_ka'$ which, by hypotheses, appear in A and therefore in A_1 become $C_1, \cdots, C_k$ in $D(0)$ and, as we see, are contained in A_1^*. At this point, we come to depend crucially on our original hypothesis that $C_1a', \cdots, C_ka'$ are the only clauses in which a' complements the occurrence of a in the clause Ba in A. This means that while a' may occur in other clauses of A, it can only occur together with the opposite of at least one of $b_1, \cdots, b_m$. Hence, any other clauses in A which contain a' get dropped in A_1(because $b_1, \cdots, b_m$ all get the value 1) and, therefore, do not occur in $D(0)$. Consequently, there cannot be any clause in $D(0)$ which does not appear in A_1^*. Hence, A_1^* is valid if $D(0)$ is; and A_1^* is valid if A_1 is. Therefore, A is valid if A^* is.

This completes the proof of the substitution theorem.

We gave some illustrations of applying this theorem together with the clause elimination theorem.

Consider a disjunction of $p\bar{q}$, $\bar{p}q$, $p\bar{r}$, $\bar{p}r$, $q\bar{r}$, $\bar{q}r$. This is a closed residue. If we apply 12.3.2 to $p\bar{q}$ and $q\bar{r}$, we can replace $p\bar{q}$ by $p\bar{r}$ which is redundant. By the clause elimination theorem, we can then also drop $q\bar{r}$ and $\bar{p}r$ since q and p in them no longer have any complements. Hence, we get a disjunction of $\bar{p}q$, $p\bar{r}$, $\bar{q}r$. Using 12.3.2 again on $\bar{p}q$ and $p\bar{r}$, we can replace $\bar{p}q$ by $q\bar{r}$, and then drop $p\bar{r}$. Hence, we get $q\bar{r} \vee \bar{q}r$. By the clause elimination theorem again, both clauses drop out, and we see that the original expression is not valid.

It is noted in [1] that every disjunctive expression can be reduced to one in which each clause contains at most three literals. Thus, let A be $C_1 \vee \cdots \vee C_m$, and C_1 be $a_1 \cdots a_t$, $t > 3$. Then A is valid if and only if the following expression A' is (with p a new variable).

$$A': pa_3 \cdots a_t \vee \bar{p}a_1a_2 \vee C_2 \vee \cdots \vee C_m.$$

Cook does not state his proof of this assertion. One proof is as follows. Thus, if A is valid, so is $pa_3 \cdots a_t \vee \bar{p}a_1a_2 \vee A$. But C_1 is the consensus of $pa_3 \cdots a_t$ and $\bar{p}a_1a_2$. It can be dropped without affecting validity. Therefore, A' is valid. On the other hand, if A' is valid, we can substitute $a_1 \cdots a_t$ for $pa_3 \cdots a_t$ by the substitution theorem and then delete $\bar{p}a_1a_2$ by the clause elimination theorem. Therefore, A is valid.

It can be seen that the natural tendency in testing validity is to move from A' to A, while the reduction to short clauses goes in the opposite direction. Of course, the reduction was not intended for the purpose of helping to test validity quickly.

By the way, it is possible to get short closed residues. But they are generally unstable. For example, $pqr \vee \bar{p}ab \vee \bar{q}\overline{bc} \vee \overline{rac}$ is a closed residue. When we substitute $qrab$ for the first two clauses, the resulting three clauses all drop by the clause elimination theorem.

The importance of the relation of complementation is further seen from the fact that when parts of an expression do not interact with one another, they can be detached from one another:

12.3.3 **Separation theorem.** *Let $A \vee B$ in the disjunctive normal form be a closed* *residue. If there is no complementation relation between A and B, then $A \vee B$ is valid if and only if either A is valid or B is. Therefore, in order to determine the validity of $A \vee B$, we need only test A and B separately.*

Proof. Of course $A \vee B$ is valid if either A or B is. To prove the other direction, we shall derive a contradiction from the assumption that neither A nor B is valid but $A \vee B$ is valid.

Let $p_1, \cdots, p_n$ be all the variables in $A \vee B$ and let us expand A and B to canonical forms A' and B' with respect to $p_1, \cdots, p_n$. It cannot be ruled out that A' and B' will have clauses in common. We assume that neither A nor B (therefore neither A' nor B') is valid. By 12.3.1, neither A' nor B' can be a closed residue. If $A \vee B$ (and therefore $A' \vee B'$) were valid and, therefore, closed residues, then A' and B' must interact. Since neither A' nor B' is valid but $A' \vee B'$ is, there must be some canonical clause C in A' not in B' and D in B' not in A'. C and D cannot differ by one negation sign only (say p_i and $\bar{p}_i$), because in that case A and B would interact for the following reason. Thus, C and D must come from C_1 and D_1 by expansion and C_1 must not be part of D and D_1 must not be part of C_1. But then C_1 must contain p_i and D_1 must contain $\bar{p}_i$, and there is a relation of complementation between C_1 and D_1, contrary to the hypothesis.

We wish to prove generally that if C and D cannot differ by $(k-1)$ or fewer negation signs, then they cannot differ by k negation signs either. The conclusion then is that there cannot be canonical clauses C exclusively in A' and D exclusively in B'. Hence $A' \vee B'$ (and therewith $A \vee B$) cannot be valid.

Return now to our assumption that $A' \vee B'$ is valid and that, therefore, there are such clauses C and D. As induction hypothesis, we assume that C and D cannot differ by less than k negation signs. We now suppose that they differ by k negation signs. Since renaming variables and inverting variables (i.e., replace p_i by $\bar{p}_i$ and $\bar{p}_i$ by p_i throughout the expression) do not affect validity, we can assume that

$$C \text{ is } \bar{p}_1 \cdots \bar{p}_k p_{k+1} \cdots p_n,$$
$$D \text{ is } p_1 \cdots p_k p_{k+1} \cdots p_n.$$

Since $A' \vee B'$ includes all canonical clauses, it includes those clauses in which exactly one of $p_1, \cdots, p_k$ is negated and $p_{k+1} \cdots, p_n$ are not negated. Call them $E_1, \cdots, E_k$. By hypothesis, these all belong to both A' and B', because they differ from C and D by less than k negation signs. Consider E_1, viz. $\bar{p}_1 p_2 \cdots p_k p_{k+1} \cdots p_n$. Since it is in A', there must be a part E_1' of E_1 which is a clause in A. Moreover, E_1' must contain $\bar{p}_1$, because otherwise E_1' would be a part of D and D would be in A'. Since D is in B', there must be a part D' of D which is in B. But D' must not contain p_1, because otherwise p_1 in D' would complement $\bar{p}_1$ in E_1', contrary to the hypothesis that A and B do not interact. By analogous argument, $E_2, \cdots, E_k$ must contain parts $E_2', \cdots, E_k'$ in A which must contain $\bar{p}_2, \cdots, \bar{p}_k$ respectively; moreover, there must be in B a part D' of D which does not contain any of $p_2, \cdots, p_k$. Consequently, D' is a part of D but does not contain any of $p_1, \cdots, p_k$. But in that case D' is also a part of C. Consequently, D must be a clause in A', which is a contradiction.

This completes the proof of the separation theorem.

There are different possible generalizations of the theorem. Roughly speaking, if we cannot quite separate an expression (in disjunctive form) into two noninteracting parts, we could still adapt the situation to apply the separation theorem. We gave two generalizations for illustration. For this purpose, we introduce the notion of "giant term" which is also useful as a means for testing validity.

12.3.4 **Definition.** Let A be an expression in disjunctive form. A disjunction G of literals is called a giant term for A if and only if G contains at least one literal from each clause of A and does not contain both p and $\bar{p}$ for any p.

12.3.5 **Theorem.** *An expression A in disjunctive form is valid if and only if there is no giant term for it.*

Suppose there is a giant term G for A, say $\bar{p} \vee q \vee r \vee \bar{s} \vee t$. We can clearly make every literal in G false (say p, s get 1, q, r, t get 0). But every clause (a conjunction) in A contains a literal from G and is falsified by the assignment. Hence, A is falsifiable and not valid.

On the other hand, if A is falsifiable, then there is an assignment which makes A (and therefore all clauses of A) false. But in that case, there is at least one literal in each clause which gets the value 0. When we pick one such literal from each clause and delete repetitions (if there are any), we get a giant term.

12.3.6 **Decomposition theorem.** *Suppose $A \vee C \vee B$ to be a closed residue, C not valid, and that A, B do not interact. Suppose $G_1, \cdots, G_m$ are all the gaint terms for C and let $A_1 \vee B_1, \cdots, A_m \vee B_m$ be the results obtained from $A \vee C \vee B$ by falsifying respectively the giant terms $G_1, \cdots, G_m$. Then $A \vee C \vee B$ is valid if and only if at least one member of each pair $\{A_i, B_i\}$ $(i = 1, \cdots, m)$ is valid.*

Clearly it is more appropriate to apply this theorem when C is short compared with the whole expression $A \vee C \vee B$.

If $A \vee C \vee B$ is valid, then of course, $A_1 \vee B_1, \cdots, A_m \vee B_m$ all are valid. By the separation theorem, at least one member of each pair is valid. Conversely, if one of each pair is valid, then of course $A_1 \vee B_1, \cdots, A_m \vee B_m$ all are valid. If $A \vee C \vee B$ were not valid, there would, by 12.3.5, be a giant term for it. But by definition, any giant term G for $A \vee C \vee B$ must contain as part a giant term for C, i.e., one of $G_1, \cdots, G_m$. Suppose now G contains G_i as part. Then the falsifying assignment for G and $A \vee C \vee B$ would also falsify $A_i \vee B_i$, contrary to the premiss that $A_i \vee B_i$ is valid.

In particular, C may be a single clause (i.e., a conjunction of literals). When C has additional properties, it is possible to achieve further simplifications. We give one example.

12.3.7 **Theorem.** *Suppose A and B are nonvalid expressions in the disjunctive form. $A \vee ab \vee B$ is a closed residue in which A and B do not interact, the literal a interacts only with A, and b only with B. The expression $A \vee ab \vee B$ is valid if and only if both $A \vee a$ and $B \vee b$ are valid.*

Proof. It is easy to verify that $CD \vee E$ is equivalent to $(C \vee E)(D \vee E)$. Hence, $A \vee ab \vee B$ is equivalent to $(a \vee A \vee B)(b \vee A \vee B)$. If $A \vee ab \vee B$ is valid, then $a \vee A \vee B$ and $b \vee A \vee B$ are valid. By the separation theorem, either $a \vee A$ or B is valid, and either A or $b \vee B$ is valid. By hypothesis, neither A nor B are valid. Therefore, both $a \vee A$ and $b \vee B$ are valid. Conversely, suppose both $a \vee A$ and $b \vee B$ are valid. Then $a \vee A \vee B$ and $b \vee A \vee B$ both are valid. Therefore, $A \vee ab \vee B$ is valid. □

We cannot directly generalize 12.3.7. Take $A \vee C \vee B$, a closed residue such that neither A nor B is valid, A and B do not interact, and every literal in C interacts with A or B but not both. Let C_a (respectively C_b) be the result obtained from C by assigning 1 to all literals interacting with B (respectively A). Then $A \vee C_a$ and $B \vee C_b$ are both valid if $A \vee C \vee B$ is, but the converse is not generally true.

Thus, if $A \vee C \vee B$ is valid, so is $A \vee C_a C_b \vee B$. So are $A \vee B \vee C_a$ and $A \vee B \vee C_b$. By the separation theorem, either $A \vee C_a$ or B is valid and either A or $B \vee C_b$ is valid. Since neither A nor B is valid, $A \vee C_a$ and $B \vee C_b$ are both valid.

To give a counterexample to the converse. Let A be $\bar{p}\bar{q}$, B be $\bar{a}\bar{b}$, C be $(pa \vee qb)$. $A \vee C \vee B$ is not valid, but $\bar{p}\bar{q} \vee p \vee q$ and $a \vee b \vee \bar{a}\bar{b}$ are valid.

12.4 Some calculations and classifications

There are a number of elementary details concerning the tautology problem which call for some elaboration. The vague general formulation is to ask whether there is a general method M for which there is a polynomial $P(x)$ such that, for every Boolean expression E and every positive integer n, if E is of length n, then the method M can decide whether E is valid or not with fewer steps or less time than $P(n)$. There are problems in fixing the universe of Boolean expressions, defining the length of an expression, and defining the time or the number of steps required for carrying out the general procedure in each case. Fortunately, these concepts are relatively stable so that we can make relatively simple choices without having to worry about detailed distinctions.

For example, we can, at least initially, confine our attention to expressions in the disjunctive normal form; the length of an expression can be identified simply with the number of literals or clauses in the expression; the time required for reaching a decision can often be identified with the number of clauses generated on the way. One reason why it is possible to use such simplified measures of complexity is that apparently large differences would not turn a polynomial function into an exponential one or vice versa. For example, multiplying by a constant factor, however large, or multiplying together a fixed number of polynomials would not lead out of the realm of polynomials.

A very essential feature of each Boolean expression is the number of variables which occur in it. Among all Boolean expressions containing n variables, we have the length ranging from the order of n to the order of 2^n. Essentially the time required for deciding an expression by the truth table method depends only on the number of variables and not additionally in any substantial way on the length of the expression. The method decides any expression with n variables in what is essentially (i.e. within polynomial transformations) 2^n steps so that all expressions whose length is of the order of 2^n can be decided in polynomial time. This leads to a simple fundamental observation that for expressions with the same number of variables, the longer ones are known to be decidable in polynomial time (by the familiar method of truth tables). Hence, the tautology problem is open only for "short" or "lean" expressions. Let us look at this observation a little more closely.

For any finite set of Boolean expressions, it makes little sense to ask whether they can be decided in polynomial time. Since these expressions can be decided in some finite time, say N, they have also a trivial polynomial bound, viz. the constant N. Moreover, once they have been decided in whatever way, we can record the (finitely many) answers and "invent" a new method which consists in looking up the table of the answers. Therefore, it is only with regard to infinite sets of expressions that we can meaningfully ask whether a set can be decided in polynomial time. Moreover, for any reasonable definition of length, there can be no finite bound on the length of an infinite set of expressions and no finite bound on the number of occurring variables. Hence, in order to exhibit counterexamples to a given method, we are expected to give an infinite set S of expressions and an exponential function $E(m)$ such that, for infinitely many values m, there are expressions is S of length m which the method M cannot decide in less than $E(m)$ steps.

Strictly speaking, there is a distinction between exponential and larger-than-polynomial functions. One standard definition of exponential function is one of the form K^m for some $K > 1$; it is never bounded by any polynomial. But $2^{\sqrt{m}}$ *and* $2^{c(\log m)^2}$ are not polynomial bounded and yet not exponential in this sense. We shall deliberately use the term "exponential" loosely to mean also just larger-than-polynomial.

An exponential function $E(m)$ increases faster than any polynomial function in the sense that for any polynomial function $P(m)$, there is some constant m_0 such that $E(m) > P(m)$ for all $m > m_0$. But there are different kinds of exponential functions: e.g.2^m, m^m, $2^{\sqrt{m}}$, $3^{\sqrt[5]{m}}$, etc. For our purpose, it seems reasonable to introduce an order of magnitude in the exponential functions.

12.4.1 **Definition.** Two exponential functions E_1 and E_2 are *of the same order of magnitude* if there are polynomial functions P_1 and P_2 such that for all m, $P_1(E_1(m)) > E_2(m)$ and $P_2(E_2(m)) > E_1(m)$.

Using this concept, we can render slightly more exact our observation that fat expressions (i. e. long expressions with few variables) are known to be decidable in polynomial time. To each infinite set S of expressions we associate a "density function" d such that $d(n)$ is the length of the shortest expression in S with n variables, and $d(n)$ is 2^n is S contains no expression of length n.

12.4.2 **Lemma.** *Let S be any infinite set of expressions which is truly fat in the sense that its density function $d(n)$ is of the same order of magnitude as 2^n. Then S is known to be decidable in polynomial time (in fact by the familiar method of truth tables).*

It seems reasonable to call a set S of expressions lean if there is a polynomial function $P(n)$ such that $P(n)$ is an upper bound of the length of the longest expression is S with n varialbes. There would then be sets of expressions which are neither lean nor truly fat, seeing that there are functions $2^{f(n)}$ or $k^{f(n)}$ such that $f(n)/\mathrm{n} \to 0$, $\log n / f(n) \to 0$. Since these remarks are only meant to elaborate a fairly simple observation, there is no need to pursue the distinction further at this point.

Consider all expressions in disjunctive form with n variables. We shall assume that no clause contains the same literal more than once and that no clause contains a variable and its negation. Moreover, the order in which different literals occur in a clause is immaterial. Clearly the longest clause would contain n literals and there are 2^n such clauses.

In fact, there are 2^n clauses of length one: $p_1, \cdots, p_n, \bar{p}_1, \cdots, \bar{p}_n$; $4n(n-1)/2$ clauses of length 2; $2^k\binom{n}{k}$ clauses of length k. Hence, if we include also one clause of length 0, there are, by the binomial theorem, 3^n possible clauses:

$$(2x+y)^n = \sum_{k=0}^{n} 2^k\binom{n}{k} y^{n-k} x^k.$$

put $x = y = 1$, $3^n = (2+1)^n = \sum_{k=0}^{n} 2^k\binom{n}{k}$.

Alternatively, we are making a selection from n triples $\{p_i, \bar{p}_i, 0\}$ $(i = 1, \cdots, n)$. The total number of possible selections is of course 3^n.

Hence, abstractly, we could have an expression D of length $3^n - 1$, viz. a disjunction of all the possible clauses. But by the familiar rule of subsumption, one could delete any clause A which contains another clause as a part. Therefore, if D contains a clause of length $k (k < n)$, the 2^{n-k} clauses containing it can be excluded.

12.4.3 **Lemma**. There are $3^n - 1$ possible clauses with n variables. Therefore, there are $2^{3n-1} - 1$ possible expressions in the disjunctive form which contain no more than n variables.

If we think of the truth table, each clause in an expression covers a number of rows. If the clause contains all n variables, it covers a single row. In general, if it contains k variables, it covers 2^{n-k} rows. The expression is valid, if all rows are covered (i.e., the expression is true under all assignments each specified by a row). The difficulty of deciding validity is to a large extent due to the fact that rows covered by different clauses overlap in various different ways.

Generally a valid expression cannot be very short relative to the length of the clauses. Roughly speaking, there must be certain minimal number of clauses to couver the whole truth table. We make an observation to illustrate this.

12.4.4 **Theorem.** *Let A be a valid expression in disjunctive form such that each clause contains k literals and there are altogether n variables, $k \leqslant n$, There must be at least 2^k clauses in A.*

When $k = n$, this is obvious, since the only valid expression in the canonical disjunction with 2^k clauses. Suppose $n > k$. There are 2^n rows in the truth table for the n variables. Since each clause contains k literals, it covers 2^{n-k} rows in the table. Therefore, there must be at least 2^k clauses because even without any overlapping at all, we need $2^n/2^{n-k} = 2^k$ clauses to cover all the 2^n rows.

In unpublished work, S.A. Cook has also obtained 12.4.4 and a generalization as well:

12.4.4g *Let A be a valid expression in disjunctive form with m clauses such that the ith clause contains k_i literals. Then* $\Sigma_{i=1}^{m} 2^{-k_i} \geqslant 1$.

Generally it is natural to expect better results when $n > k$, because it is inevitable that there are overlappings.

A natural question to ask is the maximum length (viz. number of clauses) of an expression in the disjunctive normal form with n variables when no clause subsumes any other clause in the expression. This question does have an exact answer.

Consider first the simpler case when all clauses in the expression are of the same length. Clearly, no clause subsumes any other clause. For any k, $1 \leqslant k \leqslant n$, the total number $u(n, k)$ of possible clauses with k literals is easily seen to be $2^k\binom{n}{k}$, which is also the maximum length for the set L_k of all expressions in which every clause contains k literals. Elementary calculation shows that, for a fixed n, the function $2^k\binom{n}{k}$ of k increases as k is roughly below two–thirds of n and then decreases.

12.4.5 *If* $2n-1$ *is not a multiple of* 3, *then* $u(n, k)$ *has the maximum value when* k *is* $[2(n+1)/3]$; *otherwise the maximum is attained when* k *is* $2(n+1)/3$ *and* $(2n-1)/3$. *In particular, if* n *is a multiple of* 3, *then the maximum is attained when* $k = 2n/3$, *and* $u(n, 2n/3) \sim 3^{n+1}/2\sqrt{(\pi n)}$. *In general,* $u(n, k) \sim 2^k n^n/k^k(n-k\sqrt{2\pi}$.

Since $u(n,k) = 2^k\binom{n}{k}$, $u(n, k+1)/u(n,k) = 2(n-k)/(k+1)$. But $2(n-k)/(k+1 \lesseqgtr 1$ when $k \lesseqgtr (2n-1)/3$. Hence, if $2n-1$ is a multiple of 3, the maximum is $u(n,(2n-1)/3) = u(n, 2(n+1)/3)$, otherwise, the maximum is attained when $k = [2(n+1)/3]$. In particular, if n is a multiple of 3, $k = [2(n+1)/3] = 2n/3$.

According to Stirling's formula,

$$
\begin{aligned}
& m! \sim m^m e^{-m}\sqrt{2\pi m}. \\
& u(n, 2n/3) = 2^{2n/3} n!/(2n/3)!\,(n/3)! \\
& \quad \sim 2^{2n/3} n^n e^{-n}\sqrt{2\pi n}/(2n/3)^{2n/3} e^{-2n/3}(2\pi(2n/3))^{1/2}(n/3)^{n/3} e^{-n/3}(2\pi(n/3))^{1/2} \\
& \quad = n^n\sqrt{2\pi n}/(n/3)^n 2\pi(n/3)\sqrt{2} \\
& \quad = 3^{n+1}/2\sqrt{\pi n}. \\
& u(n,k) \sim 2^k n^n e^{-n}\sqrt{2\pi n}/k^k e^{-k}\sqrt{2\pi k}(n-k)^{n-k} e^{k-n}(2\pi(n-k))^{1/2} \\
& \quad = 2^k n^n\sqrt{n}/k^k(n-k)^{n-k}(2\pi k(n-k))^{1/2} \\
& \quad \sim 2^k n^n/k^k(n-k)^{n-k}\sqrt{2\pi}.
\end{aligned}
$$

Generally, an expression in the disjunctive form may contain clauses with different numbers of literals. If, however, no clause is to subsume any other clause, the longest expression consists of clauses all with the same number of literals and therefore the maximum value of $u(n, k)$ given in 4.5 is the answer sought for. In order to establish this, we prove a more general theorem about partially ordered sets which is of some independent interest.

Consider any partially ordered finite set P. We can always define the level of each node in the set in a natural manner. If a node is not greater than any node, its level is 0. If a node is greater than certain nodes and the maximal level of these nodes is n, then its level is $n + 1$. As usual, a chain is a simply ordered subset of P, and an antichain is a subset in which no two nodes are comparable. A maximal chain (or antichain) is one to which no more nodes can be added to get a larger chain (or antichain). A chain is said to be complete if it includes a node from every level L_i. Clearly, every complete chain is maximal. Our interest is to determine how large a (maximal) antichain could be. In general, an antichain A contains 0 or more nodes from each level. Let b_i be the fraction of L_i that belongs to A; in other words, if L_i contains t_i nodes and a_i of them belong to A, then $b_i = a_i / t_i$.

12.4.6 **General Combinatorial Lemma.** *Let P be a partially ordered set with $n + 1$ levels $L_0, \cdots, L_n$ such that (1) every node in every L_i is smaller than some node in L_{i+1} if $i < n$ and greater than some node in L_{i-1} if $i > 0$, and (2) for every i, each node in L_i belongs to a same number of maximal chains. Let A be any antichain in P, then $\Sigma b_i \leqslant 1$.*

Proof. By condition (1), every maximal chain is complete. Since no chain can contain two nodes on the same level, any two sets of chains going through any two nodes on the same level have no common members. Since every maximal chain goes through all levels, the value of b_i is, by (2), the same as the ratio of the number of maximal chains going through the a_i nodes of A and L_i to the total number of maximal chains. Moreover, since no two nodes of an antichain A can belong to a same maximal chain, the sum of the numbers of maximal chains going through members of A cannot be larger than the total number of maximal chains. Hence, Σb_i cannot be greater than 1. More explicitly, we can consider $b_0, b_1, \cdots$ successively. Clearly $(\Sigma b_i) - b_0$ must be no greater than $1 - b_0$ and generally, $(\Sigma b_i) - b_0 - \ldots - b_j$ must be no greater than $1 - b_0 - \ldots - b_j$. Therefore $\Sigma b_i \leqslant 1$.

Given 12.4.6, it is easy to derive as a corollary:

12.4.7 **Corollary.** *Any expression in the disjunctive form with n variables and without any clause subsumed by another can have at most $u(n, k)$ clauses, where $k = [2(n + 1)/3]$.*

The partially ordered set is the set of all the 3^n clauses (including the empty clause) ordered by the relation of subsumption. These are obviously the $n + 1$ levels 0, 1, $\cdots$, n. Conditions (1) and (2) are easily seen to be satisfied. Let A be any expression in the disjunctive form construed as an antichain which is the set of clauses in A. Clearly, $t_i = u(n, i)$. By
12.4.6,

$$1 \geqslant \sum b_i = \sum (a_i / t_i) = \sum (a_i / u(n, i)).$$

By 12..45, $u(n, i) \leqslant u(n, k)$, where $k = [2(n + 1)/3]$.
Therefore, $1 \leqslant \Sigma(a_i / u(n, i)) \geqslant \Sigma(a_i / u(n, k)) = \Sigma a_i / u(n, k)$.
Hence, $u(n, k) \geqslant \Sigma a_i$.
In other words, A cannot contain more than $u(n, k)$ clauses.

Incidentally, 12.4.6 yields also as a corollary the familiar result on the largest antichain in the set of all subsets of $\{1, \cdots, n\}$ partially ordered by the relation of inclusion. Since the largest level is $L_{[n/2]}$, the largest antichains contain $\binom{n}{[n/2]}$ nodes.

Corollary 12.4.7 shows that among expressions with n variables which cannot be simplified by subsumption, the largest number of clauses an expression can have is $u(n, k) = 2^k \binom{n}{k}$, where $k = [2(n+1)/3]$. Such an expression is of course valid.

If we take away (say) all the absolutely positive clauses from such an expression, we obtain a nonvalid expression which contains $(2^k - 1)\binom{n}{k}$ clauses.

If we are interested in counting not the number of clauses, but the number of literals, we would need some addtional calculation. And we would be considering $2^i i\binom{n}{i}$ instead of $2^i\binom{n}{i}$.

$$2^i i\binom{n}{i} \lesseqgtr 2^{i+1}(i+1)\binom{n}{i+1} \text{ if and only if } i \lesseqgtr 2n/3.$$

If $2n/3$ is not an integer, then $[2n/3] + 1$ gives the maximum. Otherwise both $2n/3$ and $(2n/3) + 1$ give the maximum.

Another relevant classification is to subdivide all clauses with i literals, $1 \leqslant i \leqslant n$, into subsets according to the number of negated variables. Since there are n variables and each clause contains i variables, there are $\binom{i}{j}\binom{n}{i}$ clauses with j variables negated, $0 \leqslant j \leqslant i$. It is easily seen that the numbers of members of the $i + 1$ subsets are the binomial coefficients of $(a + b)^i$. For example, consider the case $i = 4$, $j = 2$, leaving $n \geqslant 4$ indefinite. Call the set $K_n(4, 2)$. In terms of the truth table, $K_n(4, 2)$ covers all the rows in which at least two variables get the value 0 and at least two variables get the value 1. But much smaller subsets of $K_n(4, 2)$ can be found to cover the same rows, because there are many overlappings among rows covered by the clauses in $K_n(4, 2)$.

Consider any disjunctive expression in n variables. Each clause B with k variables $(1 \leqslant k \leqslant n)$ covers 2^{n-k} canonical clauses (i.e., rows in the truth table with 2^n clauses). We shall call these 2^{n-k} clauses the 'range' of the clause B. Clearly, if we sum up the size of the ranges of all clauses in an expression A and get a value smaller than 2^n, then A cannot be valid. On the other hand, if the value is 2^n or more, A may be valid or not because the ranges may and generally do overlap.

It is fairly easy to adapt familiar considerations on the combination of events in probability theory to give a theoretically elegant but practically often inefficient general method for deciding whether an expression A in the disjunctive form with n variables is valid.

Suppose the clauses in A are $C_1, \cdots, C_m$. Each clause C_i is associated with a range t_i as just explained. The expression A is valid if and only if the ranges $t_1, \cdots, t_m$ together cover all the 2^n rows in the truth table. To determine whether the latter is the case, we can calculate all the possible intersections of the ranges $t_1, \cdots, t_m$.

Given any pair of clauses C_i and $C_j (i \neq j, 1 \leqslant i, j \leqslant m)$, we can calculate the intersection t_{ij} of their ranges in the following manner.

Case 1. $t_{ij} = 0$, if C_i and C_j are incompatible, i.e., there is at least one literal in C_i with its opposite in C_j.

Case 2. C_i and C_j are consistent, i.e. Case 1 does not hold. Then they contain 0 or more common literals. If C_i contains a literals and C_j contains b literals and there are c literals in common, then there are $2^{n-(a+b-c)}$ rows which are covered by both C_i and

C_j. In other words, $t_{ij} = t_i t_j / 2^{n-c}$.

We have to calculate then the intersection t_{ijk} of three clauses C_i, C_j, and C_k, etc, up to the intersection of all m clauses. The calculation is generally the same. If there is any incompatibility, the intersection is 0. Otherwise, they are all consistent, and the intersection is 2^{n-d}, where d is the total number of distinct variables in the clauses. The following holds.

12.4.8 **Lemma.** *Let $S_1 = \Sigma t_i$, $S_2 = \Sigma t_{ij}$, $S_3 = \Sigma t_{ijk}$, etc. Then the range for the whole expression A, or, in other words, the number of rows covered is given by the familiar formula:*

(1) $$t(A) = S_1 - S_2 + S_3 - \dots - (-1)^m S_m.$$

To compute $t(A)$, we should include all rows which are contained in at least one clause C_i, but each row should be taken only once. Let R be a row which appears in the ranges of exactly k clauses. Without loss of generality, we may take these to be $C_1, \cdots, C_k$. Clearly R makes k *contributions to* $C_1, \cdots, C_k$, $\binom{k}{2}$ contributions to S_2 (the pairs), etc. Hence, the total contribution in the above calculation is $k - \binom{k}{2} + \binom{k}{3} - \dots \pm \binom{k}{k}$. The value of this is 1, for $k \geqslant 1$, as is seen from the binomial formula

$$0 = (1-1)^k = 1 + \binom{k}{1}(-1) + \binom{k}{2}(-1)^2 + \dots + \binom{k}{k}(-1)^k$$
$$= 1 - \left(k - \binom{k}{2} + \binom{k}{3} - \dots + (-1)^{k-1}\binom{k}{k}\right).$$

Therefore, each row is counted exactly once, and the formula (1) is justified.

It may be noted that the number of terms t_i, t_{ij}, etc. add up to $2^m - 1$:

$$\binom{m}{1} + \binom{m}{2} + \dots + \binom{m}{m} = (1+1)^m - 1.$$

Therefore, in the worst case, this method in its naked form is certainly exponential. The likely situation is often that for sufficiently large j, no j clauses have all their ranges intersect. When that happens, there will be no need to calculate S_k for $k > j$, since they will all be 0.

If we are only interested in estimates, we can to some extent avoid the complicated interactions. For example, we may make any partition of all the clauses into disjoint subsets (e.g. pairs or triples etc.). We can then calculate the local overlapping in each subset and add up all the possibilities of these ranges. It is then also clear that the value must be $\geqslant 2^n$ for an expression with n variables to be valid.

12.5 Hard examples and negative results

We have described previously the consensus method and its dual, the resolution method. Each method can be used either as a proof procedure or as a decision procedure. In either case, there is a choice as to whether to add another rule by which a clause subsuming some other clause is to be deleted. We shall, to simplify matters, confine our attention to the consensus method as a decision procedure and do not include the rule of subsumption.

More explicitly, we begin with an expression A in the disjunctive form (i.e., a finite collection of clauses which are conjunctions of literals) and use the single consensus rule to create new clauses. It is known that A is valid if and only if we can generate from A by the consensus rule both p and $\bar{p}$ for some variable p.A natural question to ask is whether we can find examples which are hard to decide by this crude method.

Since the only rule generates AB from Aq and $B\bar{q}$, it is clear that we can often generate longer clauses from shorter ones. It is therefore highly plausible that, by suitable choice, we can find expressions which are hard to decide. For example, it seems likely that we can choose suitable valid expressions A_i so that only by going through many longer clauses can we reach p and $\bar{p}$ for some variable p. Similarly, it seems likely that we can find nonvalid expressions with a large number of possible consequences by the consensus rule.

Consider for example the class $K_n(4,2)$ mentioned before. Suppose, for each n, we have found some small subset A_n of $K_n(4,2)$ which contains all the n variables and covers all the rows of the truth table covered by $K_n(4,2)$. The consensus rule creates, among other things, all the prime implicants and there are many of them (compare [4]). Moreover, in the process of arriving at these short clauses, we would generally generate a lot more longer clauses. Since the possible applications of the consensus rule are quite uniform and rigid in each case, the problem of finding examples which would require exponential time to decide by the consensus method appears to be a manageable combinatorial problem that is highly relevant to a better understanding of the tautology problem. We plan to look more closely at this problem in continuing our studies beyond the present paper. S.A. Cook and Imre Simon have obtained certain results along this general line.

Meanwhile, it is of interest to look at an ingenious set of special examples constructed by Tseitin [10]. These examples are contradictory expressions in the conjunctive form. And Tseitin shows that exponential time is necessary in order to decide them by the resolution method even as a proof (oor rather refutation) procedure, subject to a restriction of "regularity". In other words, for these examples, even the shortest paths leading to p and $\bar{p}$ for some variable p are of exponential length relative to the length of the original expressions. The notion of regularity is to exclude the reintroduction in another branch of a literal eliminated by the cut rule in one branch. More exactly, a resolution refutation is irregular if there is some sequence of expressions $A_1, \cdots, A_k$ from the refutation such that each A_i is one of the parent clauses in generating the resolvent A_{i+1}, $1 \leqslant i \leqslant k$, and such that there is some literal which appears in both A_1 and A_k but not in all the expressions between them. The meaning of the restriction to regular refutations is not well understood.

For our present purpose, it is more perspicuous to describe Tseitin's examples is

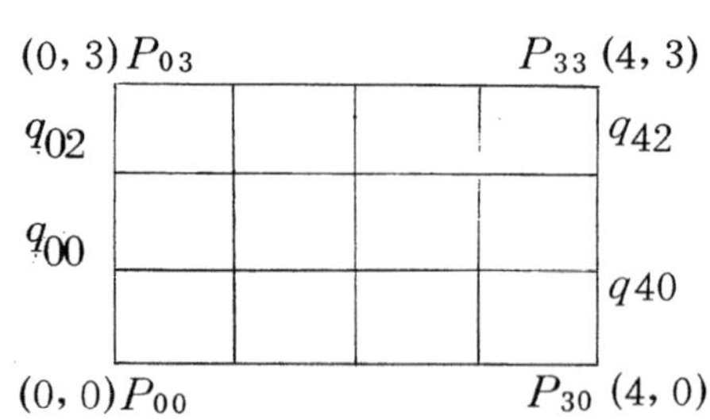

terms of biconditionals. His graphic representation defines a rectangular grid for each pair of integers n and k. Take the simple case when $n = 4, k = 3$.

The four conerns are at $(0,0), (n,0), (0,k), (n,k)$. The edges on the grid are associated with propositional variables in an obvious manner. The two edges starting at (i,j) are associated with p_{ij} (the horizontal) and q_{ij} (the vertical). There are, therefore, $2nk + n + k$ variables (edges) and $(n+1)(k+1)$ nodes. To each node is associated an expression involving the variables on the edges which meet at the node.

12.5.1 *General form of the grid problem*

For each pair n and k, $2^{(n+1)(k+1)}$ expressions are defined by choosing for each node one expression from a pair in the following manner:

(1) For the four corners, we choose in each case one clause from the pair $\{p$ iff q, p iff $\bar{q}\}$, where p, q are the variables meeting at the node;

(2) For the four boundaries, we choose for each node one clause from the pair $\{p$ iff q iff s, p iff q iff $\bar{s}\}$, where p, q, s meet at the node;

(3) For each internal node, we choose one clause from the pair $\{p$ iff q iff s iff t, p iff q iff s iff $\bar{t}\}$, where p, q, s, t meet at the node. For each total choice of $(n+1)(k+1)$ clauses, we take their conjunction as one expression.

It turns out that the choices correspond in a simple way with the status of the resulting expression. Let each choice from a pair get the index 1 or 0 according as whether it contained a negated variable or not. Then the expression is contradictory if and only if the sum of the indices is odd.

12.5.2 A special set of grid problems $\{G_i\}$

It is sufficient to consider the special case when $n = k$ and when all indices except one (say the one at the origin) are 0. These form an infinite set of contradictory expressions hard to refute by the resolution method. For example, G_3 would be the conjunction of the following 16 clauses:

(1) a iff $\bar{t}c$ iff d, p iff q, f iff g;

(2) s iff t iff y, a iff u iff b, s iff z iff q, b iff v iff c,
d iff w iff e, e iff n iff f, g iff j iff h, h iff i iff p;

(3) u iff y iff k iff x, x iff r iff w iff r, r iff n iff j iff m,
m iff k iff i iff z.

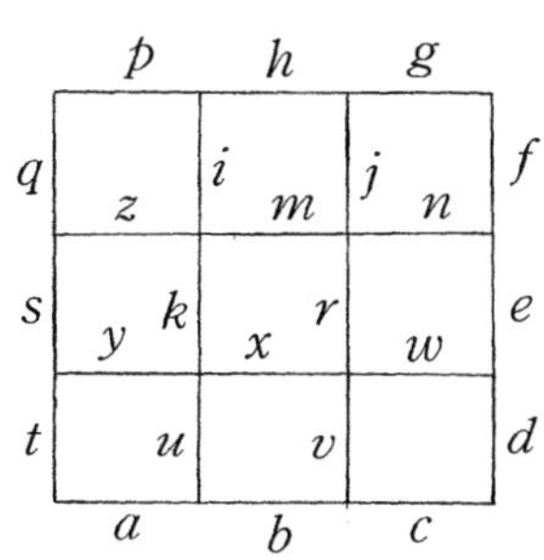

It should be noted that the clauses in each G_i are closely connected together. When the clauses are written as conjunction of disjunctions of literals and the cut rule is applied, we keep on getting longer expressions. In terms of the biconditional form, the cut rule essentially leads from p iff A and p iff B to A iff B. And we get no simplifications until we come to the end of the whole process. These vague remarks are of course only meant to suggest plausibility but they could be worked out more exactly to yield an alternative proof of Tseitin's result. We shall not undertake this task and the reader is referred to [10] for a proof that the evaluation of the set G_i of grid problems by the "regular" resolution method requires exponential time. In the next section, we shall give an efficient decision procedure for a natural class of expressions which includes all of Tseitin's examples.

Another set of interesting examples has been suggested in [8] and [1]. We shall refer to this class as the "occupancy problems". For each n, the occupancy problem of level n says intuitively that n objects cannot occupy all of $(n + 1)$ holes. We express this fact by a Boolean expression in disjunctive form as follows.

12.5.3 *The occupancy problems*

Let p_{ij} say intuitively that hole i contains object j $(1 \leqslant i \leqslant n + 1, 1 \leqslant j \leqslant n)$. For each n, the expression O_n is a disjunction of the following clauses:
For $1 \leqslant i \leqslant j \leqslant n + 1$, $1 \leqslant k \leqslant n$, $p_{ik}p_{jk}$ (hole i and hole j both contain object k). For $1 \leqslant i \leqslant n + 1$, $\bar{p}_{i1} \cdots \bar{p}_{in}$ (no object is in hole i). (Intuitively, $\exists i \exists j \exists k p_{ik}p_{jk} \vee \exists i \forall n \bar{p}_{in}$).

In other words, either some object occupies two holes or there is an unoccupied hole. It is easy to calculate that O_n contains $n(n + 1)$ variables and $\binom{n+1}{2}n + (n + 1) = (n + 1)n^2/2 + (n + 1) = (n + 1)(n^2 + 2)/2$ clauses. It has been calculated [1] that by the routine application of the resolution or consensus method, the "length" for deciding O_n is exponential: $n(n + 3)2^{n-2}$.

12.5.4

The special case O_3 is, for example, a disjunction of the following twenty–two clauses:

(1) $\bar{a}\bar{b}\bar{c}$, $\bar{d}\bar{e}\bar{f}$, $\bar{u}\bar{v}\bar{w}$, $\bar{x}\bar{y}\bar{z}$; (2) ad, au, ax, du, dx, ux;
(3) be, bv, by, ev, ey, vy; (4) cf, cw, cz, fw, fz, wz.

12.6 A feasible decision procedure for biconditional expressions

The main purpose of this section is to give a quick decision procedure for the class of expressions which are conjunctions of chains of biconditionals that join literals by the biconditional connective iff.

12. 6. 1 The biconditional is commutative and associative; hence, we can use the notation A_1 iff ... iff A_n for a chain of biconditionals. Of course, this is analogous to disjunction and conjunction rather than $=$; for example, A iff B iff C does not mean that A iff B and B iff C.

12. 6.2 In any chain of biconditionals, we can move any negation sign governed

directly by iff to govern the whole chain; hence, any even number of negation signs can be dropped (for example $\bar{p}$ iff $\bar{q}$ iff $\bar{n}$ is equivalent to p iff q iff $\bar{n}$ or p iff $\bar{q}$ iff n or $\bar{p}$ iff q iff n) so that at most one term of the chain begins with a negation sign.

12. 6. 3 Whenever a literal occurs twice in a chain of biconditionals, both occurrences can be dropped; hence, no term has to occur more than once (in a chain of length $\geqslant 3$).

In view of these and related properties of iff, it is desirable to single out iff in a given expression. For example, if a disjunctive form contains the clauses pqn, $p\bar{q}\bar{n}$, $\bar{p}q\bar{n}$, $\bar{p}\bar{q}n$, we can replace them by p iff q iff n. Before considering more general ways of using biconditionals to speed up decisions, we consider first an elegant special case.

12. 6.4 An expression is in (single literal conjunctive) biconditional normal form if it is a conjunction of clauses each of which is single literal or a chain of biconditionals in which each term is a literal.

We do not know whether each expression can be turned quickly into such a normal form, but once an expression is in this form, a general method can be applied to decide whether it is satisfiable. Take any expression A, in this biconditional form. Suppose it contains n variables $p_1, \cdots, p_n$. By 12. 6.2 and 12.6.3, we can simplify A so that each clause contains no more than n terms with no more than one variable negated. Let B be the simplified expression. If A or B contains any clause p_i iff p_i, it can be dropped; if A or B is simply p_i iff p_i, then B and A are of course satisfiable. If A or B contains a clause p_i iff $\bar{p}_i$. then B (and therewith A) is not satisfiable and we are through. Therefore, we can assume there are no such clauses and each clause contains each variable at most once. A or B may also contain clauses of the form p_i. or $\bar{p}_i$. If for some i, A or B contains p_i and $\bar{p}_i$, we are through because B(and therewith A) is not satisfiable.

In general, B contains chains of biconditionals and possibly single literal clauses. Consider all chains containing p_1, if there are any. They are (say) p_1 iff $A_1, \cdots, p_1$ iff A_m.

12. 6. 5 The conjunction of p iff $A_1, \cdots, p$ iff A_m is satisfiable if and only if the conjunction of A_1 iff $A_2, \cdots, A_{m-1}$ iff A_m is.

Suppose an assignment is given which makes p iff $A_1, \cdots, p$ iff A_m all true. p is assigned 0 or 1. In either case, $A_1, \cdots, A_m$ all get the same value. On the other hand, suppose A_1 iff $A_2, \cdots, A_{m-1}$ iff A_m has a true assignment. Then $A_1, \cdots, A_m$ all get the same value. If we give p that value, then $p, A_1, \cdots, A_m$ all get the same value. If we give p that value, then p iff $A_1, \cdots, p$ iff A_m are all true.

Hence, we can replace the m clauses p_1 iff $A_1, \cdots, p_1$ iff A_m by the $(m-1)$ clauses A_1 iff $A_2, \cdots, A_{m-1}$ iff A_m, which can again be simplified by applying 12.6.2 and 12.6.3. Observe that the resulting clauses are each at most of length $n-1$, and therefore the increase in length is well under control. In the process, we may generate single literal clauses or p_i iff p_i or p_i iff $\bar{p}_i$. These are treated as befors. Now that we have eliminated p_1, except possibly in a single literal clause, with little expansion, we can continue with the other variables in the same manner. In the final result (if we have not reached a decision earlier), we can have only clauses of the following four forms: p_i, $\bar{p}_i$, p_n iff p_n, p_n iff $\bar{p}_n$. If there is a clause of the form p_n iff $\bar{p}_n$, then the original expression A is not satisfialbe; similarly if there are two clauses p_i and $\bar{p}_i$ with the same i. Otherwise, we can assign 1 to p_j for each clause of the form p_j and 0 to p_k for each clause of the form $\bar{p}_k$. It is then seen that A is satisfiable. Therefore, we get:

12. 6. 6 Every expression in the biconditional form (as defined) can be decided quickly for satisfiability.

This simple case can be seen to apply in a natural way to the examples in [10], including those described under 12.5.1 and 12.5.2.

We turn now to the question of using biconditionals in more general situations. For this purpose,we shall begin with expressions in disjunctive normal form.

We note first that any valid expression A in disjunctive form with n variables $p_1, \cdots, p_n$ could be turned into a disjunction of p_1 iff ... iff p_n (call it B) and its negation by fairly simple rules. Clearly B and $\bar{B}$ each covers exactly half the truth table. By expanding A into the canonical normal form, we can easily arrive at a disjunction of B and $\bar{B}$. As a universal method, such a procedure does not assure speedy decisions. The relevance of the above remark is only that we could try to create biconditionals from any given expression, if there are suitable reasons for doing so. For example, if all or nearly all clauses making up a chain of biconditionals are explicitly present.

Analogous to the method of variable elimination, we have also a possibility of transmuting an expression:

12. 6. 7 The transmutation rule. Given any expression $A(p)$ containing a variable p and any expression B, $A(p)$ is valid if and only if the conjunction of $A(B)$ and $A(\bar{B})$ is valid, where $A(B)$ or $A(\bar{B})$ is obtained from $A(p)$ by substituting B or $\bar{B}$ for all occurrences of p.

Suppose $A(p)$ is valid, and $q_1, \cdots, q_n$ are all the other variables in $A(p)$. Then $A(0) \wedge A(1)$ is valid, i.e., whatever values the other variables take, $A(0)$ and $A(1)$ are always true. Consider now any assignment to $A(B)$ and $A(\bar{B})$, i.e., a choice of truth values for $q_1, \cdots, q_n$ as well as the additional variables, if any, in B. For any such assignment, B gets the value 0 or 1. If B gets 0, the value of $A(B)$ is the same as $A(0)$ and that of $A(\bar{B})$ is the same as $A(1)$, for that assignment of values to $q_1, \cdots, q_n$. Similarly, if B gets 1.

Conversely, suppose $A(B) \wedge A(\bar{B})$ is valid. Then every assignment to $q_1, \cdots, q_n$ and additional variables in B makes $A(B)$ and $A(\bar{B})$ true. Each assignment gives B the value 0 (and therefore $\bar{B}$ the value 1) or the value 1 (and therefore $\bar{B}$ the value 0). Therefore, for that particular assignment to $q_1, \cdots, q_n$, $A(0)$ and $A(1)$ are both true. Since this is true of all assignments to $q_1, \cdots, q_n$, $A(0) \wedge A(1)$ is valid. Therefore, $A(p)$ is valid.

In particular, it is not excluded that B may contain p.

In the general situation, the transmutation offers little advantage, since the possible elimination of a variable p is paid by doubling (or worse) the size of the expression. There are however cases where we could gain by the rule. One example would be when we have a biconditional p iff B as a clause in an expression $A(p)$ in disjunctive form, say (p iff B) $\vee R(p)$. In that case, $A(B)$ becomes (B iff B) $\vee R(B)$ and can be dropped, while $A(\bar{B})$ becomes ($\bar{B}$ iff B) $\vee R(\bar{B})$, which can be simplified to $R(\bar{B})$.

In particular, if an expression in disjunctive form contains two clauses pq and $\bar{p}\bar{q}$, say $pq \vee \bar{p}\bar{q} \vee R(p)$, we can simplify it to (p iff q) $\vee R(p)$ and then to $R(\bar{q})$, which is a desirable way of eliminating the variable p.

12.7 Two partial methods and an indication of two generic methods

We consider in this section a method of inversion and a method of counterterms that involve parallel eliminations of variables. We shall also briefly indicate two generic methods which we plan to investigate systematically in a future paper, viz. the method of size calculations and the method of composite expressions with symbolic abbreviations.

12.7.1 *A method of inversion*

We observe that if an expression in disjunctive form does not contain any absolutely positive clauses (or similarly negative ones), then it cannot be valid. For example, if there is no clause in which no variable is negated, then by assigning every variable the value 1, the whole expression must get the value 0. In any expression A if we "invert" any variable by substituting a variable p for $\bar{p}$ and $\bar{p}$ for p, then the result is valid if and only if A is. Hence, we can sometimes try to invert suitable variables to eliminate absolutely positive clauses (or absolutely negative ones). Of course, if the expression is valid, we cannot find such inversions. And it is possible to establish that such inversions do not exist. This yields a decision method which is not efficient in the general case but is effective for certain interesting special cases.

The inversion method is closely related to the concept of giant terms introduced in 12.3.4.

12.7.1.1 Let A be an expression in the disjunctive form. There is a giant term for A(i.e. A is not valid) if and only if there is an inversion of A which contains no absolutely positive clause.

Suppose there is a giant term G for A, say $\bar{p} \vee q \vee \bar{s} \vee t$. That means A contains no clause which makes $p\bar{q}s\bar{t}$ true. Suppose a, b, c, d are all the other variables in A. Then A does not cover the row in the truth table which makes $abcdp\bar{q}s\bar{t}$ true. Therefore, if we invert q and t, the result obtained can contain no absolutely positive clause, because any such clause would cover the missing row.

Conversely, suppose there is an inversion of A, say q to $\bar{q}$ and t to $\bar{t}$, which yields a disjunction B that contains no absolutely positive clauses. Suppose a, b, c, d, p, q, s, t are the only variables in A. Then $\bar{a} \vee \bar{b} \vee \bar{c} \vee \bar{d} \vee \bar{p} \vee \bar{q} \vee \bar{s} \vee \bar{t}$ must be a giant term for B, since every clause in B contains at least one negative literal. But then $\bar{a} \vee \bar{b} \vee \bar{c} \vee \bar{d} \vee \bar{p} \vee \bar{q} \vee \bar{s} \vee t$ would be a giant term for A.

By taking advantage of certain special features of the occupancy problem, we can decide all its cases efficiently with the method of inversion and the related concept of giant terms. Consider, for illustration, the simple case O_3 mentioned above. A first mechanically recognizable feature is that the clauses in the expression can be separated into groups containing different literals:

O_3. A_1: ad, au, ax, du, dx, ux;
A_2: be, bv, by, ev, ey, uy;
A_3: cf, cw, ca, fw, fz, wz;
B_1: $\bar{a}\bar{b}\bar{c}$; B_2: $\bar{d}\bar{e}\bar{f}$; B_3: $\bar{u}\bar{v}\bar{w}$; B_4: $\bar{x}\bar{y}\bar{z}$.

No two clauses in any two different groups have any literal in common.

We now ask whether there is any giant term for O_3. There are altogether 12 variables. Therefore, a giant term must contain no more than 12 literals. Consider A_1 first. It is easy to see that at least three variables must occur in a giant term. Similarly with A_2 and A_3. Hence, we need 9 unnegated variables for A_1, A_2, A_3. But each of B_1, B_2, B_3, B_4 needs one additional literal since all literals in them are negative. Therefore, we need at least 12 literals to make a giant term which is impossible. Hence, O_3 is valid.

Observe that if we delete any clause, say ux, we would have a giant term, viz. the disjunction of a, d, b, v, y, f, w, z, $\bar{c}$, $\bar{e}$, $\bar{u}$, $\bar{x}$. Therefore, the result obtained from O_3 by deleting any single clause is not valid.

It is not hard to convince oneself that the decision procedure can be mechanized and it applies efficiently to all cases of the occupancy problem. The natural objection is that the success depends too heavily on the special features of the occupancy problem. But our motive in including this solution is rather to illustrate a general point that metalogical arguments can be mechanized. In fact, we are currently developing a generic metamethod which has a similar flavor.

The basic idea is to find efficient ways of calculating how many rows of the truth table can be covered by the clauses in a given expression. Such calculations are possible when speed is not the issue, according to 4.8. But our goal is to find systematic shortcuts to eliminate repeated calculations which add nothing new.

We give here a few simple examples to illustrate the type of considerations we intend to study in a systematic way.

12.7.2 *Examples of size calculations*

12.7.2.1 *The disjunction of*: $p\bar{q}$, $\bar{p}r$, $q\bar{r}$. The three clauses are mutually inconsistent, and there are three variables. Each clause covers two rows of the truth table. Hence, exactly six of the eight rows are covered. Hence, the expression is not valid.

12.7.2.2 Consider the two clauses pqr and pqs. They cover the same rows as $pqr\bar{s}$ and pqs because $pqrs$ is subsumed by pqs and gives no new rows. This is a standard way of eliminating intersections of two ranges.

12.7.2.3 The expression is the disjunction of:

(1) $p\bar{q}$, (2) $\bar{p}r$, (3) $q\bar{r}$, (4) $\bar{p}q$, (5) $p\bar{r}$, (6) $\bar{q}r$.

We replace (1) by $p\bar{q}r$, $p\bar{q}\bar{r}$, (2) by $\bar{p}rq$, since $\bar{p}r\bar{q}$ is seen to be redundant by the presence of (6). Replace (3) by $q\bar{r}\bar{p}$, since $q\bar{r}p$ contains (5). Replace (4) by nothing, since $\bar{p}qr$ and $\bar{p}q\bar{r}$ are present (from (2) and (3)). Replace (5) by $p\bar{r}q$. (6) by $\bar{q}r\bar{p}$. We get altogether six clauses, and the expression is not valid. Actually, the replacement of (2) by $\bar{p}rq$ can also be avoided by appealing to (5).

12.7.2.4 The expression is the disjunction of:

(1) rs, (2) $\bar{r}s$, (3) $\bar{q}r$, (4) $\bar{p}\bar{s}$, (5) pq, (6) $p\bar{r}$.

As we compare (1) with (3), we see their ranges overlap on $rs\bar{q}$. Therefore, we can replace rs by rsq. We then compare it with (5) and see that it can be replaced by $rsq\bar{p}$. No other clause is consistent with this, and we get (1′) $rsq\bar{p}$. Next, (2) is consistent only with (6) and (5). Because of (5), we can omit the possibility of including p. We have therefore:

(2′) $\bar{r}s\bar{p}$($\bar{r}spq$ dropped by (5), $\bar{r}s\bar{p}q$ by first clause).

The clause (3) is only consistent with (4). It gives way to (3′) $\bar{q}rs$, $\bar{q}rp\bar{s}$.

Now clause (4) is consistent with none of (1′), (2′), (3′), (5), (6). So it remains unchanged.

Clause (5) is only consistent with (6). It gives way to (5′) pqr.

Clause (6) is consistent with none. So we have (1′), (2′), (3′), (4), (5′), (6). The calculation gives $1 + 2 + 2 + 1 + 4 + 2 + 4 = 16$. Hence the expression is valid.

12.7.2.5 The example O_3 again.

(1) $\bar{a}\bar{b}\bar{c}$ $\bar{d}\bar{e}\bar{f}$ $\bar{u}\bar{v}\bar{w}$ $\bar{x}\bar{y}\bar{z}$

(2) ad au ax du dx ux

(3) be bv by ev ey vy

(4) cf cw cz fw fz wz.

We have 12 variables. In (2), 4 variables occur. Of the 16 possibilities, one leaves nothing negated, 4 with one negated, 6 with two negated, 1 with all negated. The six clauses cover all possibilities when at most two are negated. We have, therefore, $(11/16) \times 2^{12}$.

In (3), we would have the same $(11/16) \times 2^{12}$ except for the fact that some of these rows have already been covered, and we need only include what is left after $11/16$ of the table is covered. Therefore, we have $(5/16)(11/16)2^{12}$ new rows covered by (3).

When we come to the clauses in (4), we do the same thing. $1 - (11/16) - (5/16)(11/16) = 25/2^8$. Hence, the number of new rows covered is: $(11/16)(25/2^8)2^{12} = 275$.

Thus far we have covered $(11)2^8 + (55)2^4 + 275 = 2816 + 880 + 275 = 3971$ rows. $4096 - 3971 = 125$.

Consider now the clauses in (1). The clause $\bar{a}\bar{b}\bar{c}$ covers only the rows in which at least three of the four variables in each of (2), (3), (4) are negated. That means $\bar{d}\bar{u}\bar{x}$, $\bar{d}u\bar{x}$, $d\bar{u}\bar{x}$ $\bar{d}u\bar{x}$ for (2) and similarly for (3) and (4). Hence, we have $4^3 = 64$. Alternatively, the only rows not covered by (2), (3), (4) are those in whcih at least three of a, d, u, x are negated, also three of b, e, v, y, and three of c, f, w, z, Hence, we get $5^3 = 125$. When we come to $\bar{d}\bar{e}\bar{f}$, we again have 64 possibilities of which 27 are covered by $\bar{a}\bar{b}\bar{c}$ because ux, vy, wz each with at least one negated is covered. Hence, we have $64 - 27 = 37$. Similarly, $\bar{u}\bar{v}\bar{w}$ cover only $64 - 27 - (27 - 8) = 18$ rows. The number 8 is the overlapping of abc and def: since we have also uvw, xyz can have any of the 8 possibilities. Finally, xyz covers only 6 rows because of the stringent conditions: (1) at least one of abc (def, uvw) must be positive; (2) at least three of $adux$ ($bevy$, $cfwz$) must be negative.

Therefore, since $64 + 37 + 18 + 6 = 125$, we get all the 4096 rows and O_3 is valid.

The treatment of Tseitin's examples by the use of biconditionals is meant to illustrate another general point which we plan to pursue extensively in working with composite expressions with the help of symbolic abbreviations. The biconditionals may be thought of as a natural existing method of abbreviation. They deviate from the familiar normal forms. The approach which we are in the process of developing refrains from turning intermediate composite expressions into standard normal form and introduces abbreviations systematically to deal with complex expressions indirectly through their symbolic representations. We have worked out a few complex examples which are rather too involved to be included here. Moreover, there are a number of theoretical points which are not yet sufficiently clear to us.

We give another partial method which is useful for many problems and use it to suggest another general approach.

12.7.3 *The method of counterterms*

Assume given an expression in disjunctive form. The method is especially appropriate when there are many short clauses. We choose a few variables to make up the "seed". Roughly we take the most frequently occurring variables, but in special cases we examine whether they occur in many different combinations of pairs, etc. For example, we may choose six variables a, b, c, d, e, f as the seed.

We try out all 64 possibilities of these variables to find out whether each possibility implies the original expression. Consider now one possibility, say $a\bar{b}cd\bar{e}f$. We compare this with each clause of the given expression and take any one which is consistent with $a\bar{b}cd\bar{e}f$ or its extension and contains one additional variable. For example, we may have a clause $a\bar{e}p$. In that case, we extend $a\bar{b}cd\bar{e}f$ to $a\bar{b}cd\bar{e}\,f\bar{p}$ This counterterm includes certain rows of the truth table not covered by $a\bar{e}p$. Moreover, all rows covered by $a\bar{b}cd\bar{e}f$ but not by $a\bar{b}cd\bar{e}f\bar{p}$ are covered by $a\bar{e}p$.

Case 1:Positive termination. After some steps, we arrive at (say) $a\bar{b}cd\bar{e}f\bar{p}q\bar{s}tn$, and there is a clause (say) $a\bar{e}n$ contained in it. This means that $a\bar{b}cd\bar{e}f$ implies the original expression, i.e., it covers no rows not covered by the original expression. To see this, we argue as follows. Certainly the whole counterterm covers no new rows not covered by $a\bar{e}n$. But $a\bar{b}cd\bar{e}f\bar{p}q\bar{s}t\bar{n}$ contains no rows not covered by $A\bar{n}$ which caused n to be added. Continuing thus, we see that $a\bar{b}cd\bar{e}f$ covers no rows not covered by clauses in the original expression. When we have a positive termination in this sense, we can add $a\bar{b}cd\bar{e}f$ to the original expression without affecting validity.

Case 2: Negative termination. We have succeeded in expanding the counterterm so that all clauses of the original expressions are involved either by interaction or by prior contradiction and yet there is no clause in the original expression contained in the counterterm. In this case, we can conclude that the original expression is not valid and stop, because we have found a row of the truth table not covered by the original expression.

Case 3: Undecided case. Neither 1 nor 2 occurs because there are clauses which are not involved by interaction. In this case, we may separate the original expression into $B \vee R$, so that R is the rest and we can ask the new question whether $A \supset R$, or $\bar{A} \vee R$ is valid. This is a simpler problem especially since $\bar{A}$ is a disjunction of single literals. It can also be treated by the counterterm method.

Generally, we do not work on the undecided cases until we have tried all 64 possibilities because any time we get a negative termination we are through.

Previously in Section 12. 4, we have remarked on the fact that "truly fat" expressions are decidable in polynomial time relative to the length of the expressions even though in absolute terms the time required is usually long. We wish to remark here on a related phenomenon about the length of the clauses in an expression. When the clauses are long, the decision is again relatively easy because a large number of clauses would then be necessary to make the expression valid. The observation 12. 4. 4 in Section 12. 4 is an illustration of this general point. The method of counterterms is specially suitable for expressions with short clauses. It does not necessarily work well, however, for expressions where clauses had been artificially shortened by the introduction of dummy variables as described earlier in Section 12. 3.

The relevance of such partial methods for the tautology problem is justified by the following envisaged possibility. We may reach a positive solution by combining a number of different methods appropriate to different types of expression. The only essential requirement is that we can determine in polynomial time the type of each expression and that the appropriate method always terminates in polynomial time.

Moreover, from the efficiency point of view, even if a certain generic method decides all Boolean expressions in polynomial time, it may handle certain types of problems much more readily than others. Therefore, a good startegy may be to have a preliminary examination of a given expression to decide which line of attack is best suited to the expression in question.

In conclusion, we have thus suggested three possible lines of attack which may on further study be shown to yield a positive solution to the tautology problem, viz. (1) the method of symbolic workthrough (dealing directly with composite expressions), (2) the method of size calculations, (3) a plurality of partial methods which taken together deals successfully with all possible expressions.

References

[1] S. A. Cook, The complexity of theorem–proving procedures, Proc. Third Annual ACM Symposium on Theory of Computing, 1971.

[2] S. A. Cook and R. Reckhow, On the length of proofs in the propositional calculus, Proc. Sixth Annual ACM Symposium on Theory of Computing, 1974 135—148 (also 8 pages of corrections).

[3] M. Davis and H. Putnam, A computing procedure for quantification theory, J. Assoc. Comp. Mach. 7(1960) 201—215.

[4] B. Dunham and R. Fridshal, The problem of simplifying logical expression, J. Symb. Logic 24(1959) 17—19.

[5] B. Dunham, R. Fridshal, and G, L. Sward, A. Nonheuristic program for proving elementary logical theorems, Proc. Int. Conf. on Information Processing (Paris: UNESCO, 1959) 282—284.

[6] B. Dunham and J. H. North, Theorem testing by computer, in: J. Fox, ed., 1962, Proc. Symposium on Math. Theory of Automata, 173—177.

[7] B. Dunham, R. Fridshal and J. H. North, Exploratory mathematics by machine, in: Recent developments in information and decision processes (Macmillan, New York, 1962) 149—160.

[8] R. M. Karp, Reducibility among combinatorial problems, in: R. E. Miller and J. W. Thatcher, eds., Complexity of computer computations, 1972, 85—103.

[9] J. A. Robinson, A machine oriented logic based on the resolution principle, JACM 12 (1965) 23—41.

[10] G. S. Tseitin, On the complexity of derivations in the propositional calculus, in: A. O. Slisenko, ed., Studies in constructive mathematics and mathematical logic, part II, 1968, 115—125.

13. RANKED MATCHING AND HOSPITAL INTERNS*

13.1 Preliminary

Many medical schools and hospitals participate in a national matching program under which each graduating medical student, besides applying directly to a number of hospitals in the group, submits to the program agency a ranked list of his preferences, and each hospital submits to it a ranked list of its preferred applicants. The assignment is worked out by a computer program on the basis of the information contained in these lists. Upon inquiry and examination, we find that the algorithm employed is 'hospital optimal'. In fact, the resulting assignment is anti-optimal for the students, in other words, every successful student gets into the hospital that is the lowest possible ('possible' in a definite and indisputable sense to be explained below) on his list of preferences. This fact is apparently not widely known.

The problem of making a judicial assignment under situations like this one is clearly a general one having many potential applications. For example, colleges could conceivably use a similar system, an election in which a candidate can run for several positions but can occupy only one could present a similar problem, the 'ideal marriage' problem amounts to the special case when each hospital has only a single vacancy. In looking at the general problem, we come across a number of simple conclusions which are surprising and have surprisingly simple proofs.

Standard matching theory deals with filling diverse positions by candidates who are generally only qualified for some of the positions (see, e.g., Liu, Chapter 11). The problem considered here may be called ranked matching: it arises whenever a multiplicity of positions or facilities are applied for by many candidates, with definite orders of preferences on both sides.

Recently we came across by chance a document (NIRMP, see reference) in which an actual ranked matching program is sketched with a simplified example. The program is to place graduates from medical schools into hospitals as interns. The sketch and example are not sufficient to reconstruct a complete algorithm; in fact, for the simple example given, different algorithms would yield the same result. We, therefore, wrote to John S. Graettinger, M. D., who is the person responsible for the program, for additional information. In the reply we were referred to the paper of Gale and Shapley (see reference) and told that 'the NIRMP algorithm is analogous to the college optimal assignment' described there.

We find that Gale and Shapley confine their attention to the separate treatments of

* Not published previously.

two extreme assignments which are the hospital (or college) optimal and the student optimal. Our approach centers around a natural reduction of all the given lists that preserves all relevant information, and continues with a study of all possible assignments including the two extreme ones considered by them. We are also able to prove additional results about the two extreme assignments, including the proof of some facts which they take for granted.

13.2 Deletion of useless names: Operations I and II

We assume given a finite number of hospitals A, B, C, ... each with a quota for interns and a fanite number of students α, β, γ, ..., together with their preference lists. Let T be the collection (or table) of these lists. The purpose is to find a judicial assignment of the students each to at most one hospital.

Gale and Shapley have introduced the basic concept of a stable assignment.
Definition. An assignment of applicants to hospitals is stable if there is no possibility of an applicant and a hospital making a deal to upset the assignment; more exactly, if the following situation does not arise. Two students α and β are assigned respectively to the hospitals A and B, but β prefers A to B and A prefers β to α.

Clearly any assignment under which the above situation occurs is not satisfactory to A and β. It is desirable to prevent such a situation and use only stable assignments.

For each hospital A, we have a ranked list L(A) which consists of two parts: D(A), the top part of L(A) up to the quota of A, is the list of desired applicants; W(A), the remainder of L(A), is the waiting list. If the number of names on L(A) is less or equal to the quota of A, L(A) = D(A). Each applicant α has a single ranked list L(α) of his preferences. Of course, an applicant α can appear on a hospital list L(A) only if A occurs in L(α).

As a first preliminary step, we can delete every ineffective applicant, i. e., one who appears on no hospital list at all, as well as any hospital with an empty list. As we revise the lists, such deletions will always be made whenever an applicant or a hospital becomes ineffective.

For individual names on a list, it is natural that we delete every hospital A from a list L(α) if α does not appear on L(A), and delete every student α from a list L(A) if A does not appear on L(α). Hence, if A is deleted from L(α), then α should be deleted from L(A); and if α is deleted from L(A), A should be deleted from L(α). For brevity, we speak of deleting the pair $\{\alpha, A\}$ to mean that both deletions are made.

The guiding principle of the systematic reductions is the obvious one: When a student or a hospital is assured of a higher choice, there is no need to retain the names of lower choices. To carry out this idea systematically, we shall introduce two (dual) operations, one for each side.

Operation I, Give a table T of lists, whenever an applicant α appears in D(A) of some hospital A, and A is not α's last choice, delete all pairs $\{\alpha, B\}$ for those hospitals B on L(α) which appear after A.

In particular, when α is deleted from some D(B), the top applicant on W(B), if

W(B) is not empty, moves onto D(B), and Operation I may be applicable again.
Definition 1. Two tables T and T′ are equivalent, $T \sim T'$, if they have the same stable assignments.
Theorem 1. Let T′ be the result obtained by applying Operation I to T. Then $T \sim T'$.

Proof. By hypothesis, α occurs in D(A). It is sufficient to prove that no stable assignment assigns α to a hospital B lower than A on L(α). Suppose the contrary. Since B accepts α and α appears on D(A), A must accept some student β on L(A) who occupies a lower place than α. But then, since α prefers A to B, α and A can make a deal. Therefore, the assignment was unstable.

Given T, there is generally a choice of singling out some student α to apply Operation I. Also, additional applications may be possible. Since, however, T is a finite collection of finite lists and each application of I reduces the size of T, we shall eventually reach a list T_I that is closed under T or a closure of T relative to I, i.e., a list T_I to which Operation I is no longer applicable.

By applying Theorem 1 repeatedly, we have:
Theorem 2. $T \sim T_I$.

A complication is to prove the intuitively obvious fact that we get the same closure of T under I no matter in what order we repeat Operation I. Following a suggestion by William Mitchell, we rearrange our results to get a direct characterization of T_I and thereby infer the uniqueness of the closure of T.

Let T_I be a closure of T under I. We consider the assignment by which to each hospital A is assigned exactly the applicants on D(A) in T_I. Clearly every student appearing in T_I is assigned to exactly one hospital. The assignment is stable in T_I since each hospital gets the students it prefers most. By Theorem 2, it is also stable in T. (In fact, this will be the hospital optimal assignment.)

Let T_I be a closure (the closure) of T under I. Consider all stable assignments for T. For each student α, there is a lowest hospital A on L(α) such that α is assigned to A under some stable assignment (but to no hospital B lower on L(α) under any stable assignment). Delete from T all pairs $\{\alpha, B\}$ for each student α and each such hospital B on L(α). Let T(I) be the result obtained from T by these deletions.
Theorem 3. $T_I = T(I)$.

We observe first that T(I) is uniquely defined because the deletions for different students are made independently of one another. By the stability of the hospital optimal assignment, clearly $T(I) \subseteq T_I$, since all the pairs deleted from T to get T_I are certainly deleted in getting T(I) from T. On the other hand, since $T \sim T_I$, $T_I \subseteq T(I)$ because all stable assignments of T are also (stable) assignments for T_I and in T_I no student can get assigned to any hospital lower on his original list than the hospital he is assigned to under the hospital optimal assignment. To see this, recall that Operation I cuts off all lower hospitals from each student's list and in T_I each student is assigned to the last hospital on his list.

From this it follows that we get the same T_I no longer in what order Operation I is repeatedly applied. Therefore, the closure T_I of T under I is uniquely defined and the hospital optimal assignment is uniquely defined.

We consider now a dual of Operation I:

Operation II. Given a table T of lists, whenever a hospital list L(A) contains at least q(A), the quota of A, applicants whose first choice is A, delete all pairs $\{\beta, A\}$ for all applicants β who appear after the applicant who is the q(A)–th applicant on L(A) with A as first choice. (of course this operation is not applicable to any hospital A with L(A) containing no more than q(A) names.)

Theorem 4. If T′ is obtained from T by applying II, then $T \sim T'$.

It is sufficient to prove that no stable assignment assigns to A any β of the type specified. Suppose the contrary. Since there are on L(A) q(A) applicants before β who have A as the first choice, at least one of them, say α, must be assigned to another hospital B. But A prefers α to β and α prefers A to B (since A is α's first choice).

Let T_{II} be a closure of T under II. We have:

Theorem 5. $T_{II} \sim T$.

We can parallel our considerations on I to prove the uniqueness of T_{II}. Consider the assignment which assigns to each hospital A exactly those applicants who are on L(A) and have A as first choice in T_{II}. The assignment is stable in T_{II} since each student gets his first choice in T_{II}. Hence, by Theorem 5, it is also stable in T. (In fact, this is the student optimal assignment.)

Theorem 6. Let T_{II} be a closure of T under II and T(II) be obtained from T by the following deletions. Consider all stable assignments T; for each hospital A, take the student α on its list who is the lowest on L(A) in T who will be assigned to A in some stable assignment. Delete every pain $\{\beta, A\}$ for each β below this α on L(A). Then $T_{II} = T(II)$.

Clearly T(II) is uniquely defined. By the stability of the student optimal assignment, $T(II) \subseteq T_{II}$. Since $T_{II} \sim T$, for reasons similar to the situation with T_I, $T_{II} \subseteq T(II)$.

Hence, T_{II} is the uniquely defined closure of T under II, and the student optimal assignment for T is uniquely defined.

The considerations so far suggest combining the two operations. This has double advantages: we end up with a simpler collection of lists which still preserves all relevant information, and we are in a better position to survey all stable assignments in order to select judicial assignments according to principles we choose to follow.

13.3 The canonical form T* of T

Definition 2. For any table T, its canonical form T* is the result obtained by making all possible applications of I and II in any order we wish. In other words, T* is the closure of T under I and II.

To justify this definition, we argue similarly as before to establish that T* is uniquely defined from T. Let T* be any closure of T under I and II. By Theorems 2 and 5, we have:

Theorem 7. $T^* \sim T$.

Let T(*) be obtained from T by making both types of deletion, viz. those to get from T to T(I) and those from T to T(II). The result T(*) is uniquely determined by T, since the individual deletions have at most trivial interdependence. If two deletions are both of the first (second) type, they involve two different students (hospitals). If two deletions are of different types, it is possible that both of them delete a same pair $\{\alpha, A\}$. But then it does not matter, since $\{\alpha, A\}$ is deleted anyhow and it has no effect on other deletions.

Theorem 8. T* = T(*)

Using the hypothesis that T* is a closure of T under I and II, we can, as before, make the two extreme assignments (hospital and student optimal) in T*. Hence, as before, $T(*) \subseteq T^*$. Moreover, since $T^* \sim T$, $T^* \subseteq T(*)$. Hence $T^* = T(*)$.

This shows that T* is the unique canonical form of T and justifies Definition 2.

The canonical table T* has some surprising properties which are, by the way, proved without using the fact that it is unique. All that is used is merely the property that T* is indeed a closure of T under I and II.

For each hospital list L(A), let f(A) be the number of applicants on L(A) whose first choice is A, and let d(A) be the number of applicants on D(A).

Theorem 9. In T*, f(A) = d(A), for every hospital A.

Proof. Since Operation II can no longer be applied to T*, we have, for every A, $d(A) \geqslant f(A)$. Let d be $\Sigma d(A)$, over all hospitals, and $f = \Sigma f(A)$, over all hospitals. Then $d \geqslant f$.

On the other hand, every applicant on some d(A) has a first choice (either A or some other hospital B) and appears on that hospital's list. Moreover, since Operation I is no longer applicable, no applicant in some D(A) can appear in any other D(B). Therefore, $f \geqslant d$. Hence, $f = d$.

Suppose for some A, $f(A) \neq d(A)$. Since, for every B, $d(B) \geqslant f(B)$, we must have $d(A) > f(A)$. But then, since $f = d$, there must be some B, $d(B) < f(B)$, contradicting $d(B) \geqslant f(B)$.

Therefore, $d(A) = f(A)$ for every A.

Corollary 1. In T*, every remaining applicant, i.e. one who appears in at least one hospital list L(B), must appear in some D(A).

If there were a student α who appears only on waiting lists, α must have a first choice A and appear on L(A), because otherwise A would no longer be a choice for α. But that would make $f > d$.

In other words, even though it is quite possible for T to contain lists such that there is some α who only appears on waiting lists, this is no longer true for T*.

Corollary 2. In T*, if for a hospital A, L(A) = D(A), i.e., the hospital A has only $\leqslant q(A)$ applicants remaining on L(A), then all applicants on L(A) have A as first choice, and exactly these applicants will be accepted by A in any stable assignment.

Since the proof of Theorem 9 does not assume that $D(A) \neq L(A)$, we have D(A) = f(A) also for such short lists. Hence, every applicant on A has A as first choice.

Since Operation I is no longer applicable to T*, these applicants do not appear on any other hospital lists. Since they have A as first choice and make up D(A), it is impossible for A to accept any other applicant in any stable assignment.

It is a little complex to try to see directly why these corollaries are true. For example, one might make explicit the following argument to show why every applicant on a short list has the hospital as first choice. If α appears on a short list A but does not have A as a first choice, then α must have some hospital B as first choice and appear on W(B). Then there must be some β on D(B) such that β does not have B as a first choice. And so on. This would lead to an unending process which is an impossibility since T* is finite.

Corollary 3. Every stable assignment for T assigns exactly the applicants appearing in T* to the hospitals appearing in T* in such a way that each student α is either assigned to the hospital A with α in D(A) or to a hospital B which α prefers to A.

Since T $\sim$ T*, they have the same stable assignments. Suppose α is assigned to C (in particular, to no hospital at all) and α prefers A to C. Since α appears on D(A), A must have accepted some applicant β lower than α on L(A). But then α and A can make a deal to upset the assignment.

13.4 The student and hospital optimal assignments

We make first a few remarks on the relations between T_I, T_{II} and T*.

13.4.1 For each hospital A, D(A) is the same in T_I and T*; hence, the hospital optimal assignment is the same in T_I and T*.

Operation II can only affect the waiting lists and affect at most the ranking of A in L(α) for α on D(A), but not the fact whether α is on D(A) or not. Therefore, for each A, D(A) in T* remains the same as in T_I. Since the hospital optimal assignment is determined entirely by the desired lists, it is the same in T_I and in T*.

13.4.2 For each hospital A, the students on L(A) with A as first choice are the same in T_{II} and T*; hence the student optimal assignment is the same in T_{II} and T*.

Applying I repeatedly to T_{II} does not affect the first choice of any student since I only eliminates $\{\alpha, B\}$ when α is on D(A) with A a higher choice for α than B.

13.4.3 $(T_I)_{II} = (T_{II})_I = T^*$.

Since T* is the uniquely determined closure of T under I and II, it is sufficient to prove that $(T_I)_{II}$ and $(T_{II})_I$ are closed under I and II. Consider first $(T_I)_{II}$, which is the closure of T_I under II. By 13.4.1, the desired lists of the hospitals in $(T_I)_{II}$ are the same as in T*, and, therefore, I is no longer applicable. On the other hand, by 13.4.2, II is no longer applicable to $(T_{II})_I$.

13.4.4 The students remaining on T* are the same as those in T_{II} and those on the desired lists of T_I.

Observe that I never eliminates any student altogether. By Theorem 9, all students remaining on T* appear on desired lists. Hence, by 13.4.1, they are exactly those who appear on desired lists in T_I. Since every remaining student has exactly one first choice, the students remaining are, by 13.4.2, the same in T* and T_{II}.

13.4.5 The hospital optimal assignment is optimal for the hospitals and 'anti–optimal' for the students; the student optimal assignment is optimal for the students and anti–optimal for the hospitals.

Thus in no stable assignment can the hospital do better than in the hospital optimal one because T ~ T* and each hospital A gets its top choices in T*. Similarly the students can do no better than using the student optimal assignment since they all get their first choices in T* and T ~ T*.

We can now see the sense that every student gets assigned to the worst possible (for him) hospital. By Corollary 3 above, in no stable assignment can a student get assigned to any hospital he prefers less.

On the other hand, in the student optimal assignment, the hospitals accept applicants they prefer least in the following sense. Let $f_\phi(k)$ be the position of the k–th candidate, accepted by a college A, on L(A) in T* in a stable assignment ϕ. Let ψ be the student optimal assignment. Then $f_\phi(k) \geqslant f_\phi(k)$ for all ϕ and all $k \leqslant q(A)$. Thus, if $f_\psi(k_0) > f_\psi(k_0)$, then there is at least one student α on L(A) before the k_0–th applicant accepted by A who has A as first choice. But then α and A can make a deal to upset the assignment.

The algorithms used by Gale and Shapley are easy to describe but awkward to carry out, because they assume that, for example, in the student optimal assignment, each student applies only to his current first choice at each stage. In other words, a student applies to his next choice only when rejected by his current first choice; this process is repeated until every student is either accepted or held as a tentative acceptance by some hospital or has been rejected by every hospital to which he is willing to apply. At this point, each hospital accepts all its tentative acceptances. (See Gale and Shapley, pp. 13–14.)

The actual current algorithm for placing interns gives, as noted above, the hospital optimal assignment. It employs only Operation I and obtains essentially T_I from T. The following advantages are claimed for the program (NIRMP, p. 7):

'Students advanced up the hospitals' preference lists only when a higher student on the list was scratched because of getting a higher choice on hier list. In other words, no student was passed by a student lower on a hospital's preference list. Each student obtained a position in the hospital highest on hier list which offered hiem a position. An important corollary is that a student does not jeopardize hier chances in a given program by ranking another program above it which s/he preferred but thought unattainable.

Of course, that no student was, in the specified sense, passed by a student lower on a hospital list is a property shared by all stable assignments. It is not a special advantage of the hospital optimal assignment.

The important point is the ambiguity of the phrase 'a hospital A offering a student α a position.' From the hospital's point of view, it would choose to mean by this phrase that α is on D(A) in T* or T_I. And this is the sense implicit in the above quotation. From the student's point of view, this ought to mean that α is on L(A) in T* or T_{II}. It is clear from our discussion so far that the misleading phrase conceals the disadvantage suffered by the students under the present system of assignment. When a hospital A is willing to accept a student α, i. e.. to put α on L(A), it is perfectly possible to proceed by giving the student's preference the first consideration, as is seen from the stability of the student optimal assignment.

We note that the hospital (in contrast to the student) optimal assignment has a further disadvantage for the students: for a student who gets assigned at the end, listing safety choices can hurt in preventing him from getting into a hospital higher on his list.

Let us look more closely at the practical question of adding or dropping choices by an applicant. Since it is difficult for a student to know how other students would alter their choices, it is simplest and reasonable to consider the situation when all other applicants have their lists fixed.

For a student α who does not get assigned, i. e. does not appear in T^*, of course adding or dropping choices can never hurt and adding choices might help. Dropping choices cannot help because a same T^* would still emerge. The interesting problem is when α does appear in T^*.

Suppose that in T^*, α has A_1 as first choice and B_1 as last choice. Generally, α lists more choices in T so that A_1 and B_1 need not be his first and last choices in T. Also it is difficult for α to predict what his first and last choices are in T^*. Suppose that initially α's first choice and last choice are A and B in T. Theorem 10. If α appears in T^*, then listing more choices initially below his last choice in T cannot help; but it can hurt under the hospital optimal assignment, and it cannot hurt under the student optimal assignment.

For definiteness, let U be the same as T except for containing also for a student α additional choices C, ..., D after his last choice B in T.

Suppose we apply I and II to T and U in all possible ways except that I is not applied to α. When we have done this, we can observe several things about the results T' and U'. Let E be the hospital such that α appears on D(E) in T^*. Then α cannot appear on $D(E_1)$ in T' or U' for any hospital E' which α prefers over E. This is true for T' because otherwise the closure under I (and II which is no longer applicable) obtained from T' would be different from T^*. This is also true for U' because the additional choices, which are certainly not α's first choice, can make no difference to II, and can at most occupy desired places to push other applicants to desired places on α's higher choices. Iterated applications of I to U' will preserve this property because otherwise α would also make a higher desired list in the closure of T' which is the same as T^* by Theorem 8.

The additional lower choices can hurt because it can happen that they will prevent another applicant β from making the desired list of a hospital which β prefers over E but is lower on α's list. On the other hand, if one uses the student optimal assignment, no difference is made because α's first choice in U^* and T^* remains the same.

This can be illustrated by a simple example in which A and B each have a quota of 1.

Example 1.	T	α	β	A	B	U	α	β	A	B
		A	B	β	α		A	B	β	α
			A	α	β		B	A	α	β

Clearly, we have $U^* = U$ and T^* is:

α	β	A	B
A	B	α	β

Hence, under T, α gets his first choice (in any stable assignment); under U, the student optimal assignment is again αA, βB, while the hospital optimal assignment is βA, αB.

A more general result is:

4.5. Suppose α appears in U* and α has A_1 as first choice, B_1 as last choice in U*. Let T be obtained from U by revising α's original list to contain only A_1. Then α will still appear in T* and will be sent to A_1 under every stable assignment.

Consider U_{II} first. Since we only apply Operation II to U, none of the choices by α below A_1 comes into play. Therefore, U_{II} and T_{II} contain the same students. By 4.4, U* and T* contain the same students. Hence, α still appears in T*. Since α has A_1 as the only choice now, he is sent to A_1 under any stable assignment.

This is surprising because in U*, α occurs on the desired list of B_1 only. It seems possible that somebody not appearing in U* might now make $D(B_1)$ in T* (now that B_1 is no longer a choice of α) and drive α out of T* altogether. In fact, 4.5 shows that such a situation cannot arise, and α is no $D(A_1)$ in T*. Here we have another result which is hard to see directly.

We note that, in general, α's original choices in U may have been A, ..., B among which $A_1, \ldots, B_1$ all appear in the same order. Let V be obtained from U by modifying α's original list to contain just A_1, ..., B_1 in that order. Then U* = V*. Hence, if we compare V with T, we obtain an alternative proof of Theorem 10 which also implies that deleting lower choices by α can help and cannot burt provided α does make V*.

Theorem 11. Adding unattainable choices by a student can help and cannot hurt.

There remains to consider the question of adding 'unattainable' choices. Of course, this may possibly help because an 'unattainable' hospital may really be eager to get the student. The harder question is rather whether this might hurt in certain cases. Suppose α lists B as first choice but he prefers A to B. A is not listed because it is thought to be hopeless. Suppose T is the table in which α lists B as first choice and U is the table in which α adds A above B. If α is on D(A), of course α is helped. Otherwise, we are only interested in the case that α makes T*.

Since α makes T*, it occurs in some D(C) in T*. Assume all possible applications of II are made to U. If α does not make L(A), the listing of A is idle. The presence of A could possibly hurt only if α remains on L(A) but fails to make D(A). In that case, some β might be driven off L(A). Since β could not have made D(A) anyway, β occupies a desired place in some other hospital even in T*. Apply I to U_{II} to get U*. It is seen that if α belongs to D(C) in T*, it also belongs to D(C) in U* since nobody has taken his place as a result of his adding A. Similarly if α adds several choices above his original first choice, he still would not get hurt. Observe that if α makes L(A), he will get into A under the student optimal assignment.

Generally the result of listing more choices can have different results in an irregular way.

Consider the following collection T of tables with A, B, C each having a quota of 1.

Example 2.

α	β	γ	A	B	C
C	A	B	α	α	γ
A		C	β	γ	α
B					

When we close under Operation I, B is deleted from L(α), α is deleted from L(B), and then γ is deleted from L(C). Hence, the only stable assignment is:

A	B	C
β	γ	α

If α has omitted A, B from his initial list, we would have α again getting his first choice. On the other hand, if α has omitted just A, we would have two stable assignments:

A	B	C		A	B	C
β	α	γ	and	β	γ	α

The first is the hospital optimal assignment, under which α fails to get his first choice C.

Notice also that even though the pair $\{\alpha, A\}$ does not occur in any stable assignment of T, deleting it creates an additional stable assignment.

13.5 Mixed assignments and a characterization of all stable assignments.

With the hospital intern problem, many of us would be inclined to favor the student optimal assignment as judicial on the ground that a student can be accepted by only one hospital while a hospital generally accepts more than one intern. When it comes to the marriage problem, the question of a judicial matching becomes more acute since it seems quite arbitrary either to favor the men or to favor the women. The question of finding assignments that would be the best possible (maximal though not maximum) for both sides is rather intricate. We consider first a characterization of all stable solutions which suggests a number of mixed (stable) assignments as candidates for the best possible assignments.

Consider a collection T of lists from students and hospitals given. We apply the * operation (that is to say, the result of making all possible applications of Operations I and II) to T to get the equivalent collection T*. We work with T*. We recall that under T*, every student appears on exactly one hospital's desired list D(A), and every hospital A has a quota number of students on L(A) with A as first choice. A terminal segment of a student's list is either empty or any unbroken part of the list including the last hospital on it but excluding the first hospital on it. Similarly a terminal segment of a hospital list L(A) is either empty or any unbroken part including the last name on it but excluding D(A).

Theorem 12. All and only stable assignments in T* are obtained by any sequence of applications of the following two operations:

(i) Delete a terminal segment from each student's list and close under Operation I; (use the hospital optimal assignment for the concluding step).

(ii) Delete a terminal segment from each hospital list and close under Operation II; (use the student optimal assignment for the concluding step).

In fact, each stable assignment in T* can be obtained by a single application of (i);

and similarly with (ii).

Proof. Let ϕ be a stable assignment in T*. Each pair of student and hospital under ϕ must occur in T*. Since every student α is paired with only one hospital A which of course appears on L(α) in T*, apply (i) by deleting from each L(α) all hospitals below his A. Since no student has deleted his first choice, the collection T′ of revised lists remains closed under II. Use now the hospital optimal assignment in T′ of revised lists remains closed under II. Use now the hospital optimal assignment in T′, and we have got ϕ, since every student gets his last choice in T′ which is, by construction, the hospital assigned to him under ϕ.

Alternatively, given ϕ, we can apply (ii) by deleting every student from each hospital list L(A) who is not accepted by A under ϕ and does not precede any student on L(A) accepted by A under ϕ. The student optimal assignment in the collection of revised lists is ϕ.

Hence, all stable assignments in T* are obtainable by (i) and by (ii).

For the other direction, we prove that given any T with stable assignments, the result T′ of applying (i) or (ii) has stable assignments and that they are all stable assignments in T. It will then follow that any sequence of applications of (i) and (ii) yields only stable assignments in T*.

Consider first T′ obtained from T by a single application of Operation (i). Take the hospital optimal assignment in T′, i.e., assign to each hospital the students on D(A) in T′. The result is a stable assignment. Suppose otherwise so that α and β are assigned to A and B, but α prefers B to A, B prefers α to β. This is impossible since β is on D(B) but α is not on D(B). Hence, T′ has some stable assignment. Moreover, any stable assignment ϕ in T′ remains stable in T since T′ results from T by the deletion of terminal segments and restoring them would not add any hospital A to a student α's list such that α prefers A to the hospital he is assigned to under ϕ. In other words, nothing new in T appears to upset ϕ which is, by hypothesis, not upsettable in T′.

Similarly, if T′ is obtained from T by a single application of (ii), no new stable assignments are generated and we can take the student optimal assignment in T′ and prove it stable.

The proof above clearly suggests different ways of generating mixed stable assignments. For example, we could take T* and delete from each student's list all but his first three choices, close under Operation I, and then use the hospital optimal assignment relative to the collection of revised lists.

A natural question to ask is whether generally we can further reduce T* to get a subtable of it which has exactly the same stable assignments as T*. Let a pair $\{\alpha, A\}$ be stable for T or T* if it appears in some stable assignments. A more definite question is: for any T (or just any T closed under I and II), is the subtable S consisting of all stable pairs equivalent to T ($S \sim T$)?

Consider the following table T which is closed under I and II, with A, B, C each having quota 1:

Example 3.

α	β	γ	A	B	C
A	C	B	γ	β	α
B	$-$A	A	$-\beta$	α	β
C	B		α	γ	

The stable assignments are: αA, βC, γB; αC, βB, γA; αB, βC, γA. For T, $\{\beta, \text{A}\}$ is the only unstable pair. Let S be the result obtained from T by deleting $\{\beta, \text{A}\}$. Then S $\sim$ T and S is stable in the sense that all pairs in it are stable.

13.5.1 Given any table T and an unstable pair $\{\alpha, \text{A}\}$, let S be the result obtained from T by deleting $\{\alpha, \text{A}\}$. Then every stable assignment of T is one of S.

Proof. Let ϕ be a stable assignment of T. If it is not stable in S, then there are β, γ, B, C such that βB, γC but β prefers C to B and B prefers β to γ. But the pairs $\{\beta, \text{B}\}$, $\{\beta, \text{C}\}$, $\{\gamma, \text{B}\}$ all appear in T in the same order, the situation would also make ϕ unstable in T.

13.5.2 A pair $\{\alpha, \text{A}\}$ is unstable in T* if and only if $\{\alpha, \text{A}\}$ is deleted after we delete all pairs $\{\alpha, \text{B}\}$ with B below A on L(α) and close under Operation I.

Proof. If $\{\alpha, \text{A}\}$ is not deleted, then it must be stable because we can use the hospital optimal assignment as in Theorem 12 under which α is assigned to A. On the other had, if $\{\alpha, \text{A}\}$ is stable, let α be assigned to A under ϕ. Obtain ϕ by applying (i) as in Theorem 12. Since $\{\alpha, \text{A}\}$ is retained in thus applying (i), it is also retained when we merely delete as in 13.5.2. Hence, $\{\alpha, \text{A}\}$ is not deleted.

Consider the following table T closed under I, II with A, B, C, D each having the quota 1:

Example 4.

α	β	γ	δ	A	B	C	D
B	A	C	D	α	β	δ	γ
A	B	A	B	γ	δ	γ	δ
		D	C	β	α		

This has three stable assignments: αA, βB, δC, γD; βA, αB, γC, δD; αA, βB, γC, δD. The unstable pairs are $\{\gamma, \text{A}\}$ and $\{\delta. \text{B}\}$. Let S be the result obtained by deleting them from T:

α	β	γ	δ	A	B	C	D
B	A	C	D	α	β	δ	γ
A	B	D	C	β	α	γ	δ

In addition to the three stable assignments for T, we have in S also; βA, αB, δC, γD. This is not stable in T because A prefers γ to β, and γ prefers A to D.

Hence, we have an example of a table T whose stable subtable is not equivalent to it.

13.6 The marriage problem

The special case of the marriage problem contains many of the essential features of the more general problem but is more symmetrical. It is customary to make simplifying assumptions such as there being an equal number of men and women with everybody ranking everyone of the opposite sex. We shall begin with a more general situation and some applications of our general results to the marriage problem.

Assume given m men α, β, .. and k women A, B, ... each with a nonempty ranking list of some members of the opposite sex. Delete from every list L(a) everybody whose list does not include a. Clearly anybody not appearing on anyone's list can just as well have his or her list deleted. Let the given collection of lists be T.

We now apply Operations I and II to get T*. The situation is more symmetrical than the matching of students with hospitals, so that the two operations are merged into one. Let a be a male or a female, if a is the first choice of somebody b on L(a), delete all names below b on L(a) and delete a from their lists. This includes I (for males) and II(for females). T* is obtained after all possible applications of I and II are made in any order.

In T*, everybody a who remains (i.e. occurs on any list at all) appears as first choice on exactly one list and has his or her list L(a) containing exactly one name (the last one) whose first choice is a. In particular, unless a person becomes the first choice of somebody after all possible applications of I and II, he or she is eliminated. It follows also that in T* the remaining men and women are equal in number, say n.

It is immediately clear that situations can arise where we have no unique best matching. Take, for example, two couples making opposite choices:

Example 5.

A	B	α	β
α	β	B	A
β	α	A	B

The only two possible solutions are equally good or equally bad. If we match A with α (and therefore B with β), the women get their first choices; if we match A with β, the men get their first choices.

In fact, this adverse pattern could occur with any number of couples. The case of three couples is sufficient to illustrate this fact:

Example 6.

A	B	C	α	β	γ
α	β	γ	B	C	A
β	γ	α	C	A	B
γ	α	β	A	B	C

Of course these examples apply also to the matching of hospitals and students, since they are just the special case when each hospital has a quota of 1.

An attractive mixed stable matching procedure is suggested by Theorem 12, beginning with T* closed under I and II. For example, delete the last choices (unless it is also the first choice) from the women's lists and close under I and II. Next delete the last choices from the men's lists and close under I and II. The two steps are repeated until everybody has only his or her (revised) first choice left. Then, by Theorem 12, we get a stable matching.

Given T* with n couples remaining, there is a natural idea of decomposing these couples into "clubs". A club is any k–club for some k, $1 \leqslant k \leqslant n$, and a k–club is a collection of k couples such that each person has his or her first k choices lying in this collection. For example, a 1–club is simply a couple who have each other as first choice. In that case, it is obvious that we can match them off and leave them out of further considerations. More generally, we have:

Lemma. Given T*, if there is any k–club, $1 < k < n$, we can consider the k couples in the k–club separately from the other n–k couples without losing any stable assignments.

Proof. If anybody a in a k–club is matched with somebody b outside the k–club, we get an unstable assignment. If a is thus matched, there must be a c of opposite sex in the k–club matched with some d outside the k–club. But a prefers c to b and c prefers a to d. Hence, the assignment was unstable.

Therefore, we can confine our attention to the case where we have n couples and we do not have any k–clubs, $1 \leqslant k < n$. This is indeed the most general case and of course the n couples form an n–club.

We consider tentatively a way of matching couples according to what seems to be a natural measure of preference. The measure does not directly apply to the student–hospital situation.

For any man α and any woman A in T*, there is a pair of numbers $(p_A(\alpha), q_\alpha(A))$, briefly $m(\alpha, A)$, where $p_A(\alpha)$ is α's place on L(A) and $q_\alpha(A)$ is A's place on $L(\alpha)$, with 1 for first choice, 2 for second choice, etc. Thus with n couples, we have at most n^2 pairs of numbers (p, q) with $1 \leqslant p \leqslant n$, $1 \leqslant q \leqslant n$. We define a partial ordering of these pairs:

$(p, q) < (p', q')$ iff $p + q < p' + q'$, or $p + q = p' + q'$ but $|p - q| < |p' - q'|$, or $p + q = p' + q'$ and $|p - q| = |p' - q|$ but $p < p'$.

The idea is that the mutual preference of A and α can be measured firstly by $p_A(\alpha) + q_\alpha(A)$, the minimum being 2 when each has the other as first choice, and that, for the same numerical sum, it is more satisfactory when the preferences for each other are not too different. The last alternative is arbitrary in putting more weight on p than on q. It is difficult to give any definite justification for the choice of such a measure.

When this idea is applied to Example 6, it is seen that, for all the nine pairs, $p + q = 4$, and that the three pairs with (2, 2) are the lowest in the above partial ordering.

The procedure proposed is the following. Given a collection T of lists, apply I and II to get T*, with say, n couples remaining. Decompose them into clubs if possible. And consider each club T* separately.

Consider first the minimal pairs according to the order defined. Let $M(\alpha, A)$ be a minimal pair. Then it is not possible that either $m(\alpha, B)$ or $m(\beta, A)$ is also minimal for some B or β. This is important because otherwise we do not know whether to match α with A or B, to match A with α or β.

Proof. Since $m(\alpha, A)$ is minimal and $p_\alpha(A) \neq p_\alpha(B)$, $m(\alpha, B)$ cannot be minimal. Similarly $m(\beta, A)$ cannot be minimal because $q_A(\beta) \neq q_A(\alpha)$.

The procedure begins by matching up A with α for every case where $m(\alpha, A)$ is minimal. For example, in the perverse case with two couples, we have:

$$m(\alpha, A) = (2, 1) = m(\beta, B),$$
$$m(\alpha, B) = (1, 2) = m(\beta, A).$$

By definition, the last line gives the two minimal pairs, and the resulting match is B with α and A with β. Because more weight is given to p, the preference of α and β, the men get their first choices.

After the couples with minimal pairs are matched off, we readjust T* by deleting these persons and renumber the choices of the remaining couples in the obvious manner. In the remaining and revised lists, we again match off those couples with minimal paris. This process is continued until all couples are matched off.

It is not clear that this procedure indeed gives a 'maximal' or optimal solution in each case. In fact, it is not even clear that it does give a stable assignment.

References

1. D. Gale and L. S. Shapley, 'College Admissions and the stability of marriage', *Am. math. monthly*, vol. 69 (1962), pp. 9—15.

2. C. L. Liu, *Introduction to combinatorial mathematics*, New York, 1968.

3. NIRMP. *NIRMP directory including hospitals and programs participating in the matching program for 1977 appointments*, National Intern and Resident Matching Program, Evanston, Illinois, October, 1976.

PART FOUR

TOPICS FROM THEORY TO PRACTICE

14. LOGICAL FRAGMENTS RELEVANT TO COMPTER SCIENCE*

14.1 Logic of many-sorted theories*

1. Introduction

Certain axiomatic systems involve more than one category of fundamental objects; for example, points, lines, and planes in geometry; individuals, classes of individuals, etc. in the theory of types or in predicate calculi of orders higher than one. It is natural to use variables of different kinds with their ranges respectively restricted to different categories of objects, and to assume as substructure the usual quantification theory (the restricted predicate calculus) for each of the various kinds of variables together with the usual theory of truth functions for the formulas of the system. An axiomatic theory set up in this manner will be called many-sorted[1]. We shall refer to the theory of truth functions and quantifiers in it as its (many-sorted) elementary logic[2], and call the primitive symbols and axioms (including axiom schemata) the proper primitive symbols and proper axioms of the system. Our purpose in this paper is to investigate the many-sorted systems and their elementary logics.

Among the proper primitive symbols of a many-sorted[3] system T_n ($n=2, \cdots, \omega$) there may be included symbols of some or all of the following kinds: (1) predicates denoting the properties and relations treated in the system; (2) functors denoting the functions treated in the system; (3) constant names for certain objects of the system. We may either take as primitive or define a predicate denoting the identity relation in T_n. In any case, it is usually desirable to include in T_n the usual theory of identity for the objects of the system. We shall assume that T_n contains the usual theory of identity[4] as a part. Then we know we can introduce descriptions by contextual definitions such as

---

* First published in *Journal of Symbolic Logic*, vol. 17, pp 105—116.

Received March 5, 1951.

[1] A term introduced in [2] as a translation of the word *mehrsortig* used in [1]. I wish to thank Professor Alonzo Church for first calling my attention to [1].

[2] A (one-sorted) elementary logic is the usual complete theory of truth functions and quantifiers (e.g. as it is formulated on p.88 of [3]) with its formulas as specified in a one-sorted axiomatic system.

[3] n refers to the number of kinds of variables in the system. We assume that n may be 2, 3, 4, $\cdots$, or ω. T_ω will be a theory with denumerably many kinds of variables.

[4] That amounts to the law of reflexivity and the principle of substitutivity for the variables of the system.

$$-(\urcorner x)\varphi x - for (\exists\ y)((x)(x=y\equiv\varphi x). - y -).$$

But we also know that once we have descriptions at hand, we can make use of additional predicates to get rid of the primitive names[5] and functors.[6] On this ground we shall assume, for simplicity, that the systems T_n which we shall consider contain neither names nor functors. In other words, we shall assume that the primitive symbols of T_n are just the truth-functional connectives, the quantifiers, the brackets, and the predicates.

We can describe each theory T_n as follows. There is at least one predicate. There are variables of different kinds: $x_1, y_1, z_1, \cdots$ (variables of the first kind); $x_2, y_2, z_2, \cdots; \cdots; x_n, y_n, z_n, \cdots$. Each k-placed ($k=1, 2, \cdots$) predicate with its places filled up by variables of the *proper* kinds is a formula (an atomic formula); and if φ and ψ are any formulas and α is a variable of any kind, then $(\alpha)\varphi$ and $\varphi\downarrow\psi$ are formulas[7]. In general, for each place of a predicate more than one kind of variable may be proper. However, to simplify our considerations, we shall always assume that each place of every predicate is to be filled up by one and only one kind of variable. Free and bound variables and occurrences will be understood as having been defined in the usual manner. A statement is a formula containing no free variables. A closure of a formula φ is a statement formed from φ by prefixing distinct general quantifiers to all the free variables of φ in an arbitrary order. We write $\vdash \varphi$ to mean that the closures of φ are theorems. Then a many-sorted elementary logic L_n is determined in the following manner. The formulas of L_n (φ, ψ, φ', etc.) are just those given above and the theorems of L_n are defined by the principles 1_n—5_n:[8]

1_n. If φ is a truth-functional tautology, then $\vdash \varphi$.

2_n. $\vdash (\alpha)(\varphi\supset\psi)\supset((\alpha)\varphi\supset(\alpha)\psi)$.

3_n. If α is not free in φ, then $\vdash \varphi\supset(\alpha)\varphi$.

4_n. If α and α' are variables of the same kind, and φ' is like φ except for containing free occurrences of α' whenever φ contains free occurrences of α, then $\vdash (\alpha)\varphi\supset\varphi'$.

5_n. If $\varphi\supset\psi$ and φ are theorems of L_n, so is ψ.

By adding certain proper axioms (or also axiom schemata) to L_n, we obtain a system T_n.

As an alternative way, we may also formulate a system involving several categories of fundamental objects using merely one kind of variables which have the sum of all the categories as their range of values. The simplest way to bring in the distinction of categories is to introduce n one-place predicates $S_1, S_2, \cdots, S_n$ such that x belongs to the i-th category if and only if $S_i(x)$. We can then set up a one-sorted theory $T_1^{(n)}$

---

[5] See [3], pp.149—152.

[6] See [4], vol. 1, pp.460—462.

[7] We shall follow [3] in using Greek letters as syntactical variables for expressions. The letters φ, ψ, χ and their accented and subscripted variants will be used to refer to formulas, and the letters α, β, γ, δ, and their variants to variables (cf. [3], p.75). Indeed, in formulating the system T_n, we are following closely the pattern set up in [3]. We shall omit the corners used in [3].

[8] These principles answer to *100, *102—*105 of [3]. A principle answering to *101 can be dropped just as in L_1; see [3], p.89.

corresponding to T_n in the following manner. In $T_1^{(n)}$ the atomic formulas are determined by the predicates of T_n plus $S_1, \cdots, S_n$ with their places all filled up by general variables. Formulas, etc. can be defined in $T_1^{(n)}$ in the usual way. And $T_1^{(n)}$ contains a usual one-sorted elementary logic L_1 determined by[9] five principles 1_1—5_1 which are similar to 1_n—5_nbut are concerned with formulas and variables of $T_1^{(n)}$. Then we under-stand by the elementary logic $L_1^{(n)}$ the system obtained from L_1 by adding the following additional principle:

6_1. For every $i (i=1, \cdots, n)$, $(\exists\, \alpha) S_i(\alpha)$ is a theorem.

And we introduce a rule for translating between statements of L_n and those of $L_1^{(n)}$:

RT. A statement φ' in $L_1^{(n)}$ and a statement φ in L_n are translations of each other if and only if φ' is the result obtained from φ by substituting simultaneously, for each expression of the form $(x_i)(-x_i-)$ in φ $(i=1, \cdots, n)$, an expression of the form (x) $(S_i(x) \supset (-x-))$ (with the understanding that differferent variables in φ are replaced by different variables in φ').

Using this rule, we see that every statement of L_n has a translation in $L_1^{(n)}$, and some (although not all) statements of $L_1^{(n)}$ have translations in L_n. In particular, the proper axioms of T_n all have translations in $L_1^{(n)}$, and $T_1^{(n)}$ is just $L_1^{(n)}$ plus the translations of these proper axioms of T_n.

The main purpose of this part is to investigate the relations between any T_n (or L_n) and its corresponding $T_1^{(n)}$ (oor $L_1^{(n)}$). By a comparative study of L_n and L_1, we shall also indicate that many known metamathematical results about a usual elementary logic L_1 have counterparts for L_n.

Preparatory to stating the results of this part, we first make a few historical remarks. In [5] Herbrand states a theorem which amounts to the following (see [5], p.64):

(I) A statement of any system T_n is provable in T_n if and only if its translation in the corresponding system $T_1^{(n)}$ is provable in $T_1^{(n)}$.

However, the proof he gives there is inadequate, failing to take into account that there are certain reasonings which can be carried out in $L_1^{(n)}$ but not in L_n. In [1], Arnold Schmidt points this out and devotes his paper to giving a careful proof of the theorem. Then Langford puts forward in [2] (a review of [1]) the problem whether the following is true in general:

(II) If a system T_n is consistent, then the corresponding system $T_1^{(n)}$ is also consistent.

This, as Professor Bernays has communicated to us in conversation, can be answered positively by the following argument. Obviously there exists a statement φ of $T_1^{(n)}$ such that both φ and $\sim\varphi$ are translatable into T_n. Assume that $T_1^{(n)}$ is inconsistent. Then every statement in $T_1^{(n)}$ is provable, and therefore φ and $\sim\varphi$ are both provable in $T_1^{(n)}$. Hence, by (I), their translations ψ and $\sim\psi$ according to RT are both provable in T_n, Hence, T_n is inconsistent.

In this part, we shall first indicate that in L_n we can easily prove counterparts of

---

9 Compare the preceding footnote. This time φ, ψ refer to formulas of T_1^n, and α, α' to variables of T_1^n.

theorems in L_1 and that about L_n we can prove counterparts of the metamathematical theorems of completeness, etc. about L_1. We shall then show that from these the theorem (I) (and therewith the theorem (II) follows. We shall also show that, conversely, given (I) and the metamathematical theorems about L_1, we can prove certain similar theorems about L_n as corollaries. In passing, we may mention here that the following converse of (II) is obviously true:

(III) If $T_1^{(n)}$ is consistent, then T_n is.

It would then seem that, merely for the purpose of proving (I), we could dispense with Schmidt's rather involved arguments. However, Schmidt actually proves in his paper the following more interesting theorem:

(IV) Given a statement of T_n and a proof for it in T_n, there is an effective way of finding a proof in $T_1^{(n)}$ for its translation in $T_1^{(n)}$; and, conversely, given a statement of $T_1^{(n)}$ which has a translation in T_n, and given a proof for it in $T_1^{(n)}$, there is an effective way of finding a proof in T_n for its translation in T_n.

Although we can prove (I) by considering the completeness of L_1 and L_n, it does not seem possible to prove (IV) similarly, for (IV) depends on syntactical considerations about the proofs in $L_1^{(n)}$ and L_n. We shall, following a suggestion of Professor Bernays, give a simpler alternative proof for (IV) by application of Herbrand's theorem. (See [4], vol.2, pp.149—163.)

From the results (I), (II), and (III), we see that for purposes of questions concerned with the consistency of T_n, we may consider $T_1^{(n)}$ instead which is simpler in that it contains only one kind of variables. However, $T_1^{(n)}$ is more complicated than T_n in that it contains new predicates $S_1, S_2, \cdots, S_n$. We contend that in many cases, given a system T_n, we can find a corresponding system which contains only one kind of variables and no new predicates, and which can serve the same purposes both for the study of consistency questions and for the development of theory. Whether we can find such a corresponding system depends on whethere we can express membership in the different categories by the following means: general variables (whose range of value is the sum of all the special domains), the quantifiers and truth-functional connectives, the brackets, plus the predicate letters of the given many-sorted theory reconstrued as having their argument places filled up by general variables. It seems that in most cases we can. The simple theory of types will afford an example of T_n for which we can give a corresponding theory relatively consistent to it, with one kind of variable and no new predicates, and essentially as rich. This example is of special interest if we want to compare the theory of types with Zermelo's set theory.

2. The many-sorted elementary logics L_n

In this section we shall sketch how theorems in and about L_n can be proved in a similar manner to theorems in and about L_1.

We first observe that in L_n we can prove from 1_n—5_n all the usual quantificational theorems of L_1 for each kind of variables. For example, we can prove in L_n all theorems which fall under principles notationally the same as *110—*171 of [3] with nearly the same proofs.[10]

Thus[11] we can define prenex normal form and Skolem normal form for L_n and prove the laws of them for L_n just as for L_1. We can prove the deduction theorem and the consistency theorem for L_n just as for L_1.

Likewise we can define valid and satisfiable formulas of L_n just as those of L_1:

2.1. A value assignment for a predicate or its corresponding atomic formula $fx_{n_1}^{(k_1)} \cdots x_{n_j}^{(k_j)}$ of L_n over a set of n non-empty domains is a function from the predicate or its corresponding atomic formula to a j-adic relation whose i-th place takes the individuals of the n_i-th domain.

2.2. A formula φ of L_n with no free variables is valid in a particular set of n non-empty domains if all value assignments for all the atomic formulas occurring in φ are such that, under the normal interpretation of the truth-functional connectives and quantifiers, φ becomes true. φ is valid if it is valid in all sets of n non-empty domains.

2.3. φ is satisfiable in a particular set of n non-empty domains if $\sim\varphi$ is not valid in it. φ is satisfiable if it is satisfiable in some set of n non-empty domains.

With these definitions we can prove the following theorems[12] for L_n just as for L_1.

2.4. If $\vdash \varphi$ in L_n, then the closure of φ is valid.

2.5. If the closure of φ is valid in a set of n denumerable domains, then $\vdash \varphi$ in L_n.

2.6. If the closure of φ is valid, then $\vdash \varphi$ in L_n.

2.7. If $\varphi_1, \varphi_2, \cdots$ are statements of L_n and the system T_n obtained from L_n by adding $\varphi_1, \varphi_2, \cdots$ as proper axioms is consistent, then $\varphi_1, \varphi_2, \cdots$ are simultaneously satisfiable in a set of n denumerable domains.

We merely outline a proof for the following theorem 2.8 from which 2.5 follows immediately.

2.8. If the statement $\sim\varphi$ is not provable in L_n, then φ is satisfiable in a set of n denumerable domains.

Suppose that the variables of the p-th kind ($p=1, \cdots, n$) in L_n are $v_p^{(1)}, v_p^{(2)}, \cdots$ and φ is the statement $(v_{n_1}^{(k_1)}) \cdots (v_{n_t}^{(k_t)})(\exists\ v_{m_1}^{(j_1)}) \cdots (\exists\ v_{m_s}^{(j_s)})\psi(v_{n_1}^{(k_1)}, \cdots, v_{n_t}^{(k_t)}; v_{m_1}^{(j_1)}, \cdots, v_{m_s}^{(j_s)})$. Let ψ_i ($i=1, 2, \cdots$) be $\psi\,(v_{n_1}^{\tau(i,1)}, \cdots, v_{n_t}^{\tau(i,t)}; v_{m_1}^{\sigma(i,1)}, \cdots, v_{m_s}^{\sigma(i,s)})$, where $(\tau(i, 1), \cdots, \tau(i, t)$ is the i-th term of

---

[10] In 4_n and theorems answering to cases of *134 of [3] which are concerned with the relation between free and bound variables, we need the condition that the variables are of the same kind.

As we come to the proofs, the only places where we need take somewhat seriously into consideration the different kinds of variables are in the proofs of the generalized modus ponens answering to *111 of [3] and the principles of generalization answering to *112 of [3]. But in both cases, proofs for these principles in L_n are easily obtainable by slightly changing the proofs of *111 and *112 in [3]. In particular, in L_n we can prove $\vdash (\alpha)(\beta)\varphi \equiv (\beta)(\alpha)\varphi$, no matter whether α, β are of the same kind or not.

[11] Cf. [6], pp.59—61, pp.68——72, pp.45—46, pp.42—44.

[12] The proof for 2.4 is easy and 2.6 follows from 2.5 as an immediate corollary. 2.7 can be proved by using arguments resembling those for 2.5. (Compare the proof of its counterpart for L_1 on pp.357—359 of [7].) The proof for 2.5 sketched below resembles that for the completeness of L_1 (cf.[6], pp.73—79) except for certain minor complications in connection with the ordering of variables and the assignment of truth values to atomic formulas.

the sequence of all the t–tuples of positive integers ordered according to the sum of the t integers and, for those with the same sum, lexicographically; and the sequence of the s–tuples $(\sigma(i, 1), \cdots, \sigma(i, s))$ $(i=1, 2, \cdots)$ is such that, if among $m_1, \cdots, m_s$, m_{r1} is identical with $m_{r2}, \cdots, m_{rq}$ and with no others, then $\sigma(1, r_1), \cdots, \sigma(1, r_q), \sigma(2, r_1), \cdots, \sigma(2, r_q), \sigma(3, r_1), \cdots$ coincide with $1, \cdots, q, (q+1), \cdots, 2q, (2q+1), \cdots$.

In order to prove 2.8, we observe first that we can prove just as in the case of L_1 the following two propositions.

2.8.1. If $\sim\varphi$ is not provable in L_n, then none of $\sim\psi_1, \sim\psi_1 \vee \sim\psi_2, \cdots$ is a tautology.

2.8.2. If none of $\sim\psi_1, \sim\psi_1 \vee \sim\psi_2, \cdots$ is a tautology, then $\psi_1, \psi_2, \cdots$ are simultaneously satisfiable.

Therefore, by correlating each variable $v_j^{(k)}$ in $\psi_1, \psi_2, \cdots$ with the j–th power of the k–th prime number, we can, similarly as in the case of L_1, provide a true interpretation for φ in the set of the domains $D_1, D_2, \cdots$ such that D_j is the set of the j–th powers of all the prime numbers. Hence, 2.8 and 2.5 can be proved.

We note in passing that we can also avoid the complications regarding the definitions of $\psi_1, \psi_2, \cdots$ and prove 2.8 more simply by treating, for any i, j, k, $v_i^{(k)}$ and $v_j^{(k)}$ as the same in our considerations. Then we can use almost completely the arguments for L_1 to give a true interpretation for φ in a set of n identical domains, each being the set of positive integers.

Since in many cases we want the different categories (e.g., points, lines, and planes, etc.) to be mutually exclusive, we might think that in such cases there should be no satisfying assignments with all the domains identical. However, the possibility just indicated shows that this is not the case. Indeed, it becomes clear that there is no means to express in L_n explicitly the requirement that the domains of any satisfying assignment for φ must be different. Such a requirement is merely one of the implicitly understood conditions which we want a normal interpretation of the theory to fulfill. But there is nothing in the definitions of the satisfying assignments of values to preclude cases where such informal conditions are not fulfilled. In a one-sorted theory we can add axioms such as $\sim(\exists\, x)(S_i(x).\, S_j(x))$ to make the demand explicit (compare Langford [2]), because in the value assignments we insist that the truth-functional and quantificational operators retain their normal interpretations.

3. The theorem (*I*) and the completeness of L_n

From the completeness of L_n, we can derive the theorem (I) stated in section 1.

Let us consider a statement φ in T_n and its translation φ' in $T_1^{(n)}$. Suppose that the variables in φ are all among the m_1-th, $\cdots$, and the m_k-th kinds. If A is a value assignment for $\sim\varphi$ in a set D of domains, then there is an associated assignment A' for $\sim\varphi'$ in the sum D' of all the domains of the set D, such that $(\exists\, x)\, S_{m1}. \cdots. (\exists\, x)\, S_{mk}(x)$ receives the value truth and that all the other predicate letters in $\sim\varphi'$ receive, for those entities of D' which belong to the proper domains of D, the same values as within A and, for all the other entities of D', receive (say) the value falsehood. Conversely, given an assignment

A' for $\sim\varphi'$ in a domain D' such that $(\exists\ x)S_{m1}(x).\ \cdots.\ (\exists\ x)S_{mk}(x)$ receives the value truth, there is an associated A for $\sim\varphi$ such that the m_i-th ($i=1, \cdots, k$) domain consists of the things x such that $S_{m_i}(x)$ receives the value truth in A' and all the predicate letters of $\sim\varphi$ receive the same values as in A'. Obviously in either case, A satisfies $\sim\varphi$ if and only if A' satisfies $\sim\varphi'$. Hence, we have: $\sim\varphi$ is satisfiable if and only if $(\exists\ x)S_{m1}(x).\ \cdots.\ (\exists\ x)S_{mk}(x).\ \sim\varphi'$ is. Therefore, we have:

3.1. φ is valid if and only if $(\exists\ x)S_{m1}(x).\ \cdots.\ (\exists\ x)S_{mk}(x).\ \supset\varphi'$ is.

Therefore, we can prove:

3.2. φ is provable in L_n if and only if φ' is provable in $L_1^{(n)}$.

Proof. If φ is provable in L_n, then, by 2.4, it is valid. Hence, by 3.1, $(\exists\ x)S_{m1}(x).\ \cdots.\ (\exists\ x)S_{mk}(x).\ \supset\varphi'$ is valid and therefore, by the completeness of L_1, provable in L_1. Hence, by 6_1, φ' is provable in $L_1^{(n)}$.

Conversely, if φ' is provable in $L_1^{(n)}$, we can assume that all the (finitely many) cases of 6_1 used in the proof for φ' in $L_1^{(n)}$ are among $(\exists\ x)S_{m1}(x), \cdots, (\exists\ x)S_{mk}(x)$, for we can so choose $m_1, \cdots, m_k$. Therefore, by the deduction theorem for L_1, $(\exists\ x)S_{m1}(x).\cdots.\ (\exists\ x)S_{mk}(x).\ \supset\varphi'$ is provable in L_1 and therefore valid. Hence, by 3.1, φ is valid and, by 2.6, provable in L_n.

From 3.2, the theorem (I) follows immediately by the deduction theorems for L_1 and L_n. Conversely, given (I) we can also derive 3.2. Moreover, as noted in section 1, the theorem (II) stated there is a corollary of (I). Now we prove that 2.6 and 2.7 can be inferred, with the help of (I), from their corresponding theorems for L_1.

Proof of 2.6. If φ is valid, then by 3.1, $(\exists\ x)S_{m1}(x).\cdots.\ (\exists\ x)S_{mk}(x).\ \supset\varphi'$ is valid and therefore, by the completeness of L_1, provable in L_1. Hence, by 6_1, φ' is provable in $L_1^{(n)}$. Hence, by 3.2, φ is provable in L_n.

Proof 2.7, Assume that the system T_n obtained from L_n by adding the statements $\varphi_1, \varphi_2, \cdots$ of L_n as proper axioms is consistent. By (II), the system $T_1^{(n)}$ corresponding to T_n is consistent. Hence, by the theorem for L_1 corresponding to the theorem 2.7 for L_n, all the axioms of $T_1^{(n)}$ are simultaneously satisfiable in a denumerable domain. But the axioms of $T_1^{(n)}$ are just those of L_1, the axioms $(\exists\ x)S_1(x),\cdots, (\exists\ x)S_n(x)$, and the translations $\varphi_1, \varphi_2, \cdots$ of $\varphi_1, \varphi_2, \cdots$. Hence we can divide the domain into n domains such that the i-th domain consists of all the individuals x such that $S_i(x)$ is true. In this way, we obtain a set of n non-empty domains each either finite or denumerable in which both φ_1, $\varphi_2, \cdots$ and the axioms of L_n are satisfiable (compare the arguments in the proof of 3.1). Consequently, we can find a set of n denumerable domains in which T_n is satisfiable. And the proof of 2.7 is completed.

4. Proof of the theorem (IV)

We may break up the theorem (IV) into two parts.

4.1. There is an effective process by which, given any proof in T_n for a statement φ of T_n, we can find a proof in $T_1^{(n)}$ for the translation φ' of φ in $T_1^{(n)}$.

4.2. There is an effective process by which, given any proof in $T_1^{(n)}$ for a statement φ' of $T_1^{(n)}$ which has a translation φ in T_n, we can find a proof in T_n for r.

First, we prove 4.1. In the proof of φ, we employ only a finite number of the proper axioms of T_n. Let the conjunction of these axioms be Φ. By the deduction theorem, we have an effective process by which, given the proof of φ in T_n, we can find a proof of $\Phi \supset \varphi$ in L_n. And if its translation $\Phi' \supset \varphi'$ has a proof in $L_1^{(n)}$, then we have immediately a proof in $T_1^{(n)}$ for φ' by modus ponens and the proper axioms of $T_1^{(n)}$, because Φ' is the translation of the conjunction of certain proper axioms of T_n. Hence, we need only prove that there is an effective process by which, given a proof in L_n for a formula ψ of T_n, we cna find a proof in $L_1^{(n)}$ for its translation ψ' in $L_1^{(n)}$.

By arguments like those used in proving *100', *102'—*105' in [8], we can prove as metatheorems in $L_1^{(n)}$ the translations of 1_n—5_n for each kind of variables in $L_1^{(n)}$. Since in each proof of L_n, we use only a finite number of special cases of 1_n—5_n, given any proof in L_n for a formula ψ of L_n, we have a proof for its translation ψ' in $L_1^{(n)}$ which consists of the proofs of the translations in $L_1^{(n)}$ of these special cases together with a translation in $L_1^{(n)}$ of the proof for ψ in L_n. Hence, 4.1 is proved.

The proof of 4.2 is more complex. We note that it is sufficient to prove the following theorem.

4.3. There is an effective process by which, given any proof in $L_1^{(n)}$ for a statement χ' of $T_1^{(n)}$ which has a translation χ in T_n, we can find a proof in L_n for χ.

Thus, let φ' be a statement of $T_1^{(n)}$ with a proof in $L_1^{(n)}$, then, by the deduction theorem for L_1, we have a proof for $\Phi' \supset \varphi'$ in $L_1^{(n)}$, Φ' being the conjunction of the proper axioms of $T_1^{(n)}$ used in the given proof of φ'. Hence, by 4.3, we have a proof in L_n for the translation $\Phi \supset \varphi$ of $\Phi' \supset \varphi'$ in L_n, and thereby also a proof for φ in T_n.

Consequently, given 4.3, we can prove 4.2. We shall prove 4.3.

By hypothesis, a proof Δ' in $L_1^{(n)}$ is given for a statement χ' of $T_1^{(n)}$ which has a translation χ in T_n. Our problem is to find a proof Δ in L_n for the translation χ of χ' in T_n. In what follows, we shall assume that χ' has been given in such a form that its translation χ is in the prenex normal form. Accordingly, since for each variable α and each formula ψ of $T_1^{(n)}$ we can substitute $(\exists\ \alpha)(S_i(\alpha).\psi)$ for $\sim(\alpha)(S_i(\alpha) \supset \sim\psi)$, each quantification in χ' is either of the form $(\alpha)(S_i(\alpha) \supset \varphi)$ or of the form $(\exists\ \beta)(S_j(\beta).\ \phi')$, where α ared β are variables in $T_1^{(n)}$, φ and φ' are formulas in $T_1^{(n)}$, and i and j are among $1, \cdots, n$. Moreover, every formula $S_i(\alpha)$ occurs, if at all in χ', in one and only one context either of the form $(\alpha)(S_i(\alpha) \supset \varphi)$ or of the form $(\exists\ \alpha)(S_i(\alpha).\varphi)$; and every variable α occurs, if at all in χ', in one unique part either of the form $(\alpha)(S_i(\alpha) \supset \varphi)$ or of the form $(\exists\ \alpha)(S_i(\alpha).\varphi)$. Such an assumption as to the form of χ' does not restrict our result in any way, because we know that each statement of $T_1^{(n)}$ which has a translation in T_n can be converted into such a form by procedures analogous to those for transforming a statement into the prenex normal form.

Therefore, if we associate each occurrence of a variable α with the number i when there is a formula $(\alpha)(S_i(\alpha) \supset \varphi)$ or a formula $(\exists\ \alpha)(S_i(\alpha).\varphi)$ occurring in χ', we see that each occurrence of a variable in χ' is associated with a unique number, and two occurrences of the same variable in χ' always have the same number.

Consider now the formula χ_1 obtained from χ' by dropping all parts of the forms $S_i(\alpha) \supset$ and $S_i(\alpha).$, or, in other words, by replacing each quantification of the form

$(\alpha)(S_i(\alpha) \supset \varphi)$ by $(\alpha)\varphi$, and each quantification of the form $(\exists\ \alpha)(S_i(\alpha).\varphi)$ by $(\exists\ \alpha)\varphi$. We see that χ_1 no longer contains occurrences of atomic formulas of the form $S_i(\alpha)$, and that χ_1 is like the translation χ of χ' in T_n except for containing occurrences of variables (say) $x, y, \cdots, z$ which are associated with the numbers $i,j,\cdots,k$ where χ contains occurrences of $x_i, y_j, \cdots, z_k$. Moreover, χ_1 is also in the prenex normal form. From now on we understand that each occurrence of any variable in χ_1 is associated with the unmber which was given to its corresponding occurrence in χ'.

Let us say that an occurrence of a variable (in a proof of L_1) is associated with the proper number if its number is exactly the number for the kind of variable which is to fill up the place in question of the predicate of T_n that occurs with the variable. For example, an occurrence of α in a context $P\alpha\beta\cdots\gamma$ is said to be associated with the proper number, if α is associated with i and the first argument place of P is to be filled up by the i-th kind of variable in T_n. From this definition and the way numbers are associated with variable occurrences, we have, since χ' has a translation in T_n, the next theorem.

4.4. Each occurrence in χ_1 of any variable is associated with the proper number.

We prove another theorem.

4.5. Given the proof Δ' in $L_1^{(n)}$ for χ', we can actually write out a proof Δ_1 in L_1 for χ_1.

Proof. In Δ' each line is either a case of 1_1—4_1 or 6_1, or a consequence by 5_1 of two previous lines. Let us replace throughout Δ' all occurrences of all formulas of the form $S_i(\alpha)$ by those of formulas of the form $S_i(\alpha) \vee \sim S_i(\alpha)$. Then, in the result Δ'', each line which was a case of 6_1 becomes an easy consequence of 1_1—5_1. If we add the easily obtainable proofs for these cases of 6_1 at the top of Δ'', tnen we obtain a proof in L_1 for a conclusion χ'' which is like χ' except for containing occurrences of formulas of the form $S_i(\alpha) \vee \sim S_i(\alpha)$ instead of those of the form $S_i(\alpha)$. But, it is then easy to see that from a proof for χ'' in L_1, we can obtain a proof in L_1 for χ_1 by 1_1 and the principle of the substitutivity of biconditionals. Hence, we obtain a proof Δ_1 in L_1 for χ_1.

Now let us apply Herbrand's theorem (see [4], vol.2, pp. 149—163, especially p. 158; cf. also bottom of p. 135) which for our purpose can be stated thus:

HT. There is an effective method which, for any given proof of L_1 for a statement ψ in prenex normal form, yields a new proof II for ψ (ψ being therefore the last line of II) whose first line is a truth-functional tautology and each of whose other lines is obtained from its immediate predecessor by applying one of the following three rules: (1) Given a formula of L_1 which has the form of an alternation (disjunction), we can replace an alternation clause $\varphi\beta$ by $(\exists\ \alpha)\varphi\alpha$ where α is an arbitrary variable; (2) Given a formula of L_1 which has the form of an alternation, we can replace an alternation clause $\varphi\beta$ by $(\alpha)\varphi\alpha$ where β is a variable not free in any other parts of the formula; (3) Given a formula of L_1 which has the form of an alternation, we can omit repetitions of an alternation clause.

It is easy to convince ourselves that the proof II for ψ as specified in HT is again a proof in L_1 or, more exactly, that from II (as given) we can easily construct a proof of L_1 with ψ as the last line. Let us refer to proofs for an arbitrary statement ψ which are of the kind as described in HT, as proofs of L_1 in the Herbrand normal form. Then the content of HT says simply that every proof of L_1 for a statement in the prenex normal form can be transformed into one in the Herbrand normal form.

By 4.5 and HT, since χ_1 is in the prenex normal form, we can actually find a proof II of L_1 for χ_1 in the Herbrand normal form. Suppose given such a proof II. Our problem is to construct from II a proof Δ of L_n with χ as its last line.

As was mentioned above, each occurrence in χ_1 of any variable is associated with a definite number, which is, more over, according to 4.4, the proper number. Using these correlations, we can now associate every occurrence in II of any variable with a definite number in the following manner.

4.6. If the occurrence is in a line φ which is followed by a line φ', then it is associated with the same number as the corresponding occurrence of the same variable in φ' except for the following special cases:

4.6.1. If φ' is obtained from φ by substituting $(\alpha)\psi\alpha$ for an alternation clause $\psi\beta$ and the occurrence in φ is one of the variable β in the clause $\psi\beta$, then it is associated with the same number as the corresponding occurrence of the variable α in the part $\psi\alpha$ of φ'.

4.6.2. Similarly for the case with a particular quantification $(\exists\ \alpha)\psi\alpha$ in φ'.

4.6.3. If φ' is obtained from φ by omitting repetitions of an alternation clause φ_1 and the occurrence in φ is in some occurrence of φ_1, then it is associated with the samenumber as the corresponding occurrence in the alternation clause φ_1 of φ'.

Let us replace every occurrence in II of a variable associated with the number i by an occurrence of a corresponding variable of the i-th kind in T_n (for instance, if an occurrence of x is associated with i in II, replace it by an occurrence of x_i) and refer to the result as Δ_2. We easily see that the last line of Δ_2 is exactly χ, the translation of χ' in T_n. Moreover, each line of Δ_2 is a formula of T_n which is either a truth-functional tautology or follows from its immediately preceding line by a quantificationally valid rule of inference (a rule of inference derivable in L_n). Therefore, from Δ_2 we can easily construct a proof Δ of L_n for the conclusion χ.

This completes the proof of 4.3. Therefore, 4.2 and theorem (IV) (using 4.1) are all proved.

5. The simple theory of types

We consider the system P which Gödel uses in [9].

Roughly, P contains as primitives the truth-functional operators, the quantifiers, the membership predicate $\in$, the symbol 0 for zero, the symbol f for the successor function, and infinitely many kinds of variables: $x_1, y_1, \cdots; x_2, y_2, \cdots; \cdots$. The predicate $\in$ occurs only in contexts of the form $x_n \in y_{n+1}$, etc. ($n=1, 2, \cdots$). The axioms and rules of inference of P may be stated as follows ($x_n = y_n$ standing for $(z_{n+1})(x_n \in z_{n+1} \equiv y_n \in z_{n+1})$).

A. The principles $1_\omega - 5_\omega$ of elementary logic for the infinitely many kinds of variables.

B. Axioms for the individuals.

1. $\vdash \sim fx_1 = 0.$
2. $\vdash fx_1 = fy_1 \supset x_1 = y_1.$
3. $\vdash 0 \in x_2 \cdot (x_1)(x_1 \in x_2 \supset fx_1 \in x_2). \supset y_1 \in x_2.$

C. Principles of extensionality ($n = 1, 2, \cdots$).

$$\vdash (Z_n)(z_n \in x_{n+1} \equiv z_n \in y_{n+1}) \supset x_{n+1} = y_{n+1}.$$

D. Principles of class existence. Let φ be any formula in which y_{n+1} is not free, then $\vdash (\exists\ y_{n+1})(x_n)(x_n \in y_{n+1} \equiv \varphi)$ ($n = 1, 2, \cdots$).

We want to show that if P is consistent, then the following system Q is also consistent. Q contains merely one kind of variable $x, y, z, \cdots$. In Q we can introduce different kinds of variables corresponding to those of P:

$x = y$	for $(z)(x \in z \equiv y \in z)$.
$t1(x)$	for $x = 0 \vee (\exists\ y)(x = fy)$.
$t(n+1)(x)$	for $(y)(y \in x \supset tn(y)). \sim t1(x)$.
	($n = 1, 2, \cdots$).
$(x_n)\varphi x_n$	for $(x)(tn(x) \supset \varphi x)$.

The axioms of Q are:

A′. The principles $1_1 - 5_1$ of elementary logic for the variables and the formulas of the system.

B′ – D′ are notationally the same as B-D of the system P.

We shall not attempt to provide a formal proof. For example, we shall retain the numeral 0 and the functor f instead of replacing them by descriptions and speak of models for them as well as those for predicates and theories. However, our arguments below, we hope, will make it clear that there is no difficulty in the way of rendering the proof more rigorous.

If the system P is consistent then, by theorem 2.7, it is satisfiable in a set of denumerably many denumerable domains. Assume that such a set $M_1 = \{D_1, D_2, \cdots\}$ is given, where D_i contains the models of the objects of the type i. Obviously D_1 must contain the models 0^*, $(f0)^*$, $(ff0)^*$, $\cdots$ of the terms $0, f0, ff0, \cdots$. Of course D_1 may also contain other things besides them. Let E_1 be the subset $\{0^*, (f0)^*, (ff0)^*, \cdots\}$ of D_1 and F_1 be the set consisting of all members of D_1 not belonging to E_1.

Let E_2 be the subset of D_2 such that if a belongs to E_2, then every member b of a(i.e., every b such that in the model $b \in^* a$ receives the value truth) belongs to E_1 and F_2 be its complement in D_2. Similarly let E_3 be the subset of D_3 consisting of all those elements of D_3 which are subsets of E_2 and F_3 be its complement in D_3. And so on.

We delete the sets $F_1, F_2, \cdots$ from M_1 and keep merely the domains $E_1, E_2, \cdots$ together with value assignments in M_1 which merely relate to these domains. It is not hard to see that the result $M_2 = \{E_1, E_2, \cdots\}$ is again a model for P. For, as we can easily check, if M_1 satisfies the axioms of groups B–D, then M_2 also satisfies them.

Moreover, since in the system P, $\alpha \in \beta$ is meaningful only when β is of one type higher than α, we may also, for instance so choose $\in^*$ that $a \in^* b$ can be true only when a and b are of two domains E_k and E_{k+1} respectively. For, since the axioms of P only involve meaningful formulas, a model for P remains one for it when we change the truth values which $a \in^* b$ may take for a and b in other domains. Let us assume that we have given such a model M_3 for the system P.

Then, if we take the sum class K of the domains $E_1, E_2, \cdots$ of M_3 as the range of values of the variables $x, y, z, \cdots$ of the system Q and use the same relations $\in^*$ as in M_3, then we have a model for Q. Thus, the variables $x_1, y_1, \cdots, x_2, y_2, \cdots$ of the various types introduced in Q by the contextual definitions can easily be seen to have the same ranges of values $E_1, E_2, \cdots$ as the variables of the system P. Hence, we obtain a model for Q because all the axioms of Q except those of group A′ remain notationally the same as in the system P, and obviously the axioms A′ are satisfied by the model. Therefore, if P is consistent, then Q is.

We want to thank Professor Bernays whose valuable suggestions have enabled us to get rid of many of the shortcomings in an earlier version of the paper.

References

[1] Arnold Schmidt, *Über deduktive Theorien mit mehreren Sorten von Grunddingen, Mathematische Annalen*, vol. 115(1938), pp.485—506.

[2] C.H. Langford, Review of [1], this JOURNAL, vol.4 (1939), p.98.

[3] W.V. Quine, *Mathematical logic*, 2nd printing, Cambridge 1947.

[4] D. Hilbert and P. Bernays, *Grundlagen der Mathematik*, vol.1, Berlin 1934; vol.2, Berlin 1939.

[5] Jacques Herbrand, *Recherches sur la théorie de la démonstration*, Dissertation, Paris 1930.

[6] Alonzo Church, *Introduction to mathematical logic*, Princeton 1944.

[7] Kurt Gödel, *Die Vollständigkeit der Axiome des logischen Funktionenkalküls, Monatshefte für Mathematik und Physik*, vol.37 (1930), pp.349—360.

[8] Hao Wang, *Existence of classes and value specification of variables*, this JOURNAL, vol.15 (1950), pp.103—112.

[9] Kurt Gödel, *Über formal unentsheidbare Sätze der Principia Mathmatica und verwandter Systeme I, Monatshefte für Mathematik und Physik*, vol.38 (1931), pp.173—198.

14.2 Ackermann's consistency proof *

1. The System Z_a

Apart from Gentzen's celebrated consistency proof of the system Z with his special formulation of the predicate calculus (*Math. Annalen*, **112**(1936), 493—565 and *Forschungen zur Logik und etc.*, no. 4, 1938, 19—44), there are also alternative proofs by Ackermann (*Math. Annalen*, **117** (1940), 162—194) and Schütte (*Math. Annalen*, **122** (1951), 369—389). In this chapter, we give an exposition of Ackermann's proof which has certain interesting applications.

Ackermann's formulation is in many ways very elegant. We quote it in full.

Basic symbols: 0, ′, —1, +, ·, =, ~, ⊃, variables x, y, etc., ε–operators εx, εy, etc.

Formulae and terms: 0 is a term, a variable is a term, if a and b are terms, so are a', $a—1$, $a+b$, ab, if $A(a)$ is a formula, $\varepsilon_x A(x)$, etc. are terms; if a, b are terms, $a=b$ is a formula, if A, B are formulae, so are $\sim A$, $A \supset B$.

* First published in *A Survey of Mathematical Logic*, by H. Wang, pp 362—375. Science Press, Beijing, 1962. Reproduced by permission.

There are three groups of axiom schemata with modus ponens as the single rule of inference. No free variables are to appear in any axioms or proofs. In particular, in the rule of modus ponens, viz. B if A and $A \supset B$, no free variables occur in A and B. We shall call a term or a formula closed if it contains no free variables, otherwise open.

I. Propositional calculus.

II. $p \supset (q \supset p)$.

I2. $(p \supset (q \supset r)) \supset ((p \supset q) \supset (p \supset r))$.

I3. $(\sim p \supset \sim q) \supset (q \supset p)$.

II. Number theory. If a, b, c are closed terms then

II1. $a = a$.

II2. $a' = b' \supset a = b$.

II3. $a \neq 0 \supset (a - 1)' = a$.

II4. $a + 0 = a$.

II5. $a + b' = (a + b)'$.

II6. $a0 = 0$.

II7. $ab' = ab + a$.

II8. $a = b \supset a' = b'$.

II9. $a = b \supset a - 1 = b - 1$.

II10. $a = b \supset a + c = b + c$.

II11. $a = b \supset c + a = c + b$.

II12. $a = b \supset ac = bc$.

II13. $a = b \supset ca = cb$.

III. The ε-operator.

III1. $A(a) \supset A(\varepsilon_x A(x))$.

III2. $A(a) \supset \varepsilon_x A(x) \neq a'$.

III3. $\sim A(\varepsilon_x A(x)) \supset \varepsilon_x A(x) = 0$.

III4. $a = b \supset \varepsilon_x A(x,a) = \varepsilon_x A(x,b)$.

The equivalence of this system with Z holds in the sense that a theorem not containing the ε-symbol is provable in Z if and only if it is provable in Z_a. For the "if" half, it is sufficient to recall the eliminability of the μ-operator in Z, and the possibility of identifying ε with μ. For the "only if" half, it is only necessary to derive the principle of mathematical induction from II3, III1, and III2 (Hilbert-Bernays II, p.85) and $(x)(x' \neq 0)$ from the definition of (x) in terms of the ε-symbol by III2 and III3.

The consistency proof aims at eliminating the ε-symbol to correlate every proof with a succession of true numerical formulae. For this purpose, a number of concepts are needed.

1.1. An ε-term is a term which begins with ε.

1.2. The principal variable of an ε-term $\varepsilon_x A(x)$ is x.

1.3. A term b is said to be subordinate to a term a, if a is an ε-term, b is a proper part of a, and the principal variable of a occurs in b. It follows that b is an open term.

1.4. A term b is said to reside in a term a, if b is a proper part of a but is not subordinate to a.

From 1.3 and 1.4, it follows:

1.5. If a is not an ε-term and b is a proper part of a, then b resides in a; if a is an ε-term and b is a proper part of a, then b either resides in a or is subordinate to a, but not both.

1.6. An occurrence of a term b is said to be a direct constituent of an ε-term a, if (i) it resides in a; (ii) it is not subordinate to any ε-term contained in a; (iii) it is not a proper part of any term residing in a.

For example, if a is $\varepsilon_y(x+2=\varepsilon_z(y+1=z))$, then $\varepsilon_z(y+1=z)$ and $y+1$ do not satisfy (i), z does not satisfy (ii), x and z do not satisfy (iii). But $x+2$ satisfies all.

1.7. The ε-category of an ε-term a, open or closed, is obtained from the term by substituting distinct free variables which do not occur in a for all direct constituents of a. When a has no direct constituent, it is its own ε-category. An alphabetic variant is regarded as the same ε-category.

For example, $\varepsilon_z(z+\varepsilon_y(y=3)=z'')$ is its own ε-category; $\varepsilon_y(u=\varepsilon_z(y+1=z))$ is the ε-category of the term a given above and also of $\varepsilon_y(\varepsilon_w(w=2)\ +3\ =\varepsilon_z(y+\varepsilon_v(v=2)=z))$. One could treat these ε-categories as functions of their free variables.

What is wanted in a direct constituent is a maximum complete unit which can be substituted without affecting other parts of the ε-term.

1.8. An ε-substitution of a set of ε-categories is an assignment of a number to each closed ε-category in the set, and a function to each open ε-category. The number or function thus assigned to an ε-category is said to be its substituent in the ε-substitution.

The number of arguments of each function is the same as the number of free variables in the original ε-category. The functions used are always recursive and indeed of the simple kind such that each takes the value 0 except for a finite number of argument values.

1.9. The resolvent of a set of closed formulae relative to an ε-substitution of its ε-categories is the result obtained from the formulae when all ε-terms are replaced by their substituents. The resolvent relative to a finite sequence of ε-substitutions is the resolvent relative to the last ε-substitution of the sequence.

The final aim is to find, for each proof, a finite sequence of ε-substitutions such that the resolvent of the set of formulae in the proof is true, i.e., all formulae in the resolvent set are true. That we can speak of true and false of the resolvent formulae follows from the fact that since the original formulae contain no free variables and since all ε-terms are replaced by their substituents, the results are numerical formulae containing no ε-terms (i.e., no quantifiers).

To get true resolvents, the main burden is to get substituents for every $\varepsilon_x A(x)$ or $\varepsilon_x B(x, a)$ such that the resolvent of $A(\varepsilon_x A(x))$ or $B(\varepsilon_x B(x, a), a)$ is true. That is to say, to make cases of III1 true.

Given a finite set of closed formulae, in particular a proof, we consider all the ε-terms occurring in it and arrange their ε-categories in a sequence such that if an ε-term a is subordinate to an ε-term b, then the ε-category of a precedes that of b.

1.10. Property P. An ε-substitution G of a set of formulae has the property P if for every ε-category in the sequence, say $\varepsilon_x B(x,y)$, and for every numeral n, the substituent of $\varepsilon_x B(x, n)$ is either 0 or else a positive m such that $B(m, n)$ is true but for no k, $k<m$, is

$B(k, n)$ true.

In other words, if G has P, then the resolvent of $A(\varepsilon_x A(x))$ can be false only when the substituent of $\varepsilon_x A(x)$ is 0.

Incidentally, each ε-substitution gives substituents for all the infinitely many ε-terms falling under an ε-category of the set, although only finitely many ε-terms occur in a proof.

1.11. The substituent of an ε-category is null if it is 0 or a function which always takes the value 0. The null ε-substitution assigns the null substituent to every category. A member of a sequence of ε-categories is said to be raw in an ε-substitution if both it and all ε-categories following it get null substituents.

Now we are to define by induction a sequence of ε-substitutions which all have the property P in the hope that we can always end up in a finite number of steps with an ε-substitution in which all resolvents of the formulae of the original proof are true.

As the initial ε-substitution, we take the null substitution. This of course has trivially the property P.

Suppose an ε-substitution G given which has the property P but the resolvent of some formula in the proof is not true. Since it has the property P, the resolvent of an axiom can be false only when it is of the form III1. Consider the first formula in the proof whose resolvent is not true, say:

(1) $$A(a, b) \supset A(\varepsilon_x A(x,b), b).$$

Suppose the value of b is n under G, and that of a is k. Then $A(k, n)$ is true, but $A(0, n)$ is false, 0 being the value of $\varepsilon_x \mathrm{A}(\mathrm{x}, \mathrm{n})$ under G. Let us determine the earliest m, $m \leqslant k$, such that $A(m, n)$ is true.

Now we define the next ε-substitution as follows. First change the substituent of the ε-category $\varepsilon_x A(x, y)$ at one place, viz. $\varepsilon_x \mathrm{A}(x, n)$ is m instead of 0 now. This does not necessarily make (1) true since the ε-category $\varepsilon_x A(x, y)$ may be subordinate to that of b so that as a result of the change, b may get a value different from n. But we do not use such a strong conclusion. We simply assign all ε-categories following $\varepsilon_x A(x, y)$ in the original sequence the null substituent; of course this change is unnecessary if they get null substituent in G already.

The new ε-substitution again has the property P. Thus, since $\varepsilon_x A(x, y)$ is not subordinate to any earlier ε-categories, their substituents remain the same. All the later ones, having null substituents, trivially possess P. With regard to $\varepsilon_x A(x, y)$ itself, if b gets the value n as before, the resolvent of (1) in the new ε-substitution is true. Otherwise, if, e.g., it has now the value j, then $\varepsilon_x A(x, j)$ has the same value as in G, and, is therefore, either 0 or the smallest number i, such that $A(i, j)$. In either case, the new ε-substitution still preserves the property P.

The problem now is to introduce suitable measures of the ε-substitutions in order to show that we are progressing toward the final goal as we continue to modify them.

2. Proof of finiteness

For this purpose we order all the closed ε-terms occurring in the original proof or in any finite set of formulae in such a way that if a resides in b, then a precedes b. Suppose there are $k+1$ such terms: $a_0, a_1, \cdots, a_k$.

2.1. The order of an ε-substitution G relative to the original finite set of formulae is given by $2^k\varphi(0)+2^{k-1}\varphi(1)+\cdots+\varphi(k)$, where $\varphi(i)$ is 1 or 0 according as a_i gets the null substituent in G or not.

Since the substituents which are not null are good according to property P, it is generally desirable to have lower orders.

2.2. The degree of an ε-substitution G relative to a finite set of formulae is its order relative to the set of formulae:

(2) $$A(0, n), \cdots. A(k, n).$$

Therein A is from (1) and (1) is the first formula in the original proof which gets a false resolvent by G, and a gets the value k in G. When all the formulae get true resolvents by G, the degree of G is taken to be 0.

2.3. The index of an ε-substitution G relative to a proof is $\omega m+n$, where m is its order and n its degree.

2.4. An ε-substitution G_i of a finite set of ε-categories is no less advanced than G_i if for every positive (i.e., not zero) substituent of a closed ε-term assigned by G_i, the same (positive) substituent is assigned by G_j. If, in addition, there is some closed ε-term which gets a positive substituent by G_i, but a zero substituent by G_i, G_j is more advanced than G_i.

Theorem 1. *if G_i is no less advanced than G_i, both of a set of ε–categories which includes all those of the ε-terms of a finite set of formulae (not necessarily a proof), then either the order of G_j relative to the set of formulae is smaller than that of G_j; or else the substituents of all the closed ε-terms in the formulae of the set are the same in G_i and G_j.*

Proof. Suppose they are not all the same in G_i and G_j. Let $\varepsilon_x A(x, b)$ be the first in the ordering of closed ε-terms defined before in 2.1. Since b precedes it, its substituents in G_i and G_j are the same, say n. Hence $\varepsilon_x A(x, n)$ must get different values by G_i and by G_j. Since G_j is no less advanced than G_i, this is possible only if it is 0 by G_i but positive by G_j. But then, by 2.1, the order of G_j must be smaller than that of G_i.

Theorem 2. *If G_i is no less advanced than G_i, then either the index of G_i relative to the proof is smaller than that of G_i,* or else G_{j+1} *(i.e. the next ε–substitution after G_i by the construction above) is no less advanced than G_{i+1} and they are obtained from G_i and G_i by adding the same positive substituents for the same closed ε-terms.*

Proof. If the order of G_j is smaller, this is proved by 2.3. Other-wise, by Th.1, all closed ε-terms in the proof get the same substituents in G_i and G_j. Consider now the set of formulae (2) under 2.2 for G_i and G_j. Since G_j is also no less advanced than G_i relative to this set, either the order of G_i relative to this set, i.e., its degree relative to the original proof, is smaller than that of G_i, or else, by Th.1, ε_x $A(x, n)$ must get the same positive value in G_{i+1} and G_{j+1}.

2.5. Given a proof and its finite sequence of ε-categories, the rank of an ε-substitution G relative to the sequence is the number of raw ε-categories in G (cf.1.11). When the last ε-category is not raw, then the rank is 0. If g is the number of ε-categories for a proof, then the ranks are always $\leqslant g$.

If it were true that G_{i+1} is always no less advanced than G_i, so that the rank of G_{i+1} is no greater than that of G_i, we would be able to establish the consistency rather simply; since every new ε-substitution corrects at least one substituent, we would come to an end soon. But, as we mentioned before, we could spoil the ε-category of a by changing the substituent of $\varepsilon_x B(x, a)$, if $\varepsilon_x B(x, y)$ happens to be subordinate to the ε-category of a. Hence, G_{i+1} may have a higher rank than G_i.

It is, however, true that in any sequence of ε-substitutions defined for a proof in the manner of 1, either the rank of G_{i+1} is always no greater than that of G_i, or else there exists some G_i of rank m followed by a finite number of ε-substitutions of rank $<m$, and then a G_{i+k} which is of rank $\geqslant m$ again.

2.6. An m–section of ε-substitutions relative to a proof is a finite sequence of ε-substitutions $G_i, \cdots, G_{i+k}$ $(k\geqslant 0)$, such that G_i is of rank $\geqslant m$, $G_{i+1}, \cdots, G_{i+k}$ are of ranks $<m$, and either G_{i+k} is the last ε-substitution (making all resolvents true), or G_{i+k+1} is of rank $\geqslant m$.

Since we never have an ε-substitution of rank less than 0, a 0–section has a single term only. Since the highest rank is g, the number of ε-categories for the proof, the null ε-substitution has rank g, and all later ε-substitutions have ranks $<g$. Hence, we need only consider m-sections for $m<g$.

Theorem 3. *If $G_1, \cdots, G_k$ and $H_1, \cdots, H_j$ are two consecutive m–sections with indices $a_1, \cdots, a_k, b_1, \cdots, b_i$, and H_1 has the rank m, $m<g$, then:* (i) *H_1 is more advanced than G_1;* (ii) *there is some number p, $1\leqslant p\leqslant k$, $1\leqslant p\leqslant j$, such that $a_p>b_p$, and for all i, if i $<p$, then $a_i=b_i$, if $1<i\leqslant p$ then G_i and H_i have the same rank.*

Proof. Let $\varepsilon_x A(x, y)$ be the $(g-m)$–th ε-category, i.e., the $(m+1)$-th from the end, of the original sequence. Since $G_2, \cdots, G_k$ all are of ranks less than m, they retain the substituents in G_1 of $\varepsilon_x A(x, y)$ and all preceding ε-categories. Therefore, since H_1 is of rank m, it must be got from G_k by giving a new positive substituent to $\varepsilon_x A(x, n)$ *for some numeral n.* Hence (i) is proved when $1<k$. This, by the way, does not necessarily mean that H_1 has a smaller index than G_1 or a new closed ε-term gets a positive substituent. It may happen that exactly the same closed ε-term $\varepsilon_x A(x,b)$ which got a positive substituent in G_1 now requires a different positive substituent on account of changes made in $G_2, \cdots, G_k$.

If, H_1 follows directly G_1, then, since the rank of H_1 is no greater than that of G_1, (i) is again true. Hence, (i) is proved.

To prove (ii), we note that if $a_1>b_1$, $p=1$. If $a_1=b_1$, then there must be some p, $a_p\neq b_p$. We assume there is no such p and consider two different cases. Suppose $k>j$. First H_j cannot be the last ε-substitution, otherwise since it has the same index as G_j, we would have stopped at G_j. But H_j is no less advanced than G_j by (i). Hence, by Th.2, H_{j+1} must be no less advanced than G_{j+1}. But this is impossible since G_{j+1} has rank less than m, while H_{j+1} has rank no less than m.

Suppose $k\leqslant j$. By hypothesis, G_k and H_k have the same index. Hence, H_1 and H_{k+1} must get the same new positive substituent. But this is impossible, since H_k being of rank less than m still preserves the positive constituent introduced at H_1.

Hence, in either case, there is some p, $a_p \neq b_p$. Take the first such p, then preceding ε-substitutions in both m–sections all yield the same resolvents, and (ii) is proved.

We now extend the definition of index to m–sections.

2.7. The index of a 0-section is the same as the index of the single ε-substitution which it contains, the index of an $(m+1)$–section is $\omega^{a_1} + \cdots + \omega^{a_k}$, where $a_1, \cdots, a_k$ are the indices of the finitely many m–sections which together make up the $(m+1)$–section.

Theorem 4. *Given two consecutive m–sections, $m<g$, such that the first ε-substitution of the second m–section has rank m. Let p be any number $0 \leqslant p \leqslant m$, and a_1, $\cdots$, a_i; b_1, $\cdots$, b_k be the indices of the p–sections out of which the two m–sections are made, j, $k \geqslant 1$. There is then a number q, such that $a_q > b_q$ and for all i, $i<q$, $a_i = b_i$.*

Proof. By Th. 3, this theorem is true when p is 0 and m is arbitrary. Assume it true for all smaller m and all smaller p.

By Th. 3, there are two earliest corresponding ε-substitutions in the two m–sections such that the index of the first is greater than that of the second. Since the index of an ε-substitution determines its rank, they must belong to two corresponding p–sections with indices a_q and b_q. Now each of the two p–sections is made up of one or more $(p-1)$–sections. By the induction hypothesis, $a_q = \omega^{c_1} + \cdots + \omega^{c_u}$, $b_q = \omega^{d_1} + \cdots + \omega^{d_v}$, and there is an earliest $c_t > d_t$. Hence, $a_q > b_q$.

From this theorem, the consistency of Z_a follows because given any proof, we can write out its ε-categories and construct ε-substitutions in the manner described before. If there are g ε-categories, then the initial ε-substitution G_1 has rank g. Since all later ε-substitutions, if there are any, have smaller rank, there is one g–section. Since the second of two consecutive p–sections in a $(p+1)$–section always has a smaller index than the first, there can be only finitely many $(g-1)$–sections. In each $(g-1)$–section, there can be only finitely many $(g-2)$–sections. Hence, for every proof, there is a finite sequence of ε-substitutions which gives numerically true resolvents for all formulae in the proof. Hence, e.g., $0=1$ is not provable.

To formalize this proof, one would need transfinite induction up to the first Cantor ε-number. Thus, if we consider all proofs each of which contains g ε-categories or less, we would at most need a transfinite induction up to $\omega(g+1)$, where $\omega(0)=\omega$, $\omega(n+1) = \omega^{w(n)}$. Since a single ε-substitution has an index of the form $\omega m+n$, a 0-section has index less than $\omega(1)$, in fact less than ω^2. If a p–section has index less than $\omega(p+1)$, then a $(p+1)$–section has index less than $\omega(p+2)$. Hence, the single g–section has index less than $\omega(g+1)$. To formalize this, we prove that every decreasing sequence of ordinals less than $\omega(g+1)$ is finite. If we are concerned with all proofs of Z_a, then g is not bounded, and we need induction through all of $\omega(0)$, $\omega(1)$, etc., up to the first ε-number. From Gödel's second theorem, such induction cannot be formalized in Z_a. Gentzen gives a direct proof of this fact in connection with his own consistency proof (see *Math. Annalen*, **119** (1943), 140—161).

3. Estimates of the substituents

To reproduce as much of the argument as is possible in Z_a, one represents ordinals by natural numbers. For each $\omega(n)$ a well-ordering R_n of natural numbers is defined by induction. Thus, R_0 is the usual natural ordering, $R_1(2^a(2b+1)-1, 2^c(2d+1)-1)$ if and only if either $a<c$ or $a=c$ but $b<d$. Given R_p, $p>0$, if $R_p(b_i, b_{i+1})$, $R_p(c_i, c_{i+1})$, $R_{p+1}(2^{b_1}+\cdots+2^{b_1}-1, 2^{c_1}+\cdots+2^{c_k}-1)$ if and only if $R_p(b_1, c_1)$ *or etc. or* $b_1=c_1$, $\cdots$, $b_i=c_i$, but $j<k$.

Now the index of each m–section can be represented by a natural number, if the index of an m–section is represented by a, then the total number of ε-substitutions in the m–section can be defined by a simple ordinal recursive function $f(a, m)$:

3.1. $f(a, 0)=1$, $f(2^{a_1}+\cdots+2^{a_k}-1, p+1)=f(a_1, p)+\cdots+f(a_k, p)$.

We have defined degree, order, rank of an ε-substitution. Now we define degree, order, rank of a proof in a different way.

3.2. The order of a proof is the number of closed ε-terms in it.

3.3. The degree of a closed term is 0 if it is 0 or an ε-term, it is $n+1$ if it is of the form a', $a \dot{-} 1$, $a+b$, or ab in which a has the degree n, or the maximum degree of a and b is n. The degree of a proof is the maximum degree of the closed terms in it.

3.4. The rank of a proof is the number of ε-categories in its associated sequence.

Suppose given a proof of degree d and an ε-substitution G. If m is the maximum value by which a closed ε-term can get under G, then $b(d, m)$ is the maximum value by which any closed term can get:

3.5. $b(0, m)=m$, $b(d+1, m)=(b(d, m))^2+1$.

This is so because with each increase in degree we only go from $b(d, m)$ to $(b(d, m))'$, $b(d, m)\dot{-}1$, $b(d, m)+b(d, m)$, $b(d, m)\cdot b(d, m)$ all less than or equal to $(b(d, m))^2+1$.

Since for the initial ε-substitution G_1, $m=0$, the greatest value is no greater than $b(d, 0)$. From the way we obtain G_{i+1} given G_i, the maximum value at each substitution G_i is $c(d, i)$ defined by:

3.6. $c(d, 0)=b(d, 0)$, $c(d, i+1)=b(d, c(d, i))$.

It therefore follows that given a bound to the number of ε-substitutions, a bound for numerical values can also be obtained.

If the degree of the proof is k, then the degree of the ε-substitution G_i is no greater than:

3.7. $e(d, i, k)=2^{(c(d,i)+1)k}$.

This is so because in the sequence $A(0, n)$, $\cdots$, $A(j, n)$ of 2.2, j is no greater than $c(d, i)$, and each formula contains no more than k terms.

The difficult part is to define a function $g(p, i, a)$ which gives an upper bound to the ordinal notation less than a which is the index of a p–section beginning with G_i. Assume the function g is given, then we can define $h(p, i, a)$ which gives an upper bound to the index of a $(p+1)$–section that begins with a p–section whose first term is G_i whose index is a:

3.8. $h(p, i, a)=2^a+h(p, i+f(a, p), g(p, i+f(a, p), a))$.

Both g and h are also functions of d and k (the degree and the order of the proof), although we are not writing out these arguments explicitly. In fact, the functions g and h are defined simultaneously.

(i) $g(p, i, 0)=0$. (ii) $g(0, i, 2^b(2c+1)-1)$ is $2^b(2c-1)-1$, if $c\neq 0$, and $2^{b-1}(2e(d, i,$

$k)+1)-1$, if $c=0$, $b\neq 0$. (iii) when $p>0$, $a>0$: (iiia) if a is even, $g(p, i, a)=a-1$; (iiib) if a is odd and of the form 2^b-1, $g(p, i, a)=h(p-1, i, g(p-1, i, b))$; (iiic) if a is $2^j+2^b+\cdots+2^c-1$, then $g(p, i, a)=2^j+g(p, i+f(j, p-1), 2^b+\cdots+2^c-1)$.

Once h is given, we may forget the function g, and again use the letter g as the rank of the proof, and rewrite the function h as $h(d, k, g, i, a)$.

Then we can estimate the total number of ε-substitutions from a given proof of Z_a whose degree, order, rank are d, k, g. We can effectively find its sequence of ε-substitutions and therefore calculate the index $2^a+\cdots+2^b-1$ of the only g–section in the sequence. Hence, the total number of ε-substitutions $f(2^a+\cdots+2^b-1, g)$ is no more than: $f(h(d, k, g-1, 1, a), g)$.

Now a satisfies $R_g(a, t(g-1, k))$, where $t(0, k)=2^{k+1}$, $t(p+1, k)=2^{c(p,k)}-1$. Hence, the total number of ε-substitutions is no more than:

3.9. $m(d, k, g)=f(h(d, k, g-1, 1, t(g-1, k)), g)$.

By combining this witb 3.7, we see that for any proof with degree, order, rank d, k, g given, the maximum numerical value used in the final resolvents is:

3.10. $c(d, m(d, k, g))$.

From this situation, some surprising consequences can be drawn. For this purpose, it is more direct to state the consequences in the notation of the equivalent system Z.

Theorem 5. *Given any theorem of the form* $(Ex)\cdots(Ey)\ A(x, \cdots, y)$ *where* A *contains no quantifiers, we can calculate from the proof a number* p *such that* $A(j, \cdots, n)$ *holds for some* $j, \cdots, n$ *all no greater than* p.

This can be directly generalized to systems which contain more recursive functions as primitive symbols. This shows that in proving a pure existence theorem, we can obtain actual examples with a finite amount of labour which has a predetermined bound yielded by the proof.

A related consequence is:

Theorem 6. *If a theorem* $(x)\ (Ey)\ A(x, y)$, A *containing no quantifiers, is given in* Z, *we can take a recursive function* $f(x)$ *of a definite type (viz., the functions actually used in defining* 3.10, *the ordinal recursive functions of Ackermann) such that for all* x, $R(x, f(x))$ *is true.*

For this purpose, we assume d to be the degree of the proof and consider the proofs of $R(0, \varepsilon_y R(0, y))$, $R(1, \varepsilon_y R(1, y))$, $\cdots$. Then we see that the following must hold:

$$(x)\ (Ey)\ (y\leqslant c(x+d,\ b(d, k, g))\ \&\ R(x, y)).$$

Hence, the theorem follows by eliminating the bounded quantifier y.

One may hope to generalize these to more complex theorems. However, Kreisel has shown that a direct generalization is impossible and introduces a program of interpretation (see *J, Symbolic Logic*, **16**, 241—267; **17**, 43—58; **23**, 155—182). We proceed to summarize some of Kreisel's results in the next section.

4. Interpretation of nonfinitist proofs

Theorem 7. *There are theorems of the form* $(x)\ (Ey)\ (z) R(x, y, z)$, R *quantifier-free, such that* $(x)\ (z) R(x, f(x), z)$ *is not true for any recursive function* $f(x)$.

Consider the theorem:

(3) $$(i)\,(Ey)\,(z)\,(B(y, s(i, i)) \vee \sim B(z, s(i, i))).$$

Therein B and s are those used previously in proving Gödel's theorems.

Since there is a notation in Z for every recursive function, we may take every proposed recursive function f and prove that there is for it some values of x and z such that:

(4) $$B(f(i), s(i, i)) \vee \sim B(z, s(i, i))$$

is false. Suppose the Gödel number of

(5) $$\sim B(f(i), s(i, i))$$

is $\bar{p}$. Then

(6) $$\sim B(f(\bar{p}), s(\bar{p}, \bar{p}))$$

is true, because otherwise, $f(\bar{p})$ would give a proof of the formula whose number is $s(\bar{p}, \bar{p})$, viz. (6) itself. On the other hand, since (6) is a numerical formula, it, being true, is provable in Z with a proof whose number is $\bar{k}$. Hence,

(7) $$B(\bar{k}, s(\bar{p}, \bar{p}))$$

is true and provable in Z. Hence (4) is false, if we substitute $\bar{p}$ and $\bar{k}$ for i and z.

Another example was used by Specker (*J, Symbolic Logic*, **14** (1949), 145—158). Classically, if $a(m)$ is a monotone bounded sequence of rational numbers, then for all x, there exists y, such that for all z and w, $2^x|a(z)-a(w)|<1$, for $z, w>y$. The usual proof gives no idea how y is to be determined from x. And Specker gives a monotone bounded recursive sequence $a(m)$ such that there exists no recursive function f for which the following holds:

$$(x)\,(z)\,)(w)\,(z,\ w>f(x) \supset 2^x|a(z)-a(w)|<1).$$

Kreisel uses a free-variable formalism F which is obtained from Z by dropping all quantifiers but adding all ordinal recursive functions of order k, for every k. That is to say, all functions definable from primitive recursive functions by addition of ordinal recursions of each order k, where

$$(0, x)=g(x),\ f(m', x)=h(x, m, f(\varphi(m'), x)),$$

g, h, φ are given functions such that $R_k(\varphi(m), m)$, for all m, according to the ordering R_k defined in 3. The system F contains also a rule of transfinite induction for each order k, viz., if there is a function φ, $R_k(\varphi(m), m)$. Then $A(n)$ follows from $A(0)$, and $A(\varphi(m')) \supset A(m')$.

To each formula A of Z is associated effectively a sequence of formulae $A_1, A_2, \cdots$ in F such that:

4.1. From a proof of A in Z, we can read off a proof of some A_i in F.

4.2. From a proof of any A_i in F, we can read off a proof of A in Z; indeed, since A_i can be expressed in Z, we can prove A_i in Z and derive A from A_i in Z.

We can easily generalize and modify Theorem **6** to get:

4.3. If $(x)\,(Ey)\,R(x, y)$ is provable in Z, R primitive recursive, then there is an ordinal recursive function g of some finite order such that we can prove in F:

(8) $$R(n, g(n)) \& (m<g(n) \supset \sim R(n, m)).$$

Let $t(n, x)$ be $t_n(x)$, where t_n is the n-th function which is 0 except for a finite number of argument places in some simple enumeration.

Let the formula A of Z be, e.g., $(Ex)(y)(Ez)\,C(x, y, z)$.

Enumerate all the proofs of Z which lead up to a conclusion with $(Ex)(y)(Ez)$ followed by a primitive recursive predicate and let the i–th such proof lead to

$$(Ex)(y)(Ez)D_i(x, y, z).$$

Then we can also prove in Z:

$$(n)(Ex)(Ez)D_i(x, t(n, x), z).$$

Hence, by 4.3, there are ordinal recursive functions a_i, b_i in some enumeration such that:

$$D_i(a_i(n), t(n, a_i(n)), b_i(n))$$

can be proved in F.

Now the sequence of formulae associated with A is simply that A_i is, for $i=1, 2, \cdots$:

(9) $$C(a_i(n), t(n, a_i(n)), b_i(n)).$$

If A can be proved in Z, C must coincide with some D_i, and therefore A_i, i.e., (9) is provable in F for some i.

Conversely, if some A_i is provable in F, it is also provable in Z and we can derive from it, again in Z, A itself:

In (9), if we choose a suitable term s, as on pp.6—7 of *Math. Zeitschrift*, **57**(1952), we can make:

$$a_i(s)=\mu_x(y)(Ez)C(x, y, z),$$
$$t(s, a_i(s))=\mu_y(z)\sim C(a_i(s), y, z),$$
$$b_i(s)=\mu_z C(a_i(s), t(s, a_i(s)), z).$$

Then A follows from A_i in Z.

14.3 Partial systems of number theory *

1. Skolem's non–standard model for number theory

In this section we summarize the work of Skolem in *Fund. Math.*, **23** (1934), 157—159, *Mathematical Interpretation of Formal Systems*, 1955, 1—14, and the related results of Ryll–Nardzewski, *Fund. Math.*, **39** (1952), 239—263.

By the famous theorem of Löwenheim, set theory has also an arithmetic model. Skolem emphasized that this leads to a relativization of the concept of set to each formal system. If one desires to develop arithmetic as a part of set theory, a definition of natural numbers in a formal set theory has a relative meaning so that an enumerable (and, therefore nonstandard) interpretation of the whole system would also yield a nonstandard interpretation of the natural numbers. From this it is natural to expect if we try to characterize the sequence of natural numbers directly by a formal system, we would not obtain a complete characterization. Skolem has succeeded in showing that this is really so.

Let us use the formal system Z. Every formula is equivalent to a prenex normal form beginning with a string of quantifiers, followed by truth–functional combinations

* First published in *A Survey of Mathematical Logic*, by H. Wang, pp 376—382. Science Press, Beijing, 1962. Reproduced by permission.

of equations. In familiar manner, we can delete the truth–functional connectives, e.g., by the following relations.

1.1. $a \neq b$ by $(Ex)((a = x + b) \vee (b = a + x))$.

1.2. $(a = b) \vee (c = d)$ by $ad + bc = ac + bd$.

Then we can drop all quantifiers and replace all variables attached to particular quantifiers by functions or ε–terms. The resulting system contains no more quantifiers but only equations with free variables. Call it Z_f.

We enumerate all the functions of one argument, i.e., all terms containing one free variable:

$$(1) \qquad f_1(t), f_2(t), \cdots$$

Let M_1, M_2, M_3 be respectively the subsets of the set N of natural numbers for which

$$f_1(t) < f_2(t), \quad f_1(t) = f_2(t), \quad f_1(t) > f_2(t).$$

Let N_1 be the M_i with least subscript which is infinite. Of course, one at least of M_1, M_2, M_3 must be infinite. This defines whether

$$f_1 < f_2, \quad f_1 = f_2, \quad f_1 > f_2$$

For the infinitely many members of N_1, there must be at least one infinite subset for which every member has a same relation between $f_1(t)$ and $f_3(t)$, and a same relation between $f_2(t)$ and $f_3(t)$. For example, if N_1 is M_1, then for every t in N_1, at least one of the following relations holds:

$$f_3(t) < f_1(t) < f_2(t), \quad f_3(t) = f_1(t) < f_2(t), \quad f_1(t) < f_3(t) < f_2(t),$$
$$f_1(t) < f_3(t) = f_2(t), \quad f_1(t) < f_2(t) < f_3(t).$$

This gives five subsets of N_1, choose the first infinite subset as N_2.

This process is continued so that we get an infinite sequence of infinite subsets of N of monotone nonincreasing size. In this way an ordering of the functions in (1) is defined.

Let $g(i)$ be the least member of N_i, then we have:

Theorem 1. *For any pair a, b, the same relation $<$, $=$, or $>$ holds between $f_a(g(t))$ and $f_b(g(t))$ for all $t > \max(a, b)$, this relation is also the ordering relation between f_a and f_b. The function $g(t)$ is monotone nondecreasing. In particular, since all constants occur in (1), viz., functions which always take a same value, $g(t)$ is not bounded.*

It is easy to see that the relations $=$ and $<$ thus defined over the sequence (1) have the usual properties of such relations. We now treat each equivalent class as a nonstandard number, and the ordering defines a sequence N^* of higher order type than N.

Among these equivalent classes, the constant numbers also occur in distinct equivalent classes with the correct ordering relation. If we reinterpret all the free variables of Z_b as ranging over N^* instead of N, we can see that all theorems of Z_f would be true in the new interpretation. In other words, if we replace variables over N by variables over N^* in theorems of Z, the results are true. For example, this can be verified for 1.2 and for the principle of induction.

Theorem 2. *The system has a model which is not isomorphic to the standard model, and the order type of the nonstandard numbers is greater than* ω.

It may be noted that the argument applies equally well to partial systems of Z. While the model defined above is not constructive, the same method can be applied to weak fragments of number theory to obtain transparent effective nonstandard models. Skolem himself has given simple examples. Related questions are also studied by Hasenjaeger, *J. Symbolic Logic*, **17** (1952), 81—97.

An application in a different direction was made by Ryll–Nardzewski. He takes the system Z and deletes all but a finite number of special cases of the schema of mathematical induction. For any such partial system S, he shows by using Skolem's model that there is some case of the induction schema not deducible in the system.

Theorem 3. *There is no finite set of theorems of Z (indeed, no finite set of formulae true in the standard model of Z) from which we can derive by the predicate calculus all axioms of Z.*

Assume given such a partial system S. In order that it be adequate at all, it must include the various special things we use below from Z. We assume that quantifiers are taken away as before.

There are only a finite number of axioms beyond the predicate calculus, and only a finite number of functions $f_1, \cdots, f_k$ in these axioms. These are taken as the basic functions.

For the arbitrary terms a and b of S, we can define in S a predicate which says:

1.3. b is of order i relative to a and $f_1, \cdots, f_k$, or, briefly, b is an i–th descendant of a: a is its own 1–descendant; b is an $(n+1)$–descendant of a if a is $f_i(c, \cdots, d)$ and some of $c, \cdots, d$ is an n–descendant of a but none is of order more than n relative to a and $f_1, \cdots, f_k$.

In addition, it is easy to define a relation R such that:

1.4. $R(x, y, z)$ if and only if x is at most of order y relative to z and $f_1, \cdots, f_k$.

Let $\Phi(y)$ be the formula that for every z there exists t such that for all x, if x is of order no more than y relative z and $f_1, \cdots, f_k$, then $x \leqslant t$:

1.5. $\Phi(y)$ is $(z)(Et)(x)(R(x, y, z) \supset x \leqslant t)$.

Then it is possible to prove in S:

1.6. $\Phi(1)$,

1.7. $\Phi(y) \supset \Phi(y+1)$.

However, the conclusion

1.8. $\Phi(y)$

is not provable in S.

Assume a Skolem model N^* for S given. The equivalent classes containing the standard natural numbers form only an initial segment of N^*. Let w be an arbitrary element of N^* not among the above. Then all descendants of w determine a model of S. Then the equivalent class (n) for each natural number n satisfies the relation: $(\mathrm{n}) < \mathrm{w}$.

If $\Phi(w)$ is true, then by 1.5, there is some c,

$$(2) \qquad (d)(R(d, w, w) \supset d \leqslant c).$$

But the successor c^* of c in the model can be obtained from w as a descendant, say a k–descendant. Hence, $R(c^*, w, w)$, contradicting (2). Hence, Theorem 2 is proved.

It seems possible to use the same argument to prove a somewhat stronger result, viz. we cannot derive all axioms of Z from any finite consistent set of formulae in Z. Thus, given a such set which can be written as a single formula A, either not all theorems of Z are derivable from A, or else we can carry out Ryll–Nardezwski's construction and get a case of the induction schema which is derivable in Z but not derivable from A, provided only A is consistent. Hence, if all theorems are derivable from A, A must be inconsistent.

There appears to be a connection between Skolem's result and Gödel's first theorem. From Gödel's theorem, we can get a nonstandard model of Z. Conversely, we can also extend Ryll–Nardezwski's argument to find an undecidable proposition of Z by the Skolem model. Thus, the above argument depends on the possibility of enumerating all terms of S in Z. If S is Z itself, the enumeration can no longer be made directly in Z, but we can make an indirect enumeration by the proof predicate. Compare Ch. II, §5 above.

2. Some applications of formalized consistency proofs

In this section we give an explanation of the paper by Kreisel and the writer, *Fund. math.*, **42** (1955), 101—110; **45** (1958), 334—335.

We consider partial systems F of Z and Z_a which are obtained from them by suppressing those proofs which are too "complex". We shall prove the consistency of such systems F in Z and Z_a; F would be demonstrably weaker since a formula expressing its consistency cannot be proved in F itself. We shall denote such a formula by $Con(F)$; it is to be understood that the formula chosen satisfies conditions sufficient to ensure the application of Gödel's second theorem.

For the first measure of complexity we use the rank of a proof in Z_a, *viz., the number of ε–categories.*

2.1. The system Z_a^n: a proof of Z_a of rank $\leqslant n$ is a proof of Z_a^n.

2.2. For each integer n, $Con(Z_a^n)$ is a theorem of Z_a.

This is direct from Ackermann's consistency proof since, for each n, transfinite induction to $\omega(n)$ can be proved in Z_a.

A second measure of complexity is obtained from the truth definition of Z given in Hilbert–Bernays II.

2.3. The system $Z^{(n)}$: a proof Z whose formulae are all of type $\leqslant n$, i.e., in each formula if we construct a graph of quantifiers governed by a given quantifier, the maximum succession is always of length $\leqslant n$, is a proof of $Z^{(n)}$.

2.4. For each n, $Con(Z^{(n)})$ is a theorem of Z.

To prove this, the truth definition has to be modified somewhat.

It is assumed that "natural" definitions of the following syntactical terms and predicates have been chosen.

$\eta_1^{(m)}(n), \cdots, \eta_m^{(m)}(n)$ as in Hilbert–Bernays II, p. 235.

$\rho(m, a, n)$ is the number of the formula got from A (with number a) by replacing the variable v_i, $i \leqslant m$, in A by $\eta_i^{(m)}(0^{(n)})$ and v_i, $j > m$, by 0. It is not assumed that all v_i, $i \leqslant m$, occur in A. All the variables we use in the proofs will be v_1, v_2, v_3, etc. Trivially, if A is a closed formula, $\rho(m, a, n) = \rho(0, a, 0)$. ($v_i$ are free variables.)

$P(a, b)$ if and only if a is a numerical proof of the formula b (*i.e.*, a proof in the elementary calculus, no variables). We recall that the consistency of numerical arithmetic can be proved in Z.

If a and b are numbers of A and B, then $t(a, b)$ is the number of $A|B$.

$U(a)$ if and only if a is the number of a formula of the form $(x)B(x)$, and then $s[u(a), y]$ is the number of $B(0^{(y)})$.

Similarly, $Q(a)$ if and only if a is the number of a formula of the form $(Ex)B(x)$, and then $s[q(a), y]$ is the number of $B(0^{(y)})$.

For each k, a truth definition $T_k(b)$ can be given by means of a formula of Z, satisfying the following conditions (compare *ibid.*, p. 334):

$$T_0[\rho(m, a, n)] \text{ if and only if } (Ey)P[y, \rho(m, a, n)]$$

or

$$(Ex)(Ey)\{x < \rho(m, a, n) \;\&\; y < \rho(m, a, n) \;\&$$
$$\&\; \rho(m, a, n) = t(x, y) \;\&\; [T_0(x) | T_0(y)]\};$$
$$T_{k+1}[\rho(m, a, n)] \text{ if and only if } T_k[\rho(m, a, n)]$$

or

$$\{U[\rho(m, a, n)] \;\&\; (y)\; T_k[\rho(m, s(u(a), y), n)]\}$$

or

$$\{Q[\rho(m, a, n)] \;\&\; (Ey)\; T_k[\rho(m, s(q(a), y), n)]\}$$

or

$$(Ex)(Ey)\{x < \rho(m, a, n) \;\&\; y < \rho(m, a, n) \;\&$$
$$\&\; \rho(m, a, n) = t(x, y) \;\&\; [T_{k+1}(x) | T_{k+1}(y)]\}.$$

It can be verified in the usual manner that $T_k(b)$ is a normal truth definition, and hence $Con(Z^{(k)})$ may be proved in Z. Note that $T_n(b)$ is a truth definition for the system $Z^{(n)}$ only, and not for Z; in particular, n is not a free variable.

Observe that the consistency proof of 2.4 is capable of various extensions: *e.g., if 3* is an extension of Z by some principle of transfinite induction, we get a consistency proof of $3^{(n)}$ in 3.

Let F be a system consisting of the predicate calculus with a single closed formula A as axiom.

Theorem 4. *If F is consistent, there is a theorem of Z not provable in F.*

Let k be the type of A, roughly the number of distinct quantifiers in it. If $\sim Con(F)$, there is a proof in the predicate calculus of $\sim A$. By Herbrand's theorem, $\sim A$ can be proved with a proof each formula of which is of type $\leqslant k$. Thus, $\sim A$ can be proved in $Z^{(k)}$. Hence, by the normal truth definition of Z^k, we can prove in Z: $\sim Con(F) \supset \sim A$. Therefore, the following is a theorem of Z:

$$(1) \qquad A \supset Con(F).$$

Now if F contains all theorems of Z, the deducibility conditions for the application of Gödel's second theorem would be applicable to F, since we are assuming the "natural" proof predicate of F; further the formula (1) and hence $Con(F)$ would be theorems of F, and F *would be inconsistent.*

It is actually possible to exhibit from the proof, for any given consistent F, a theorem of Z not provable in F.

This result and the result in the preceding section shows that there are infinitely many axioms, which cannot be reduced to a finite number. Kleene proves in *Memoirs of Am. Math. Soc.*, no. 10, 1952, 27—68, that if we introduce auxiliary predicate symbols, such reduction is always possible.

These considerations can also be applied to deal with Gödel's result on the length of proofs (*Ergebnisse Math. Kolloquiums*, 7 (1936), 23—24). Thus since $Con(Z^{(n)})$ is provable in Z but not provable in $Z^{(n)}$, there is also some formula $(x)C(x)$ which is provable in Z but not in $Z^{(n)}$, *although* $C(1)$, $C(2)$, $\cdots$, all are provable in $Z^{(n)}$. Then for large constants k, we would seem to need a proof of unbounded length to prove $C(k)$ in Z^{n}, although in Z it is always an immediate consequence of $(x)C(x)$.

In view of 2.2. and 2.4, the measure of length or complexity of a proof is for us more easily given by the rank and type of a proof, i.e., the number of ε*–categories or the number of distinct quantifiers.*

Theorem 5. *If the length of a proof of* Z_a *is measured by the number of* ε*–categories in it, and* F' *is an extension of* Z_a *in which* $Con(Z_a)$ *is a theorem, then for each n,* $Con(Z_a^n)$ *can be proved in* F' *by proofs of bounded length, but its shortest proof in* Z_a *is longer than n. Similarly, if we use the type as a measure of length, an analogous theorem holds for* Z *and the theorems* $Con(Z^{(n)})$.

Finally, according to the Bernays' lemma, every S is translatable into Z_s, viz. Z plus a new axiom $Con(S)$. Now for systems with finitely many axioms, we can prove a converse to it:

Theorem 6. *If a finite axiom system F is translatable into Z, then* $Con(F)$ *can be proved in Z.*

Let the translation of the axiom A of F be the theorem B of Z. If $\sim Con(F)$, there would be a proof of $\sim$ F in the predicate calculus, and hence a proof of $\sim B$ in $Z^{(m)}$ for some fixed number m, the type of A. Hence, we have in Z, $\sim Con(F) \supset \sim T_m(b)$, b being the number of B.

Since $T_m(a)$ is a normal truth definition, $\sim T_m(b) \supset \sim B$. Hence, we have in Z: $B \supset Con(F)$. Hence, $Con(F)$ is a theorem of Z.

14.4 The calculus of partial predicates and its extension to set theory I*

The usual predicate calculus and set theory deal with completely defined predicates and sets. One natural extension would seem to be a predicate calculus and a

---

* First published in *Zeitschrift f. Mathematische Logik und Grundlagenforschung*, vol, 7, pp 283—288. © VEB Deutscher Verlag, 1961. Reproduced by permission.

1) Presented at the Logical Colloquium on July 7, 1961 at Cambridge, England. Part I deals only with the predicate calculus, set theory and other related matter will be considered in Part II.

set theory in which partially defined predicates and sets are also permitted, using, besides the values t (true) and f (false), a third value u (undefined). Among other things, a different interpretation of the paradoxes of set theory would ensue. Various people including SKOLEM, BEHMANN, BOCHVAR, ACKERMANN, FITOH, SCHÜTTE have considered the set–theoretical paradoxes from such an approach and reached rather diverse results. It appears that the full implications of such a generalized outlook on logic have not been fully clarified yet and many avenues remain to be explored. It is, therefore, proposed here to reexamine some of the alternative sets of underlying assumptions and discuss the questions of setting up formal systems to correspond.

The two main items of indefiniteness are the interpretation of implication and the mixing of complete sets with partial sets. On account of the need to give implication a special place, we cannot reduce our problems to those which have been dealt with by standard works on many–valued logics such as [10]. If implication is not iterated, it is easy to set up a natural calculus of partial predicates. This part will be developed quite thoroughly below. In addition, the questions relevant to set theory will be discussed, possible interpretations of iterated implications will be suggested, and comments on the system A* in SCHÜTTE'S book [15] will be made. Because of the presence of the third value u, the general law of excluded middle is not a part of the logic of partial predicates. While this is a feature in common with the usual constructive theory, the study of partial predicates can be combined with either a constructive or a classical position. In fact, the considerations will be based on a classical viewpoint, and the question of a constructive theory of partial predicates will not be investigated.

1. The basic truth tables

The tables for negation, conjunction, and disjunction are those originally introduced by LUKASIEWICZ ([8], p. 94):

1.1. *Negation, conjunction,* and *disjunction* satisfy the following tables.

p	$\neg p$
t	f
u	u
f	t

$\vee$	t	u	f
t	t	u	f
u	u	u	f
f	f	f	f

$\vee$	t	u	f
t	t	t	t
u	t	u	u
f	t	u	f

Given the intuitive meaning of $\neg$, $\wedge$, $\vee$ and the values t, u, f, the above choice seems natural. BOCHVAR and SMILEY use a different set in which, for example, $u \wedge f$ and $f \vee u$ are u rather than f; such tables are suitable when u is taken to mean "meaningless". Our first assumption is 1.1.

The interpretation of implication presents serious dificulties, even if we are willing to follow PHILO and FREGE to accept material implication when we are only concerned with complete predicates, using only the two values t and f. The minimum requirement on $p \rightarrow q$ is that if p is t, q must not be f, or that $t \rightarrow f$ is f. If, in the 2–valued case, we further require that the value of $p \rightarrow q$ is determined uniquely by the values of p and q, then the only possible choice is to give t to $t \rightarrow t$, $f \rightarrow t$, and $f \rightarrow f$ because in each case we can find some propositions p and q such that p implies q is intuitively true: if Yü was a man, then Yü was not a fish; if George Sand was a man, George Sand was a human being; if Algeria is in France, Algeria is in Europe. It is familiar that this interpretation of implication does not give the full meaning but is, nonetheless, very useful. If we now try to extend this interpretation to the 3–valued case, we are led fairly naturally to the following incomplete tables:

1.1.1.

$\to$	t	u	f
t	t		f
u			
f	t		t

$\to$	t	u	f
t	t		f
u	t		
f	t	t	t

$\to$	t	u	f
t	t	u	f
u	t		u
f	t	t	t

We may wish to give the value u to $u \to u$. In that case, $p \to p$ is no longer a universal logical law. Or, we may wish to give the value t to $u \to u$. In that case, $p \to \neg p$ gets the value t when p gets the value u. Such consequences do not constitute conclusive reasons for rejecting either of these choices. But they serve to illustrate that we do not yet possess either a sharp enough notion of $\to$ as applied to propositions with the value u, or any reliable guiding principle to enable us to decide between contrary alternatives. If, e.g., we try to evaluate $((p \to \neg p) \to p) \to p$ for the case when p gets the value u, we may feel that $u \to u$ gets different values in different contexts. Hence, another course suggests itself: use more than 3 values, perhaps an infinite number of values.

The alternative of giving the value t to $u \to u$ is adopted by LUKASLEWICZ and has the advantage of preserving the law $p \to p$. According to such an interpretation, we of course can no longer identify $p \to q$ with $\neg p \vee q$ since $\neg p \vee p$ is not generally valid although $p \to p$ is. The other alternative of giving the value u to $u \to u$ is suggested, e.g., by KLEENE ([7], 64). By this interpretation $\neg p \vee q$ is the same as $p \to q$ and one can develop a three–valued logic without including $\to$ as a primitive. However, in either case, we would have to depend on a notion of $\to$ that is quite beyond the complete control of our rather limited intuitive conception.

We are inclined to regard $p \to q$ as a logical law (which gets the value t) if under all interpretations of the occurring predicates, whenever p is true, q is also true. If we adopt this natural interpretation of $\to$, the value of $p \to q$ is no longer a function of the specific values of p and q in each case but depends on the relations between all possible values of p and q. As a result, iterated applications of $\to$ in the antecedent introduce new elements not taken care of by such an explanation. For example, although we can quite reasonably regard $p \to (q \to p)$ as $(p \wedge q) \to p$, we are at a loss with regard to $(\neg p \to p) \to p$. In order that it be t, we require that for all evaluations E, if $v_E(\neg p \to p) = t$, then $v_E(p) = t$. But we have given no maening to $v_E(\neg p \to p)$, because we have only given a value to $\neg p \to p$ relative to all evaluations of p, and have not explained what it means for $\neg p \to p$ to be t under each single evaluation E.

If we are content not to iterate $\to$, the present definition of $p \to q$, when $\to$ occurs neither in p nor in q, is satisfactory. If we are concerned with a two–valued logic, then such a definition of $p \to q$ agrees with $\neg p \vee q$, and this is the reason why in GENTZEN'S formulation, $\to$ is dispensable, and, for example, SCHÜTTE (in [12]) uses

$\neg p_1 \vee \cdots \vee \neg p_m \vee q_1 \vee \cdots \vee q_n$ in place of GENTZEN'S $p_1, \cdots, p_m \rightarrow q_1, \cdots, q_n$. In the three–valued case we are interested in here, $\rightarrow$ thus defined is an indispensible high–level connective and we do not define truth of an implication but validity relative to sets of evaluations. Our second assumption is:

1.2. $p \rightarrow q$, with p and q free of $\rightarrow$, is valid if and only if for every E, $v_E(q) = t$ when $v_E(p) = t$.

On account of the presence of the third value u, it is also possible to make a more stringent requirement on $\rightarrow$:

1.2.* $p \rightarrow q$ is valid if and only if for every evaluation E, (a) if $v_E(p) = t$, then $v_E(q) = t$, and (b) if $v_E(p) = u$, then $v_E(q) \neq f$.

In contrast with $\rightarrow$, the quantifiers present no serious problems but can be explained in the natural manner as generalized conjunctions and disjunctions:

1.3. The value of $\forall xFx$ is (a) t, if all cases get the value t; (b) u, if no cases are f but at least one case is u; (c) f, if at least one case is.

The value of $\exists xFx$ is (a) t, if at least one case is t; (b) u, if no cases are t but at least one case is u; (c) f, if all cases are f.

On the basis of the interpretations of $\neg, \vee, \wedge, \rightarrow, \forall\ \exists$ as given under **1.1, 1.2, 1.3,** we can develop both a partial predicate calculus and the usual predicate calculus. The presence of a third value enables us to make some distinctions more sharply than in the ordinary two–valued approach.

2. The partial predicate calculus PP

It is possible to give several different formulations of **PP** by modifying alternative formulations of the classical predicate calculus. The following is along the line of HERBRAND'S system in [6].

2.1. The formal system **PP.**

2.1.1. The symbols are $\vee, \wedge, \neg, \rightarrow, \forall\ \exists$, parentheses, variables, predicate letters, proposition letters, possibly certain other terms and constant predicates.

2.1.2. The atomic propositions are the proposition letters, and the predicate letters or constant predicates followed by suitable numbers of terms (variables or other terms).

2.1.3. The formulae are the smallest set S such that (a) all atomic formulae are in S, (b) if x is a variable, p, q are in S and do not contain $\rightarrow$, $\neg p$, $p \vee q$, $p \wedge q$, $p \rightarrow q$, $\forall xp$, $\exists xp$ are in S.

2.1.4. The axioms and rules are of four kinds.

2.1.4.1. $p \rightarrow q$ is an axiom if p, q are quantifier–free and for every E, $v_E(q) = t$ if $v_E(p) = t$. In particular, $\rightarrow q$ or q is an axiom if for every E, $v_E(q) = t$. Clearly, given any quantifier–free formula $p \rightarrow q$, we can decide whether it is an axiom by testing all possible truth distributions, i.e., assignments of the truth values t, u, f to all the occurring atomic propositions.

2.1.4.2. If x is a variable, y, is a variable not free in p, and t is a term, then:

$$\frac{p \rightarrow F_y}{p \rightarrow \forall xFx} \qquad \frac{p \rightarrow F_t}{p \rightarrow \exists xFx} \qquad \frac{F_y \rightarrow p}{\exists xFx \rightarrow p} \qquad \frac{F_t \rightarrow p}{\forall xFx \rightarrow p}$$

2.1.4.3. Contraction rules:

$$\frac{p \wedge p \wedge q \rightarrow r}{p \wedge q \rightarrow r} \qquad \frac{p \rightarrow q \wedge q \wedge r}{p \rightarrow q \wedge r}.$$

2.1.4.4. Rules for shifting quantifiers. If x is not free in p and $q \equiv r$ means within each theorem, replacing a part q by r or a part r by q, we again get a theorem; then: $\neg \exists xFx \equiv \forall x \neg Fx$, $\quad \neg \forall xFx \equiv \exists x \neg Fx$, $\quad \forall x(Fx \wedge p) \equiv (\forall xFx \vee p)$, $\exists x(Fx \vee p) \equiv (\exists xFx \vee p)$, $\forall x(Fx \wedge p) \equiv (\forall xFx \wedge p)$, $\exists x(Fx \wedge p) \equiv (\exists xFx \wedge p)$.

2.2. Definition of completeness.

A formula free of $\rightarrow$ is true under a given interpretation of the occurring predicate and proposition letters (and the constant predicates and terms) with a domain as the range of variables, if and only if it comes out true according to the interpretation of $\wedge$, $\vee$, $\neg$, $\rightarrow$, $\forall$, $\exists$ in **1.1, 1.2** and **1.3**. A system is complete if and only if every valid formula $p \rightarrow q$ is a theorem.

2.3. Proof of completeness of **PP**.

It is easy to verify in familiar manner, that all theorems of **PP** are valid by the above definition:

2.3.1. *Every theorem in* **PP** *is valid.*

Conversely, we can also adapt familiar arguments for the classical predicate calculus to prove that every valid formula of **PP** is a theorem of **PP**.

Theorem 1. *The system* **PP** *is complete; in other words, if for every* E, $v_E(p) = t$ *implies* $v_E(q) = t$, *then* $p \rightarrow q$ *is a theorem of* **PP**.

Thus, given any formula $p \rightarrow q$, by 2.1.4.4, we can assume that p and q are both in the prenex form, e.g.:

$$\forall x \exists y \forall zFaxyz \rightarrow \forall u \exists v \forall w\ Gauvw. \tag{1}$$

This is not valid if and only if there is some interpretation E such that, for suitable a:

$v_E(p) = v_E(\forall x \exists y \forall zFaxyz) = t$, $\qquad v_E(\neg q) = v_E(\exists u \forall v \exists w \neg Gauvw) = t$ or u.
(2)

We can adapt familiar methods of SKOLEM and HERBRAND to associate with p, q two sequences of quantifier–free formulae $p_1, p_2, \cdots, q_1, q_2, \cdots$ so that if for some n, $p_1 \wedge \cdots \wedge p_n \rightarrow q_1 \vee \cdots \vee q_n$ is a theorem, then $p \rightarrow q$ (in particular, (1)) is; and if this is true for no n, $p \rightarrow q$ is not valid (in particular, (2) is true). In the example (1), the sequence $p_1, p_2, \cdots$ is, if we write (h, i, j, k) for $x_h x_i x_j x_k$:

$$(1, 1, 2, 1), \quad (1, 1, 3, 2), \quad (1, 2, 4, 1), \quad (1, 2, 5, 2), \quad (1, 1, 6, 3) \cdots$$

Similarly, the sequence $q_1, q_2, \cdots$ is

$$(1, 2, 1, 3), \quad (1, 2, 2, 4), \quad (1, 2, 3, 5), \quad (1, 2, 4, 6), \cdots$$

From Theorem 1, it follows that if we add standard rules governing $\rightarrow$ (i.e., rules which can be seen to be correct from the meaning of $\rightarrow$) to **PP**, we do not change the body of theorems. In particular, the cut rule is permissible:

2.4. The generalized cut rule. *If* $p_1 \wedge \cdots \wedge p_m \rightarrow q_1, \cdots, p_1 \wedge \cdots \wedge p_m \rightarrow q_n$, *and* $q_1 \wedge \cdots \wedge q_n \rightarrow r$ *are theorems of* **PP**, *then* $p_1 \wedge \cdots \wedge p_m \rightarrow r$ *is a theorem too.* Thus, if $p_1 \wedge \cdots \wedge p_m$ is t, then $q_1, \cdots. q_n$ are t, hence $q_1 \wedge \cdots \wedge q_n$ is t, *and* r is also t.

It is also possible to parallel familiar arguments to give finitist proofs of the eliminability of cuts.

Such results can be extended to systems obtained from **PP** by adding new axioms of the form $p \vee \neg p$ with atomic formulae p, or also of the form p or the form $\neg p$ with atomic formulae containing only constants.

If we replace 2.1.4.1 by:

2.1.4.1.* $p \rightarrow q$ is a theorem if for all E, $v_E(p) = t$ implies $v_E(q) = t$, and $v_E(p) = u$ implies $v_E(q) \neq f$,

we obtain a system **EP** that is complete relative to the interpretation **1.2*** (instead of 1.2) of $\rightarrow$. In particular, $p \wedge \neg p \rightarrow q$ is always a theorem in **PP**, but not always one in **EP**.

3. Relation of PP to the classical predicate calculus

If we add the law of excluded middle to **PP**, we get the classical predicate calculus. Alternatively, we can give a system **KP** which differs from **PP** only in having a quantifier–free rule stronger than 2.1.4.1:

3.1. $p \rightarrow q$ is a theorem if for all E, $v_E(p) = t$ implies $v_E(q) \neq f$; in particular, when p is empty, if for all E, $v_E(q) \neq f$. Hence, if r is quantifier–free $\rightarrow r \vee \neg r$ is a theorem.

We can prove the completeness of **KP** relative to either the classical definition of validity, or equivalently:

3.2. $p \rightarrow q$ is valid if and only if for all E, $v_E(p) = t$ implies $v_E(q) \neq f$; in particular, q or $\rightarrow q$ is valid if and only if for all E, $v_E(q) \neq f$.

3.3. KP *is complete;* in fact $\vdash_{KP} p \rightarrow q$ if and only if $p \rightarrow q$ is valid by 3.2.

Hence, it follows that:

3.4. *The cut rule is permissible in* **KP**.

3.5. $\vdash_{KP} \rightarrow p \vee \neg p$, *for every formula* p.

Theorem 2. $\vdash_{KP} p \rightarrow q$ *if and only if* $\vdash_{PP} p \wedge (q \vee \neg q) \rightarrow q$; in particular, $\vdash_{kp} \rightarrow q$ if and only if $\vdash_{PP} q \vee \neg q \rightarrow q$.

Since every theorem of **PP** is one of **KP**, if $\vdash_{PP} p \vee (q \vee \neg q) \rightarrow q$, then, $\vdash_{KP} p \wedge (q \vee \neg q) \rightarrow q$. Hence, by 3.5 and 3.4, $\vdash_{KP} p \rightarrow q$.

Conversely, if $\vdash_{KP} p \rightarrow q$, then for all E, $v_E(q) \neq f$ if $v_E(p) = t$. If $v_E(q \vee \neg q) = t$, then $v_E(q) = t$ or f, but not u. Hence, for all E, if $v_E(q \vee \neg q) = t$ and $v_E(p) = t$, then $v_E(q) = t$. Hence, by the completeness of **PP** (Theorem 1), $\vdash_{PP} p \wedge (q \vee \neg q) \rightarrow q$.

Alternatively, we can also prove Theorem 2 directly without appeal to Theorem 1 and deduce Theorem 1 from Theorem 2 and 3.3.

Corresponding to **EP**, we have also a complete system **CP** such that $p \rightarrow q$ is valid if and only if for all E, $v_E(p) \neq f$ implies $v_E(q) \neq f$. Here again, we need only modify the quantifier–free rule, and we have:

$$\vdash_{CP} p \to q \ if \ and \ only \ if \vdash_{EP} p \wedge (q \vee \neg q) \to q.$$

In the classical approach, there is no sharp distinction between **KP** and **CP** since they would contain the same theorems (not containing →).

References

[1] ACKERMANN, W., Widerspruchsfreier Aufbau einer typenfreien Logik II, Math. Z. **57** (1953), pp. 155—166.

[2] BEHMANN, H., Zu den Widersprüchen der Logik und der Mengenlehre, Jahresber. d. Dt. Math.-Ver. **40** (1931), pp. 37—48.

[3] BOCHVAR, D. A., On a Three-valued Logical Calculus and its Application to the Analysis of Contradictions (Russian), Recueil Mathématique, N.S. **4** (1939), pp. 287—308.

[4] —, On the Consistency of a Three-valued Logical Calculus (Russian), ibid, **12** (1943), pp. 353—369.

[6] HERBRAND, J., Recherches sur la Théorie de la Demonstration, Warsaw, 1930.

[7] KLEENE, S. C., Introduction to Metamathematics, New York, 1952.

[8] LUKASIEWICZ, J., Die Logik und das Grundlagenproblem, Les Entretiens de Zurich (1941), pp. 82—108.

[9] MOH SHAW-KWEI, About the Rules of Procedure (Chinese), Journal of Nanking University, **1** (1953), pp. 801—809.

[10] ROSSER, J. B. and TURQUETTE, A. R., Many-valued Logics, Amsterdam, 1952.

[11] SCHMIDT, H. ARNOLD, Mathematische Gesetze der Logik I, Berlin, 1960.

[12] SCHÜTTE, K., Schlu*β*weisen-Kalküle der Prädikatenlogik, Math. Annalen **122** (1950), pp. 47—65.

[13] —, Ein System des verknüpfenden Schlie*β*ens, Archiv f. Math. Logik u. Grundlagenfor-schung **2** (1956), pp. 55—67.

[14] —, Aussagenlogische Grundeigenschaften formaler Systeme, Dialectica **12** (1958), pp. 422—442.

[15] —, Beweistheorie, Berlin, 1960.

[16] SKOLEM, TH., Über einige Grundlagenfragen der Mathematik, Oslo (1929), 49 pp.

14.5 Model theory*

A. Background and typical problems.

Model theory studies the interpretations (models) of theories formalized in the framework of formal logic, especially that of the (first-order) predicate calculus with equality (briefly, elementary logic). A (first-order) language is given by a collection S of symbols for relations, functions and constants which, in combination with the symbols of elementary logic, single out certain combinations of symbols as sentences. Thus, for example, in the case of the system N, the formation rules give the language which is determined in accordance with a uniform procedure by the set of (uninterpreted) extralogical symbols:

$$S = \{+, \cdot, 0, 1\}.$$

A (first–order) theory is determined by a language and a set of sentences of the language (the distinguished or "true" sentences of the theory). In the particular case of the system N, one theory T_a is determined by the language and the set of theorems of N, and another theory T_b is determined by the true sentences of N according to the natural interpretation. In general, we can use the language of N and any set of sentences of the language to make up a theory.

A *realization* of a language (for example, the one based on S) is a structure $\mathfrak{A}$ of the form

$$\mathfrak{A} = < A, +_{\mathfrak{A}}, \cdot_{\mathfrak{A}}, 0_{\mathfrak{A}}, 1_{\mathfrak{A}},$$

where A is a nonempty set (called the domain of $\mathfrak{A}$), $0\,\mathfrak{A}$ and $1\mathfrak{A}$ are members of A, + and $\cdot$ are functions from AxA (i.e. the set of ordered pairs $\langle a, b\rangle$ such that a,b belong to A) into A. The structure $\mathfrak{A}$ satisfies or is a *model* of the theory T_a (or T_b) if all the distinguished sentences of T_a (or T_b) are true in $\mathfrak{A}$ (or satisfied by $\mathfrak{A}$). Thus, if $\mathfrak{A}$ is the structure of the ordinary nonnegative integers $\langle\omega, +, 0, 1\rangle$ then it is not only a realization of the language based on S but also a model of both T_a and T_b. Gödel's incompleteness result permits nonstandard models of T_a which contain more objects than ω but in which all the distinguished sentences of T_a (viz. the theorems of the system N) are true. Skolem's constructions (related to ultraproducts, see below) yield nonstandard models for both T_a and T_b.

The use of the relation of satisfaction or model–of between a structure and a theory (or a sentence) can be traced back to the book *Wissenschaftslehre* (published in 1837) by B. Bolzano (1781—1848), and, in a more concrete context, to the introduction of models of non–Euclidean geometries around that time. In the mathematical treatment of logic, these concepts can be found in works of E. Schröder (1841—1902) and L. Löwenheim (in particular, his paper of 1915). Basic tools and results in model theory such as the Löwenheim–Skolem theorem, the completeness theorem of elementary logic, and Skolem's construction of nonstandard models of arithmetic were developed during the period 1915—1933. A more general and abstract study of model theory began after 1950 by Tarski and others.

One group of new results may be classified as refinements and extensions of the Löwenheim–Skolem Theorem. A rather direct generalization says that if a theory has any infinite model, then for any infinite cardinal number θ, it has a model of cardinality θ. It follows that no theory with any infinite model can be categorical (i.e., such that any two models of the theory are isomorphic), since models of different cardinalities can obviously not be isomorphic. A natural question is whether a theory T can be categorical in certain (infinite) cardinalities, i.e., whether there are cardinal numbers θ such that any two models of T of cardinality. θ are isomorphic. A central result due to M. Morley (1963) says that if a theory is categorical in any uncountable θ, then it is categorical in every uncountable cardinality. On the other hand, examples are known for all four combinations of countable and uncountable cardinalities: there are theories which are categorical (1) in every infinite cardinality, (2) in the countable cardinality but in no uncountable cardinality, (3) in every uncountable cardinality but not in the countable, (4) in no infinite cardinality. In another direction, there are "twocardinal problems" which ask about the possibilities of changing, from one model to another, the cardinality, not only of the domain of the first model, but also of some chosen property (such as being a prime number). There are various results on these questions including independence results (pom. ordinany axioms of set theory) and conditional nesultos proved under the assumption of certain faniliar hypotheses of set theory.

An area of perhaps more philosophical interest is concerned with the nature of elementary logic itself. On the one hand, the completeness result seems to show in some sense that elementary logic is what we naturally wish to have. On the other hand, one is still inclined to ask whether there might be some uniqueness result to the effect that elementary logic is the only solution to satisfy certain natural requirements on what a logic should be. The development of model theory has led to a more general outlook that enabled P. Lindström to prove a general result along this direction (1969): roughly speaking, within a broad class of possible logics, elementary logic is the only one which satisfies the requirements of axiomatizability and the Löwenheim–Skolem theorem. While this theorem does not settle satisfactorily whether elementary logic is the right logic, it does seem to suggest that mathematical results can help us to clarify our concept of logic and logical truth.

A particularly useful tool for getting new models from given models of a theory is the construction of the ultraproduct of a family of structures (in particular, the ultrapower when the structures all are copies of the same, just as the product of $a_1, \cdots, a_n$ is the same as the power a^n if $a_i = a$, for each i). The intuitive idea is to establish that a sentence is true in the ultraproduct if and only if it is true in "almost all" ("almost everywhere") the given structures: an idea which was present in a different form in Skolem's construction of a nonstandard model of arithmetic in 1933. It then follows that if the given structures are models of a theory T, then their ultraproduct is one also since every sentence in T is true everywhere (which is a special case of "almost everywhere" in the technical sense employed). For example, ultraproducts have been applied to providing a foundation for nonstandard analysis which yields an unambiguous interpretation of the classical concept of infinitesimals. They have also been applied by Ax and Kochen to problems in the field of algebra (oon p–adic fields).

There are also studies developing model theory of nonelementary logic: such as second order logic and infinitary logics. Second order logic contains a second kind of variable ranging over sets of objects so that the model $\mathfrak{a}$ of a second order sentence or theory involves, beyond the basic domain A, also (the set of) all subsets of A. Infinitary logics may include functions or relations with infinitely many arguments, infinitely long conjunctions and disjunctions, infinite strings of quantifiers. From studies on infinitary logics, Hanf was able to define certain cardinals some of which have been studied in connection with large cardinals in set theory. Yet another direction is the development of model theory for modal logics and the intuitionistic logic.

There is a big gap between the general theory of models and the construction of interesting particular models such as those which are employed in the proofs of independence (and consistency) of special axioms and hypotheses in set theory. It is natural to look for further developments of model theory which will yield more systematic methods of constructing models of axioms with interesting particular properties, especially along the line of deciding whether certain given sentences are derivable from the axioms. Relative to the present state of our knowledge, such goals appear fairly remote. The gap is not unlike that between the abstract theory of computers and the basic properties of actual computers.

B. Characterizations of (the first order) logic.

We have outlined above a proof of the completeness of elementary logic without including equality. The proof can be extended to the full elementary logic in a fairly direct manner. Thus, if F is a sentence containing equality, we can adjoin a sentence G which embodies the special properties of = relevant to the sentence F. Then we can treat the conjunction C of F and G as a sentence not containing = (i.e., = is treated as an arbitrary relation symbol). Hence, C has a model in the sense of logic without equality, if and only if F has a model in the sense of logic with equality, and we can infer the completeness of elementary logic (with equality).

A concept more general than validity is the relation of logical entailment between a (possibly infinite) set X of sentences and a single sentence P, which holds if and only if p is true in every model of X, or, for all M, $M \models p$ if $M \models X$. In particular, p is valid if the emty set logically entails P. This suggests a stronger requirement on a formal system of logic: $X \vdash p$ (p is derivable from X by the system) whenever X logically entails P. Theusual systems of logic do satisfy this requirement because, besides the completeness theorem. there is also a compactness theorem: A theory X has a model if every finite subset of X has a model.
Roughly speaking, this enables us to reduce an infinite set X to a finite subset X_1 in each individual case, and the case of entailment when X_1 is finite is taken care of by the completeness result.

These results show that the ordinary systems of elementary logic are the correct formulation, provided we assume that the actual choice of the truth functions (say negation and disjunction), the quantifiers and equality as the "logical constants" is the correct one. There remains the question of justifying the particular choice of logical constants. For example, one might ask whether "for most x" or "for finitely many x" should not be counted as a logical constant. Lindström gives a general concept of logic and shows that logics which apparently extend the first order logic all end up being the same as it, provided they satisfy the Löwenheim–Skolem theorem and they either have the compactness property or are formally axiomatizable. There remains the question whether or why these requirements (especially the one on the Löwenheim–Skolem theorem) are intrinsic to the nature of logic.

C. Generalizations and extensions of the Löwenheim–Skolem theorem.

A generalized theorem can be proved using basically the same ideas as in the more special case discussed above:

Generalized Löwenheim–Skolem Theorem. If a theory T has any infinite model, then, for any infinite cardinality α, T has a model of cardinality α. More explicitly, this contains two parts. (a) If a theory T has a model of infinite cardinality β, then, for each cardinal $\alpha > \beta$, T has a model of cardinality α, (b) If a theory T has a model of infinite cardinality β, then, for each infinite cardinal $\alpha < \beta$, T has a model of cardinality α.

It follows immediately that any theory having an infinite model has two nonisomorphic models, and is, therefore, not categorical. This applies, in particular, to the theories T_a and T_b of arithmetic (based on the language of N) mentioned before, whose natural models are countable, as well as theories dealing with real numbers and arbitrary sets whose natural models are uncountable. Both kinds of theory have both countable and uncountable models. There is much philosophical discussion about this phenomenon.

It is not excluded that a theory may be categorical in some infinite cardinality. For example, the theory T_d of dense linear ordering is categorical in the countable cardinality. One application of the Löwenheim–Skolem theorem is: If a theory T has no finite models and is categorical in some infinite cardinality α, then T is complete (i.e., for every closed sentence p in the language of T, either p or its negation belongs to T). An immediate consequence of this is that the theory T_d is complete.

A result which is generally regarded as one of the hardest to prove in model theory is:

Morley's Theorem. A theory, which is categorical in one uncountable cardinality, is categorical in every uncountable cardinality.

The two–cardinal theorems deal with languages with some distinguished preducate U. We say that a theory T admits the pair $\langle\alpha, \beta\rangle$ of cardinals if T has a model (with its domain) of cardinality α in which the value of U is a set of cardinality β. The central two–cardinal theorem says:

If a theory T admits the pair $\langle\alpha, \beta\rangle$ of infinite cardinals with $\beta < \alpha$, then, for each regular cardinal γ, T admits $\langle\gamma^+, \gamma\rangle$ where γ^+ is the next larger cardinal after γ. The most interesting case is when γ is the least infinite cardinal (The general theorem is established only under the assumption of the "generalized continuum hypothesis").

D. Ultraproducts and ultrapowers.

An ultrafilter on a nonempty set I is a set D of subsets of I such that (1) the empty set does not belong to D; (2) if A, B are in D, so is $A \cap B$; (3) if $A \subseteq B$ and A is in D, then B is in D; (4) for every subset A of I, either A is in D or I–A is in D. To use a rough terminology, each ultrafilter of a set I gives a notion of large subsets of I so that any property applying to all members of a member of D applies to I "almost everywhere".

Let $\{\mathfrak{a}_i\}$, $i\varepsilon I$ be a family of structures indexed by I, and D be an ultrafilter on I. We can pass from the direct product B of this family to a new structure U, with the help of D. We consider the equivalence relation $\equiv$ (relative to D) on B such that for a, bεB, $a \equiv b$ if and only if $\{i|a_i = b_i\}\varepsilon D$. This structure U, viz. the quotient set of B by the relation $\equiv$, is the ultraproduct of the original family of structures. For example, $R_D abc$ if and only if $\{i|Ra_ib_ic_i\}\in D$. We get the ultrapower of a structure $\mathfrak{a}$ (relative to I and D) when every $\mathfrak{a}_i$, $i\varepsilon I$, coincides with $\mathfrak{a}$.

The central theorems are the following:

(1) If $\mathfrak{a}_i$, $i\varepsilon I$ are realizations of the same language, then a sentence p is true in the ultraproduct U, if and only if the set of i, such that p is true in $\mathfrak{a}_i$, belongs to D. In particular, if each $\mathfrak{a}_i$ is a model of a theory T, then U is also a model of T.

(2) Two realizations of the same language are elementarily equivalent if they have the same set of true sentences. A necessary and sufficient condition for two realizations to be elementarily equivalent is that they admit ultrapowers which are isomorphic.

One application is to the introduction of nonstandard analysis, which was initially set up by corpactness considerations. By using a suitable ultrapower of the structure of the field $\mathbb{R}$ of real numbers, we get a nonarchimedean (i.e., permitting numbers a and b, such that no n can make na greater than b) real closed field which is elementarily equivalent to $\mathbb{R}$. This supplies an unexpected exact foundation of the classical differential calculus using infinitesimals. The result has considerable historical, pedagogical, and philosophical interest.

A widely known application to the area of algebra deals with the fields Q_p (p–adic completion of the rational numbers). A famous conjecture says that every form of degree d over Q_p, in which the number of variables exceeds d^2, has a nontrivial zero in Q_p. Ax and Kochen showed, using ultraproducts, that the conjecture is true for arbitrary d with the possible exception of a finite set of primes p (depending on d). Subsequently, it was found that the original conjecture is not true in full generality.

Another useful tool in model theory is the pigeon hole principles. The basic principle is that if a set of large cardinality is partitioned into a small number of classes, some one class will have large cardinality. Those elements of the set which lie in the same class cannot be distinguished by the property defining that class. A related idea is that of "indiscernibles" which also has rather extensive applications in set theory.

An ordered subset X of the domain of a model $\mathfrak{a}$ of a theory is a (homogeneous set or) set of indiscernibles for $\mathfrak{a}$, if $\mathfrak{a}$ cannot distinguish members of X from one another. More exactly, given any $x_1 < \cdots < x_n$, $y_1 < \cdots < y_n$ in X, for any sentence $F(a_1, \cdots, a_n)$ of the language of the theory, $\mathfrak{a}\ F \models (x_1, \cdots, x_n)$ if and only if $\mathfrak{a} \models F(y_1, \cdots, y_n)$. A first theorem on this notion says that given a theory T with an infinite model and a linearly ordered set X, there is a model $\mathfrak{a}$ of T such that X is a set of indiscernibles for $\mathfrak{a}$.

15. COMPUTERS AND MATHEMATICAL ACTIVITY*

15.1 Remarks on machines, sets and the decision problem[1]

1. Machines and production systems

1.1 The basic distinction between monogenic and polygenic systems corresponds to the contrast of calculations with proofs, functions with relations, and machines with production systems. In calculations, we generally have a fixed procedure such that the answer is completely determined by the question. In looking for a proof of a given statement in a given formal system, we have in general an unbounded number of choices at each stage since, for example, there are infinitely many p's such that $p \supset q$ together with p would yield q. If there is a fixed number n such that at each node, there are only n or less choices, then clearly we can get a monogenic system in the search for proofs. A monogenic proof procedure, such as the Herbrand expansion procedure for the predicate calculus, need not give a decision procedure. On the other hand, a monotone system, such that by some criterion the conclusion is always longer or more complex than the premisses, is always decidable when there are finitely many rules only. Thus, given a statement p, the total number of statements which can enter in a proof of p is finite since every rule has a fixed number of premisses.

Hence, it is of interest to inquire when a polygenic system is equivalent to a monogenic one, and when either is equivalent to a monotone one.

1.2. A machine which halts on every finite input corresponds to a function from the input to the output. If, on the other hand, we allow, e.g., that the machine can do either of two things at each moment, then for each input we can get many outputs, and we get, in general, a relation Rxy such that y is an output of the machine for the input x. It seems somewhat unnatural to speak of a polygenic machine, but with a Post production system, the distinction between monogenic and polygenic is perfectly natural.

In Turing machines, we are usually interested in tapes which are blank for all but a finite number of squares. The consecutive minimum portion containing all marked squares and the square presently under scan could be taken as the string of symbols in a production system. In that case, a machine corresponds to a monogenic production system except for the fact that the former has a scanned square at each moment and has different states.

---

* First published in *Formal Systems and Recursive Functions*, by Crossley et al. © 1963, North–Holland Publishing Company. Reproduced by permission.

1) Work for this paper was supported in part by NSF grant GP–228 and in part by Bell Telephone Laboratories, Inc., Murray Hill, New Jersey.

Definition 1: A labeled rewriting system is a finite set of rules $P_i \to Q_i$ such that in each P_i and Q_i exactly one symbol has an arrow above it (the label indicating the square under scan).

Theorem 1: *There is an effective method by which, given any Turing machine, we get a corresponding monogenic labeled rewriting system in which each P_i (also each Q_i) contains exactly two symbols, one of which is labeled.*

To prove this, we use a Turing machine formulation such that in each state, a machine prints, shifts, and changes state according to the symbol newly under scan. In other words, if there are m states $q_1, \ldots, q_m$, n symbols $S_1, \ldots, S_n$, a machine is given by $q_a S_i \pm 1 S_j q_b$ ($a = 1, \ldots, m$; $i, j = 1, \ldots, n$), so that if the machine is in state q_a scanning symbol S_i, it shifts right ($+1$, or left, -1) and then scans the next square, ending up in a state q_b determined by the newly scanned symbol S_j. It is not hard to verify that this formulation is equivalent to the usual one in the sense that they can simulate each other.

With this formulation, we can always use an alphabet with $(m+1)n$ symbols and one state only. Thus, instead of the given state q_a and the symbol S_i, we have the symbol (a, i). This is changed to $(0, i)$. After the shift, the scanned symbol is $(0, j)$ which is now changed into (b, j). In other words, for $c = 1, \ldots, m$ and $d = 1, \ldots, n$, (c, d) is a symbol indicating state c and symbol d, when the square is under scan; a symbol d in other squares is represented by $(0, d)$. This makes it easy to give a 1–state universal machine and yields a measure of the complexity of Turing machines solely by the size of the alphabet (using always 1 state only). This also gives Theorem 1 immediately, since the rules are simply of the forms

$(a\downarrow i)(0, j) \to (0, i)(b\downarrow j)$ for right shift,

$(0, j)(a\downarrow i) \to (b\downarrow j)(0, i)$ for left shift.

1.3. Multiple tapes naturally make it possible to simulate each $m \times n^k$ one–tape machine by an (m, n, k) (m states, n symbols, k tapes) machine; but the full force is not used in the simulation and it is desirable to find more accurate measures than these.

Recently, P. K. Hooper [7] proved:

Theorem 2: *There is a (2, 3, 2) universal Turing machine; there is a (1, 2, 4) UTM, having a fixed loop for one of its four tapes.*

In the realm of "real time computation," Michael Rabin has recently proved that there are calculations which can be performed by two tapes but not by one tape. The whole area of efficient calculations (as against theoretical computability) is wide open and promises much interesting work.

Although there are various elegant formulations of Turing machines, they are still radically different from existing computers. To approach the latter, we should use fixed word lengths, random access addresses, accumulator, and permit internal modification of the programs. Alternatively, we could, for example, modify computers to allow more flexibility in word lengths. Too much energy has been spent on oversimplified models so that a theory of machines and a theory of computation which have extensive practical applications have not been born yet.

1.4. There are a number of conceptually neat results on the theoretical side. We mention a few recent ones at random.

The most elegant formulation of Turing machines is perhaps the *SS*–machines of Shepherdson and Sturgis [16]. An SS–machine is a finite sequence of instructions, each of which is of the following two types.

P_0, P_1:print 0(or 1) at the right end of the string S and go to the next instruction.

SD(k):scan and delete the leftmost symbol of S; if it is 0, go to the next instruction, otherwise, go to instruction k; if S is null, halt.

They have proved:

Theorem 3: *Every Turing machine (in particular, a UTM) can be simulated by an SS–machine.*

It is particularly easy to simulate these machines by Post production systems (see [22]).

1.5. A combinatorial system in the most general sense would be any finite set of rules, each of which effectively produces a finite set of conclusions from a finite set of premisses. The most intensively studied case is the one in which each rule has a single premiss and a single conclusion. Such a system is called monogenic if the rules are such that for any string at most one rule is applicable.

From this broad class of monogenic systems, Post chooses to consider the tag systems. A tag system is determined by a finite set of rules:

$$T_i: \qquad s_i \rightarrow E_i, \qquad i = 1, \ldots, \rho,$$

such that if the first symbol of a string is s_i, then the first P symbols are removed and the string E_i is appended at the end. Since the system is monogenic, $s_i \neq s_j$ when $i \neq j$. If the alphabet contains σ symbols, then $\rho = \sigma$.

Another natural class is, for want of a better name, the lag systems. A lag system is a set of σ^P rules:

$$L_i: \qquad s_{i1} \ldots s_{iP} \Big(\rightarrow E_i,$$

such that if the first P symbols of a string are $s_{i1} \ldots s_{iP}$, the first symbol, viz, s_{i1}, is deleted and E_i is appended at the end of the string.

In either case, E_i may be the null string. If S_i is the length of E_i and S is the maximum among S_i, then each system has a prefix number P and a suffix number S.

In [11] and [12], Minsky has proved the following remarkable result:

Theorem 4: *There is a tag system with prefix number $P = 2$ and suffix number $S = 4$, whose halting problem is unsolvable.*

This is improved slightly in [22] to get the suffix number down to $S = 3$, and then the result is shown to be best possible because every tag system with $P = 1$ or $P \geqq S$ is always decidable (i. e., both its halting problem and its derivability problem). More recently, Cocke and Minsky gave an improved proof of Theorem 4, from which the simplification to $S = 3$ follows directly. In these considerations, attempts to use the *SS*–machines have not been possible.

A similar result for lag systems is proved in [22] by using *SS*–machines:

Theorem 5: *There is a lag system with $P = S = 2$, whose halting problem is unsolvable; moreover, when $P = 1$ or $S \leqq 1$, every lag system is decidable.*

The tag systems are a subset of Post's monogenic normal systems, each of which has rules of the form

$$B_i \rightarrow E_i$$

such that a given string B_iQ becomes QE_i by the rule. It is quite easy to use *SS*–machines to get a normal system with $P = S = 2$ (P the maximum of the lengths of B_i) whose halting problem is unsolvable (see [22]).

A specially interesting subcase of the normal systems is the 1–normal systems in which B_i is always a single symbol. The 1–normal systems include all tag and lag systems with $P = 1$. It is obvious from [22] that the halting problem for every 1–normal system is decidable. S. Cook and S. Greibach have strengthened the result, with two radically different proofs, to get also:

Theorem 6: *The derivability problem (i.e., whether one string is deducible from another) of every 1–normal system is decidable.*

1.6. It has been known for quite some time that for Turing machines erasing is dispensible (see [19]). In theory, this result has the practical application that, e.g., paper tapes can be used in place of magnetic tapes.

The dispensibility of erasing is understood in the sense that every calculation can in theory be done without erasing. Recently, the consequence problem is considered and it is proved [15]:

Theorem 7: *If T ranges over nonerasing T. M., W ranges over words in their history, I ranges over (finite) inputs, then the relation $P(W, T, I)$ (i.e., W belongs to the history of T with input I) is recursive; on the other hand, for a fixed initial (finite) input, we can find a T. M. with erasing permitted such that the set of words in its history is not recursive.*

2. The decision problem and its reduction problem

2.1. In this part, we consider recent results on the decision and reduction problems of the (restricted) predicate calculus.

Since all mathematical theories can be formulated within the framework of the predicate calculus (quantification theory, elementary logic), Hilbert spoke of *the* decision problem when he was referring to the problem of finding a general algorithm to decide, for each given formula of the predicate calculus, whether it is satisfiable in some nonempty domain (or, has a model). He called this the main problem of mathematical logic. It is familiar today that this problem in its general form is unsolvable in a technical sense which is widely accepted as implying unsolvability according to the intuitive meaning. An interesting problem is to investigate the limits of decidable subdomains and the underlying reasons for the phenomenon of undecidability.

Recently, the general problem has been reduced to the formally simple case of formulas of the form $AxEx'\ Ay\ Mxx'y$, where M is quantifier–free and contains neither the equality sign nor function symbols. In fact, one can further restrict the class to those AEA formulas in which all predicates are dyadic, and each dyadic predicate G_i occurs only in some of the nine possible forms G_ixx, G_ixx', $G_ix'x$, $G_ix'x'$, G_iyy, G_ixy, G_iyx, $G_ix'y$, G_iyx'. The following is proved in [9].

Theorem 8: *Any AEA class including all formulas which contain only atomic formulas in three of the four forms* $(xy, yx, x'y, yx')$ *is undecidable; the class of all AEA formulas of the form* $Wxx \wedge U(xy, x'y) \wedge V(yx, yx')$, *that of the form* $U(xy, x'y) \wedge V(xy, yx)$, *that of the form* $U(yx, yx') \wedge V(xy, yx)$, *are all undecidable, where* W, U, V *are truth–functional expressions. Moreover, all these classes are reduction classes.*

This completely settles the question of decidable and undecidable prefix subclasses of the predicate calculus. This is true even if we allow formulas in the extended prenex forms, i.e., formulas which are conjunctions of formulas in the prenex normal form. (Compare [9] and [21]).

Theorem 9: *An extended prefix form class is a reduction type (and undecidable) if and only if either the prefix of at least one conjunct contains AEA or AAAE as an (oorder–preserving but not necessarily consecutive) substring, or there are two conjuncts of which the prefixes contain AAA and AE respectively. Moreover, it is decidable if and only if it contains no axioms of infinity. i.e., formulas which have only infinite models.*

2.2. In [21], a simpler alternative proof of Theorem 8 is given which has two additional properties: (a) only a small fixed finite number of dyadic predicates are needed, together with arbitrarily many monadic predicates; (b) finite models are preserved in the reduction procedure so that a formula has a finite model if and only if its corresponding AEA formula has a finite model.

Definition 2: Consider classes of formulas of the predicate calculus. For any class X, let $N(X)$, $I(X)$, $F(X)$ be the subclasses of X which contain all formulas in X which have respectively no model, only infinite models, finite models. If R is a reduction procedure which reduces a given class Y to Y^* and every subclass Z of Y to Z^*, then R is said to be a conservative reduction procedure for Y, if $(F(Y))^* = F(Y^*)$.

The following two theorems are proved in [21]:

Theorem 10: *If K is the class of all formulas of the predicate calculus and R is a conservative reduction procedure for K, then no two of the three classes $N(K^*)$, $I(K^*)$, $F(K^*)$ are recursively separable.*

Theorem 11: *If Z is the class of AEA formulas (or some suitable subclass of this, such as Δ_1 given below), then no two of the three classes $N(Z)$, $I(Z)$, $F(Z)$ are recursively separable.*

In another direction, Kahr (see [8]) extends Theorem 8 to the following:

Theorem 12: *A reduction class for the predicate calculus is the set Δ_1 of formulas with prefix AEA such that each formula of the set contains only monadic predicates and a single dyadic predicate.*

This proof can be modified as in [21] to get theorem 11 for Δ_1 and to give a corresponding result for the prefix $AAA \wedge AE$, and therewith an alternative proof of Suranyi's similar result [18] for the more complex prefix $AAA \wedge AAE$.

2.3. *In studying the AEA* case, "dominoes" were first introduced in [20], and are found to be useful for the study. They are also of some independent interest and are reviewed here mainly for the remaining open problems.

We assume there are infinitely many square plates (the domino types) of the same size (say, all of the unit area) with edges colored, one color on each edge but different edges may have the same color. The type of a domino is determined by the colors on its edges and we are not permitted to rotate or reflect any domino. There are infinitely many pieces of every type. The game is simply to take a finite set of types and try to cover up the whole first quadrant of the infinite plane with dominoes of these types so that all corners fall on the lattice points and any two adjoining edges have the same color.

Definition 3: A (finite) set of domino types is said to be solvable if and only if there is some way of covering the whole first quadrant by dominoes of these types.

It is natural to use ordinary Cartesian coordinates and identify each unit square with the point at its lower left hand corner. Then we can speak of the origin (0, 0), the main diagonal $x = y$, etc.

The following general questions on these games have been considered:

Definition 4: The (unrestricted) domino problem. To find an algorithm to decide, for any given (finite) set of domino types, whether it is solvable.

The origin–(diagonal–, row–, column–) constrained domino problem. To decide, for any given set P of domino types and a subset Q thereof, whether P has a solution with the origin (the main diagonal, the first row, the first column) occupied by dominoes of types in Q.

Theorem 13: *All the constrained domino problems are unsolvable (see* [9] *and* [21]*).*

The unrestricted domino problem remains open. In fact, as discussed in [20], there are two related open questions.[1]

Problem 1. Is the unrestricted domino problem solvable?

Problem 2. Does every solvable domino set have a periodic solution?

A positive solution of the second problem would yield also a positive solution of the first problem, but not conversely.

The unrestricted domino problem is related to a special subclass of the *AEA* formulas with dyadic predicates only, viz., those of the form

$$(1) \qquad U(G_1xy, \ldots, G_kxy; G_1x'y, \ldots, G_kx'y) \wedge V(G_1yx, \ldots, G_kyx; G_1yx', \ldots, G_kyx'),$$

or briefly,

$$(1) \qquad U(xy, x'y) \wedge V(yx, yx'),$$

where U and V are truth–functional combinations of the components.

Theorem 14: *Given a domino set P we can find a formula F_p of the form (1) such that P has a solution if and only if F_p has a model; conversely, given a formula F of the form (1), we can find a domino set P_F such that F has a model if and only P_F has a solution. Hence, the unrestricted domino problem is undecidable if and only if the decision problem of the class of all formulas of the form (1) is unsolvable.* (See [21].)

2.4. Results on the degree of complexity of *AEA* formulas are announced in the preliminary report [10]. The whole paper has not been completed because of the unwieldy construction of the simulation. An outline with proofs of the less combinatorial part is reproduced here.

The method of simulating Turing machines by domino sets with diagonal constraints, as developed in [9], can be extended to obtain a simulation of each Turing machine X with all its numerical inputs by a single domino set P_X *such that* when X is viewed as a function from inputs to outputs, every diagonal-constrained solution of P_X

[1] Recently (May 1964) Robert Berger has settled both questions in the negative.

satisfies the condition: if $X(n) = 1$, then K occurs at the point $(\alpha(n), \alpha(n))$; and if $X(n) = 0$, then K does not occut at $(\alpha(n), \alpha(n))$, where K is a domino type, α is a fixed monotone increasing recursive function. Expressing the solvability condition for P_X by an *AEA* formula, we can establish the following:

Lemma 1: *For every Turing machine X, there is an AEA formula* $F_X = (x)(Eu)(y)\, Jxuy$ *which contains a monadic predicate M, such that every model of* F_X *in the domain of natural numbers has the property that the model* M^* *of M separates the sets* $\hat{n}(X(n) = 0)$ *and* $\hat{n}(X(n) = 1)$, *and any such set can be used as* M^*, *with models of other predicates being recursive.*

More specifically, identify $M^*(\alpha(n))$, with $K^*(\alpha(n), \alpha(n))$, and, for all k, if for non, $\alpha(n) = k$, $M^*(k)$ is true. In this way, we shall be able to choose M^* which is recursive in $\hat{n}(X(n) = 0)$.

We shall leave the proof of the lemma out and discuss what consequences we can derive from it.

In the intended model, all other predicates of F_X are recursive. We use the fact that if F has a model, then $Jxx'y$, x' being short for $x + 1$, has a model in the domain of natural numbers. It can be shown that the formula has no finite models. A nonstandard model must also contain all the natural numbers. This seems sufficient for showing that any RE (recursively enumerable) predicate A is recursive in every model of M when $\hat{x}(X(x) = 0)$ and $\hat{x}(X(x) = 1)$ are suitably chosen (see below). "Recursive in" is defined for natural numbers, but if M* also includes other objects, we seem to require a generalization of the concept, which can be done in the natural manner. In any case, it is true that A is recursive in M* because A is recursive in the standard part of M* already. Further, the restriction to *RE* models also requires a definition for M* to be *RE*, one possibility is that its standard part is *RE*.

It may be pointed out, incidentally, that if we require that F_X has a unique model relative to the domain of natural numbers and the successor function, then all the predicates must have recursive models by the infinity lemma.

Alternatively, we may also wish to relativize the definition of the given *RE* predicate A. Then we have to define A by a quantificational schema. Hence, we have to begin with all possible models.

Another way of proceeding is to confine our attention to models in the domain of natural numbers since otherwise recursive and *RE* are not defined. This last alternative seems the most natural way.

In other words, we are only concerned with models of F_X in which the domain is the set of natural numbers and the existential quantifier is replaced by the successor function. This is not regarded as a weakened condition because otherwise we cannot talk about recursive and *RE* models. This is indeed the practice followed by earlier authors.

Lemma 2: *If A is RE, then there are disjoint RE sets B, C, which are* $\leqslant_T A$, *i.e., recursive in A, such that if an RE set D separates B and C, i.e.,* $B \subset D$, $C \subset \bar{D}$, *then*

$A \leqslant_T D$, *i.e., A is recursive in D; in particular,* $A \leqslant_T B$, *and, hence,* $A =_T B$.

This follows from the proof (though not the statement) of Theorem 1 in [17].

Observe that unlike recursive separability, we cannot infer from the existence of an *RE* set *D* separating *B* and *C*, *that there is an RE set E* separating *C* and *B*, i.e., $C \subset E$ and $B \subset \bar{E}$, since $\bar{D}$ is not *RE* unless *D* is recursive.

The condition that *D* is an *RE* set is essential. Thus, if *A* is of degree $0'$, then *D* must be of degree $0'$ too. In an unpublished work, Dana Scott shows that there is a degree $d <_T 0'$ such that any two disjoint *RE* sets are separable by a set (not necessarily *RE*) of degree $< d$.

Theorem 15: *For every RE set A, there is an AEA formula* F_A *which contains a monadic predicate M among its predicates such that (1) the model M* of* $M =_T A$ *and all other predicates have recursive models, and that (2) in every RE model of* F_A, $M^* \geqslant_T A$.

Proof. Given the set *A*, take the sets *B* and *C* as given in Lemma 2. It is familiar that for any disjoint *RE* sets *B* and *C*, there is a corresponding Turing machine *X* such that $x \varepsilon B \equiv X(x) = 0$, $x \varepsilon C \equiv X(x) = 1$. Use the machine *X* corresponding to the given sets *B* and *C*, and apply Lemma 1. Hence, in every model of F_X, $M^* \geqslant_T A$. If now we choose $M^* = B$, we have $M^* =_T A$.

It appears likely that if we do not want the stronger result with the restriction to *AEA* formulas, we can combine Lemma 2 with familiar considerations to get a weaker form of Theorem 15. Thus, for example, we can write a more complex formula characterizing the machine *X* in Lemma 1.

Proof. of Lemma 2. By definition of *RE*, there exists *g*:

$$x \varepsilon A \equiv (Eu)(Ey)[x = U(y) \wedge T(g, u, y)] \equiv (Eu)(Ey) R(x, u, y).$$

Hence, there is *f*:

$$x \varepsilon A \equiv (Ey)\, T(f, x, y).$$

Let

$$x \varepsilon B \equiv (Ey)[T(f, (x)_0, y) \wedge \neg (Ez)_{z \leqslant y}\, T((x)_1, x, z)]$$

$$x \varepsilon C \equiv (Ey)[T(f, (x)_0, y) \wedge (Ez)_{z \leqslant y}\, T((x)_1, x, z)].$$

Clearly *B* and *C* are disjoint. Since

$$[(Ey)(Fy \wedge \Box\, Gy) \vee (Fy)(Ey \wedge Gy)] \equiv (Ey) Fy,$$

$$(x \varepsilon B \vee x \varepsilon C) \equiv (x)_0 \varepsilon A.$$

It is easy to see that *B* and *C* are recursive in *A*. Thus, if $(x)_0 \notin A$, then $x \notin B$, $x \notin C$. If $(x)_0 \varepsilon A$, we can determine the unique *y* such that $T(f, (x)_0, y)$. Hence,

$$x \varepsilon B \equiv \neg (Ez)_{z \leqslant y}\, T((x)_1, x, z)$$

$$x \varepsilon C \equiv (Ez)_{z \leqslant y} T((x)_1, x, z).$$

Suppose now $B \subset D$, $C \subset \bar{D}$, and D is RE. Choose e so that $D = \hat{x}(Ey)\, T(e, x, y)$. To determine whether $w \varepsilon A$, we ask just whether $x = 2^w 3^e$ belongs to D. i.e., whether $(Ey)\,[y < \mu_z T(e, x, z) \wedge T(f, w, y)]$.

Case 1. $x \notin D$. We have then

Hence, $w \notin A$, because otherwise, if $w \varepsilon A$, then $x \varepsilon B$. But $x \varepsilon B$ implies $x \varepsilon D$, contrary to hypothesis.

Case 2. $x \varepsilon D$. We can then find unique z, $T((x)_1, x, z)$. Since $x \varepsilon D$ implies $x \notin C$,

$$w \varepsilon A \equiv x \varepsilon B \bigcup C$$

$$\equiv x \varepsilon B.$$

But then,

$$x \varepsilon B \equiv (Ey)_{y<z} T(f, (x)_0, y),$$

because otherwise, i.e., if $z \leqslant y$, then the second half of the condition for $x \varepsilon B$ cannot be satisfied.

Since B can serve as a D, $A =_T B$.

2.5. Since the class U of AEA formulas with dyadic predicates only is unsolvable and a reduction type, it is of interest to consider what subclasses are solvable. The following is proved in [5] and a more "geometrical" alternative proof is given in [21].

Consider the four forms xy, yx, $x'y$, yx'. First take any three of them. From Theorem 8 above we know that any subclass of U which includes all formulas whose atomic formulas are in just these three forms is a reduction class and hence is undecidable. Now take any two of the four forms. Combining them with the other five forms yields a subclass of U. In this way we obtain six subclasses of U which divide into three pairs:

$$J = \{xy,\ x'y\}, \qquad J^* = \{yx,\ yx'\},$$
$$L = \{xy,\ yx\}, \qquad L^* = \{x'y,\ yx'\},$$
$$Q = \{xy,\ yx'\}, \qquad Q^* = \{yx,\ x'y\}.$$

Theorem 16: *With the exception of subsets of Q and those of Q^*, a class, determined by the forms of atomic formulas occurring, is decidable if and only if it contains at most two of the four forms $xy, yx, x'y, yx'$; it contains an axiom of infinity if and only if it contains three forms including either xy and $x'y$, or yx and yx'.*

Problem 3. Is the class of AEA formulas with dyadic predicates only which occur only in contexts with xy, yx', xx, xx', $x'x$, $x'x'$, yy decidable? Is the class decidable when dyadic predicates occur only in the contexts with xy, yx', xx?

Problem 4. Does either case contain an axiom of infinity, i.e., a formula that has only infinite models?

If the answer to the latter question is no for either class, then, by familiar arguments, that class is solvable.

2.6. Surányi has applied his reduction classes with the prefix $AAA \wedge AAE$ to obtain reduction classes with more complex prefixes but fewer predicates. Denton has undertaken to study similar consequences of the AEA reduction:

Theorem 17: *The following classes are reduction classes*
(a) $Ey_1 \dots Ey_n AxEyAzM (n = 1, 2, \dots)$ *with one predicate only which is dyadic;*
(b) $AxEy(Pxy \wedge (pxx \not\equiv Pyy)) \wedge Az_1 \dots Az_n M$ *(and therewith* $AxEyAz_1 \dots Az_n M$*)* *with only the predicate* P.

Part (a) follows from Theorem 12 in exactly the same way as Surányi's Theorem IV follows from his reduction class with prefix $AAA \wedge AAE$. Part (b) was announced in [3] and afterwards also proved by another argument using only Theorem 8.

In [6], Gegalkine claims that the class of formulas $AxEyFxy \wedge Az_1 \dots Az_n M$ with M containing any number of monadic and dyadic predicates is decidable for finite satisfiability. Denton shows in [4] that this would contradict Theorem 11 for Δ_1, and singles out the mistake in Gegalkine's paper. Further, he proves, by extending Ackermann's work [1], the following:

Theorem 18: *The class* $AxEyPxy \wedge Az_1 \dots Az_n M$, *where* M *contains only the dyadic* P *and monadic predicates, is decidable for finite satisfiability.*

Problem 5. Is the class $AxEy_1 \dots Ey_n AzM$, with only a single predicate (dyadic), a reduction class?

Problem 6. Is the class in Theorem 18 (or even without the monadic predicates) a reduction class?

3. Sets

3.1. The basic axioms of the system ZF of set theory are extensionality, infinity (unconditional existence) and four axioms of conditional existence: (a) pairs, (b) sum set, (c) power set, (d) replacement (a schema). It is noted in [23] that these axioms (a)–(d) are equivalent to a single axiom (schema): if A is a one–many correlation, x is a set, and $A``t = \hat{u}(Ev)(t \varepsilon v \wedge Auv)$, *then there is a set* $y$, $y = \Sigma A``\pi x$ *(*Σ is sum set, π is power set*)*.

Thus, if Auv *is* $v = \{u\}$, then $\Sigma x = \Sigma A``\pi x$. If $Auv$ is $u = \{v\}$, then $\pi x = \Sigma A``\pi x$. If Auv *is* (Ez) (Ew) $(Gzw \wedge u = \{z\} \wedge v = \{w\})$, then $G``x = \Sigma A``\pi x$. If 0 is $(u = v \wedge v \neq v)$ $``x$, $Guv$ is $(u = a \wedge v = 0) \vee (u = b \wedge y = \{0\})$, then $\{a, b\} = G``\pi\pi 0$. The part about $\{a, b\}$ is familiar from the literature.

Previously, Bernays ([2], p. 65) had employed a similar schema $y = \Sigma A``x$ to get (b) and (d). Ono ([13]) had introduced a schema which would yield (a), (b), (c), (d) if we add another axiom: $(x)(Ey)(y = \{x\})$. As it turns out, in a less explicit way, the same axiom is also needed for the result stated in [23], although Ono's schema is different.

3.2. A different procedure is followed in [24] to get a schema which would yield all the axioms including infinity and extensionality.

A partial hull is a transitive set closed with respect to power sets, i.e., *PH(x)* if (1) $\Sigma x \subseteq x$, (2) $y \varepsilon x \subset \pi y \varepsilon x$. A natural hull is closed also with respect to sum sets, i.e., *NH(x)* if PH(x) and (3) $y \varepsilon x \subset \Sigma y \varepsilon x$. The natural holl ηa (or paritla hull ξa) of a set a is the intersetion of all natural (partial) hulls x such that $a \varepsilon x$.

Theorem 19: *In ZF, ηa can be shown to exist for each a, and to satisfy the conditions (1)—(4), as well as that of being the minimum; similarly for ξa with the conditions (1), (2), (4).*

Let *SE* be obtained from the usual axiom of replacement by substituting ηx *or* ξx for the given set x, viz., the schema: If *Huv* is many–one, then (Ey) $(y = H``\eta x)$ (or $(Ey)(y = H``\xi x)$). Let *UE* be obtained from *SE* by adding uniqueness, i.e., by substituting (E!y) for (*Ey*). Both *SE* and *UE* can be expressed in the primitive notation of *ZF*.

Theorem 20: *In the predicate calculus with equality, SE yields all existence axioms of ZF, UE is equivalent to all these axioms plus extensionality.*

For some purposes, it is also useful to define other closures, e.g., τa, the transitive closure of a, would be the smallest x, $a \subseteq x$ and $\Sigma x \subseteq x$.

3.3. The usual definitions of the class On of the Zermelo–Neumann ordinals do not reveal the intuitive picture of how the ordinals are obtained successively. It is possible to use a "genetic" definition roughly in the tradition of Frege and Dedekind. This approach has the advantage that we can also use different successor functions, e.g., $x' = \pi x$ rather than $x' = x \cup \{x\}$. Such differnt successor functions are useful, e.g., in studying the natural models of von Neumann and Bernays.

Two definitions of the Zermelo–Neumann ordinals are found to be adequate:

DI. $\mathrm{On}_1(x)$ when x belongs to every set u such that (1) $v' \varepsilon u$, if $v \varepsilon u$ and $v' \varepsilon x'$; and (2) $\Sigma w \varepsilon u$, if $w \subseteq u$ and $\Sigma w \varepsilon x'$.

DI*. $\mathrm{On}_2(x)$ when for every set u, there is w, $\Sigma w \in u$, $w \subseteq x$, and $u \cap w = 0$; if $x \in u$ and for all v, $v \varepsilon u$ if $v' \varepsilon u$.

In fact, a general theorem on these ordinals is proved:

Theorem 21: *If for a predicate* On(x) *we can prove that for every F, Fy if* (1) On(y), (2) $(v)(Fv \supset Fv')$, *and* (3) $(w)(w \subseteq F \supset F(\Sigma w))$; *then every x which satisfies* On(x) *is a genuine ordinal. Hence, if* On(x) *is a property known to hold for all ordinals, then the definition is adequate.*

When specialized to finite ordinals, we get:

DF. $Nn_1(x)$ when x belongs to every set u such that (1) $0 \; \varepsilon \; u$ if $0 \varepsilon x'$; (2) $v' \varepsilon u$, if $v \varepsilon u$ and $v' \varepsilon x'$.

DF*. $Nn_2(x)$ when for every u, $0 \varepsilon u$, if (1) $x \varepsilon u$, (2) $v \; \varepsilon \; u$ if $v' \; \varepsilon \; u$.

These are also adequate definitions. In fact, DF* is one which had previously been studied by W. V. Quine and K. R. Brown.

In all these developments, weak axioms are enough, viz., extensionality, Aussonderung, and self-adjunction $(x)(Ey)(y = x \cup \{x\})$. To get also recursive definitions or transfinite recursions, some strengthening of the axioms in the standard manner is necessary.

References

[1] W.Ackermann, Beitrage zum Entscheidungsproblem der mathematischen Logik. Math. Annalen **112** (1936) 418–432.

[2] P. Bernays and A. Fraenkel, Axiomatic Set Theory. (Amsterdam 1958).

[3] J. S. Denton, A Reduction Class with a Single Dyadic Predicate. Notices AMS **10** (1963) 124—125.

[4] J. S. Denton, A False Decision Procedure for the Halting Problem. Notices AMS **10** (1963) 125.

[5] B. Dreben, A. S. Kahr, and Hao Wang, Classification of AEA Formulas by Letter Atoms, Bulletin AMS **68** (1962) 528—532.

[6] I. Gegalkine, Problema razresimosti na konecnik klassah. Ucenye Zapiski Mosk. Gos. Univ. **100** (1946) 155—212.

[7] P. K. Hooper, Some Small Multi–tape Universal Turing Machines. Notices AMS **10** (1963) 584.

[8] A. S. Kahr, Improved Reductions of the Entscheidungsproblem to Subclasses of AEA Formulas, Symposium on the Mathematical Theory of Machines, Brooklyn Polytechnic Institute (April 1962); Proceedings (New York 1963) 57—70.

[9] A. S. Kahr, Edward F. Moore, and Hao Wang, Entscheidungsproblem reduced to the AEA Cace. Proc. Nat. Acad. Sci., U.S.A. **48** (1962) 365—377.

[10] A. S. Kahr and Hao Wang, Degrees of RE Models of AEA Formulas. Notices AMS **10** (1963) 192—193.

[11] M. L. Minsky, Recursive Unsolvability of Post's Problem of Tag. Annals of Mathematics **74** (1961) 437—455.

[12] M. L. Minsky, Universality of (p = 2) Tag Systems. A. I. Memo No. 33 (Cambridge, Mass. 1962).

[13] K. Ono, A Set Theory founded on Unique Generating Principle. Nagoya Mathematical Journal **12** (1957) 151—159.

[14] E. L. Post, Formal Reduction of the General Combinatorial Decision Problem. American Journal of Mathematics **65** (1943) 197—215.

[15] M. O. Rabin and Hao Wang, Words in the History of a Turing Machine with a Fixed Input. Journal ACM **10** (1963) 526—527.

[16] J. C. Shepherdson and H. E. Sturgis, Computability of Recursive Functions. Journal ACM **10** (1963) 217—255.

[17] J. R. Shoenfield, Degrees of Formal Systems. Jorunal of Symbolic Logic **23** (1958) 389—392.

[18] J. Surányi, Reduktionstheorie des Entscheidungsproblem (Budapest 1959).

[19] Hao Wang, A Variant to Turing's Theory of Computing Machines. Journal ACM **4** (1957) 53—92.

[20] Hao Wang, Proving Theorems by Pattern Recognition, II. Bell Systems Technical Journal **40** (1961) 1—41.

[21] Hao Wang. Dominoes and the AEA Case of the Decision Problem. Symposium on the Mathematical Theory of Machines, Brooklyn Polytechnic Institute (April 1962); Proceedings (New York 1963) 23—55.

[22] Hao Wang, Tag systems and Lag Systems. Math. Annalen (1963) 65—74.
[23] Hao Wang, A Universal Axiom of Conditional Set Existence. Notices AMS **10** (1963) 588.
[24] Hao Wang, Natural Hulls and Set Existence. Nitices AMS **10** (1963) 594.

15.2 Logic and computers *

1. Historical and philosophical background

Familiar connections between mathematical logic and automatic computers are the possibility of representing basic building blocks of computers by (sequential) Boolean functions and the close resemblance between programming languages and symbolisms of logic. As a result, both for the construction and for the use of computers, a certain degree of acquaintance with logic becomes indispensable. As early as 1656, Leibniz dreamed of a universal scientific language in his first published work; and many people today are actively seeking for a universal language for computers. Gottlob Frege wished to reduce arithmetic to logic from 1879 on, and in an oblique way the performance of arithmetic operations in computers by means of electronic circuits which are essentially logical functions may be said to accomplish the task in a particularly down-to-earth manner.

A more basic link between logic and computers is perhaps the common interest in algorithms. Although Charles Babbage conceived of and started to build in the 1830's the Analytical Engine which possessed most of the basic characteristics of modern computers, it was only in the 1940's that automatic computers began to appear througn the efforts of Howard Aiken, John von Neumann, and others. The logicians had, on the other hand, made not only a highly successful abstract study of algorithms but even clarified the relation between machines and algorithms, largely through A. M. Turing's theory of idealized machines, all in the 1930's.

Traditionally the study of algorithms falls outside the domain of logic. The deepest source of the affinity of logic and computers is the preoccupation of logicians with formalization. The long evolution of attempts to formalize mathematical proofs, from Euclid to *Principia*, finally led to mechanizability as the ultimate criterion of complete success. This desire to make arguments formally precise and exact is concerned more with the product rather than with the theory of formalization. Only in the 1920's, D. Hilbert, P. Bernays and others began to study metamathematics: the theory of proofs, the theory of formal systems. The distinction may be illustrated by the method of "discarding 9's" for checking multiplications. It is a metamathematical result on the usual technique of multiplying particular integers that a number is divisible by 9 if the sum of its digits is. This example also brings out the fact that the distinction between mathematics and metamathematics is not sharp. For we can easily state and prove the above result as a simple theorem in number theory by the easy relation $10^n \equiv 1 \pmod 9$.

---

* First published in *American Mathematical Monthly*, vol, 72, pp 135—140.

It was the concern with a theory of proofs which at first led J. Herbrand to an abstract definition of calculation processes, as a particularly simple type of proofs. Although Turing was said to have had formulated his theory of machines before he was familiar with much of the achievements in logic, he certainly did his homework quickly and soon put his work in the main stream. The surprisingly simple solutions to the question of giving a general definition of algorithms were undoubtedly an important cause of the rapid developments in the area of abstract studies of calculations.

2. Between engineering and mathematics

Logic was a bastard of mathematics and philosophy; while actual computers first came into being as a great feat of engineering. This divergence in their ancestry presents serious sociological and scientific difficulties for those who are interested in the vaguely defined region referred to as "logic and computers." This is not the place to digress into sociology.

The trouble on the scientific side is that most ambitious people find dreary piecemeal engineering and idle intellectual gymnastics equally repulsive. And it seems as though there is little else to offer at the present stage, except it be irresponsible speculations. The origin of the problem goes further back. Every branch of applied mathematics embodies an intrinsic dilemma: each piece of work is either not sufficiently applied or not sufficiently mathematical. We have to present a patient defense in each case.

As Turing machines and actual computers werer studied more or less independently of one another, there slowly developed a desire to bring about a marriage of theory and practice. This has proved to be an exceedingly difficult task. True, there are a number of basic results in very general terms. Turing machines are equivalent to actual computers if we disregard speed and the question of a potentially infinite supply of tape. In fact, there are alternative formulations of Turing machines which are more similar to actual computers, e.g., a representation of Turing machines by programs with a small number of basic instructions. Hence, since there are problems (e.g., the halting problem) which are unsolvable on Turing machines, the corresponding problems are unsolvable on actual computers. Another example is the result that in theory erasing is dispensable on Turing machines. Hence, magnetic tapes (in contrast with, say, paper tapes) are in theory not necessary for building computers.

Along with Turing machines, a simpler model of computers under the name "finite automata" has been extensively investigated and sometimes compared, in an expansive and speculative mood, to the human brain. This elegant model has also given rise to a number of amusing mathematical results.

In both cases, however, decisive steps remain to be taken before more significant applications can be made, probably with more realistic and less neat models. One basic difficulty is the transition from theoretical possibility to practical feasibility. For example, even though finite automata are closer to actual computers in being finite, the large size needed to represent an actual computer makes it seem likely that Turing machines will be a more useful model than finite automata even for practical purposes.

It must be emphasized that the mathematical theory of machines is a young discipline and, as such, it is doing very well so far. Moreover, it has the important advantages that little equipment beyond native wits is required for its pursuit and that it promises great things to come. But great things are rare. What is needed at present is not quantity but rather pursuers of good quality.

On the more theoretical level, the study of impractical algorithms and abstract machines, not as isolated idealizations, but in relation to other parts of logic and mathematics, has led to many significant mathematical results. Moreover, these results, although not often their proofs, can usually be stated in quite simple terms. In terms both of their intrinsic intellectual merit and of their potential applications, they would seem to have as wide an appeal as, say, molecular biology.

3. Unsolvable problems

While engineering is primarily the study of how to make things, mathematics is more often concerned with showing that under certain general conditions, certain things can or cannot be done. There is a special appeal to show that certain things cannot be done, because such results involve, in a negative way, all the available resources of a given method. For example, in bisecting an angle, we use only a small part of the resources of ruler and compass; while in proving the impossibility of trisecting an arbitrary angle, we have to possess a clear conception of all the possible constructions which we can make with ruler and compass. It is in this area of demonstrating unsolvability that the abstract study of idealized machines has produced results of the greatest mathematical interest. In particular, the interplay of logic and the theory of computers is striking.

It is familiar to logicians that all mathematical theories can be formulated in the framework of elementary logic. Hence, if we could decide whether in general a statement is a theorem of logic, we would also be able to decide whether a statement is provable in any given mathematical theory. This situation explains why the Hilbert School regarded the Entscheidungsproblem, i. e., the problem of deciding whether a statement in logic is a theorem, as the main problem of logic.

From 1920 on, Post tackled this problem by formulating a more general one that deals with derivability in arbitrary production systems, of which the system of elementary logic is a special case. This turned out to yield a general concept of formal systems, and indirectly, one of calculation processes. Moreover, the abstract formulation renders possible experimentations on apparently simple cases, with a view to discovering some common pattern useful for the handling of the general case.

Unfortunately, as a means to get positive results, this attractive approach is, on the whole, quite powerless. One of the very first examples which Post studied in 1920–21, and reported publicly in 1943, remains unsettled today. Consider all (finite) strings made out of 0's and 1's and use two very simple rules: if a string begins with 0, delete the three symbols at the beginning and add 00 at the end; if it begins with 1, delete the three initial symbols and append 1101 at the end; (stop if a string contains less than three symbols). The problem is simply: do we have a general method of deciding, for any two strings, whether the second can be obtained from the first by the above two rules?

On the other hand, Post's approach can be employed to establish negative results, once the step is taken to identify solvability with that by a production system or some other equivalent method, say by a Turing machine. In fact, in 1936 Turing proposed and argued for such an identification, and applied results on Turing machines to prove the unsolvability of the Entscheidungsproblem. Quite recently, this result has been sharply refined to a certain degree of finality, with the help of a picturesque auxiliary tool of "dominoes." This type of work exemplifies the rich possibilities of applying the theory of machines to establish basic results about logic.

Applications in other branches of mathematics include P. S. Novikov's proof (1955) that the word problem for groups is unsolvable, a result which has been applied by A. A. Markov to prove (1958) that the 4–dimensional homeomorphy problem is unsolvable. The 3–dimensional homeomorphy problem remains open. Impressive partial results have been obtained on Hilbert's tenth problem: whether tere is a general method of deciding the question of solvability in integers of every polynomial equation with integer coefficients. To establish the unsolvability of this problem would be considered a major mathematical result.

The (formally) simplest example of an unsolvable problem is probably the following word problem formulated by G. S. Tsentin and D. Scott in 1955. Consider strings (words) made out of the five symbols, a, b, c, d, e and the following seven rules for mutual substitution: $ac \leftrightarrow ca$, $ad \leftrightarrow da$, $bc \leftrightarrow cb$, $bd \leftrightarrow db$, $adac \leftrightarrow abac$, $eca \leftrightarrow ae$, $edb \leftrightarrow be$. It is an unsolvable problem to decide whether any two words are equivalent by these rules.

4. Formalization

From the rich domain of mathematical logic, we have selected two aspects as specially relevent to computers, viz. the theory and the practice of formalization. The unsolvability results belong to the theory side, while formalizing individual mathematical proofs or "deriving mathematics from logic" belongs to the practice side. In this latter aspect the interplay of logic and computers is significant on a more concrete level: developments of logic combined with the great power of actual computers give rise to the hope of mechanizing mathematical arguments, not just in principle, but in practice as well.

As in engineering, there is little likelihood of general results in this positive enterprise. But, unlike an engineer, we are not concerned with actually making things and we do get exact results in each individual case.

The interest in mechanization implies a reorientation of formal logic with a view toward greater efficiency. In particular, this means that the need for economy of axioms and primitive concepts is to be supplemented with an exact formulation of a large body of concepts and rules, which make up the average mathematician's stock of trade.

This can best be illustrated by an example from elementary number theory.

Suppose we wish to prove:

(I) $$x > 1 \rightarrow (Ey)[Py \wedge (y \mid x)],$$

i.e., every integer greater than 1 has a prime divisor.

We assume given an organized stock of information *SF* with properties of $+$, $\cdot$, $<$ listed first, and then properties of P and $\mid$, which may involve the more basic concepts $+$, $\cdot$, $<$. The list *SF* is organized so that not too much searching is necessary to look up required properties.

The basic strategy is to assume the theorem false and try to derive a contra-diction from the least counterexample, which embodies a mechanically convenient form of the principle of mathematical induction. The imagined least counterexample provides an "ambiguous constant" which possesses not only general properties true of all integers but also unusual properties arising from the assumption that it is a counterexample. In simple cases such as (I), after we draw on *SF* and use simple truth–functional deductions, we quickly get an ambiguous constant with contradictory properties. It should be emphasized that the proof below is merely a sketch of an illustration. More elaborate strategies of roughly the same type are necessary in order to prove more complex theorems.

To prove (I), we first assume it false and let m be the least counterexample:

(1) $$m > 1$$

(2) $$Pb \rightarrow b \nmid m$$

(3) $$1 < a < m \rightarrow Py_a \wedge (y_a \mid a).$$

The only ambiguous constant thus far is m. Substitute m for a and b above in order to get further properties of m.

(4) $$Pm \rightarrow m \nmid m$$

(5) $$1 < m < m \rightarrow Py_m \wedge (y_m \mid m).$$

Look up *SF* and try to derive simple consequences from (1), (4), (5) with the help of *SF*. Find $m \nless m$ in *SF* and delete the trivially true (5). Find $m \mid m$ in *SF* and infer from (4):

(6) $$\sim Pm.$$

Now (4) may be deleted since it is a direct consequence of (6). We have now (1), (2), (3), (6). Look up *SF* and get the defining property of P as applied to m:

$$\sim Pm \leftrightarrow (Ex)[1 < x < m \wedge (x \nmid m)].$$

By (6), we get:

(7) $$1 < x_m < m$$

(8) $$x_m \mid m.$$

Since x_m is a new ambiguous constant, it is desirable to substitute it for the free variables in the general statments obtained so far, viz. (2) and (3) only.

(9) $$Px_m \rightarrow x_m \mid m$$

(10) $$1 < x_m < m \rightarrow Py^x{}_m \wedge (yxm|_{xm}).$$

Derive truth-functional consequences (first without appealing to *SF*) from (1), (2), (3), (6)—(10).

(11) $$\sim Px_m, \text{ by (8) and (9).}$$

(12) $$Py_{xm}, \text{ by (7) and (10).}$$

(13) $$y_{xm} \mid x_m, \text{ by (7) and (10).}$$

Now appeal to the list *SF* and use: $(a \mid b) \wedge (b \mid c) \rightarrow (a \mid c)$.
(14) $y_{xm} \mid m$, by (8) and (13).

Substitute the ambiguous constant $y^x{}_m$ for the free variables in (2) and (3):

(15) $$Py_{xm} \rightarrow y^x{}_m \mid m.$$

(16) $$1 < y_{xm} < m \rightarrow Py_{yx_m} \wedge (y_{yx_m} \mid y_{xm}).$$

By (14) and (15), we get:

(17) $$\sim Py_{xm}, \text{ contradicting (12).}$$

Obviously we have not listed all the blind alleys and the method has to be specified much more exactly before a machine program can be written. But, it is thought, the above outline makes it plausible that a fairly natural program can be written on existing machines to prove theorems like (1) and, for example, also $2x^2 \neq y^2$ (x, y range over positive integers).

A Short Reading List

1. B. A. Trakhtenbrot, Algorithms and automatic computing machines, D. C. Heath, Boston, 1963.
2. Marthin Davis, Computability and unsolvability, McGraw–Hill, New York, 1963.
3. American Mathematical Society, Experimental arithmetic, high speed computing and mathematics, Proc. Symposia in Appl. Math., 15 (1963).
4. Brooklyn Polytech. Inst., Mathematical theory of automata (Proc. of Symposium held April, 1962), 1963.

15.3 Remarks on mathematics and computers *

1. Introduction

The main body of this paper is devoted to suggestions on mechanical mathematics

* First published in *Theoretical Approaches to Nonumerical Problem Solving* (Lecture Notes Oper. Res. & Math Systems 28), pp. 152—160. © Springer Verlag, New York, 1970. Reproduced by permission.

(sections 3 and 4) and an analysis of the relations between mathematics and physically executable procedures (section 5). The more general comments in the first two sections are to round off the picture.

2. New uses of computers

The eventual goal of studying new uses of computers must be practical in a broad sense. They may be used to do familiar things in order to eliminate drudgery, to reduce cost, to increase reliability, or to speed up operations. The greater accuracy and speed alone may also make possible hitherto unachievable aids such as space projects or weather forecasting. The practical goal could also be the advance of knoweledge and understanding. Much of the unorthodox experiments and speculations on new uses of computers has to be justified in such terms. And it can be frustrating to remind oneself that much of the theoretical work on computers may turn out to be pointless in the long run.

There is a sort of conservation law. The immediately practicable applications such as airline reservations or recognition of characters typed by a given kind of machine are, though financially profitable, intellectually less challenging. While the more exciting problems are, almost by definition, much harder.

For example, computers are useful as a model of "thinking machines" in that we can now experiment with hardware models of program simulations there of which will perform certain mental acts. It is not so much (not in the foreseeable future anyway) that we aim at duplicating the brain but rather, we can try to ioprove existing computers, both in their use and in their structure, to perform zore and more sophisticated tasks. On account, however, of the radical novelty of qualitatively new applications, we are mostly at a loss as to how to proceed. In fact, this area shares with many new things serious and interrelated drawbacks: no solid foundation (such as Newtonian mechanics) to rely upon, no heritage to fall back on, cumulative advance not easy, standard of evaluating results less objective, vulnerability to exaggeration and deception.

There are also safer uses which are not practical in the narrow sense. For example, the very concept of computers lends a new dimension to discussions on Philosopnical problems such as mind and body, the nature of consciousness. In the area of mathematics, we can also list a few rather noncontroversial examples. Computers have been used as heuristic aids to deal with nonlinear problems. The complex data are not only useful in themselves but may suggest solutions to abstract mathematical problems in more general cases. There have also been work to prove general theorems in number theory by reducing them to some special numerical cases manageable on large computers. In numerical analysis, it is desirable to mechanize the sequencing of connecting steps between different procedures in order to take advantage of the automatic aspect of computers.

3. Influence of mathematics on the development of computers

Rather surprisingly, the influence of specific mathematical theories and results on the development of computers is quite limited. Perhaps we can mention only the two elementary things: Boolean algebra for circuit design and the binary notation of numbers. The abstract theory of idealized computers has had little practical impact.

In a more general way, the abstract theory has a course of good deal of educational value for users of computers. Moreover, however pure mathematicians may say, programming is quite typically a mathematical activity in so far as it involves a lot of "thought experiments" with characters and numerals. On the whole, a sort of mathematical spirit is crucial to the use of computers. In fact, with the current shift of emphasis from hardwares to softwares, one would expect the influence of mathematics to increase.

The mathematical study of computers is attractive but not easy since it often calls for new conceptual tools to achieve the correct formulations of right theorems to be proved. Some of the directions under development are: (1) to find more realistic idealized models of computers and programs; (2) to relate computer programs to more standard logical and mathematical formulas in order to assist simplification and debugging of programs; (3) to establish a natural framework for proving that multiplication is in general more complex than addition; (4) to formulate the appropriate notion of effective method and prove that the travelling salesman problem is unsolvable; (5) to develop a mathematical theory of pattern recognition.

Of course, it is possible that a higher level of abstraction may impose some order and uniformity on how to use computers. One might think of the examples familiary from high school mathematics: clever word problems in arithmetic become a matter of routine in algebra, and ingenious proofs in elementary geometry can be treated systematically in analytic geometry.

4. Logical mathematics

In general, formalization or rendering exact and explicit vague procedures is of practical interest in extending the range of application of computers. This is perhaps the most basic link between logic and computers. It is in this direction that a large scale revolution of mathematics is likely to be achieved in the long run. As more and more of our mathematical arguments get mechanized, the human contribution to the mathematical activity will have to be less and less routine and more and more imaginative or creative.

The initial experiment with and limited success at automatic demonstration came from an appreciation of the fairly advanced state which mathematical logic had arrived at with respect to formalization. Further attempts at progress revealed the limitations of the achievements of logic as a formal and systematic treatment of mathematics. Very roughly speaking, what one needs is not just formalization in principle of mathematical textbooks but rather formalization in practice of mathematical activities. The goal is to enrich logic (or mathematics) so that computers can aid pure mathematicians at least as much as they assist the applied scientists at present. It calls for the mechanization of two related aspects: the formalization of proofs after discovery, and the abstraction of general methods to guide the search for proofs of new theorems. There seems to be a need to develop a sort of "logical mathematics", the idea of which must be quite repulsive to pure mathematicians who would think of a hybrid of mathematicians and librarians. It is most likely that such a discipline will be more relevant to automatic demonstration than "mathematical linguistics" is to mechanical translation. Moreover, it may even be the most promising avenue in the near future that will lead to general progress on the study of the potentialities and limitations of "artificial intelligence".

Formalization is obviously central to all uses of computers. The very existence of computers depends on the basic fact that we have exact rules for numerical calculations. Arguing by analogy, we may contend that the great expansion of the uses of computers for mental acts will be achieved first in the area of mechanizing mathematical arguments. Compared with game–playing, this area is much richer and more central to all works of the intellect.

5. Reductionism, reflectionism, and the dialectic method

Typically the reductionist is struck by the power or beauty of certain modes to proceed and wish to build up everything on them. Logical positivism is the most recent historical example. A reflectionist takes the data of existing human knowledge more seriously and often is not able to come up with as sweeping answers. In its extreme form. we arrive at phenomenology which is serious philosophy but hardly of immediate relevance to technical advances. For example, inconclusive arguments have been put forward to contend that it is intrinsically impossible to use computers to perform mental tasks such as making perspicuous grouping, tolerating ambiguities, distinguishing essence from accident, and appealing to fringe consciousness. While these discussions help to focus certain long–renge issues, we do not at present possess sharp enough concepts of realizable computers and feasible algorithms to prove, or even to conjecture, such impossibility results.

Although such extreme positions do not seem promising, it does seem highly desirable to coordinate reduction (synthesis) with reflection (analysis) in the area of automatic demonstration, and, in particular, at the present stage. The preoccupation with Herbrand's theorem illustrates for me a reductionist tendency, and should, in my opinion, be balanced by more reflections on the data (viz. existing mathematics). For example, in number theory, we should obviously make use of *least* counterexamples rather than just counterexamples. In each branch of mathematics, we should bring in, besides general features common to all branches, also distinguishing characteristics of the particular branch. In addition, we are no longer interested in the economy of axioms but rather lean heavily on derived rules (metatheorems). As we progress, what is known at each stage has to be more carefully digested and organized in order that mechanical retrieval be feasible. More concretely, I feel that an extensive and systematic examination of a large body of existing proofs is of value at the present stage.

If we reflect on the mathematical activity, one striking feature is man's ability to operate simultaneously on different levels. It is not necessary to perfect the lower levels in a hierarchy in order to be able to act on a higher level. And it is hard to see how machines can be made to do the same. As a result, one often finds it easier to adapt oneself to take advantage of what machines can currently do (such as checking numerical instances after the man himself has reduced a general theorem to these crucial special cases). But the primary objective of automatic demonstration is certainly to extend the general power of computers to take over new types of work.

6. Finite computations and infinite mathematics

Physical Limitations

It seems unquestionable that we cannot have arbitrarily small or arbitrarily fast computer components (say for switching). Physics should be capable of calculating lower or upper bounds to these quantities. This kind of limitation does not affect in any inevitable way the meaning of infinite mathematical procedures. Of course, if there were no such bounds, we might be able to justify mathematical infinity simply by physically actual infinity. I see no reason to delay over, this unrealistic assumption.

The problem of noise and the nonexistence of infallible components can to some extent be treated by means of redundancy. For example, von Neumann asserted that if the probability of basic units to malfunction is no more than $\varepsilon = .005$ (half of one per cent), then one can arbitrarily improve the reliability by majority organs (fiduciary levels $\Delta = 0.07$ is favored). There are other complications not considered in von Neumann's scheme, but it seems reasonable to accept that for moderately long computations, we can, with enough efforts, improve reliability to as high a degree as we wish.

In short, we wish to distinguish two kinds of problem: the scientific problems of physical limitations of speed, reliability, size, and length of computation on the one hand; the epistemological problem of arbitrarily long computations on the other. The scientific problems are important and contain different interrelated aspects each calling for careful attention. But, for the purpose of our present discussion at least, the epistemological problem is essentially one, viz. the apparent fact that there can be no physical machinery to carry out arbitrarily long computations, either without error or just without appreciable probability of error. For this epistemological problem, I do not view the distinction between certainty and high probability as the central issue. I shall leave aside the challenging problem of a theory of physical computations and confine myself to considering the philosophical implications of the finite nature of actual computations. The main features of the basic problem are fully present in the simple matters of adding or multiplying large integers.

There is indeed a distinction between one machine to do arbitrarily long computations and each long computation to be done by some machine. It is logically possible that there is no machine M which deals with all lengths n, yet for each length n there is a machine M to do it. But we shall not speculate on whether such a logical possibility is actual. Rather we shall take for granted that neither is physically possible; therefore, in particular, that there is some large N such that we can never do a

computation of length N with reasonable accuracy. For those who do not like the assumption, we may base our discussion on the weaker postulate:
(*) There can be no physical machine which does correctly arbitrarily long computations.

Does it follow from this that there exists no procedure for calculating the digits of π? The problem of mathematical existence is notoriously controversial. We are accustomed to saying that there exist infinitely many prime numbers, that there exists (indeed, we have) an effective method by which we can, for each n, calculate the n–th digit of π, that there exists a relatively simple effective function f such that f(n) gives the n–th digit of π. To say that there exists no such procedure invokes not only the postulate (*), but also, more seriously, the stipulation:
(#) Existence of a mathematical procedure can only be established by the existence of a physically constructable automaton to carry out the procedure arbitrarily far.

Even in applying infinite mathematics, physics possess a closer contact with reality and executable procedures. Experimental confirmation of a physical theory has to go through performable measurements and calculations. Mathematics supplies a detour through the nonexecutable. If applications of nonexecutable mathematics are to be accepted at all, the physical scientist can also, no less than the mathematician, legitimately work on such material in order to help complete the detour.

Mathematics and application

The stipulation (#) presents serious problems to both mathematics and physics. It may be thought that mathematics could go on as "purely formal systems", but physics cannot hide behind such formalities. This can at best serve to evade the issue. It is certainly not an arbitrary matter that we choose to emphasize the "formal systems" of natural numbers and real numbers. Why do we favour some formal systems over others?

Application is a distinguishing characteristic of mathematics, in contrast with mere games. One does not justify the study of pure mathematics exclusively or primarily in terms of applicability. Mathematics in its advanced stage also lives a life of its own. For example, the criterion of beauty and elegance, that of depth, all are commonly employed in judging works in mathematics.

But it is an undeniable fact that infinite mathematics has been applied in a most spectacular way in the study of natural phenomena. In terms of applications, infinity has thus far proven to be a highly useful detour. One might ask whether it may not further improve matters if we eliminate this detour altogether. We have no guide line as to how to accomplish this. In fact, the mathematical way of thinking in terms of infinities is so deeply rooted, it is hard to see why we should wish to give up such a powerful tool.

Less drastic new directions would be to retain what we have but look more closely at infinities as detours and try to extract as much executable content as we can, as well as to justify infinite mathematics in terms of experiencible facts and more concrete intuitions.

Attempts along these directions are not unfamiliar, but usually less drastic than eliminating infinities altogether. Rather they represent a domestic affair for mathematicians who wish to eliminate or justify higher infinities (the actual infinite) in favor of or in terms of simple infinities (the potential infinite). Thus we have intuitionism, finitism, as well as various efforts to rebuild classical analysis in terms of recursive functions or constructive sets in some suitable sense of "constructive". On the whole, there has been no definitive success in the sense of actually changing the common practice in mathematics. But a suitable rough–edged recursive approach may turn out to be a wholesome way of looking at mathematics.

There are also a few scattered discussions of strict finitism and ultraintuitionism, which reject numbers which are not "executable". In particular, A. S. Esenine–Volpine ("Le programme ultra–intuitioniste des fondements des mathematiques", *Infinitistic Methods*, Pergamon Press, 1961, pp. 201—233.) attempts to prove the consistency of current set theory on this basis. (Compare also D. van Dantzig, "Is $10^{10^{10}}$ a finite number?" *Dialectica*, vol. 9 (1956) pp. 273—277).
The proposed proof is rather obscure and some people regard it as an elaborate joke. There is, however, no doubt that the author is quite serious about his program.

Mathematical activity

Mathematical activity is a phenomenon in nature and is, as such, like all mechanical and mental activities, finite. This undeniable fact does not in itself exclude infinite mathematics. Rather, it excludes, for example, any alleged proof that is too complex to be digestable. For example, even though whether "the billionth digit of π is 7" is a problem decidable in principle, we do not possess at present a digestable proof either of this proposition or of its negation.

What we have here is not something controversial but rather an aspect of mathematics grossly neglected in foundational discussions. One can accept mathematics as is commonly practiced or choose some different outlook on mathematics. But in any case, a mathematical theorem is established only if it is somehow accepted by the relevant mathematical community and somebody must have understood the proof. Execution is central to mathematics, but not in the restricted sense of exhibiting the billionth digit of π, rather more in the extended sense of actual understanding (a mental activity) by some human mind. Attention to this aspect of mathematics can even resolve the deep–rooted conflict on the question of how central applications are for mathematics. The pursuit of elegance is central to mathematics perhaps for the reason that mathematics, as a mental activity, has to be perspicuous and surveyable. And elegance generally extends the range of complexities which we can command.

15.4 On the long-range prospects of automatic theorem-proving *

There is a false contrast between the algorithmic and the heuristic approaches.

* First published in *Synposium on Automatic Demonstration* (Lecture Notes Math 125), pp. 101—111. © Springer Verlag, New York, 1970. Reproduced by permission.

Every program has to embody some algorithm and for serious advances, partial strategies or heuristic methods are indispensible. Hence, no serious program could avoid either component. Perhaps the contrast is more between anthropomorphic and logicist, as typified by the general problem solver on the one hand and elaborate refinements of the Herbrand theorem on the other. This polarization appears to me to be undersirable and to represent what I would call the reductionist symptom.

Typically the reductionist is struck by the power or beauty of certain modes to proceed and wish to build up everything on them. The two extremes seem to share, in practice if not in theory, this reductionist preoccupation. In my opinion, there should be more reflective examination of the data, viz. the existing mathematical proofs and methods of proof. It is true that what is natural for man need not be natural or convenient for machine. Hence, it will not be fruitful to attempt to imitate man slavishly. Nevertheless, the existing body of mathematics contains a great wealth of material and constitutes the major source of our understanding of mathematical reasoning. The reasonable course would be to distill from this great reservoir whatever is mechanizable. In other words, we should strive for an interplay between reduction and reflection which, for lack of a better name, may be called the dialectic method.

In a previous survey ([8], 1965), I have set forth a few vague suggestions which are buried in the examples. I should like now to list these suggestions explicitly and use them to make a few remarks on the current scene. (1) It is recommended that powerful methods with restricted ranges of application be explored. (2) Crude strategies are sketched for selecting lemmas in proving theorems of number theory. (3) An example in the predicate calculus is given to illustrate possibilities of directly exploiting special properties of $\equiv$ and local quantifiers (to reduce $\exists$ x(Fx $\wedge$ x = y) to Fy). (4) The need for an adequate treatment of equality is emphasized for both proof procedures and decision procedures in the predicate calculus.

With regard to (4), there have been several proposals during the last few years for adjoining equality to proof procedures of the predicate calculus. In connection with decision procedures, it has turned out that there is a major open theoretical problem, viz. no proof exists in the literature for the belief that there is a decision procedure for the Gödel case with equality. More exactly, the belief is that there is a decision procedure for satisfiability for the class of prencx formulas with equality whose prefix is $\forall x_1 \ldots \forall x_m \exists y_1 \exists y_2 \forall z_1 \ldots \forall z_n$, and, more, that any formula in the class, if satisfiable at all, has a finite model.

With regard to (2), there have been work to carry out the examples from number theory on computers, but only in a weakened form. No strategies are included to select lemmas. Rather, the lemmas are taken as given and a conditional theorem to the effect that the theorem follows from the lemmas is proved as a theorem of the predicate calculus. It is clear that this is not making use of special properties of particular branches in mathematics but rather continuing to "logicize mathematics".

In connection with (3), the second proof of ExQl ([8], p.55) is intended to give examples of mechanizable special strategies which are suggested by human deductions. The following features are present in the example. (a) Substitute given constants for variables to get stronger conclusions. (b) To eliminate local. quantifiers when possible, i.e. strive to introduce a condition $x = y$ to yield $\exists x(x = y \wedge Fx)$ or $\forall x(x = y \supset Fx)$ in order to reduce the quantified expression to Fy. (c) Substitute equivalences freely (if $A_1 \equiv A_2, \ldots, A_{n-1} \equiv A_n$, then A_i can be substituted for A_j). (d) Apply implication chains: $A_1 \supset A_n$ if $A_1 \supset A_2, \ldots, A_{n-1} \supset A_n$. The features (a) and (d) can be incorporated into Herbrand type proofs fairly directly. But features (b) and (c), though mechanizable and familiar, seem to be destroyed when the problem is transformed into a normal form suitable for obtaining proofs of the Herbrand type. It is thought that by studying examples of human proofs, one may come up with a fair number of useful special strategies such as (b) and (c).

In connection with (1), we may mention the use of least counterexamples in number theory and strategies like (b) and (c) above. In general, it seems desirable to consider directly, besides Skolem functions obtained from dropping quantifiers, also descriptive functions with predetermined meaning such as addition and multiplication in number theory, pair and power set in set theory. It seems desirable to be miserly in the use of quantifiers. In dealing with set theory, it seems desirable to view every axiom of relative existence

$$\exists y \forall x (x \epsilon y \equiv Fxu..v)$$

as defining a function $f_F(u, \ldots, v) = \hat{x}Fxu..v$. In this way, we may operate with constants (such as 0 and ω), functions, and extensionality in form:

$$A \equiv B \supset f_A = f_B.$$

If one reviews the literature on automatic demonstration during the last few years, one gets the impression that the whole field consists of variations on Herbrand's theorem. Often a slight modification is given with full details in a somewhat new dress, accompanied by an elaborate completeness proof. Alternative procedures are offered for alternative advantages. It is hard either to compare the relative efficiency or to accumulate different advantages into one procedure. Hence, some people are looking for a theoretical criterion of relative efficiency.

In the direction of formalization, there are two major successes in modern logic. First, the fairly well established conclusion that all of mathematics is reducible to axiomatic set theory and that, if one takes enough trouble, mathematical proofs can be reproduced in this system completely formally in the sense of mechanical checkability. Second, the results of Skolem and Herbrand according to which we can, by construing mathematical theorems as conditional theorems (viz. that the axioms imply the theorem) in the predicate calculus, search for each mathematical proof in a mechanical (in principle) way to determine whether a related Herbrand expansion contains a contradiction. Impressive as these results are, and encouraging as they are for the project of mechanizing mathematical arguments, they are only theoretical results which do not establish the strong conclusion that mathematical reasoning (or even a major part of it) is mechanical in nature.

What is exciting in the unestablished strong conclusion is that we are facing an altogether new kind of problem which crys out for a totally new discipline and which has wide implications on the perennial problem about mind and machine. We are invited to deal with mathematical activity in a systematic way. Even though what is demanded is not mechanical simulation, the task requires a close examination of how mathematics is done in order to determine how informal methods can be replaced by mechanizable procedures and how the speed of computers can be employed to compensate for its inflexibility. The field is wide open, and like all good things, it is not easy. But one does expect and look for pleasant surprises in this requirement of a novel combination of psychology, logic, mathematics and technology.

It is highly likely that there are different levels of mathematical activity which can be measured by the ease of mechanization. For example, Euler told of how his theorems were often first discovered by empirical and formalistic experimentations. While these experimentations are probably easy to mechanize, the steps of deciding what experimentations to make and of finding afterwards the correct statement and proof of the theorems suggested, are of a higher level and much harder to mechanize. Ramanujan is reported to have commented on the taxicab number 1724 that it is the smallest number expressible as a sum of two cubes in two different ways. The memory and powers of calculation exemplified in this anecdote are probably not hard for a computer, but it would be less easy to have a computer prove most of his theorems. One suspects, however, it would be easier for a computer to prove his theorems than many of the more famous theorems in number theory which are more "conceptual" and further removed from calculations. Axiomatic set theory has in more recent years become much more mathematical, and one gets the impression that long formal proofs of relatively simple results are much easier to discover mechanically than advanced neat proofs which can be communicated succinctly between experts.

On the highest level, Poincaré compares Weierstrass and Riemann. Riemann is typically intuitive while Weierstrass is typically logical. In this case, it is natural to believe that it is easier to reach results of Weierstrass mechanically. Hadamard contrasts his impression of the great works of Poincaré and Hermite and states that he finds Hermite's discoveries more mysterious ([4], p. 110). By stretching greatly one's imagination, one might wish to claim that Hadamard would have found it easier to design a program to discover Poincaré's results than to get one for Hermites.

G. Wallas (*Art of Thought*, 1926, pp. 79—107) suggests that there be four stages in the process of bringing about a single achievement of thought: (1) preparation, (2) incubation, (3) illumination, (4) verification. This fits in well with Poincaré lecture on mathematical discoveries (*Science and Method*). Hadamard ([4]) and Littlewood ([5]) discuss these four stages at great length. The first and the last stages are done consciously. The preparation stage contains two parts: the long–range education of the individual, and the immediate task of learning and digesting what is known about the problem under study. The verification stage consisting of making vague ideas precise and filling in gaps (in particular, carrying out calculations). To mechanize these stages appear formidable enough, but incubation leading to illumination would seem in principle a different kind of process from the operation of existing computers. Since incubation implies an element of rest (an abstention from conscious thought on the initial problem), we may perhaps claim that the importance of this stage comes from a weakness on the part of man, and that machines do not need the period of rest or abstention.

To come back to the current scene, I venture to make some general comments on a few specific aspects. It is appealing to think of an interaction between man and machine, so that computers may become research assistants. In fact, an example of man–machine programs has been written by Guard and others ([3]). It seems that human interventions would be able to improve more substantially the end results if we move from Herbrand proofs to programs with more varied data and strategies.

Practical applications of computers are mainly concerned with repetitions of simple steps rather than individualized long sequences of simple steps such as mathematical proofs. It is natural to think of applying mechanical inference to cases where a lot of short deductions are made. For example, it has been suggested that we can retrieve simple consequences of stored information on individual persons (e. g., Darlington, [2]).

Suggestions have been made to extend automatic demonstration to higher–order logic. It is, however, not clear to me why this could be considered more promising than looking directly at, say, number theory or axiomatic set theory which, in my opinion, is more suggestive and closer to real life. Usable examples in set theory can be found in [8], 1967.

The central idea of automatic demonstration during the last few years appears to be the observation that in order to derive a contradiction from the Herbrand expansion of a formula, it is sufficient to examine mechanically all possible substitutions to obtain potential contradictions. It was noted by Prawitz ([6]) that we can devise an algorithm to a decide whether, given a conjunction C of finitely many clauses and a recursive set of terms, there exists a substitution of terms for variables in C such that the result contains a contradiction. Moreover, given any partition of all terms in C into equivalence classes, there is a least or most general substitution, if there is any, that yields the partition: α is the least if for any β yielding the same partition, we can find γ, $\gamma\beta C = \alpha C$. This idea was applied independently by Robinson ([7]) and Aanderaa ([1]) to introducing what is called resolution (by Robinson) or generalized cut (by Aanderaa). Various generalizations and refinements of the "resolution method" have been proposed.

Elsewhere, I have stressed the advantage of "miniscope" form. In this way, the Skolem functions resulting from existential quantifiers in general get fewer argument variables than in the usual prenex form (compare reference number 10 of [8]). This is adopted in Aanderaa's algorithm. Aanderaa also uses "generalized contraction" and a priority function to govern the order in which different clauses are "confronted" to yield generalized cuts. Unfortunately, I am not able to follow all his intricate steps to give a reasonable sketch of his detailed methods.

References

[1] S. Aanderaa, *A deterministic proof procedure* (manuscript of a term paper), 61 pp., Harvard, May, 1964.

[2] J. L. Darlington, "Theorem proving and information retrieval", *Machine intelligence*, vol. 4 (1969), Edinburgh.

[3] J. R. Guard, J. H. Bennett, W. B. Easton, L. G. Settle, "CRT–aided semi–automated mathematics", AFCRL–67–0167, 1967.

[4] J. Hadamard, *Psychology of invention in the mathematical field*, Princeton, 1945.

[5] J. E. Littlewood, "The mathematician's art of work", *The Rockefeller University Review*, September–October, 1967, New York.

[6] D. Prawitz, "An improved proof procedure", *Theoria*, vol. 26 (1960), pp.102—139.

[7] J. A. Robinson, "A machine–oriented logic based on the resolution principle", *J. ACM*, vol. 12 (1965), pp. 23—41.

[8] H. Wang, "Formalization and automatic theorem–proving" *Proc. IFIP Congress*, 1965, vol. 1, pp. 51—58; "Examples in set theory", *Z. f. Logik u. Grundl. d. Math.*, vol. 13 (1967), pp. 175—188, 241—250.

16. ON INFORMATION PROCESSING OF THE CHINESE LANGUAGE*

The development of science and technology has changed our way of life and work in many respects. Some people are always looking for whatever is novel. They collect and use varieties of newlyinvented gadgets. In their minds, technology can take the place of thought and therefore, their work is often flashy and superficial. Others are conservative by nature. They decline to make use of the new "bizarre technology and excessive ingeniousness", and prefer to ignore the tools that can improve efficiency, which results in diminished potential. But more offen there are few opportunities or suitable conditions in which to apply the technology which saves both time and effort. For instance, xerox machines and telephones are much more popular in the United States than in China.

Let us take writing as an example. In the West, people usually use a word processor to write articles. What's more, in recent years, some people write with the help of a tape-recorder. It was said that when Churchill wrote, he used to typeset and print it out first before he polished it, then typeset it again, polished it again, until it was finalized. This convenient method, formally avaible only to the privileged, is nowadays enjoyed by a larger number of people due to the development of computer technology. After necessary deletions and modifications have been programmed with appropriate instructions, the computer could be used to provide a new printed proof sheet.

For different modes of mechanical manipulation of information, there are a certain number of problems in common and others which are specific to particular languages. For instance, even now a Chinese word processor is not as convenient as an English one. One of the basic reasons lies in the fact that the structure of Chinese characters is not so evidently regular as that of English words. Each English word is composed of a sequence of letters of the same size -- a linear structure. The development of computer and mimeograph technology means that the complicated structure of Chinese characters is no longer an insurmountable barrier, and in addition, it facilitates the development of a more efficient and easy-to-learn Chinese word processor. The most important concept is that it is possible to type out simultaneously several parts of a character, such as its letters, strokes, radicals or phonetic symbols. This sort of scheme has been put forward several times since 1973, although word processors built according to this design are not

---

* **First published in Chinese in *The State of the Art Report of Computer Technology*, no. 98 (June 1979), pp 1—4. Reproduced by permission of the author.**

* This is the English version of an article written by Wang, Hao and published in "Dian Zi Ji Suan Ji Dong Tai" ("The State of Art Report of Computer Technology") June issue of 1979. This English version is done by Fan, Lanying of Beijing Institute of Computing Technology.

available yet. It might be the case that there are still misunderstandings about them. Thus, further explanation of a few ideas, which are actually very simple, is given here in passing along with a digression on the significance and prospects of mechanical manipulation of information.

A survey of information processing

The modes of information processing can be divided into two catagories: replication and transformation. Strictly speaking, transformation includes replication, because the mechanical process cannot produce something genuinely novel. It can only preserve and transmit the original information or, at most, transform the original information into a more proper form. For instance, in computing, what is put in is the question and what comes out, the answer. The essential content has been preserved.

Information processing in the replication form consists of the telephone (in other words, an ear that hears voices a long way off), television (that is to say, an eye that sees a long way off), photolithograph, video, optical plate-making and so on. In general, the above mentioned forms of information processing are mainly concerned with the traditional engineering field based on applied physics, in which people do not participate in the transformation of information and there is, essentially, no problem of "coding".

Information processing in the form of transformation consists of telegraph, computing, editing, word processing, recording etc. By comparing word processing with photolithograph, we can distingquish between the two types. The difference between a xeroxed document and a retypewritten one is negligible. To re-type one takes a little bit more effort but it can be done by using a computer, yet the conceptual procedure is far more complicated than that of xeroxing. If the original document is handwritten, it then, relates to the problem of "pattern recognition" in a deeper way. This is a question in the field of artificial intelligence. To transform a document written by hand into a printed proof sheet still remains a difficult problem.

Recently, computer development has made amazing progress in terms of increased speed. reduction of size, reduced cost and increased reliability. On the other hand, computer applications have lagged behind the leap in hardware technology. Therefore, the full utilization of these technological advances leads to a lot of new questions. Here, I will not tackle such a large area but try to deal with the most simple question -- the reduction of size. Microfiche can store a very large book on a tiny film, bu it needs a machine specially make to read it. A pocket computer can be as small as input and output permit. If it is too small, human hands can no longer accomplish the work of input and human eyes can no longer see the answer clearly either. The major difficulty of using a computer is the fact that programs need to be worked out beforehand. In order to do that, one needs to learn one or more artificial languages, a fact which proves to be a sheer nuisance. A type of computer graphics has recently been developed which could help communication between people and the computer through the use of pictures.

Language is a very important tool in communication among people. The development of the written language from the oral one has made possible the preservation of information free from the limitations imposed by space and time. The invention of printing has extended this possibility one step further. The widespread use of the taperecorder has raised an interesting question: in terms of storage and the speed transmission only, tape-recorders can fulfill the function of writing and printing. Then what is the basic distinction in effectiveness between the two different transmiting modes? First, there is naturally the distinction derived from the difference between the sense of vision and the sense of hearing. To people who are used to reading and writing, they are more at ease with printed matter or written material. It is hard to grasp mathematical formulae, chemical notation and concise classical Chinese through the sense of hearing only. Here is a hypothetical question: If tape-recorders had been invented before writing, then would writing have been invented at all? Another question is: Will writing become obsolete in the future? Since the widespread use of the telephone, letter writing has become less popular. Many youngsters spend much more time on watching TV than on reading. There are peculiar cases in which tapes have served as substitutes for books in recent years. Will this become more common in the future? TV teaching has already replaced part of the function of both text books and teachers. To what extent will this develop?

To determine numerical codes for chinese characters

We will neither deal with the questions of how to transform speech into printed words, or vice versa, nor the topic of recognition of characters by machine. These are still problems beset with difficulties and no noticeable progress has been made. What we will discuss here are several simple, yet necessary, tasks of writing and printing, and input and output on a computer. Conceptually, the process of dictionary-consulting, which is familiar to everybody, is to be analysed to compare and ascertain the similarities and differences between these methods and to serve as an example of the common phenomenon outlined below: certain simple ideas, however, usually turn out to be crucial questions and difficult to deal with in the development of the technology concerned.

In achieving these tasks, Chinese and other languages (English for instance) have their common problems and specific difficulties. The peculiar properties of Chinese, which appear as either barriers or conveniences in the course of mechanical processing, are the main topics to be discussed here. By and large, because several selected tasks in the mechanical processing of English have been fulfilled, a brief discussion of the common problems will be enough.

There are some evident differences between Chinese and English. In English, the ingredients of words are letters, whereas the ingredients of Chinese characters are strokes and radicals. Each English word is a sequence of letters of similar size and fixed position in a linear sequence, while the composition of each Chinese character from its components is a two-dimensional graph, with variety in size and position of the strokes and radicals of different characters. Even ordering is not always definitely standard. In

Chinese homonyms are more frequent than in English, so if one wants to recognize a character by listening to its sound, it can be recognized only in context or by knowing its composition of radicals. The pronunciation is much more complicated in Chinese than in English which is, in turn, considered to be more difficult in terms of pronunciation than German.

Since English words are composed of a sequence of letters, there is a natural solution -- the general method of dictionary-consulting and word processing. What one needs is only the ability to recognize individual letters. So long as individual letters are represented in forms recognizable by machine, the machine will be able to perform the work of dictionary-consulting and word processing. However this is rather complicated in the case of Chinese characters. An appropriate means of classification for complex Chinese characters should be chosen to create a kind of systematic numerical coding which is applicable to mechanical processing. For the computer, it can compute only if the individual steps are definite. Due to the high speed, a large number of steps matters little.

To start with dictionary-consulting, first of all one has to pick out the numerical code to find the exact character in turn. Thousands of characters are catalogued in a dictionary and the numerical code is the coordinate of these categories. I remember that when in secondary school, a lot of time was spent in consulting the dictionary by means of radicals. In doing so, strokes should be counted twice and then looking up the character should be done among the characters with the same number of strokes and certain radicals. Sometimes a wrong radical was selected. I did not learn the fourcorner coding system method of consulting the dictionary very well either, though some of my classmates claimed to be able to use the method easily. Later when I came abroad, the Romanization system was used. There were still difficulties though it proved to be generally faster. Characters which were unknown could not be pronounced. Homophones are good, for there are certain different pinyin systems. If using the latter two methods of dictionary consulting, there are many characters with the same codes, while if using the first method -- radical -- to the question of what the code of a certain word is there are various explanations: If the numbers of strokes of the radicals, or the strokes of a whole character are used as the numerical code, there will be many characters with the same code. If the original character is used as the numerical code, there will not be many characters with the same code of course.

The problem concerning the word processing of Chinese text can be stated as below. A dictionary is stored in a word processor in a suitable way. (This point will be discussed later.) When the character wanted is found from the dictionary, there will be no technical problem in printing that character out. So in theory, in order to create a high speed and easily-monitored word processor of Chinese text, the key point lies in selecting the appropriate numerical code to shorten and simplify the procedure of consulting. Leaving aside the basically solved step of immediate-print-by-character instruction stored inside, the differences between word processing and dictionary-consulting are as follows:

The step of selecting a character from the code is executed by a computer in the process of typewriting. So there might be some complex computation if only the steps are explicit. The step of selecting the code from a character is executed by man both in word processing and dictionary-consulting. The distinctive point is that when consulting a dictionary, the code is often borne in mind, while in word processing, the code is external -- printed on the keyboard. The criteria for determining whether a code for word processing and dictionary-consulting is good or poor are not all the same because of the distinction mentioned above. Now, let us focus specially on the topic of selecting codes for making word processors.

The following four standards are quite precise:

(1) It is better that the number of characters and codes are the same. The main idea of this is that it is hoped that different characters would have different codes. It does not matter if there is more than one code for a single character.

(2) It is hoped that the step of selecting a code from the character should be easy for the human subject to use. This means two points: First, it should be easy to learn. Secondly, as the method is mastered initially, increased experience should lead to the procedure in externalizing the code being increased.

(3) It is hoped that the typist can externalize the codes of the characters speedily since this step determines essentially the speed of word processing. Other steps that are much faster and are based on existing technology, are operated by machines.

(4) The process of selecting the character from the code must be explicit though the steps may be tedious.

When these four standards are recognized, many people will be able to get involved in researching, discussing and experimenting in the search for better codes. It is clear that NOs.(1) and (3) might be in conflict. Usually an increase in the content of the codes and selection of more information from the characters are needed to avoid the free quent occurence of the different characters with the same code, then the speed of the externalization of the codes would be reduced. When this sort of conflict occurs, a choice might be made. For instance, if there are a few characters with the same code included in a kind of codes then the number of the codes should be increased to eliminate this case. Therefore the speed of keypressing is greatly affected. So a display such as a screen could be adopted. When characters with the same code occur, a bell rings and these characters are shown on the display. The typist could choose a proper key, for instance, 1, 2 or 3 to get the character required.

Three types of codes —— radicals, phonetical symbols and the four-corner coding system are mentioned above. There is another type that ought to be mentioned, that is, the use of the strokes of a character by standard ordering or canonical ordering. Other suggestions, such as a three-corner coding system have also been Put forward. In fact, different coding systems could be combined to generate a new one. It depends on whether the way of combining is appropriate. The four standards mentioned above are also applicable to the determination of whether these compound systems are good or not. Take the following as an example: Radicals and strokes can be combined naturally. It is said that there are 200-odd different radicals, which, if all were exploited, would make the keyboard too large. As to the strokes, there are about 20. Some people complain that stroke-counting is a method too tedious to be desirable. The answer could be found only through experiment. However, if 20—30 commom radicals are provided, then only the left strokes need to be typed when the radicals have been eliminated. To decide what radicals should be chosen, it is necessary to check the structure of commonly used characters in every detail.

A recipe for chinese word processors

Mr.B.Dunham and I put forward in 1973 a recipe which was published in "Dousou Magazine" (a magazine published in Hong Kong) March 1979. In recent years, we have had discussions with specialists concerned, and found that some passages of that article were not explicit or clear enough. So here in this article, we hope to elaborate and supplement it in the hope that this type of word processor, which would be fast and efficient, or at least a type that partially adopts the idea will be built soon.

There is a new idea that seems not to have been mentioned before. Whatever the codes used might be, there are two ways of looking at them. In one way the sequencial order of occurence of the ingredients of each code is taken into consideration, while in the other this is not the case. When the strokes contained in a character are used as codes, if the stroke occurrences are condidered, the code could be the sequence of strokes in the order as indicated, otherwise, the code would be a set of strokes. The new idea is to adopt the latter one -- to use a set of strokes as the code instead of a sequence of strokes. The great advantage of this lies in the fact that several keys can be pressed simultaneously -- a feature which improves the speed tremendously. This idea can be considered and used to speed up whatever type of code is chosen. Let us take the character "明" as an example: If the selected radicals are "日" and "月" then, both of them can be pressed at the same time. Word processing is similar to piano playing and a skillful typist can do it by using several of her fingers simultaneously.

We suggest calling them unordered and ordered codes. Note that the unordered form should be extensively used no matter which type of coding is to be adopted. It is certain that if the codes of two characters are two different sequences of the same class of ingredients then the use of the unordered coding would increase the number of characters with the same code. But since the method of simultaneous key pressing greatly improves the speed, it is worthwhile paying the price of increasing the number of characters with the same code and the ingredients of the codes. Generally speaking, gains prevail over losses.

The original suggestion is to use unordered strokes mainly and radicals as a subsidiary means. Now I think multiple-varient compound codes should be investigated and what is important is the preservation of the advantage of the unordered coding. Probably, by choosing the advantages from various codes mentioned above, the best codes satisfying the above four standards can be devised. There is one thing worth mentioning here, that is, the existing simple unordered coding is adequate for the production of Chinese word processors which are much faster than English ones by using existing technology.

In the following we present a specified recipe that is similar to the original suggestion and consists of unordered coding of radicals primarily and strokes subsidiarily.

The original idea of using the radicals as the basis of classification is to indicate each Chinese character as a combination of strokes which should satisfy certain constraints (such as the positions of the strokes and radicals) but we would not impose the constraints concerned here. Therefore once a radical commonly in use is selected, each time the stroke combination of this radical occurs in a character, it can be processed as a unit, that is to say, it can be processed on the keyboard by a single key. One of the tasks is to analyse and choose the stroke combinations —— which most frequently occur —— they are called radicals for simplification. So there exists a problem of how many radicals should be chosen. The keyboard would become unwieldy if the number of distinct radicals is unnecessarily large, whereas efficiency would be compromised if that number is too small. At first, it was thought that without using radicals, it was already fast enough, especially if several keys could be pressed at one time. But, it was said that the average number of strokes of a commonly used character is just over nine. So now it is thought that if two to three dozen radicals are added, the average number of strokes can be reduced by half and usually several strokes can be pressed simultaneously.

With characters being coded into a word processor, the idea of storing a complete dictionary therein would not be prohibitively expensive nor prohibitively large because of the availability of IC which have been developed to a new level characterized by high speed, small size and low cost. To find a character by using codes is much faster than people using a keyboard. Characters can be stored in different ways. Our idea is to store each character in the word processor by using a short program by means of which the selected character is printed or written out through the included instruction -- the controlling printing part of the word processor.

Specifically, an unordered code for a character is entered through the keyboard, and the controlling part of the word processor changes the code into a form with canonical order.

Example, change:

(N, 7, A, 3, C, N)

into 3 7 A C N N.

The Arabic numerals represent radicals and the letters represent strokes. Then for the code concerned the corresponding chain of "branches" should be looked for on the "tree" composed of canonical ordered codes of all characters. This is called "TREE SEARCH" in the technical jargon. First, to find out the lowest branch that corresponds to 3, then the branch of this sub-branch -- 7, continue in this way until the whole chain of branches corresponding to the six symbols is found, that is, the subprogram or short program which represents the original character is found. The controlling part will start to execute the instructions included and finally prints out the required character.

Different alternatives of printing techniques might be considered. For example, a "matrix of dots" can be used to output a character either by punching small holes or dotting with small spots. The size of the matrix of dots can be variable. The larger size is more accurate because of the concentrated dots, but it costs more. There is another new technique called ink jet printing which has the advantage of both high speed and flexibility, but the cost is high at present. By using this technique, characters in different sizes and styles (such as running style, typeface of imitation Song-Dynasty-style or seal style) can be printed. It is said that the ink jet printing technology is quite simple. A sprinkler head sprays out drops of ink at a fixed speed successively and it moves in a topdown way like writing out lines of crowded dots. The paper moves at a fixed speed from left to right. The main point is that there should be a program controlling the magnetic field or electric field, to make each ink drop either fall on the paper or drop down into a storage bottle for later use as required. Since each character has its corresponding sub-program, when the controlling part of the word processor has chosen the character or corresponding subprogram through the code, the subprogram starts the sprinkler head and controls the acceptance or rejection of the ink, thus writing the relevant character.

To sum up, the procedure of word processing can be divided into the following steps:

(1) The character is coded and typewritten on the keyboard by human agent.

(2) The code is put into canonical order by machine and the character located, i.e. corresponding subprogram, through the "Tree Search".

(3) The subprogram obtained by (2) is executed by the machine and the character required is printed on the paper.

The speed of typewriting roughly equals the speed of step (1), because the other two steps are so fast that they will have already been finished before step (2) for the next character is carried out.

The printing of books and newspapers and the input/output of a computer pose no extra problems conceptually and the author neither intends nor is he able to discuss the concrete technical questions here. As to the typesetting, two points should be noted. Firstly, once the word processor works out a satisfactory original document, the techniques of laser and optics can be utilized to make a photographic plate for printing. Of course, direct replication from the original document is also feasible. Secondly, the construction of the word processor can be adjusted a bit to make it into a typesetting machine. Compared with word processing the required speed of printing is lower and it does not matter if the machine is heavy. The only difference between a word processor and a typesetting machine lies in the fact that changes in the subprogram representing the character are needed. The subprogram of a typesetting machine links up directly with the typeface corresponding to the character and is moved to the required position by instructions. Usually there are several spare parts for the typeface of each character.

Usually tens or hundreds of characters are used to input and output on a computer, so the codes mentioned can still be in use or be simplified in accordance with the characters less frequently in use. If a great number of characters are to be input and output later, all the methods of input and output by a word processor discussed above can be applied. Generally speaking, to process the characters on a computer, numerical codes are used, while for output a separated machine is often used to translate the codes into the characters to be printed out.

These are the main points I wanted to talk about. But to express my ideas more explicitly herewith is a concrete proposal to serve as an example. The example might not be the most ideal recipe since I have not analysed carefully the construction of the characters commonly in use. However, it might be a stimulus to experienced research workers in this field to make further studies.

The Example of the Recipe and Other Questions

The following concrete recipe must be rather sketchy because of my lack of knowledge in philology.

As was said before, the so-called radicals indicate the ingredients. It is feasible for any characters that contain one of these ingredients:

1	氵,水	10	月
2	扌,手	11	疒
3	土	12	辶,辶
4	口,囗	13	衤,衣
5	忄,心,⺗	14	石
6	艹,廾	15	钅,金,釒
7	纟,糸	16	虫
8	木,木	17	竹,⺮
9	日	18	讠,言

The list of 24 atoms (strokes):

Examples of Strokes

A	一	厂,上,口,列,支	M	㇀,㇀	汁,冶,扎,纠,红
B	一	二,土,王	N	㇜	去,红,参
C	丨	共,工,土,干	O	㇙	比,饥,针,良,长,仰
D	丨	十,木,中	P	乚,㇄	札,匕,四,心,志
E	丿,㇀	木,欠,人,矢,戈,四,水,猝,比	Q	㇖	完,宅,了,危
F	丶,㇀,㇀,㇏	心,主,卞,杰,材,共	R	㇂	戈,式
G	㇕,㇕,㇖	巴,田,日,尸,局,与	S	㇉	与,弓,马,传
H	㇏,㇏	人,入,大,矢	T	㇗,㇜,㇙	凶,幽,山,出,纠,壮
I	㇆,㇆,亅	月,刀,即,方,豕,狗,马	U	㇏	走,廷,建,边,这,送
J	㇇,㇇	又,水,迅,夕	V	㇈,乙	杭,抗,乙,亿,气
K	亅	打,刊,了,事	W	㇌,㇋	廷,邦,那,防,陂
L	𡿨	涇,巡,巡,災	X	㇊	计,认,讨,讲

Strokes resembling each other are grouped together to alleviate difficulty in recognition. The horizontal and vertical strokes occur most frequently, being 33 percent and 18 percent respectively according to some scholors. Let us divide the two strokes respectively into two cases -- the long ones and the short ones. The long ones are those which have both ends unbound.

There are six-four keys altogether in which forty-two are radicals and strokes, ten are Arabic numerals and about twelve are punctuation marks. Besides these, a few extra keys might be added, such as: one key for the input of two horizontal strokes simultaneously and another one, for three at a time. Therefore if these three keys are pressed at the same time, six horizontal strokes can be input.

The numerical codes mentioned above are in unordered form. It is desirable to investigate the following questions: How to introduce the four-corner coding system and pronunciation codes, how to choose the more appropriate radicals and how to further distinguish the strokes. The critical problem is that there are many different characters with the same code according to the above recipe, for example: 吕 and 回; 田, 由, 甲 and 用; 申 is not included, since it possesses a long vertical stroke. The original suggestion was touse a display unit. When characters with the same code occur, the Arabic numeral keys are used for the purpose of selecting the proper character. Though in this case no special technique is needed, and the display unit helps to increase the efficiency of word processing in other ways. I think it is completely possible to have an unordered code with no characters with the same code, which is easy to learn and command. It is also good for the speed.

On word processors for both Chinese and Western languages you are not able to press more than one key at a time, so some people hold that to press more simultaneously must be a difficult technique. As a matter of fact, it is quite easy. For instance, the board can leave a fixed position for each key, supply electricity for the corresponding position when several keys are pressed, so those positions are open and others closed.

When a character is entered into a typewriter by using an unordered code, it is easy to turn it into a canonical ordered code with numbers preceding the letters, and in which both numbers and letters are arranged according to their size and order. As for using this code to search for the subprogram corresponding to the character to exccute Tree Search. it is easy to experiment on a computer. I have heard that some people have already exprimented with this.

To sum up: I hope people with adequate training in philology do some further research to find appropriate unordered codes, especially those which would not result in producing the same code for distinct characters. And I also hope hardware and software engineers will give more thorough consideration and discuss the techniques concerned.

THE LIST OF THE PUBLICATIONS OF THE AUTHOR

Hao Wang

1952(50). Logic of many-sorted theories, *JSL* (i.e., *Journal of symbolic logic*), vol.17, pp. 105—116.

1953(52). Quelque notions d'axiomatique, *Revue philosophique de Louvain*, vol.51, pp. 409—443. English version entitled 'The axiomatic method' is included in *Survey* [1962(59)] as chapter 1.

1955(53). On formalization, *Mind*, vol. 64, pp. 226—238. Reprinted in *Contemporary readings in logical theory*, edited by I. Copi and J. Gould, 1967, pp. 29—39; also included as the opening essay in their *Contemporary philosophical logic*, 1978, pp. 2—13.

(1953). The concept of computability. This essay was first written in 1953. It was then revised in 1954; but it has not been published before.

1962(53). Ackermann's consistency proof. These notes were written in 1953 and first published in *Survey* (pp. 362—375).

1955(54). On denumerable bases of formal systems. Invited hour lecture at the International Congress of Mathematicians, Amsterdam, 1954; published in *Mathematical interpretation of formal systems*, pp. 57—84.

1957(54). A variant to Turing's theory of computing machines, *Journal ACM* (i.e., of the Association for Computing Machinery), vol.4, pp. 63—92. The paper was presented to the meeting of ACM in June, 1954.

1974(55). On formalizing mathematical concepts. Six essays delivered as the second series of John Locke Lectures at the University of Oxford in spring 1955; parts were published in revised form in 1974(72), chapters 1 and 2.

(1956). Elementary philosophy of mathematics. An uncompleted typescript of 450 pages written during 1955—56; only some fragments have been published.

1957(56). (With A. W. Burks). The logic of automata, *JACM*, vol.4, pp. 193—218 and pp. 279—297. Reprinted in *Survey* as chapter 8.

1957(57). Universal Turing machines: an exercise in coding, *ZMLGM* (i.e., Zeitshrift für Mathematische Logik und Grundlagen der Mathematik), vol. 3, pp. 69—80. Reprinted in *Survey* as chapter 7.

1958(57). Eighty years of foundational studies, *Dialectica*, vol. 12, pp. 466—497. Reprinted in *Survey* as chapter 2.

1960(58). Toward mechanical mathematics, *IBM journal of research and development*, vol. 4, pp. 2—22. Reprinted in *Survey* as chapter 9; also reprinted in *The modelling of mind*, edited by K. Sayre and F. Crosson, 1963; Russian translation in *Problems of cybernetics*.

1959(59). Circuit synthesis by solving sequential Boolean equations, *ZMLGM*, vol.5, pp. 216—239. Reprinted in *Survey* as chapter 10.

1962(59). *A survey of mathematical logic* (*Survey*), Science Press, 1962, 652 pp. + x; also distributed by North-Holland Publishing Company, 1963. Reprinted by Chelsea, New York, 1970 under the title *Logic, computers and sets*. The manuscript was completed and submitted in June 1959 (at Oxford).

1960(59). Proving theorems by pattern recognition, I, *Communications ACM*, vol.3, pp. 220—234. Inited lecture at the ACM meeting in May 1960.

1961(60). Process and existence in mathematics, *Essays on the foundations of mathematics*, pp. 328—351. This was read to the Philosophical Club of Harvard University in spring 1960; a Russian translation came out in 1965.

1961(60)a. Proving theorems by pattern recognition, II, *Bell System technical journal*, vol. 40, pp.1—41.

1961(60)b. An unsolvable problem on dominoes, The Computation Laboratory, Harvard University, *Report BL-30, I,* July 1961, 5 pp.

1975(60). Notes on a class of tiling problems, *Fundamenta mathematicae*, vol. 82, pp. 295—305.

1961(61). The calculus of partial predicates and its extension to set theory, I, *ZMLGM*, vol. 7, pp. 283—288. It was read to the Logic Society in England in spring 1961; the second part (extension to set theory) has not been written.

1963(61). Mechanical mathematics and inferential analysis, *Computer programming and formal systems*, edited by P. Braffort and H. Hirschberg, pp. 1—20. This is a revised version of an invited lecture at a seminar, spring 1961, sponsored by IBM in Holland.

1962(61). (With A.S. Kahr and E. F. Moore). Entscheidungsproblem reduced to the AEA case, *Proceedings of the National Academy of Science, U. S. A.*, vol. 68, pp. 528—532.

1963(62). Dominoes and the AEA case of the decision problem, *Mathematical theory of automata*, pp. 23—55. Invited lecture given April 1962 in New York.

1963(62)a. The mechanization of mathematical arguments, *Experimental arithmetic, high speed computing and mathematics*, pp. 31—40. This was an invited lecture at a meeting of the American Mathematical Society in 1962.

1963. Tag systems and lag systems, *Mathematische Annalen* (*MA*), vol. 152, pp. 65—74. This was included in *Popular lectures* [1981(78)] as appendix C5.

1963a. (With M. O. Rabin). Words in the history of a Turing machine with a fixed input, *Journal ACM*, vol. 10, pp. 526—527.

1964(63). Remarks on machines, sets and the decision problem, *Formal systems and recursive functions*, pp. 304—320. An invited lecture given at Oxford, England in summer 1963.

1964. Critique of logic for the computer sciences, *Communications ACM*. vol. 7, p. 218.

1964a. (with W. V. Quine). On ordinals, *Bulletin of American Mathematical Society*, vol. 70, pp. 297—298.

1965(64). Formalization and automated theorem proving, *Proceedings of the IFIP Congress 65*, pp. 51—58. This was an invited lecture to the Congress; Russian translation, *Problems of cybernetics*, vol. 7 (1970), pp. 180—193.

1965. Logic and computers, *American mathematical monthly*, vol. 72, pp. 135—140.

1965a. Games, logic and computers, *Scientific American*, vol. 213, no.5 (November), pp. 98—106. There is a Swedish translation in *Modern Datateknik*.

1965b. Note on rules of inference, *ZMLGM*, vol.11, pp. 193—196.

1966. (With S. A. Cook). Characterizations of ordinal numbers in set theory, *MA*, vol. 164, pp. 1—25.

1966a. (With K. R. Brown). Finite set theory, number theory and axioms of limitation, ibid., pp. 26—29.

1966b. (With K. R. Brown). Short definitions of ordinals, *JSL*, vol. 31, pp. 409—414.

1971(66). Logic, computation and philosophy, *L'âge de la science*, vol.3, pp. 101—115.

1967(66). On axioms of conditional set existence, *ZMLGM*, vol.13, pp. 183—188.

1967(66)a. Natural hulls and set existence, ibid., pp. 175—182.

1967(66)b. A theorem on definitions of the Zermelo-Neumann ordinals, ibid., pp. 241—250.

1970(67). Remarks on mathematics and computers, *Theoretical approaches to nonnumerical problem solving*, pp. 152—160. An invited lecture given at Cleveland, Ohio in 1967.

1970(68). A survey of Skolem's work in logic, *Selected logical works of Th. Skolem*, pp. 17—52.

1970(68)a. On the long-range prospects of automated theorem-proving, *Symposium on automatic demonstration*, pp. 101—111. Invited lecture given at Versailles, France in December 1968.

1974(71). Metalogic, *Encyclopaedia Britannica*, vol. 11, pp. 1078——1086. All except the part on model theory is reprinted in 1974(72) as chapter 5.

1974(72). *From mathematics to philosophy*, Routledge and Kegan Paul, 413 pp. +xiv. Italian translation *Dalla mathematica alla filosofia*, Boringhieri, 1984, by Alberto Giacomelli.

1976(73). (With B. Dunham). A recipe for Chinese typewriters, IBM report RC4521, September 5, 1973. Chinese translation appeared in *Dousou bimonthly*, no. 14, March 1976, pp. 56—62.

1976(74). (With B. Dunham). Toward feasible solutions of the tautology problem, *Annals of mathematical logic*, vol. 10, pp. 117—154. (Originally issued as IBM report RC4924 on July 9, 1974).

(1977). (With D. A. Martin). Ranked matching and hospital interns. Some of the results are mentioned in 1981(78) under chapter 3.6.

1981(78). *Popular lectures on mathematical logic*, Science Press and van Nostrand Reinhold. 273 pp. + x. Chinese translation appeared about the same time, Science Press, 257 pp. +vii.

1979. On information processing of the Chinese language (in Chinese), *The state of the art report of computer technology*. no.98 (June 1979), pp. 1—4.

1981(80). Specker's mathematical work from 1949 to 1979, *L'enseignement mathématique*, vol. 27, pp. 85—98.

1984(82) Computer theorem proving and artificial intelligence, *Automated theorem proving: after 25 years*, pp. 47—70. Lecture to accept the first Miletone Prize in automated theorem proving, awarded January 1983 at the annual meeting of the American Mathematical Society.

1984(83). The formal and the intuitive in the biological sciences, *Perspectives in biology and medicine*, vol. 27, pp.525—542. Opening lecture at the Ninth International Congress of Thrombosis and Haemostasis, Sweden, on July 3, 1983.

GPSR Compliance

The European Union's (EU) General Product Safety Regulation (GPSR) is a set of rules that requires consumer products to be safe and our obligations to ensure this.

If you have any concerns about our products, you can contact us on ProductSafety@springernature.com

In case Publisher is established outside the EU, the EU authorized representative is:

Springer Nature Customer Service Center GmbH
Europaplatz 3
69115 Heidelberg, Germany

Batch number: 10607220

Printed by Printforce, the Netherlands